OXFORD LOGIC GUIDES

For a full list of titles please visit
http://www.oup.co.uk/academic/science/maths/series/OLG/

21. C. McLarty: *Elementary Categories, Elementary Toposes*
22. R.M. *Smullyan: Recursion Theory for Metamathematics*
23. Peter Clote and Jan Krajícek: *Arithmetic, Proof Theory, and Computational Complexity*
24. A. Tarski: *Introduction to Logic and to the Methodology of Deductive Sciences*
25. G. Malinowski: *Many Valued Logics*
26. Alexandre Borovik and Ali Nesin: *Groups of Finite Morley Rank*
27. R.M. Smullyan: *Diagonalization and Self-Reference*
28. Dov M. Gabbay, Ian Hodkinson, and Mark Reynolds: *Temporal Logic: Mathematical Foundations and Computational Aspects, Volume 1*
29. Saharon Shelah: *Cardinal Arithmetic*
30. Erik Sandewall: *Features and Fluents, Volume I: A Systematic Approach to the Representation of Knowledge about Dynamical Systems*
31. T.E. Forster: *Set Theory with a Universal Set: Exploring an Untyped Universe, Second Edition*
32. Anand Pillay: *Geometric Stability Theory*
33. Dov M. Gabbay: *Labelled Deductive Systems*
34. R.M. Smullyan and M. Fitting: *Set Theory and the Continuum Problem*
35. Alexander Chagrov and Michael Zakharyaschev: *Modal Logic*
36. G. Sambin and J. Smith: *Twenty-Five Years of Martin-Löf Constructive Type Theory*
37. María Manzano: *Model Theory*
38. Dov M. Gabbay: *Fibring Logics*
39. Michael Dummett: *Elements of Intuitionism, Second Edition*
40. D.M. Gabbay, M.A. Reynolds and M. Finger: *Temporal Logic: Mathematical Foundations and Computational Aspects, Volume 2*
41. J.M. Dunn and G. Hardegree: *Algebraic Methods in Philosophical Logic*
42. H. Rott: Change, Choice and Inference: *A Study of Belief Revision and Nonmonotoic Reasoning*
43. Johnstone: Sketches of an Elephant: *A Topos Theory Compendium, Volume 1*
44. Johnstone: Sketches of an Elephant: *A Topos Theory Compendium, Volume 2*
45. David J. Pym And Eike Ritter: *Reductive Logic and Proof Search: Proof Theory, Semantics and Control*
46. D.M. Gabbay And L. Maksimova: *Interpolation and Definability: Modal and Intuitionistic Logics*
47. John L. Bell: Set Theory: *Boolean-Valued Models and Independence Proofs, Third Edition*
48. Laura Crosilla and Peter Schuster: *From Sets And Types to Topology and Analysis: Towards Practicable Foundations for Constructive Mathematics*
49. Steve Awodey: *Category Theory*
50. Roman Kossak and James Schmerl: *The Structure of Models of Peano Arithmetic*
51. André Nies: *Computability and Randomness*
52. Steve Awodey: *Category Theory, Second Edition*
53. Byunghan Kim: *Simplicity Theory*
54. Ettore Casari: *Bolzano's Logical System*

Bolzano's Logical System

ETTORE CASARI

OXFORD
UNIVERSITY PRESS

OXFORD
UNIVERSITY PRESS

Great Clarendon Street, Oxford, OX2 6DP,
United Kingdom

Oxford University Press is a department of the University of Oxford.
It furthers the University's objective of excellence in research, scholarship,
and education by publishing worldwide. Oxford is a registered trade mark of
Oxford University Press in the UK and in certain other countries

First Edition published in 2016

Published in the United States of America by Oxford University Press
198 Madison Avenue, New York, NY 10016, United States of America

British Library Cataloguing in Publication Data
Data available

Library of Congress Control Number: 2016932192

ISBN 978–0–19–878829–4

To the memory of my teachers Giulio Preti, Ludovico Geymonat, Hans Hermes, Wilhelm Ackermann, Gisbert Hasenjaeger

PREFACE

1. Studies concerning Bernard Bolzano's thought have seen remarkable growth in recent decades. As is evident even by simply considering the catalogue of the excellent series of the *Beiträge zur Bolzano-Forschung*, such studies have examined all of the multifaceted aspects of his thought: from logic and mathematics to ontology, epistemology, and philosophy of language; from moral philosophy to philosophy of religion, theology, political and social philosophy, and aesthetics. In addition, a great deal of attention has been devoted to the processes through which his thought matured.

Certainly, an important motivation for the recent interest in Bolzano's work is the large-scale project to produce a complete edition of his works. This substantial enterprise began at the end of the 1960s and has already reached some 90 of the roughly 130 volumes planned in total.

While such scholarly interest has succeeded in bringing to light Bolzanian thought as a whole, as well as continuing to provide new insights into his many-sided work, there is no doubt that the most important aspect of his thought was that concerned with logical problems. The wide scope of research carried out on this aspect of Bolzano's work, especially after the publication in 1962 of Jan Berg's *Bolzano's Logic* (still a fundamental text for this subject), confirms Husserl's judgement that: "we must count him as one of the greatest logicians of all time". One of the foundations for this assessment no doubt is Bolzano's monumental *Wissenschaftslehre* of 1837. In this work, especially the second and third parts, we can appreciate the mature core of his reflection on logic in its broadest sense, including the theory of knowledge.

As is well known, the *Wissenschaftslehre* is divided into five parts. The first part, 'Theory of Fundamentals' (*Fundamentallehre*) focuses on the conceptual basis of the logical system and its justification. It does this, on the one hand, by introducing the fundamental distinction between the sphere of *logical objectivity* (truths in themselves and, more generally, propositions in themselves) and the sphere of *logical subjectivity* (judgements and knowledge), and, on the other hand, by refuting the respective sceptical objections, by firstly demonstrating that truths in themselves, even an infinity of them, really are there, and, secondly, by arguing that truths of this kind can also be recognized.

The second part, the 'Theory of Elements' (*Elementarlehre*), develops the theory of logical objectivity, first of all by analysing objective representations and then gradually propositions in themselves, followed by truths in themselves, and concluding with logical truths. Husserl observed that this part of Bolzano's writing represents "a work which, in his treatment of the logical 'theory of elements', far surpasses everything that world literature has to offer in the way of a systematic sketch of logic".

The third part of the *Wissenschaftslehre,* the 'Theory of Knowledge' (*Erkenntnißlehre*), on the other hand, develops the theory of logical subjectivity, discussing first of all subjective representations, then judgements, then the truth of judgements, and, finally, the trust we place in our judgements.

The last two parts, which deserve far greater attention than has been given them to date, by contrast, have an explicitly didactic aim. The fourth section, the 'Heuristics' (*Erfindungskunst*), aims to demonstrate "how one must proceed in the search for truth" (as far as this may depend on logic). The fifth part of his work, instead, explores what Bolzano regarded as the 'Proper Theory of Science' (*Eigentliche Wissenschaftlehre*), that is, the presentation of the "general rules according to which one must proceed in the determination of the domains of the various sciences as well as in the working out of such sciences in treatises".

2. As already mentioned, many specialist studies have analysed many of the questions that arise from the first three parts of Bolzano's work, as well as providing comprehensive expositions of them, often very successfully. Yet, it seems to us that there remains room for a more systematic reconsideration of Bolzano's logical thought.

This book is concerned precisely with this aim. In undertaking this task, the book is intended as an exploration, not so much of the more specifically discursive aspects of Bolzano's logical thought—already amply studied—as much as one aimed at identifying the singularly coherent and systematic nature of the logic presented in the *Wissenschaftslehre.*

In order to render as visible as possible the systematic nature of that logic, I have decided to present it within a formal system. Despite being surprising even to me, it has become clear that in pursuing this aim, it is sufficient to adopt the approach of the predicate calculus with identity and choice operator, that is, enlisting the well-known *Hilbert's epsilon calculus.* As this book reveals, the formalization of Bolzano's logic in this calculus emerges quite effortlessly.

3. As is normal for a reconstructive work of this nature—a process that I began some 30 years ago—the state of scholarship has evolved such as to enable better access to a number of relevant questions. A few particularly significant examples of this development are in order here.

The first relates to a remark that I made in my earliest study of Bolzano many years ago, which at the time gave rise to some objections, but now seems to be accepted generally: I argued that the objective representations of quality may be not only empty or singular, as traditionally agreed, but also *plural*, that is, referring to more than one quality. A relevant consequence of this observation is that Bolzano's authentic definition of the truth of a proposition of the form *A has B*, results as: each object falling under *A* enjoys at least one of the qualities falling under *B*. Another important consequence is that in this way Bolzano's extraordinary insight into the strict parallelism between the non-emptiness of an objective representation and the truth of a proposition in itself becomes much clearer.

Another relevant example here concerns Bolzano's view on the interpretation of the logical complexity of propositions. We are accustomed to analysing this complexity in terms of operations: as operations on propositions or propositional schemata (as in the usual logics of the *Fregean tradition*) or, alternatively, as operations on concepts (as in the logics of the *algebraic tradition*: for example, in *Quine's functorial logic*). Bolzano, on the contrary, analyses this concept in terms of semantic predication: emptiness or non-emptiness of an objective representation; truth or falsity of a proposition in itself. A brief example illustrates this point more clearly: what we analyse as a proposition obtained from the two propositions *p* and *q* by means of the operation of disjunction is seen by Bolzano as the assertion of non-emptiness of the representation of a true proposition between *p* and *q*. Among the main consequences of accepting this daring and singular viewpoint is that it is possible to understand the reason of Bolzano's *prima facie* surprising conviction that all propositions have the form *A has B*, or at least are equivalent to propositions of this form.

Another observation worthy of mention concerns a point that Bolzano raises—while not explicitly in the *Wissenschaftslehre*, quite clearly in his *Athanasia*—whereby he outlines a particularly interesting treatment of those singular qualities that today often are called 'tropes'; indeed, he considers tropes of tropes, tropes of tropes of tropes, and so on, and discusses reduction procedures in relation to them.

Finally, it seems apposite to mention the apparently ignored fact that Bolzano developed a theory of signs, which includes, in particular, a theory of linguistic signs, where he provides, with great accuracy, the first formulation in the modern age of the distinction between *sense* and *reference*.

4. In accordance with the key themes to be explored in this study, the book is divided into nine chapters.

The first chapter defines the Bolzanian logical universe as being inhabited by *objects* and *qualities*, and introduces the basic concepts that determine its structure: *belonging; part, variation; existence; objective representation* (abbreviated to *'idea'*), *referring, proposition in itself* (abbreviated to *'proposition'*), *objectuality, truth; consecutivity, causality; having as material, subjective representation* (abbreviated to *'representation'*), *judgement, knowledge.*

The subsequent chapters provide elaborations on the theories of these concepts. In particular, the second and third chapters are concerned with the properties and relations of ideas and, respectively, of propositions: they do this from a perspective that we call 'elementary'; such a perspective neglects both the fact that such entities are, in themselves, objects, and hence possible references of ideas, and the fact that qualities are, in themselves, again objects and hence possible bearers of qualities.

The fourth chapter removes this first limitation and reconsiders ideas and propositions from the point of view that we call 'symbolic', to use a Bolzanian term: that is, from the perspective that acknowledges that such entities are also objects and hence possible references of ideas. In this context, particular attention is given to Bolzano's conviction, mentioned above, that all propositions have the form *A has B*, or at least are equivalent to propositions of that form.

The fifth chapter discusses Bolzano's theory of *inferences*, that is, the theory of the propositions enunciating those relations among propositions that we would call 'of logical consequence', but which Bolzano called 'of derivability'.

The sixth chapter removes the second limitation of the elementary level and takes into consideration qualities of qualities, and so on; it also clarifies, besides the theory of 'tropes', Bolzano's parallelism between the objectuality of an idea and the truth of a proposition. In this chapter, a further question is discussed concerning whether an empty quality is present or absent in Bolzano's universe.

The seventh chapter deals with the objective relation that may subsist among truths, according to which some truths are reasons for other truths. This relation is known as 'consecutivity' and is also, in particular, the relation to which Bolzano reduces the causality relation.

The eighth chapter discusses the theory of knowledge, as that concerned with representing and judging: this is what Bolzano regarded as the subjective counterpart of the sphere of entities in themselves. This theory relates to those psychic phenomena that have as material ideas and, respectively, propositions.

Following from this analysis, the final ninth chapter is devoted to the Bolzanian theory of signs.

After this book was completed, R. George and P. Rusnock published their excellent English translation of the *Wissenschaftslehre*. So it seemed appropriate to replace with it all original German quotations occurring in the footnotes. As the book and the translation do not always strictly coincide with regard to terminology, there may be some small discrepancies between the main text and the footnotes, discrepancies that will, however, in no way disturb the comprehension of the exposition.

ACKNOWLEDGEMENTS

My first encounter with the *Wissenschaftlehre* was at a seminar that Heinrich Scholz conducted on the work, shortly before his death (1956), with the collaboration of Friedrich Kaulbach. A second encounter with Bolzano's work arose in 1963, through the Einleitung that my old fellow student Friedrich Kambartel posed at the beginning of his anthology of the first two parts of Bolzano's book, titled *Grundlegung der Logik.* Only at the beginning of the 1980s, however, did I begin my close reading of the *Wissenschaftlehre.* On this score, I am very grateful to Justus Diller: his invitation for me to present a paper in Münster for the Scholz centenary in 1984 forced me to write down my initial ideas and responses to the work. Since then, my study of Bolzano's thought has continued to grow, gradually developing into this book.

Over the course of such a long period, then, I have received assistance and support from so many people that it is impossible to list them all. Nevertheless, to the following I extend my particular gratitude: Jan Berg, Sergio Bernini, Arianna Betti, Andrea Cantini, Stefania Centrone, Maria Luisa Dalla Chiara, Lorenzo Fossati, Rolf George, Rudolf Haller, Frank Krickel, Wolfgang Künne, Paolo Mancosu, Pierluigi Minari, Edgar Morscher, Massimo Mugnai, Francesco Paoli, Gianni Rigamonti, Paul Rusnock, and Jan Sebestik. In addition, I am indebted to the students who studied my courses on this topic at the University of Florence and the Scuola Normale Superiore of Pisa, for their many stimulating questions and remarks. I am grateful, furthermore, to the participants in my seminars at the Universities of Innsbruck, Graz, and Salzburg. Finally, I am indebted to Laurence Steinman for his careful revision of my first English draft.

In the final part of the preparation for publication, I was particularly supported by the already-mentioned Centrone, Minari, and Mugnai, and, besides them, by Riccardo Bruni. To all of them my warmest thanks. Particular thanks must I also address to Daniel Taber of Oxford University Press for his unlimited availability and patience.

CONTENTS

1 **Introduction** **1**
1.1 The Bolzanian Conceptual Universe 1
1.2 The Logic and the Language 5
1.3 General (Elementary) Principles of Ontology, Lectology, and Epistemology 6
1.4 Restricted Quantifiers 11
1.5 Variation 12
1.6 Ordered *n*-tuples 15
1.7 Relations (Relational Qualities) 17

2 **Elementary Level (A): Ideas** **20**
2.1 Preamble 20
2.2 Properties of Ideas 20
2.3 Relations Between Two Ideas 24
2.4 Relations Among Several Ideas 28
2.5 The Idea of Something 37
2.6 Concretization 38
2.7 Concrete and Abstract Ideas 39
2.8 Intuitions 43
2.9 Sum of Quality Ideas 45
2.10 Another Lectological Operation on Quality Ideas 51
2.11 Negation 52

3 **Elementary Level (B): Propositions** **62**
3.1 Atomic Propositions 62
3.2 Aristotelian Universal Propositions with Existential Import 66
3.3 Material Proprieties and Relations of Propositions 67
3.4 Complex Propositions 70
3.5 Variation 74
3.6 Formal Properties and Relations of Ideas 76
3.7 Validity of Propositions 76
3.8 Analytic and Synthetic Truth and Falsity 79
3.9 Formal Relations Among Propositions 81
3.10 Formal Connectives 110
3.11 Probability 111
3.12 Probability of Events 120

4 Symbolic Level 124
4.1 Preliminary Remark: Properties of Things and Quality of Objects 124
4.2 Bolzano's Conjecture 125
4.3 Restricted Symbolic Level—Ideas of Ideas 128
4.4 The Traditional Categorical Propositions 132
4.5 Extended Symbolic Level: Ideas of Propositions 133
4.6 Material Connectives 136
4.7 Formal Properties and Relations 140
4.8 Formal Properties of Collections of Propositions 143
4.9 Formal Relations Among Propositions 144
4.10 Generalization to a Finite Set of Ideas 147
4.11 Formal Connectives Among Finite Sets of Propositions 149
4.12 'Arbitrary Sets' of Ideas and Propositions 151
4.13 Generalization of Properties and Material Relations among Ideas 152
4.14 Two Operations on Ideas of Object Ideas 154
4.15 New Formulation of the Material Relations among Ideas of Ideas 158
4.16 Bolzano's Thesis for Material Relations Among Sets of Ideas 160
4.17 Material Conjunctions and Disjunctions 161
4.18 Formal Properties of 'Arbitrary Sets' of Propositions 166
4.19 Formal Relations among 'Arbitrary Sets' of Propositions 167
4.20 Formal Connectives among 'Arbitrary Sets' of Propositions 170

5 Inferences 173
5.1 Preliminaries 173
5.2 Rules to Obtain Inferences from Inferences 176
5.3 Inferences with One or More Universal Premises 177
5.4 Inferences with at Least One Negated Universal Premise 185
5.5 Inferences with One or More Existential or Inexistential Premises 188
5.6 Syllogistic Inferences 202
5.7 Inferences from Premises Containing Width Determinations 206
5.8 Inferences from Premises Enunciating Relations among Ideas 210
5.9 Inferences from Premises Enunciating Relations among Propositions 214

6 Higher Level 233
6.1 Preamble 233
6.2 The Problem of Vacuous Qualities 233
6.3 Truth as Enunciation of a Situation 239
6.4 Qualities of Qualities 243

7 Etiology **250**
7.1 Consecutivity 250
7.2 Dependence 259
7.3 Consecutivity and Derivability 261
7.4 Causality (A): Ontological–Metaphysical Preliminaries 267
7.5 Causality (B): the Relations of Causality 270
7.6 Alethic Modalities 272

8 Theory of Knowledge **275**
8.1 Having as Material: Representations and Judgements 275
8.2 Logical Properties of Representations and Judgements 278
8.3 Cognitive and Physical Properties of Representations and Judgements 280
8.4 The Formation of Judgements 282
8.5 Knowledge and Error 289
8.6 Trust 293
8.7 Knowing and Believing 294

9 Semiotics **296**
9.1 Semiotic Relations 296
9.2 Hermeneutics 302

A The Epsilon Calculus with Definitory Expansions 307
B Specific Axioms 309
C Aggregates 315
1 Preamble 315
2 The Notion of Aggregate 316
3 Ideas of Aggregates 323
4 Existence of Aggregates 325
D Unity and Multiplicity 328
E Antinomies 331
F Real numbers 337
References 339
Author Index 347
Subject Index 349

1

Introduction

1.1 The Bolzanian Conceptual Universe

As is particularly clear in the *Wissenschaftslehre,* we may distinguish four basic conceptual areas in Bolzano's logical system.

(a) The first is *ontology*. As is well known in the tradition of this discipline, quite generally intended as the 'theory of being as such', we may further distinguish three main issues. The first issue centres on the question of the possible *types* or *categories* of *things* [*Dinge*] and the relationships among them. Among the various names that may be used to denote this, '*categorical ontology*' is possibly the preferable term and anyhow the one that will be used in this book. '*Mereology*' is the name that has become widely established to refer to the second issue, which centres on the *structure* of *things*, the ways things are composed. The third issue concerns the problem of the *reality* of *things* or, more properly, the problem of the possible ways a thing may exist; instead of the often-used names '*existence ontology*' (as opposed to the first one as '*essence ontology*') or '*onticology*', we prefer the name '*metaphysical ontology*'.

(a1) The distinctive notions of categorical ontology are *object* [*Gegenstand*] and *quality* [*Beschaffenheit*], as well as the notion of the relation that ties an object and a quality when the former *has* [*hat*] the latter or, equivalently, the latter *belongs* [*kommt zu*] to the first. As we will later discuss in greater detail, it may be asserted that for Bolzano there are neither '*Musilian*' objects (i.e. objects without qualities) nor '*vacuous*' qualities (i.e. qualities that belong to no object). Therefore, the true basic and primary notion of categorical ontology is the relation of *belonging among things* and, consequently, *an object may be identified with something to which something belongs and a quality with something that belongs to something.*

In this book, we will write '$x \in y$' for '(the thing) x has (the thing) y' and '$y \ni x$' for '(the thing) y belongs to (the thing) x', on the understanding that $x \in y$ iff $y \ni x$; furthermore, we will write '$O(x)$' for '(thing) x is an object' and '$Q(x)$' for '(the thing) x is a quality'.

(a2) Insofar as Bolzanian *mereology* is concerned, the situation is not so clear. There is of course the notion of part, which does not always interact in a univocal way with the notion of *aggregate* [*Inbegriff*]. Moreover, another mereological notion plays a fundamental role in the Bolzanian system: the relation of *variation* [*Veränderung*] that obtains among two things when the second differs from the first by having certain things exactly in those places in which the first one has certain other things—roughly, when the second is obtained from the first by uniformly replacing in it some parts for other parts.

We will write '$x < y$' for '(the thing) x is (proper) part of (the thing) y'; we will come back later to the problems related to aggregates and replacements.

(a3) Finally, insofar as *metaphysical ontology* is concerned, Bolzano's position is very clear: among the things occurring in the universe of discourse, among the things '*that there are* [*die es gibt*]', as Bolzano says, or that '*subsist*', as we will also say, *some and only some* have the property to *exist*, to *have existence* [*Daseyn haben*], to *have reality* [*Wirklichkeit haben*]. As will be seen subsequently, besides God, such things are all and only all those things that have spacial–temporal determinations.

We will write '$D(x)$' for '(the thing) x exists'.

(b) The second main area of the Bolzanian system deals with the possible objective contents of our cognitive activities that have the fundamental metaphysical characteristic of not having reality. Given the close affinity between this conceptual area and the Stoic doctrine of 'what may be said' or '*lektòn*' we will use the name of '*lectology*' for it. Within lectology, we may distinguish a '*(lectological) morphology*', which deals with the types and structures of lectological entities, and a '*(lectological) semantics*', which investigates the relationships between such entities and those entities that are investigated by the categorical ontology.

(b1) The main notions of *(lectological) morphology* are those of *representation in itself* or *objective representation* [*Vorstellung an sich* or *objective V.*] and of *proposition in itself* [*Satz an sich*]. The representations in themselves, on their part subdivide into *object representations in themselves* [*Gegenstandsvorstellungen an sich*] and *quality representations in themselves* [*Beschaffenheitsvorstellungen an sich*]. Following the widespread use in the literature, we will use '*idea*' as synonymous with '*representation in itself*' and '*proposition*' as synonymous with '*proposition in itself*'.

We will write '$V(x)$' for '(the thing) x is an object idea', '$B(x)$' for '(the thing) x is a quality idea' and '$S(x)$' for '(the thing) x is a proposition'.

(b2) The main notion of lectological semantics is the relation of *referring to* [*sich beziehen auf*] that may subsist between an object idea and an object or, respectively, between a quality idea and a quality. This relation induces a basic partition among the object (quality) idea, according to their referring or not referring to at least one object (one quality). In the first case, the ideas are said to be *objectual* [*gegenständlich*]; in the second case, they are said to be *objectless* or *object-free* [*gegenstandlos*]. Analogously (and later on we will see that this analogy is truly very close), the propositions divide into *true* [*wahr*] and *false* [*falsch*]. A true proposition is also said to be a *truth* (*in itself*) [*Wahrheit* (*an sich*)].

We will write '$x \,\mathrm{E}\, y$' for '(the thing) x underlies [*steht unter*] (or is subordered [*untergeordnet o subordinirt*] to or is subsumed [*subsumirt*] under or is contained [*enthalten*] in or is comprehended [*umfasset*] in or is included [*eingeschlossen*] in) (the thing) y' as well as, and especially, for '(the thing) y refers to (the thing) x';[1] we will also write '$G(x)$' for '(the thing) x is objectual' and '$W(x)$' for '(the thing) x is true'.

(c) The third main conceptual area we will refer to as '*etiology*'. It has to do with the idea, inspired by the Leibnizian–Wolffian tradition, that among the truths (in themselves) there is an ordering according to which some of them are the *reason* [*Grund*] of others and the latter are the *consequence* [*Folge*] of the former. The basic relation occurring between two truths—when the first is the reason for the second and hence the second consequence of the first—is referred to as a relation of *consecutivity* [*Abfolge*].[2] In fact, this relation occurs not only between two truths, but more generally also among sets of truths; how to account for this additional peculiarity will be discussed in due course.

For the relation of consecutivity, we will use the notation '$\triangleright$'.

(c1) The Bolzanian theory of *causality* arises from the interaction between the relation of consecutivity and the metaphysical property of existence.[3] Bolzano,

[1] Cf. TS §66, I 215.

[2] In the WL, Bolzano did not use the term 'etiology' for this problematic area; he had, however, done so in the preliminary works for what was to be the second volume of the *Beyträge*, where he even distinguished a 'general etiology [*allgemeine Aetiologie*]' as the theory of reason and consequence [*Grund und Folge; ratio et sequela*] from a 'particular or special or chronic etiology [*besondre, specielle oder chronische Aetiologie*]' resulting from the interaction of the preceding with a 'chronology [*Chronologie*]', that is, a theory of time. Cf. BGA II, A, 5, pp. 77–112.

[3] This takes the place of the 'special etiology' of the *Beyträge* II.

indeed, believes that the thing x is *cause* [*Ursache*] of the thing y (henceforth the *effect* [*Wirkung*] of x) if and only if the truth asserting the existence of x is the reason for the truth asserting the existence of y.

For '*(the thing) x is the cause of (the thing) y*' or, equivalently, '*(the thing) y is the effect of (the thing) x*', we will write '$x \succ y$'.

(d) Finally, the fourth main conceptual area is the *theory of knowledge*. The basic notion is the relation of *having as content* or *material* [*als Inhalt oder als Stoff haben*] that may occur between certain psychic phenomena [*psychische Erscheinungen*] (i.e. some existent singular qualities of thinking beings) and lectological entities. In particular, the *subjective representations* [*subjective Vorstellungen*], which we will simply call '*presentations*', have ideas as their materials and are *objectual* iff such are their materials, whereas the *judgements* [*Urtheile*] have propositions as their materials and are *true* or *knowledges* [*Erkenntnisse*] iff their materials are true.

> We will write '$x \lessdot y$' for '(the thing) x is material or content of (the thing) y', '$R(x)$' for '(the thing) x is a presentation', '$U(x)$' for '(the thing) x is a judgement', and '$E(x)$' for '(the thing) is a knowledge'.

(d1) The Bolzanian *theory of signs* is developed through the interaction between causal relations and cognitive psychic phenomena.

As a simple inspection of the preceding remarks shows, the real 'extension' of the notion of object remains open. Indeed, they only indicated that *objects are all and only all things that have qualities*. It is worth emphasizing, as we will see subsequently, how it is possible to acquire a rich, initial, and organic view of the Bolzanian system from this simple information. For convenience, we will refer to this context as the *elementary level of the system*. For the development of other, very important features of the system, it will later become necessary to cope with two questions, which are, by the way, completely independent of each other. The first question is: *Are qualities also objects*? That is to say, are there in the universe things that belong to qualities (i.e. things that belong to things that belong to things)? The richer view obtained from a positive answer to this question will constitute what we call the *higher level of the system*. The second question is: *Are the lectological things objects*? That is to say, do ideas and propositions have qualities (i.e. are there things in the universe that belong to ideas or to propositions)? For reasons to be explained later, we will call the richer view obtained from a positive answer to this question the *symbolical level of the system*. Clearly, at this point, the question may arise: Is *every thing an object*? We cannot exclude this. It is even probable that Bolzano would have answered this question positively. Nonetheless, we prefer to avoid this question for two reasons. The first is that a positive answer exposes the system to risks of inconsistencies that are not easily avoided, while the second is that, besides the risks, such an answer does not seem to enrich the system.

1.2 The Logic and the Language

For the exposition of the Bolzanian system, we will use the logical framework constituted by the '*epsilon calculus with identity and (conditional) definitions*'. The exact description of this system is given in Appendix A; here we just point out that, besides the usual apparatus of elementary logic with identity and definitions, we have, for each formula $\alpha[x]$, the individual term $\varepsilon x\alpha$, which, freely speaking, denotes a thing chosen among those that satisfy $\alpha[x]$, if such things exist. Indeed, the *First epsilon axiom* states $\alpha[x] \rightarrow \alpha[\varepsilon x\alpha]$. We will, however, opt for the 'intensional' character of the choice, that is to say that the element chosen by the epsilon among those that satisfy $\alpha[x]$ strictly depends on the 'nature' of $\alpha[x]$ and not simply on its extension; in other words, we will not accept the '*Second epsilon axiom*' (or '*Ackermann's axiom*'), which says $\forall x(\alpha[x] \leftrightarrow \beta[x]) \rightarrow \varepsilon x\alpha = \varepsilon x\beta$.

As far as the *descriptive part* of our language is concerned, for the moment we simply accept that we have a range of predicates and precisely:

1. The six monadic predicates: O [to be an object], Q [to be a quality], D [to exist], V [to be an object idea], B [to be a quality idea], S [to be a proposition].
2. The four dyadic predicates: $\in$ [to have], E [to underlay to], $<$ [to be (proper) part of], $\lessdot$ [to be material of].
3. For each $n > 1$, the $(2 + 2n)$-adic predicate H^{2+2n} [to differ from with respect to the n components].
4. The dyadic predicate $\triangleright$ [to be reason of].

Many other symbols (individual, functorial, or predicative) will be introduced, usually by means of defining axioms (briefly, *definitions*).[4] These definitions will be enumerated in the 'Dm.n' form, where m is the chapter and n is the progressive number within the chapter in which they are introduced. Following the same rule, the theorems will be enumerated in the form 'Tm.n'. Often, instead of proper definitions, we will have to make do with *definition schemata* (i.e. metalinguistic formulations synthesizing an infinity of definitions, each of which depends on one or more numerical indices). They will be listed as simple definitions, but prefixed

[4] See, however, the expansion of the primitive stock of symbols discussed in Section 1.6.

by an S (i.e. 'SDm.n'); analogously, *theorem schemata* will be indicated as 'STm.n'. The full name of definitions and theorems will only be indicated in other chapters; within a chapter, the index m will be omitted. Sometimes, it will be convenient to use abbreviations (i.e. metalinguistic definitions) of the form ... := ..., and these will be enumerated as 'MDm.n' (or 'SMDm.n').

As far as the *specific axioms are concerned*, we distinguish between *general* or *frame axioms* that govern the general behaviour of primitive symbols, and *existential* or *comprehension axioms* that ensure the presence in the universe of particular things (qualities, object or quality ideas, and propositions). General axioms will be enumerated in the form 'GAm' and, in the case of schemata (for instance with H^{2+2n} or $\triangleright^{m,n}$), in the form 'SGAm'. Comprehension axioms will be enumerated respectively in the forms 'CQm', 'CVm', 'CBm', and 'CSm', with an S prefix for the schemata.[5]

1.3 General (Elementary) Principles of Ontology, Lectology, and Epistemology

As we have already remarked, the basic principle that connects the main notion of *categorical ontology* is that the domain of the relation of *having* ($\in$) consists of the *objects* (O), whereas its codomain consists of the *qualities* (Q). This can be reduced to the following three axioms:

(a) *having may hold only between objects and qualities*

$$x \in y \to O(x) \wedge Q(y) \qquad \text{GA1}$$

(b) *every object has at least one thing*

$$O(x) \to \exists y(x \in y) \qquad \text{GA2}$$

(c) *every quality belongs to at least one thing*

$$Q(x) \to \exists y(y \in x) \qquad \text{GA3}$$

These axioms lead directly to the following theorems

$$O(x) \leftrightarrow \exists y(x \in y) \qquad \text{T1.1}$$

$$Q(x) \leftrightarrow \exists y(y \in x) \qquad \text{T1.2}$$

[5] Appendix B will make their finding possible in every moment.

which may be regarded as definitions of O and Q in terms of $\in$ and show that the true primitive notion of categorical ontology is in fact the latter. Indeed, by assuming these two theorems as defining axioms of O and Q, the three preceding axioms become theorems. However, we prefer the other formulation, as it is more apt to illustrate (as will be done in due time) the criticisms that have been raised against this point of view. These are related to the fact that we can deduce from these axioms not only that there are no 'Musilian objects',[6] i.e.

$$\neg\exists x(O(x) \wedge \forall y(Q(y) \rightarrow x \notin y)) \qquad \text{T1.3}$$

which does not seem to disturb anyone, but also that there are no 'vacuous properties', i.e.

$$\neg\exists y(Q(y) \wedge \forall x(O(x) \rightarrow x \notin y)) \qquad \text{T1.4}$$

In agreement with the remarks at the end of §1.1, our axioms neither affirm nor exclude that *also qualities are objects*. The assumption $Q(x) \rightarrow O(x)$ will be the principle that elevates us from the '*elementary level*' to the '*higher level*'.

As far as mereology is concerned, here we will limit ourselves to the assumption that the proper part relation ($<$) is a 'strict ordering' (i.e. irreflexive[7] and transitive)

$$\neg(x < x) \qquad \text{GA4}$$

$$x < y \wedge y < z \rightarrow x < z \qquad \text{GA5}$$

and to introduce by definition the more general notion of part ($\leq$) by

$$x \leq y \leftrightarrow x < y \vee x = y \qquad \text{D1.1}$$

as well as the notion of aggregate [*Inbegriff*] (Ag) as any thing having at least two distinct proper parts:[8]

$$Ag(x) \leftrightarrow \exists yz(y \neq z \wedge y, z < x) \qquad \text{D1.2}$$

The notion of replacement will be discussed in Section 1.5.

As far as *metaphysical ontology* is concerned, we accept that

(a) *only objects or qualities may exist*

$$D(x) \rightarrow O(x) \vee Q(x) \qquad \text{GA6}$$

[6] TS §80, I 274: "It is easy to see that every object will have its own attributes".

[7] TS §19, I 70: "[...] which seems as absurd as to claim that a part of a whole constitutes the whole itself".

[8] See, however, Appendix C.

(b) *existent qualities may be only qualities of existent things*

$$D(x) \wedge Q(x) \rightarrow \forall y(y \in x \rightarrow D(y)) \qquad \text{GA7}$$

(c) *parts of existent things are existent*[9]

$$D(x) \wedge y < x \rightarrow D(y) \qquad \text{GA8}$$

At this point, we introduce the predicates *St* [*being a substance*] and *Ad* [*being an adherence*] by stating that *substances are existent things that are not qualities*, whereas *adherences are existent things that are qualities*; i.e.

$$St(x) \leftrightarrow D(x) \wedge \neg Q(x) \qquad \text{D1.3}$$
$$Ad(x) \leftrightarrow D(x) \wedge Q(x) \qquad \text{D1.4}$$
$$D(x) \rightarrow Sost(x) \vee Ad(x) \qquad \text{T8.1}$$

To be more precise, when Bolzano speaks of 'adherences', he does not seem to think in general about existent qualities, but rather about *singular* existent qualities. Hence we introduce the new predicate Ad° [*being a singular adherence*] by stating such is any adherence that belongs to one and only one thing; i.e.

$$Ad^{\circ}(x) \leftrightarrow Ad(x) \wedge \exists y \forall z(z \in x \leftrightarrow z = y) \qquad \text{D1.5}$$
$$Ad^{\circ}(x) \leftrightarrow D(x) \wedge Q(x) \wedge \exists y(D(y) \wedge \forall z(z \in x \leftrightarrow z = y)) \qquad \text{T8.2}$$

As far as *lectological morphology* is concerned, we begin with the introduction by definition of the predicate *L* [*being a lectological thing*], stating that *lectological things are ideas (of object or of quality) as well as the propositions*; i.e.

$$L(x) \leftrightarrow V(x) \vee B(x) \vee S(x) \qquad \text{D1.6}$$

and, for the moment, we will limit ourselves to the assumption that *the notion of idea (of object or of quality) and the notion of proposition are disjoint*; i.e.

$$V(x) \vee B(x) \rightarrow \neg S(x) \qquad \text{GA9}$$

It is important to point out that this axiom allows us neither to affirm nor to exclude that *qualities ideas are also object ideas*; obviously, the principle $B(x) \rightarrow V(x)$ will become true at the '*higher level*'.

Concerning the relationships between these and the ontological notions, we may assume two fundamental principles:

[9] It is worth remarking that, in general, it does not hold that a quality of an existent exists, or that an existent be only part of existents. Nevertheless, in TS I §80, 278, Bolzano says: "*Note* 2. . . . In the same case in which we can and must say that a property exists, namely, when the object to which it belongs itself has existence".

(a) *lectological things do not exist, they have no reality*

$$L(x) \rightarrow \neg D(x) \qquad \text{GA10}$$

(b) *lectological things are not qualities*

$$L(x) \rightarrow \neg Q(x) \qquad \text{GA11}$$

Again, in agreement with the remarks at the end of §1.1, our axioms neither allow us to affirm nor to exclude *that lectological things are also objects*; the assumption $L(x) \rightarrow O(x)$ will be the principle that elevates us from the '*elementary level*' to the '*symbolic level*'.

As already indicated, *lectological semantics* is centred on the relation of *underlying* (E).

The first assumption about it is that only *objects or qualities may underlie and only ideas may refer*, and, more precisely, *object ideas to objects and quality ideas to qualities*; i.e.

$$x \mathsf{E} y \rightarrow (O(x) \wedge V(y)) \vee (Q(x) \wedge B(y)) \qquad \text{GA12}$$

Bolzano was also convinced that *every object and every quality should underlie at least one idea of the suitable category*; i.e.

$$O(x) \rightarrow \exists y(V(y) \wedge x \mathsf{E} y) \qquad \text{GA13}$$

$$Q(x) \rightarrow \exists y(B(y) \wedge x \mathsf{E} y) \qquad \text{GA14}$$

It is important to notice that, in contrast to what happens in the case of *belonging* and *being a quality*, the notion of *being an idea (of object or of quality)* is not definable in terms of *referring*. Indeed, the realm of things determined by the notion of being an idea (of object or of quality) is not exhausted by the domain of the referring relation, because there are vacuous ideas, ideas that do not refer to anything; in other words, the following two principles hold:

$$\exists y(V(y) \wedge \forall x \neg(x \mathsf{E} y)) \qquad \text{and} \qquad \exists y(B(y) \wedge \forall x \neg(x \mathsf{E} y))$$

which, however, do not need to be postulated because, as we will see later, they follow from other axioms. Obviously, these two principles lead to the next two theorems:

$$\neg(V(y) \leftrightarrow \exists x(x \mathsf{E} y)) \qquad \text{and} \qquad \neg(B(y) \leftrightarrow \exists x(x \mathsf{E} y))$$

We should also point out that at the higher level, where both $Q(x) \rightarrow O(x)$ and $B(x) \rightarrow V(x)$ hold and hence GA12 reduces to $x \mathsf{E} y \rightarrow (O(x) \wedge V(y))$, we have $V(y) \wedge x \mathrel{\text{E}} y \rightarrow O(x)$ and consequently, remembering GA13, also $O(x) \leftrightarrow \exists y(V(y) \wedge x \mathrel{\text{E}} y)$. In other words, *at the higher level, being an object is equivalent to underlying an idea.*

At the *higher level,* we will have the possibility to consider a different form of this group of axioms, through the intervention of another more complex ontological notion: that of *situation* (Sit) as the reference of a proposition. So GA12 will become

$$x \mathrel{\text{E}} y \rightarrow (O(x) \wedge V(y)) \vee (Q(x) \wedge B(y)) \vee (Sit(y) \wedge S(x)) \qquad \text{GA12}'$$

while GA13 e GA14 will become

$$Sit(x) \rightarrow \exists y(S(y) \wedge x \mathrel{\text{E}} y) \qquad \text{GA14}'$$

Insofar as *epistemology* is concerned, we have already pointed out that its basic notion is that of *being material* $(\lessdot)$, in terms of which the other characteristic notions—*object representation* (R), *quality representation* (R'), and *judgement* (U)—are definable. Indeed, we set

$$R(y) \leftrightarrow \exists x(x \lessdot y \wedge V(x)) \qquad \text{D1.7}$$

$$R'(y) \leftrightarrow \exists x(x \lessdot y \wedge B(x)) \qquad \text{D1.8}$$

$$U(y) \leftrightarrow \exists x(x \lessdot y \wedge S(x)) \qquad \text{D1.9}$$

We assume the left-univocity of $(\lessdot)$: *the materials of a thing, if ever, are unique:*

$$x \lessdot y \wedge z \lessdot y \rightarrow x = z \qquad \text{GA15}$$

Object or quality presentations, as well as judgements, are existent singular qualities of existent things, i.e. *singular adherences of existents.*

We have the following two principles:

(a) *only lectological things may be materials*

$$x \lessdot y \rightarrow L(x) \qquad \text{GA16}$$

(b) *only singular adherences have materials*

$$x \lessdot y \rightarrow Ad^{\circ}(y) \qquad \text{GA17}$$

As in some contexts it is convenient to have particular symbols for the converses of the predicates $\in$ [*to have*], E [to *underlay to*], $<$ [*to be (proper) part of*], and $\lessdot$ [*to be material of*], we take for them, respectively, $\ni$ [*to belong to*], $>$ [*to have as part*], $\reflectbox{$\mathsf{E}$}$ [*to refer to*], and $\succ$ [*to contain as material*], together with the following definitions:

$$y \ni x \leftrightarrow x \in y \qquad \text{D1.10}$$

$$y > x \leftrightarrow x < y \qquad \text{D1.11}$$

$$y \mathrel{\reflectbox{E}} x \leftrightarrow x \mathrel{\mathsf{E}} y \qquad \text{D1.12}$$

$$y \succ x \leftrightarrow x \prec y \qquad \text{D1.13}$$

1.4 Restricted Quantifiers

In general, a monadic predicate P circumscribes, within the universe of the discourse, a certain subclass: the class of those x for which $P(x)$ holds. A useful convention to study the behaviours of those things that lie in the class isolated by P is referred to as the 'restriction of the quantifiers to', characterized by two metalinguistic definitions (i.e. simple notational abbreviations):

$$\underset{P(x)}{\forall x}\, \alpha[x] := \forall x(P(x) \to \alpha[x]) \qquad \text{DM1.1}$$

$$\underset{P(x)}{\exists x}\, \alpha[x] := \exists x(P(x) \land \alpha[x]) \qquad \text{DM1.2}$$

Naturally, this restriction may also be applied to the epsilon by the further definition

$$\underset{P(x)}{\varepsilon x}\, \alpha[x] := \varepsilon x(P(x) \land \alpha[x]) \qquad \text{DM1.3}$$

It is even more advantageous, especially when we need to simultaneously consider subclasses determined by various predicates, to introduce 'special variables' (i.e. variables that vary over such subclasses). In the following sections, we will use the following special variables, while others will be introduced further on:

$$\text{for objects } i, j, \ldots \begin{cases} \forall i\alpha[i] := \underset{O(x)}{\forall x}\, \alpha[x] \\ \exists i\alpha[i] := \underset{O(x)}{\exists x}\, \alpha[x] \\ \varepsilon i\alpha[i] := \underset{O(x)}{\varepsilon x}\, \alpha[x] \end{cases}$$

$$\text{for qualities } r, s, \ldots \begin{cases} \forall r\alpha[r] := \underset{Q(x)}{\forall x}\, \alpha[x] \\ \exists r\alpha[r] := \underset{Q(x)}{\exists x}\, \alpha[x] \\ \varepsilon r\alpha[r] := \underset{Q(x)}{\varepsilon x}\, \alpha[x] \end{cases}$$

$$\text{for object ideas } a, b, \ldots \begin{cases} \forall a \alpha[a] & := \underset{V(x)}{\forall x} \, \alpha[x] \\ \exists a \alpha[a] & := \underset{V(x)}{\exists x} \, \alpha[x] \\ \varepsilon a \alpha[a] & := \underset{V(x)}{\varepsilon x} \, \alpha[x] \end{cases}$$

$$\text{for quality ideas } \bar{a}, \bar{b}, \ldots \begin{cases} \forall \bar{a} \alpha[\bar{a}] & := \underset{B(x)}{\forall x} \, \alpha[x] \\ \exists \bar{a} \alpha[\bar{a}] & := \underset{B(x)}{\exists x} \, \alpha[x] \\ \varepsilon \bar{a} \alpha[\bar{a}] & := \underset{B(x)}{\varepsilon x} \, \alpha[x] \end{cases}$$

$$\text{for propositions } p, q, \ldots \begin{cases} \forall p \alpha[p] & := \underset{S(x)}{\forall x} \, \alpha[x] \\ \exists p \alpha[p] & := \underset{S(x)}{\exists x} \, \alpha[x] \\ \varepsilon p \alpha[p] & := \underset{S(x)}{\varepsilon x} \, \alpha[x] \end{cases}$$

$$\text{for lectological things } l, l', \ldots \begin{cases} \forall l \alpha[l] & := \underset{L(x)}{\forall x} \, \alpha[x] \\ \exists l \alpha[p] & := \underset{L(x)}{\exists x} \, \alpha[x] \\ \varepsilon l \alpha[p] & := \underset{L(x)}{\varepsilon x} \, \alpha[x] \end{cases}$$

It should be noted that the use of special variables requires a certain degree of goodwill by the reader, insofar as they present a degree of ambiguity, although this should be resolved by the context. One particular case, for example, is before '$\neg G(a)$', which usually stands for '$V(x) \wedge \neg G(x)$' rather than '$\neg(V(x) \wedge G(x))$'.

The reading of the last sections of this chapter may be deferred until it becomes explicitly necessary.

1.5 Variation

As has already been remarked, a fundamental characteristic of Bolzanian mereology is the presence, for $n > 0$, of a $2(n+1)$-ary relation of *variation* [*Veränderung*] that holds between the thing x, the things $x_1, \ldots, x_n$ (each one of which is different from the others), the things $z_1, \ldots, z_n$, and the thing z iff the thing z differs from the thing x by having the things $z_1, \ldots, z_n$ exactly in the places in which x has the things $x_1, \ldots, x_n$; in other words, when we may consider that z was obtained from x by (simultaneously and uniformly) substituting the thing z_1 for the thing $x_1, \ldots,$ the thing z_n for the thing x_n. Let us remark that when setting some parts in the places of some other parts in an x, one must choose between two different and reciprocally independent characteristics of the other parts. The first characteristic concerns their mereological simplicity or atomicity (i.e. their having no proper parts). The second characteristic concerns their possible 'multiplicity' (i.e. their occurring in more 'copies', having more 'occurrences', being 'equiform' to other parts). As is well known, the most common transformational operations of logical practice concern linguistic things and are two in number. The first, usually called '*substitution*', is characterized by the fact that it may only concern *atomic* parts, but in *all their occurrences*, while the second, usually called 'replacement', is characterized by the fact that it may concern *any type* of parts (i.e. not necessarily atomic), but, if so, some and not necessarily all of their occurrences. Bolzanian *variation* appears to take an intermediate stance between substitution and replacement. In fact, it may concern *every* type of parts, but, in such cases, every occurrence of them.

The exact determination of the conditions that completely characterize this relation is still a *desideratum*. Nonetheless, for our present aims, it will suffice to introduce a suitable notation and agree on handling it according to certain principles intuitively suggested by the preceding metaphor. Thus we will write '$H(x, \genfrac{}{}{0pt}{}{x_1,\ldots,x_n}{z_1,\ldots,z_n}, z)$' for 'the thing z differs from the thing x by having the things $z_1, \ldots, z_n$ exactly in the places the thing x has the things $x_1, \ldots, x_n$ (each one of which is different from the others)'. Moreover, we will accept that this relation is many–one in z, that is to say, for given $x, x_1, \ldots, x_n$ and $z_1, \ldots, z_n$, there can be at most one z such that the relation holds. So, besides establishing that the *things to be replaced must all differ from one another*, i.e.

$$H(x, \genfrac{}{}{0pt}{}{x_1,\ldots,x_n}{z_1,\ldots,z_n}, z) \rightarrow \bigwedge_{i \neq j \leq n} x_i \neq x_j \qquad \text{SGA1}$$

we also accept

$$H(x, \genfrac{}{}{0pt}{}{x_1,\ldots,x_n}{z_1,\ldots,z_n}, z) \wedge H(x, \genfrac{}{}{0pt}{}{x_1,\ldots,x_n}{z_1,\ldots,z_n}, u) \rightarrow z = u \qquad \text{SGA2}$$

The question of the existence of such a z is more complex. Given an x, some $x_1, \ldots, x_n$ and some $z_1, \ldots, z_n$, let us consider a generic x_i $(1 \leq i \leq n)$. It is clear that if x_i is not a part of x, there is no problem. Indeed, in this case, its substitution with the corresponding z_i does not modify x and is therefore completely harmless. So, as far as it depends on it, the z may exist. If, however, x_i is a part of x, its substitution with the corresponding z_i will be acceptable only in the case that it does not interfere with the other parts of x to be exchanged. In other words, *z may exist only if all those things among the $x_1, \ldots, x_n$ that are really part of x have no common parts*. When this is the case, however, it seems reasonable to accept that z exists.

Therefore

$$\bigwedge_{i\neq j\leq n}(x_i \leq x \wedge x_j \leq x \rightarrow x_i \perp x_j) \rightarrow \exists z H(x, \substack{x_1,\ldots,x_n \\ z_1,\ldots,z_n}, z) \qquad \text{SGA3}$$

where $x \perp y$ (*x is disjoint from y*) is of course defined by

$$x \perp y \leftrightarrow \neg\exists z(z \leq x \wedge z \leq y) \qquad \text{D1.14}$$

We may now introduce the definition

$$x\begin{bmatrix} x_1,\ldots,x_n \\ z_1,\ldots,z_n \end{bmatrix} = \varepsilon z H(x, \substack{x_1,\ldots,x_n \\ z_1,\ldots,z_n}, z) \qquad \text{SD1.1}$$

so that

$$\bigwedge_{i\neq j\leq n}(x_i \leq x \wedge x_j \leq x \rightarrow x_j \perp x_i) \rightarrow H(x, \substack{x_1,\ldots,x_n \\ z_1,\ldots,z_n}, x\begin{bmatrix} x_1,\ldots,x_n \\ z_1,\ldots,z_n \end{bmatrix}) \qquad \text{ST1.1}$$

Moreover, we may also set the aforesaid harmlessness of the x_i that are not parts of x by assuming

$$\begin{aligned} \neg(x_i \leq x) \rightarrow (H(x, \substack{x_1,\ldots,x_n \\ z_1,\ldots,z_n}, z)) & \qquad \text{SGA4} \\ \leftrightarrow H(x, \substack{x_1,\ldots,x_{i-1},x_i,x_{i+1},\ldots,x_n \\ z_1,\ldots,z_{i-1},x_i,z_{i+1},\ldots,z_n}, z) & \\ \leftrightarrow H(x, \substack{x_1,\ldots,x_{i-1},x_{i+1},\ldots,x_n \\ z_1,\ldots,z_{i-1},z_{i+1},\ldots,z_n}, z)) & \end{aligned}$$

This leads to the obvious corollary that if the replacement only concerns things that are not part of x, then nothing happens, i.e.

$$\bigwedge_{i\leq n}\neg(x_i \leq x) \rightarrow H(x, \substack{x_1,\ldots,x_n \\ z_1,\ldots,z_n}, x) \qquad \text{ST1.2}$$

or, equivalently, simply writing '$x\left[\overset{n}{x}/\overset{n}{z}\right]$' for '$x\begin{bmatrix} x_1,\ldots,x_n \\ z_1,\ldots,z_n \end{bmatrix}$',

$$\bigwedge_{i\leq n}\neg(x_i \leq x) \rightarrow x\left[\overset{n}{x}/\overset{n}{z}\right] = x \qquad \text{ST1.3}$$

It is important to note that nothing prevents the $z_1,\ldots,z_n$ from interfering. In fact, they may even all be equal.

Last, but not least, if the replacement concerns a unique part u, then this part is trivially disjoint from every other part involved in the replacement and so, by SGA3, $x[u/w]$ exists; if, on the contrary, u is not part of x, then, by SGA4, $x[u/w]$ is x and so it exists again; i.e.

$$H(x, u, w, x[u/w]) \qquad \text{T1.5}$$

$$x < y \leftrightarrow \exists z(y[x/z] \neq y) \qquad \text{T1.6}$$ [10]

1.6 Ordered n-tuples

Bolzanian mereology is a rather complex subject that contains not only obscurities and ambiguities but also some possible incongruities. Nevertheless, for our aims, we need only two things, the 'Bolzaneity' of which, apart from terminology and notation, may hardly be questioned: the notion of an *ordered n-tuple of objects* and the notion of an *n-ary relational quality*.

As far as the first notion is concerned, it does not seem possible to really question that Bolzano accepted that, for any n ($n \geq 2$) objects, their *aggregate* [*Inbegriff*] could be considered, and this in such a way that it would be possible to distinguish in it a first, a second, ..., an n-th *component* [*Bestandtheil*]. At this point of our exposition, however, it appears difficult to account for this idea while maintaining close adherence to Bolzanian views about aggregates. So we prefer to directly accept the presence, among our primitive symbolical fund, for every $n \geq 2$, of the following:

(a) an n-adic functor 'π^n' denoting the 'construction of an (ordered) n-tuple', i.e. the operation that associates with the n objects $i_1, \ldots, i_n$ the object called the '(ordered) n-tuple of $i_1, \ldots, i_n$';

b) n monadic functors 'π_{n1}', ..., 'π_{nk}', ..., 'π_{nn}' denoting the n operations 'projection of (ordered) n-tuple', i.e. those operations that with the tuple of $i_1, \ldots, i_n$ associate the first, the second, ..., the n-th component, respectively.

On the basis of this notation, the formal behaviour of these functors is ruled by the following axioms.

[10] A satisfying systematization of the relation H should make it possible to use this equivalence as a definition of the part relation.

As we intend these operations to take arguments and values only among the objects, and, in particular, we will consider the domain of the construction of n-tuples to be the realm of all objects but the domain of the n-projections to be only the codomain of the construction of n-tuples, we must first state that $\pi^n(x_1, \ldots, x_n)$ *is an object iff such are all the* $x_1, \ldots, x_n$, i.e.

$$\bigwedge_{k \leq n} O(x_k) \leftrightarrow O(\pi^n(x_1, \ldots, x_n)) \qquad \text{SGA5}$$

and, moreover, after defining the predicate '$O^n(x)$' [x is *an n-tuple of objects*] by

$$O^n(z) \leftrightarrow \exists x_1 \ldots x_n ((\bigwedge_{k \leq n} O(x_k)) \wedge z = \pi^n(x_1, \ldots, x_n)) \qquad \text{SD1.2}$$

we also state that $\pi_{nk}(z)$ *is an object iff* z *is an n-tuple of objects*, i.e.

$$O^n(z) \leftrightarrow \bigwedge_{k \leq n} O(\pi_{nk}(z)) \qquad \text{SGA6}$$

It should be noted that we have

$$O^n(z) \rightarrow O(z) \qquad \text{ST1.4}$$

Proof By SD19, $O^n(z)$ implies that z is $\pi^n(x_1, \ldots, x_n)$ for some $x_1, \ldots, x_n$ that are objects; therefore, by SGA5, $\pi^n(x_1, \ldots, x_n)$, that is to say z, is also an object.

As far as the interactions among the operations are concerned, it will of course suffice to accept

$$O^n(x) \rightarrow \pi^n(\pi_{n1}(x), \ldots, \pi_{nn}(x)) = x \qquad \text{SGA7}$$

$$\bigwedge_{k \leq n} O(x_k) \rightarrow \pi_k^n(\pi(x_1, \ldots, x_k, \ldots, x_n)) = x_k \qquad \text{SGA8}$$

or

$$\pi^n(\pi_{n1}(i), \ldots, \pi_{nn}(i)) = i \qquad \text{SGA7}'$$

$$\pi_k^n(\pi(i_1, \ldots, i_k, \ldots, i_n)) = i_k \qquad \text{SGA8}'$$

from which the *characteristic property of the notion of ordered n-tuple* follows:

$$(\pi^n(i_1, \ldots, i_n) = \pi^n(i'_1, \ldots, i'_n) \leftrightarrow \bigwedge_{k \leq n} i_k = i'_k) \qquad \text{ST1.5}$$

Proof If $\pi^n(i_1, \ldots, i_n) = \pi^n(i'_1, \ldots, i'_n)$, then clearly also

$$\pi_{nk}(\pi^n(i_1, \ldots, i_n)) = \pi_{nk}(\pi^n(i'_1, \ldots, i'_n))$$

holds for all k; however, by SGA7′, $\pi_{nk}(\pi^n(i_1,\ldots,i_n)) = i_k$ and $\pi_{nk}(\pi^n(i'_1,\ldots,i'_n))= i'_k$; thus $i_k = i'_k$. This is so for all $k \leq n$. The inverse is just a logical axiom of identity.

In the following, instead of '$\pi^n(x_1,\ldots,x_n)$', we will use the familiar '$< x_1,\ldots,x_n >$', and, if convenient, '$< \overset{n}{x} >$'; i.e.

$$< x_1,\ldots,x_n > := \pi^n(x_1,\ldots,x_n) \qquad \text{SDM1.4}$$

$$< \overset{n}{x} > := < x_1,\ldots,x_n > \qquad \text{SDM1.5}$$

1.7 Relations (Relational Qualities)

Moving on to the second question, a preliminarily remark is that even if not completely ambiguity-free, the Bolzanian treatment of the relations appears worthy of consideration. In fact, the essential traits are far closer than any other antecedent to the treatment current in today's set theory.

First of all, Bolzano is convinced, against a long and well-established tradition, that there is no difference between the ontological status of the properties and the ontological status of the relations. In particular, therefore, with respect to existent objects, relations are as real as the properties and not at all simply 'imaginary [*eingebildete*]'.[11]

Second, referring to Leibniz, Bolzano accepts that the relations are not simply binary, but may concern an *arbitrary number of objects*:

> "That moreover the objects required for a relation should not always be just two but may sometimes be more, has already been remembered by Leibniz [...]."[12]

Third, Bolzano is persuaded that *either the notion of relation may be defined by some reduction to the notion of quality or it is undefinable and must be taken as a primitive notion*:

> "If one would not accept that the notion of relation could be reduced to the notion of quality, one should also give soon up the hope to a decomposition of it and so immediately affirm its simplicity."[13]

In this situation, Bolzano decides, not without hesitation, to go along with the first possibility and conceive *relations* [*Verhältnisse*] as *qualities of aggregates* [*Beschaffenheiten von Inbegriffen*]:

[11] Cf. TS §80, I 278, *Note* 2.

[12] TS §80, I 276.

[13] TS §80, I 276.

> "Because a whole which is made up of more objects *A*, *B*, *C*, *D*, ... as its parts is, as such, a particular object that is essentially distinct from each of its parts, it is clear that to every whole, qualities may belong which do not belong also to its parts; if I am not mistaken, these are the qualities which we call relations among these parts."[14]

This definition is adopted again in the *Grössenlehre*, where we find the following:

> "Definition. If a quality *m* does not separately belong to the several objects *A*, *B*, *C*, *D*, ... , but to the whole they make up in their union, we say that *m* is a relation which holds [*obwaltet*] among the objects *A*, *B*, *C*, *D*, ... The objects *A*, *B*, *C*, *D*, ..., to the aggregate of which the quality adheres, which is called a relation among them, are called one by one the members [*Glieder*] of this relation."[15]

It is worth pointing out that Bolzano seems to be aware, if confusedly, of the ambiguity of the current use of the term 'relation' and, in particular, of the fact that it is used to denote both a relation in abstract and a singular instance of it—in set-theoretical perspective, to denote both an *n*-ary relation and a particular *n*-tuple among which it holds.[16]

It should also be pointed out that Bolzano appears to be well aware of the difference between a *relation* and a *parametric quality*, which he calls '*external* [*aüßere*] *quality*'. It is the difference occurring, for instance, between the *relation* of fatherhood between Philip and Alexander the Great and the *two qualities*, first, that of Philip being the father of Alexander and, second, that of Alexander having Philip as father:

> "Although [...] a relation *x* holding among *A*, *B*, *C*, *D*, ... is a quality which in principle only belongs to the whole made up by the parts *A*, *B*, *C*, *D*, ...; nevertheless of every part, for instance of *A*, we may at least truely affirm, that the fact that *A* together with the things *B*, *C*, *D*, ..., makes up a whole to which the quality *x* belongs, is a quality belonging to *A*. It is this quality of *A* which we call external. Thus, under an external quality of an object we understand a quality of it consisting just in the fact that it entertains a certain relation with another object."[17]

[14] TS §80, I 274.

[15] BGA, IIA7, p.108: "*Erkl[ärung]*. Wenn eine Beschaffenheit *m* keinem der mehreren Gegenstände *A*, *B*, *C*, *D*, ... für sich allein, wohl aber dem Ganzen, das sie in ihrer Vereinigung bilden, zukommt; so sagen wir, *m* sey *ein zwischen den Gegenständen A*, *B*, *C*, *D*, ... *obwaltendes Verhältniß*. [...] Die Gegenstände *A*, *B*, *C*, *D*, ... an deren Inbegriffe die Beschaffenheit *m*, die ein Verhältniß zwischen denselben gennant wird, haftet, werden im Einzelnen die *Glieder* dieses Verhältnisses gennant."

[16] See the central part of TS §80.3, I 274.

[17] Cf. TS §80, I 274. Textor [1996, pp.70–3] seems to have misunderstood the question; his misunderstanding is inherited by Krause [2004, pp.101–4].

The not-external qualities are called '*internal* [*innere*]' or also '*proprieties* [*Eigenschaften*]'.[18]

Leaving aside all other aspect of the Bolzanian theory of relations, we may thus easily accept that *n-ary relations* are qualities of objects of the form $< x_1, \ldots, x_n >$ and introduce, by definition, the monadic predicate '$Q^n(x)$' ['*x is an n-ary relation*', or, better, '*x is an n-ary (relational) quality*'], stating

$$Q^n(r) \leftrightarrow \forall i(i \in r \rightarrow O^n(i))) \qquad \text{SD1.3}$$

and hence

$$Q^n(r) \leftrightarrow \forall i(i \in r \rightarrow \exists i_1 \ldots i_n(i = < i_1, \ldots, i_n >)) \qquad \text{ST1.6}$$

Using 'r^n' as a special variable for n-ary qualities, we may introduce 'parametric qualities' by the comprehension axiom

$$\forall r^n \forall i \exists s \forall i_1 \ldots i_{k-1} i_{k+1} \ldots i_n (i \in s \leftrightarrow < i_1, \ldots, i_{k-1}, i, i_{k+1}, \ldots, i_n > \in r^n) \qquad \text{SCQ1}$$

which we solve as follows:

$$r^n_{i_1 \ldots i_{k-1} i_{k+1} \ldots i_n} = \varepsilon s \forall i_1 \ldots i_{k-1} i_{k+1} \ldots i_n (i \in s \leftrightarrow < i_1, \ldots, i_{k-1}, i, i_{k+1}, \ldots, i_n > \in r^n \qquad \text{SD1.4}$$

Thus we have

$$Q(r^n_{i_1 \ldots i_{k-1} i_{k+1} \ldots i_n}) \qquad \text{T1.7}$$

$$i \in r^n_{i_1 \ldots i_{k-1} i_{k+1} \ldots i_n} \leftrightarrow < i_1, \ldots, i_{k-1}, i, i_{k+1}, \ldots, i_n > \in r^n \qquad \text{T1.8}$$

[18] TS §80, I 275: "4) [...] Attributes which are not external, [...] we call internal *attributes* or *properties*".

2

Elementary Level (A): Ideas

2.1 Preamble

The main points of the Bolzanian theory of ideas are as follows:

1. Two fundamental distinctions are introduced: the mereological distinction between simple and composite ideas and the semantic distinction among ideas that refer to something and ideas that do not refer to anything.
2. A whole set of relations (compatibility, inclusion, equivalence, etc.) are introduced between two ideas that refer to something.
3. The same relations are extended to situations involving more than two ideas.
4. The presence or some simple ideas and idea-generating operations is postulated.
5. The relations of point 2 are extended to non-referring ideas.

This chapter addresses points 1, 2, and 4 on the elementary level. Moreover, we provide an initial version of point 3 that will be expanded upon and in part superseded by the treatment developed on the symbolic level (where lectological things are also regarded as objects). Point 5 will be treated in the following chapter in the context of the Bolzanian theory of variation.

2.2 Properties of Ideas

As just mentioned, from the mereological point of view, ideas (of objects or of qualities) are subdivided into *simple* [*einfach*] and *composite* [*zusammengesetzt*]

Bolzano's Logical System. Ettore Casari.

according to the occurrence—or not—in them of other ideas as their parts. Bolzano distinguishes two aspects in a composite idea: the set of its proper parts, which he calls the '*content* [*Inhalt*]' of the idea,[1] and the way in which these parts are combined together, which he calls the *composition mode* [*Zusammensetzungs*— or *Verbindungsart*] of the idea.[2] It is not completely clear, however, when Bolzano refers to the 'content' of an idea, whether he intends solely the set of simple ideas occurring in it or also some of its composite sub-ideas. This is not a negligible issue, as it is related to the *theory of definition*, a fundamental chapter of Bolzanian Philosophy of Science.

From the semantic point of view, on the other hand, the fundamental distinction existing among ideas (of objects or of qualities) is their *objectuality* [*Gegenständlichkeit*] or *objectlessness* [*Gegenstandlosigkeit*], respectively. An idea is *objectual* [*gegenständlich*] if it refers to at least one thing of the suitable type (i.e. an object, if it is an object idea, and a quality, if it is a quality idea), while an idea is *objectless* [*gegenstandlos*] in the opposite case (i.e. when it does not refer to any object or any quality).[3]

We thus introduce a new monadic predicate 'G' [$G(x)$ :: x *is objectual*], the behaviour of which is fixed by the conditional defining axiom

$$\forall x(V(x) \vee B(x) \rightarrow (G(x) \leftrightarrow \exists z(z \mathrel{\text{E}} x))) \qquad \text{D2.1}$$

from which, of course,

$$\underset{V(x)}{\forall x}(G(x) \leftrightarrow \exists z(z \mathrel{\text{E}} x)) \qquad \text{T2.1}$$

$$\underset{B(x)}{\forall x}(G(x) \leftrightarrow \exists z(z \mathrel{\text{E}} x)) \qquad \text{T2.2}$$

and so also

$$\underset{V(x)}{\forall x}(\neg G(x) \leftrightarrow \neg\exists z(O(z) \wedge z \mathrel{\text{E}} x)) \qquad \text{T2.3}$$

$$\underset{B(x)}{\forall x}(\neg G(x) \leftrightarrow \neg\exists z(Q(z) \wedge z \mathrel{\text{E}} x)) \qquad \text{T2.4}$$

[1] TS §56, I 178: "The sum of the parts of which a given idea consists is usually called its content."

[2] TS §96, I 320: "[...] provided that when we say that ideas have the same content we only mean that their ultimate constituents are the same, *not the way these constituents are combined* (§56).'

[3] TS §50, I 162: "When I say of an idea that it has objectuality I mean nothing but that there are objects which are subsumed under it, where I want the *are* to be interpreted in the same way as in the phrase there *are* truths (§30)."

TS §66, I 215: "2) I call ideas that have one or more objects *objectual*, or say that they are *object-ideas*. On the other hand, I call those that have no object corresponding to them *objectless*."

that is, using the special variables,

$$G(a) \leftrightarrow \exists i(i \mathrel{\varepsilon} a) \qquad \text{T2.5}$$

$$G(\overline{a}) \leftrightarrow \exists r(r \mathrel{\varepsilon} \overline{a}) \qquad \text{T2.6}$$

$$\neg G(a) \leftrightarrow \neg\exists i(i \mathrel{\varepsilon} a) \qquad \text{T2.7}$$

$$\neg G(\overline{a}) \leftrightarrow \neg\exists r(r \mathrel{\varepsilon} \overline{a}) \qquad \text{T2.8}$$

An important point, which is widely and critically discussed in the Brentanian school (see Twardowski) is the Bolzanian conviction that there are objectless ideas. The apparent paradoxical nature of 'a representation that does not represent anything' is solved by Bolzano by means of the distinction between subjective representation (presentation) and objective representation (idea). Every presentation has *material* and therefore a presentation like 'nothing' or 'round square' has it too: the very ideas nothing or round square, respectively; however, these ideas have no reference. Bolzano also makes a distinction (we will return to this later)[4] between ideas that have no reference for logical reasons (*nothing, round square, etc.*), which he calls '*imaginary* [*imaginär*] *ideas*', and ideas that lack of reference for contingent reasons (*golden mountain*). Surprisingly, he includes among the imaginary ideas those such as *green virtue*, which Husserl will later wisely distinguish as *Unsinne*, as opposed to the former, *Widersinne*.[5]

According to Bolzano, each objectual idea is associated with what he calls the *field* or the *extension* or the *sphere* [*Gebiet, Umfang, Sphäre*] of the idea. This is constituted, he says, by "that quality of an idea in virtue of which it represents these and not other objects"[6] and the exact determination of which, he insists, cannot by given by simply determining the "set of underlying objects" but, if they are more than one, also by the indication of which they " singularly are"[7]—a notion that is rather complex and the whole significance of which is not easy to grasp. In turns out, however, that in the later developments of the theory, such complexity does not play a significant role; what counts is just the "set of underlying objects". Let us

[4] In Chapter 8.

[5] TS §67, I 220-1: "*There are also objectless ideas.* It is true that most ideas have some, or even infinitely many, objects. Still, there are also ideas that have no object at all, and thus do not have an extension. [...] Somebody might in turn find it absurd that an idea or representation should have no object at all, and thus represent nothing, but the reason for this is in all likelihood that he means by ideas merely mental ideas, i.e., thoughts, and that he identifies the matter of these mental ideas (i.e., the ideas in themselves) with their objects."

[6] TS §66, I 215: "2) [...] that particular attribute of an idea by virtue of which it represents only those and no other objects".

[7] TS §66, I 215: "3) In order to determine completely the extension of an idea that has several objects, it is not sufficient [...] merely to determine the number of these objects. It must be indicated precisely which objects they are."

remark, however, that, strictly speaking, Bolzano reserves the term 'aggregate' and its various specifications 'set', 'sum', etc. for the cases in which there is more than one element.

Moreover, each field of an idea is associated with what he calls the *width (of the field)* [*Weite des Gebietes*]: the *size of the set* [*Größe der Menge*] of underlying objects.[8]

Bolzano considers, for every object idea a and every natural number n, the following three properties:

(1) a refers to at least n objects (we write '${}_nA(a)$');
(2) a refers to at most n objects (we write '${}^nA(a)$');
(3) a refers (exactly) to n objects (we write '${}^n_nA(a)$').

Analogously, the same is done for quality ideas. So we may set

$$ {}_nA(a) \leftrightarrow \overset{n}{\exists} i(i \mathrel{\text{E}} a) \qquad \text{SD2.2} $$

$$ {}^nA(a) \leftrightarrow \overset{n}{\exists}! i(i \mathrel{\text{E}} a) \qquad \text{SD2.3} $$

$$ {}^n_nA(a) \leftrightarrow \overset{n}{\exists}!! i(i \mathrel{\text{E}} a) \qquad \text{SD2.4} $$

$$ {}_nA(\overline{a}) \leftrightarrow \overset{n}{\exists} r(r \mathrel{\text{E}} \overline{a}) \qquad \text{SD2.5} $$

$$ {}^nA(\overline{a}) \leftrightarrow \overset{n}{\exists}! r(r \mathrel{\text{E}} \overline{a}) \qquad \text{SD2.6} $$

$$ {}^n_nA(\overline{a}) \leftrightarrow \overset{n}{\exists}!! r(r \mathrel{\text{E}} \overline{a}) \qquad \text{SD2.7} $$

Clearly,

$$ G(a) \leftrightarrow {}_1A(a) \qquad \text{T2.9} $$

$$ \neg G(a) \leftrightarrow {}^0A(a) \qquad \text{T2.10} $$

Ideas having width 1 are labelled *singular ideas* [*Einzelvorstellung*], whereas objectual but not singular ideas (which may also have infinite width) are called *common* or *universal ideas* [*Gemeinvorstellung* or *allgemeine Vorstellung*].

$$ Cm(a) \leftrightarrow G(a) \wedge \neg {}^1_1A(a) \qquad \text{D2.8} $$

$$ Cm(\overline{a}) \leftrightarrow G(\overline{a}) \wedge \neg {}^1_1A(\overline{a}) \qquad \text{D2.9} $$

[8] TS §66, I 215: "3) [...] The multitude of these objects, as any multitude, has a certain *magnitude*, which is called the *width of the extension*."

Last, but not least, this theorem implicitly states that also *quality ideas may be common*, meaning that *such ideas may refer to more than one quality*. We will return to this later when we address the Bolzanian notion of truth, where these possibilities play a very important role.[9]

2.3 Relations Between Two Ideas

In §§ 94–7 and 103–5 Bolzano introduces a series of relations among objectual ideas that depend on their objects. They are essentially relations that, in one form or another, belong to the logical tradition. The distinctive features of the Bolzanian presentation are, on the one hand, rigorous systematics and, on the other, an emphasized restriction on objectual ideas.

The first basic distinction is between ideas that are *compatible* or *in agreement* or *consonant* [*verträglich, einstimmig, einhellig*] —meaning ideas that have at least one object in common—and ideas that are *incompatible* or *dissonant* [*unverträglich, mißhellig*], ideas that have no common elements.[10]

It is important to emphasize that the ideas involved in the relations that are treated here are always—and not only in the case of compatibility— objectual;[11] as mentioned in the preamble, the extension of these relations to the case of objectless ideas will be discussed in the following chapter. Moreover, although, for simplicity, we will limit ourselves to the treatment of object ideas, everything we say holds also for quality ideas.

Thus, writing '$a \top b$' for '*a and b are compatible*' and '$a \perp b$' for '*a and b are incompatible*', we set

$$a \top b \leftrightarrow \exists i(i \sqsubset a \wedge i \sqsubset b) \qquad \text{D2.10}$$

$$a \perp b \leftrightarrow G(a) \wedge G(b) \wedge \neg\, a \top b \qquad \text{D2.11}$$

[9] TS§60, I 191: "Note 2. The distinction that obtains between abstracta and their corresponding concreta with respect to their extensions seems not to have been observed by certain logicians. They have assigned the same objects to the former as well as to the latter. In my opinion, the objects that stand, e.g., under the abstract idea 'virtue" are quite different from those that are comprised under the corresponding concrete idea 'a virtuous person'. The abstractum is an attribute-idea; hence it can comprise nothing but attributes. Thus, the attributes veracity, charity, etc., stand under "virtue', while Socrates, Aristides, etc. stand under the concretum 'a virtuous person'."

[10] TS §94, I 316: "1) Concerning the objects themselves which are represented by ideas, it appears that two ideas either have common objects or they do not. Both cases are important enough to deserve a special name. I therefore call ideas that have one or several objects in common *compatible* or *concordant*, and those that have not even a single common object I call *incompatible*."

[11] §108 I 371: "According to their definitions, the relations that we have considered in §93 ff. only hold between ideas that *have objects*."

Bolzano also distinguishes between two special cases of compatibility: (a) *inclusion* [*Umfassung*],[12] in which two compatible ideas are such that all objects underlying one of them also underlie the other, and (b) its opposite, *entanglement* or *concatenation* or *disparateness* [*Verschlungenheit, Verkettung* or *Disparation*],[13] in which each one of the two compatible ideas refers to an object to which the other does not refer. Writing '$a \sqsubseteq b$' for '*a is included in b*' and '$a \dot{\top} b$' for '*a and b are disparate*', we therefore assume

$$a \sqsubseteq b \leftrightarrow a \top b \wedge \forall i(i \in a \rightarrow i \in b) \qquad \text{D2.12}$$

$$a \dot{\top} b \leftrightarrow a \top b \wedge \exists i(i \in a \wedge \neg i \in b) \wedge \exists i(\neg i \in a \wedge i \in b) \qquad \text{D2.13}$$

Thus

$$a \dot{\top} b \leftrightarrow a \top b \wedge a \not\sqsubseteq b \wedge b \not\sqsubseteq a \qquad \text{T2.11}$$

Among the properties of the inclusion, Bolzano mentions explicitly the transitivity of $\sqsubseteq$:[14]

$$a \sqsubseteq b \wedge b \sqsubseteq c \rightarrow a \sqsubseteq c \qquad \text{T2.12}$$

Moreover, Bolzano also distinguishes special cases of inclusion: (a) *equivalence* [*Gleichgültigkeit*] or *reciprocal inclusion* [*wechselseitiges Umfassen*], which occurs precisely when inclusion is reciprocal[15] and (b) its opposite, *subordination* [*Unterordnung*], which occurs when the inclusion is *proper*.[16] Writing '$a \equiv b$' for '*a and b are equivalent*' and '$a \dot{\sqsubseteq} b$' for '*a is subordinated* to *b*', we thus assume

$$a \equiv b \leftrightarrow a \sqsubseteq b \wedge b \sqsubseteq a \qquad \text{D2.14}$$

$$a \dot{\sqsubseteq} b \leftrightarrow a \sqsubseteq b \wedge b \not\sqsubseteq a \qquad \text{D2.15}$$

[12] TS §95, I 319: "1) [. . .] If a pair of ideas stand in the relation of compatibility to one another, then it may be the case that not only one, but all of the objects that stand under *A* also stand under *B*. If it is not assumed that this relation is mutual, i.e., if it is undetermined whether *B* has some objects that do not stand under *A*, then I wish to call this relation between *A* and *B* a relation of *inclusion*. I wish to say that the extension of the idea *B* or, more briefly, that *B* includes *A*, and I call *B* the *including* and *A* the *included* idea."

[13] TS §98, I 327: "1) [. . .] when it can be said of neither of two compatible ideas that all its objects stand under the other, we encounter a relation that I call *overlapping* or *linking*, or, using a term that others have already employed, *disparateness*". Bolzano gives as an example of disparate ideas 'erudite' and 'virtuous'.

[14] TS §95, I 319: "4) If an idea *A* is included in *B*, and *B* is included in *C*, then *A* is also included in *C*".

[15] TS §96, I 320: "1) The definition of the concept of inclusion in the preceding section allows this relation to be mutual between two ideas *A* and *B*. *A* can be included in *B* and *B* in *A*. [. . .] This relation I call a *mutual* or *precise inclusion* or *equivalence*, and the ideas themselves I call *equivalent* or *interchangeable*."

[16] TS §97 I 324: "1) The second case that can occur with a relation of inclusion (§95) is that the relation is not *mutual*. [. . .] This relation is called *subordination* and we say that *B* is *higher*, *A* is *lower*, that *A* is *subordinated* to *B*, or *stands under A*."

Bolzano answers the question whether distinct equivalent ideas may also have the same content affirmatively. Here, by 'identity of content' he means the identity of the last components and not also to the way of composition (examples are an intelligent virtuous and a virtuous intelligent; 2^4 and 4^2). However, he rejects this possibility if both ideas are simple.

Bolzano explicitly admits the *transitivity* of $\equiv$,[17] but—a very interesting fact—he rejects its *substitutivity*. In fact, he points out that ideas that are generated "*in the same way* [*auf eine gleiche Weise*]" out of equivalent ideas are not necessary equivalent. The surprising example he provides is the '*extraction of the root of a power*' starting from the equivalent ideas 2^4 and 4^2; in the first case, we get 2, in the second 4.[18]

As far as species of incompatibility are concerned, he distinguishes between *contradictoriness* [*Widerspruch*] and *contrariety* [*Widerstreit*]. Two incompatible ideas are *contradictory* [*widersprechend* or *contradictorisch*] if they are also *exhaustive*, in the sense that any object falls under the one or the other, whereas they are *contrary* [*widerstreitend* or *conträr*] when this does not occur. Writing '$a \mathrel{\dot{\perp}} b$' for '*a and b are contrary*' and '$a||b$' for '*a* and *b* are contradictory', we therefore assume

$$a \mathrel{\dot{\perp}} b \leftrightarrow a \perp b \wedge \exists i(\neg i \mathrel{\varepsilon} a \wedge \neg i \mathrel{\varepsilon} b) \qquad \text{D2.16}$$

$$a \,||\, b \leftrightarrow a \perp b \wedge \forall i(i \mathrel{\varepsilon} a \vee i \mathrel{\varepsilon} b) \qquad \text{D2.17}$$

In summary, the semantic relations between two (objectual!) ideas are as follows:

- *compatibility* [$\top$]
  - *inclusion* [$\sqsubseteq$]
    - *equivalence* [$\equiv$]
    - *subordination* [$\dot{\sqsubseteq}$]
  - *disparateness* [$\dot{\top}$]
- *incompatibility* [$\perp$]
  - *contradictoriness* [$||$]
  - *contrariety* [$\dot{\perp}$]

At the end of his discussion, Bolzano provides in §105 a "*List of some herein pertaining theorems*" that he claims to have derived mainly from the *Grundrisse der Logik* (1823) of J. G. E. Maaß. We will quote only 12 of them here; the other 11

[17] TS §96, I 322: "6) If *A* is equivalent to *B* and *B* is equivalent to C, then *A* and C are equivalent."

[18] TS §96.5, I 322.

will be provided in the final section of this chapter, after we have addressed the notion of *negation of an object idea*:

$$\exists c(c \mathrel{\dot{\sqsubseteq}} a \wedge c \mathrel{\dot{\sqsubseteq}} b) \rightarrow a \mathrel{\dot{\top}} b \qquad \text{T2.13}$$ [19]

$$a \perp b \rightarrow \neg\exists c(c \mathrel{\dot{\sqsubseteq}} a \wedge c \mathrel{\dot{\sqsubseteq}} b) \qquad \text{T2.14}$$ [20]

$$a \mathrel{\dot{\top}} b \wedge a \mathrel{\dot{\sqsubseteq}} c \wedge b \mathrel{\dot{\sqsubseteq}} d \rightarrow c \mathrel{\dot{\top}} d \qquad \text{T2.15}$$ [21]

$$a \perp b \wedge c \mathrel{\dot{\sqsubseteq}} a \wedge d \mathrel{\dot{\sqsubseteq}} b \rightarrow c \perp d \qquad \text{T2.16}$$ [22]

$$a \perp b \wedge a \mathrel{\dot{\top}} c \wedge b \mathrel{\dot{\top}} c \rightarrow \neg c \mathrel{\dot{\sqsubseteq}} a \wedge \neg c \mathrel{\dot{\sqsubseteq}} b \qquad \text{T2.17}$$ [23]

$$a \mathrel{\dot{\top}} a \mathrel{\dot{\top}} b \wedge c \mathrel{\dot{\sqsubseteq}} a \rightarrow c \mathrel{\dot{\sqsubseteq}} b \vee c \mathrel{\dot{\top}} b \vee c \perp b \qquad \text{T2.18}$$ [24]

$$b \wedge a \mathrel{\dot{\sqsubseteq}} c \rightarrow b \mathrel{\dot{\sqsubseteq}} c \vee b \mathrel{\dot{\top}} c \qquad \text{T2.19}$$ [25]

$$a \mathrel{\dot{\top}} b \wedge a \perp c \rightarrow c \mathrel{\dot{\sqsubseteq}} b \vee c \perp b \qquad \text{T2.20}$$ [26]

$$a \,||\, b \rightarrow \neg\exists c(c \perp a \wedge c \perp b);\ \ a \,||\, b \rightarrow (c \perp a \rightarrow c \sqsubseteq b) \qquad \text{T2.21}$$ [27]

$$a \mathrel{\dot{\perp}} b \rightarrow \exists c(a \mathrel{\dot{\sqsubseteq}} c \wedge b \mathrel{\dot{\sqsubseteq}} c) \qquad \text{T2.22}$$ [28]

$$a \mathrel{\dot{\perp}} b \rightarrow \exists c(a \perp c \wedge b \perp c) \qquad \text{T2.23}$$ [29]

$$a \mathrel{\dot{\perp}} b \rightarrow \exists c(a \perp c \wedge b \perp c) \qquad \text{T2.24}$$ [30]

[19] TS §105, I 354: "1) If two ideas *A* and *B* have one and the same idea *X* under them, then they are concordant. For [. . .]."

[20] TS §105, I 354: "2) Conversely, when two ideas are mutually exclusive, there is no third idea subordinate to both of them."

[21] TS §105, I 354: "3) If two ideas are compatible, then all ideas that are higher than them are also compatible. For [. . .]."

[22] TS §105, I 354: "4) Conversely, if two higher ideas are mutually exclusive, then so too are any ideas subordinate to them. Since [. . .]."

[23] TS §105, I 354–5: "5) If two mutually exclusive ideas *A* and *B* are each compatible with a third *X*, then *X* is subordinate neither to *A* nor to *B*. For [. . .]."

[24] TS §105, I 355: "7) If two ideas overlap, then there is no idea *X* which is lower than one of them, say *A*, and higher than the other, *B*; rather, *X* is either lower than *B*, or overlaps *B*, or is incompatible with *B*. [. . .]."

[25] TS §105, I 355: "6) If two ideas *A* and *B* overlap, then there is no idea *X* which is higher than one of them, say *A*, and lower than the other, *B*: either *X* is higher than *B*, too, or overlaps *B*. For [. . .]."

[26] TS §105, I 356: "8) If two ideas *A* and *B* overlap, and a third *X* is incompatible with one of them, e.g., *A*, then it cannot be higher than the other, *B*; rather *X* must be lower than, or incompatible with, *B*. [. . .]."

[27] TS §105, I 357: "10) No idea can be incompatible with two contradictory ideas. If it is incompatible with one, then it must be equivalent or subordinate to the other. For [. . .]."

[28] TS §105, I 357: "11) Two ideas which are merely contrary can both be subordinate to one and the same higher idea. For [. . .]."

[29] TS §105, I 357: "12) Two merely contrary ideas can both be incompatible with one and the same idea. For [. . .]."

[30] TS §105, I 358: "13) One and the same idea can be compatible with each of two contrary ideas, and also with each of two contradictory ideas. For [. . .]."

2.4 Relations Among Several Ideas

An important fact is that after introducing a binary relation between two ideas, Bolzano extended this relation to situations in which more ideas are involved. *Unfortunately, he did this without making the question explicit,* thereby paving the way to misunderstandings. In fact, he reached these generalizations in two very different ways. The first was to extend the binary relation to ternary, quaternary, etc. ones. This was done *usually* by employing locutions of the type '*the more ideas a, b, c, . . . [die mehren Vorstellungen a, b, c, . . .]*'. The second way was to extend the binary relation between ideas to a binary (or also ternary) relation among *aggregates* of ideas, in which case the *preferred* locution is '*the aggregate of the ideas a, b, c, . . . [der Inbegriff der Vorstellungen a, b, c, . . .]*'. Actually, in the first case, when Bolzano speaks of an object underlying the $a, b, c, \ldots$, he means an object that underlies *each one* of the $a, b, c, \ldots$, whereas, in the second case, when referring to an object underlying the $a, b, c, \ldots$, he intends an object that underlies *at least one* of the $a, b, c, \ldots$[31]

The first type of extension, *in so far as it involves only a finite number of ideas,* does not create any particular problem. For its treatment, it will suffice to extend the predicate already introduced for the dyadic case to the 3-adic, 4-adic, etc.. and, in general, to introduce, using a schema of defining axioms, an n-adic predicate, for each $n \geq 2$. Hence, for example, we will generalize the notion of compatibility to the case of the ideas $a_1, a_2, \ldots, a_n$, by stating that they are compatible iff there is an object that underlies each one of them, i.e. by setting

$$\mathsf{T}^n(a_1, a_2, \ldots, a_n) \leftrightarrow \exists i(i \mathsf{E} a_1 \wedge i \mathsf{E} a_2 \wedge \ldots \wedge i \mathsf{E} a_n).$$

In order to simplify the notation, we will agree to use '$\overset{n}{a}$' for '$a_1, a_2, \ldots, a_n$'; moreover, as the adicity of the predicate is already deducible from the number of arguments and its explicit mention is therefore redundant, we will agree to omit it. Hence, instead of writing '$\mathsf{T}^n(a_1, a_2, \ldots, a_n)$', we will simply write '$\mathsf{T}\,(\overset{n}{a})$'; furthermore, by using the current logical notation for the iterated conjunction, the above equivalence will also be written as

$$\mathsf{T}\,(\overset{n}{a}) \leftrightarrow \exists i(\bigwedge_{j \leq n} i \mathsf{E} a_j)$$

[31] In contemporary set-theoretical terminology, if we think of the '$a, b, c, \ldots$' as sets of their objects, we could say that in the first case he thinks of their intersection and in the second case their union.

In the case of the second type of extensions, the situation appears more complex. Indeed, they seem to require an enlargement, through a notion of set, of the logical framework in which we have decided to formalize the system. However, a satisfactory solution to this issue that respects the limits of our logical framework, will only be reached by going over to what we have called the 'symbolic level', i.e. the level characterized by the fact that ideas and propositions are also objects. Nonetheless, for the moment, we can be content with the following solution.

If we want to extend a binary relation R between object ideas to a relation R^* between two arbitrary *finite and not-empty* sets of ideas, we could replace this relation R^* by a (double) infinity of $m+n$-adic predicates $R^{m,n}$ $(m, n \geq 1)$, each of which satisfies the condition

$$R^{m,n}(a_1,\ldots,a_m; c_1,\ldots,c_n) \leftrightarrow \alpha[a_1,\ldots,a_m, c_1,\ldots,c_n]$$

In principle, as the intended relation should be *among sets*, it is important that the unavoidable order among the arguments does not play a significant role. Hence, we must also assume, that, for all permutations π of $\{1,\ldots,m\}$ and ρ of $\{1,\ldots,n\}$, the following holds:

$$R^{m,n}(a_1,\ldots,a_m; c_1,\ldots,c_n) \leftrightarrow R^{m,n}(a_{\pi(1)},\ldots,a_{\pi(m)}; c_{\rho(1)},\ldots,c_{\rho(n)})$$

This remark—which, by the way, also applies to the generalizations of the first type—should not cause problems. Indeed, the relations to be considered are such that this permutability will be ensured, each time, by the logic ruling the defining condition (in the preceding example by the commutativity and associativity of $\wedge$).

In this way, the place of the R^* intended by Bolzano is actually taken by a (double) infinity of relations among *multisets* of every finite and non- empty cardinality. Taking into account the fact that, as already alluded to, such relations will be superfluous at the 'symbolic level', it appears reasonable to be content with what has been established so far and avoid the technical complications arising from a further attempt to reduce the *multisets* to *sets* (i.e. from the attempt to eliminate the possibility of repetition of an argument).

Notational tricks similar to those discussed above, will allow us to simply write '$R(\overset{m}{a}; \overset{n}{c})$', and even often '$\overset{m}{a}R\overset{n}{c}$', instead of '$R^{m,n}(a_1,\ldots,a_m; c_1,\ldots,c_n)$'.

On the basis of these premises, let us proceed to the rigorous formulations.

2.4.1 Compatibility and Incompatibility

We will begin by introducing, by means of suitable defining axioms, the extensions to '*more object ideas*' for both of these relations.

Bolzano holds that *several ideas* $a_1, \ldots, a_m$ are *reciprocally compatible* iff there is an object that simultaneously underlies all of them, whereas, on the contrary, they are *reciprocally incompatible* iff there is no such object. Bolzano's use of the term 'compatibility of several ideas' with this meaning is clearly proven by the following statement:

> "If the relation of reciprocal compatibility must hold amongst the *several* ideas $A, B, C, D, \ldots$, then the same relation must also hold amongst the fewer ideas $A, B, \ldots$, the aggregate of which constitutes only a part of the first aggregate. Indeed, if it were not so, if there were no object that is represented by all the $A, B, \ldots$, even less there would be an object which is represented by all the ideas $A, B, C, D, \ldots$"[32]

Hence we may establish

$$\top\,(\overset{m}{a}) \leftrightarrow \exists i(\bigwedge_{j\leq m} i \sqsubset a_j) \qquad \text{SD2.18}$$

and

$$\bot\,(\overset{m}{a}) \leftrightarrow (\bigwedge_{j\leq m} G(a_j)) \wedge \forall i(\bigvee_{j\leq m} \neg i \sqsubset a_j) \qquad \text{SD2.19}$$

respectively.

We choose not to formalize the two generalizations of these relations that were also considered by Bolzano and may be called '*m-compatibility*' and '*m*-incompatibility', respectively, and that obtain among the several ideas $a_1, \ldots, a_n$ (with $m \leq n$) iff, taking m of them, they and only they turn out to be compatible or respectively incompatible.[33]

As far as the extension of *compatibility* and *incompatibility* to 'aggregates of ideas' is concerned, Bolzano says that the aggregate of the $a_1, \ldots, a_m$ is *compatible* with the aggregate of the $c_1, \ldots, c_n$ iff there is at least one object that simultaneously underlies *one* of the $a_1, \ldots, a_m$ and *one* of the $c_1, \ldots, c_n$, whereas it is *incompatible* iff there

[32] TS §94, I 317.

[33] TS §94, I 317: "Sometimes we find among a given set of ideas $A, B, C, D, \ldots$ the noteworthy situation that only a certain number of them, e.g., any n, are compatible."

TS §246, II 345: "[...] that any n of the ideas $A, B, C, D, \ldots$ are compatible [...] any n of the ideas $A, B, C, D, \ldots$ stand in the relation of *incompatibility* [...]."

is no such object.[34] Hence we may set

$$\overset{m}{a}\top\overset{n}{c} \leftrightarrow (\bigwedge_{j\leq m} G(a_j)) \wedge (\bigwedge_{k\leq n} G(c_k)) \wedge \exists i((\bigvee_{j\leq m} i \,\mathrm{E}\, a_j) \wedge (\bigvee_{1\leq k\leq n} i \,\mathrm{E}\, c_k)) \qquad \text{SD2.20}$$

$$\overset{m}{a}\perp\overset{n}{c} \leftrightarrow (\bigwedge_{j\leq m} G(a_j)) \wedge (\bigwedge_{k\leq n} G(c_k)) \wedge \forall i((\bigvee_{j\leq m} i \,\mathrm{E}\, a_j) \rightarrow (\bigwedge_{k\leq n} \neg i \,\mathrm{E}\, c_k)) \qquad \text{SD2.21}$$

In order to make clear the difference between the two notions *ad abundantiam*, let us remark that we have, obviously, $\top(\overset{m}{a}, \overset{n}{c}) \rightarrow \overset{m}{a}\top\overset{n}{c}$, but its converse does not hold true. Indeed, $\top(\overset{m}{a}, \overset{n}{c})$ says that there is an object that underlies simultaneously *all* $\overset{m}{a}$ and *all* $\overset{n}{c}$, whereas $\overset{m}{a}\top\overset{n}{c}$ only says that there is an object that underlies simultaneously at least *one* $\overset{m}{a}$ and *one* $\overset{n}{c}$.

The incompatibility between two aggregates of ideas implies the incompatibility between any two of their elements; the compatibility, on the other hand, implies the compatibility of at least two of their elements,[35] i.e.

$$\overset{m}{a}\perp\overset{n}{c} \rightarrow \bigwedge_{j\leq m,\, k\leq n} a_j \perp c_k \qquad \text{ST2.1}$$

$$\overset{m}{a}\top\overset{n}{c} \rightarrow \bigvee_{j\leq m,\, k\leq n} a_j \top c_k \qquad \text{ST2.2}$$

In fact

$$\overset{m}{a}\perp\overset{n}{c} \leftrightarrow \bigwedge_{j\leq m,\, k\leq n} a_j \perp c_k \qquad \text{ST2.3}$$

$$\overset{m}{a}\top\overset{n}{c} \leftrightarrow \bigvee_{j\leq m,\, k\leq n} a_j \top c_k \qquad \text{ST2.4}$$

It should be noted that

$$a \top c \leftrightarrow \top(a, c) \leftrightarrow \overset{1}{a}\top\overset{1}{c} \qquad \text{ST2.5}$$

[34] TS §94, I 317-8: "4) Just as an individual idea can stand in the relation of compatibility to another idea, a whole collection of ideas *A, B, C, D,* ... can stand in this relation or in the relation of incompatibility either to some other collection *M, N, O,* ... or to a single idea *M*. The first is the case if there is some object that stands under any one of [*unter einer der*] the ideas *A, B, C, D,* ... and at the same time under *M* or under one of these ideas *M, N, O,* ... ; the second if this is not the case."

[35] TS §94 I 318: "5) If a whole collection of ideas *A, B, C, D,* ... stands in the relation of *incompatibility* to a whole collection of ideas *M, N, O,* ... then any one of the ideas *A, B, C, D,* ... must stand in this relation to any one of the ideas *M, N, O,* .... But when both collections stand in the relation of *compatibility*, then it is not necessary that each, but only that one of the ideas *A, B, C, D,* ... stand in this relation to one of the *M, N, O,* ... "

$$a \perp c \leftrightarrow \perp(a, c) \leftrightarrow \overset{1}{a} \perp \overset{1}{c} \qquad \text{ST2.6}$$

The relations of inclusion, equivalence, and subordination are generalized only with respect to aggregates of ideas.

2.4.2 *Inclusion*

According to Bolzano, the aggregate of the $a_1, \dots, a_m$ is *included* in the aggregate of the $c_1, \dots, c_n$ iff each object falling under the first (i.e. under one of the $a_1, \dots, a_m$) falls also under the second (i.e. under at least one of the $c_1, \dots, c_n$).[36] In graphic terms, this is true iff the union of the areas of the extensions of the $c_1, \dots, c_n$ completely covers the union of the areas representing the extensions of the $a_1, \dots, a_m$.[37] From the whole context in which this definition is presented, we can draw the obvious conclusion that some hypothesis of compatibility is involved; however, it is not at all clear which type of compatibility is meant. Is it the compatibility among the $a_1, \dots, a_m$, which also automatically involves one of the $c_1, \dots, c_n$? Or it is just the compatibility between the two aggregates? By also taking into account what will later occur in the case of propositions that, as Bolzano explicitly says, behave similarly to ideas, we are inclined to opt for the first hypothesis, while also remarking that such a choice does not apparently influence later developments. Hence we set

$$\overset{m}{a} \sqsubseteq \overset{n}{c} \leftrightarrow \mathsf{T}(\overset{m}{a}) \wedge \forall i((\bigvee_{j \leq m} i \mathrel{\mathsf{E}} a_j) \rightarrow (\bigvee_{k \leq n} i \mathrel{\mathsf{E}} c_k)) \qquad \text{SD2.22}$$

Bolzano also considers two special cases, which he calls *subdivision* [*Eintheilung*] cases and which are given when $m = 1$ or, respectively, $n = 1$. In the first case (i.e. when $a \sqsubseteq \overset{n}{c}$), he speaks of *distribution* [*Vertheilung*] of the a among the $c_1, \dots, c_n$; in the second case (i.e. when $\overset{m}{a} \sqsubseteq c$), he speaks of *enumeration* [*Aufzählung*] of c by the $a_1, \dots, a_m$.[38]

2.4.3 *Equivalence*

According to Bolzano, the aggregate of the $a_1, \dots, a_m$ and the aggregate of the $c_1, \dots, c_n$ are *equivalent* iff the same objects fall under them, that is, iff every object that falls under one of the $a_1, \dots, a_m$ falls also under one of the $c_1, \dots, c_n$, and vice versa, or also iff the aggregates of the $a_1, \dots, a_m$ and of the $c_1, \dots, c_n$ are reciprocally

[36] TS §95, I 319: "2) It is clear how this relation can be extended so that we have a collection of ideas on each side, or a collection of ideas on one, and a single idea on the other. I shall say that *A, B, C, D*,... are included in the ideas *M, N, O*,... if every object of one of the idea *A, B, C, D*,... also stands under one of the ideas *M, N, O*,..."

[37] TS §95.5, I 319.

[38] TS §140, II 41: "3) [...] If, however, one term of this relation is a sum of several ideas, i.e., when [...]"

included.[39] In graphic terms, this is true when the total area covered by the union of the areas representing the extensions of the $a_1, \ldots, a_m$ coincides with the area determined by the extensions of the $c_1, \ldots, c_n$:[40]

$$\overset{m}{a} \equiv \overset{n}{c} \leftrightarrow \exists i(\bigwedge_{j\leq m} i \,\mathrm{E}\, a_j) \wedge \forall i((\bigvee_{j\leq m} i \,\mathrm{E}\, a_j) \leftrightarrow (\bigvee_{k\leq n} i \,\mathrm{E}\, c_k)) \qquad \text{SD2.23}$$

$$\overset{m}{a} \equiv \overset{n}{c} \leftrightarrow \overset{m}{a} \sqsubseteq \overset{n}{c} \wedge \overset{n}{c} \sqsubseteq \overset{m}{a} \qquad \text{ST2.7}$$

$$\overset{m}{a} \equiv \overset{n}{c} \leftrightarrow \overset{n}{c} \equiv \overset{m}{a} \qquad \text{ST2.8}$$

Bolzano explicitly remarks that for the equivalence of $a_1, \ldots, a_m$ and $c_1, \ldots, c_n$, it is not necessary that at least one of the $a_1, \ldots, a_m$ be equivalent to at least one of the $c_1, \ldots, c_n$.[41]

2.4.4 *Subordination*

According to Bolzano, the aggregate of $a_1, \ldots, a_m$ is *subordinated* to the aggregate of $c_1, \ldots, c_n$ iff the second includes the first but not vice versa.[42] Graphically, this is obvious. Hence

$$\overset{m}{a} \mathrel{\dot{\sqsubseteq}} \overset{n}{c} \leftrightarrow \overset{m}{a} \sqsubseteq \overset{n}{c} \wedge \neg \overset{n}{c} \sqsubseteq \overset{m}{a} \qquad \text{SD2.24}$$

2.4.5 *Concatenation (or Entanglement or Disparateness)*

In this case, we must distinguish between concatenation among several ideas $a_1, \ldots, a_m$ and concatenation between the aggregate of $a_1, \ldots, a_m$ and the aggregate of $c_1, \ldots, c_n$. The first relation holds when the $a_1, \ldots, a_m$ are concatenated in pairs. The second occurs when there is at least one object that simultaneously falls under one of the $a_1, \ldots, a_m$ and one of the $c_1, \ldots, c_n$, but there also is another object that falls under one of the $a_1, \ldots, a_m$ without falling under any one of the $c_1, \ldots, c_n$,

[39] TS §96, I 322: "8) If we want to extend the concept of equivalence to whole collections of ideas, then we shall have to say that the collection of ideas *A, B, C, D, . . .* *is equivalent* to the collection of ideas *M, N, O, . . .* if every object of one of the ideas *A, B, C, D, . . .* also stands under one of *M, N, O, . . .* and conversely."

[40] TS §96.10, I 322–3.

[41] TS §96, I 448–9: "9) In order for the two collections *A, B, C, D, . . .* and *M, N, O, . . .* to be equivalent, it is not necessary that any one of the ideas *A, B, C, D, . . .* taken by itself, should be equivalent to any one of *M, N, O, . . .* [. . .]."

[42] TS §97, I 325: "2) Several ideas *A, B, C, D, . . .* stand in the relation of *subordination* to one or several other ideas *M, N, O, . . .* if each of the objects represented by one of the former is also represented by one of the latter, but not conversely. The collection *A, B, C, D, . . .* will be called the lower and *M, N, O, . . .* the higher one."

as well as a still another object that falls under one of the $c_1, \ldots, c_n$, without falling under any of the $a_1, \ldots, a_m$.[43].Hence we set

$$\dot{\top}(\overset{m}{a}) \leftrightarrow \bigwedge_{j \neq k \leq m} a_j \dot{\top} a_k \qquad \text{SD2.25}$$

$$\overset{m}{a} \dot{\top} \overset{n}{c} \leftrightarrow \overset{m}{a} \top \overset{n}{c} \wedge \overset{m}{a} \not\sqsubseteq \overset{n}{c} \wedge \overset{n}{c} \not\sqsubseteq \overset{m}{a} \qquad \text{SD2.26}$$

2.4.6 *Omnilateral Concatenation*

A special case of concatenation to which Bolzano pays particular attention is what he calls *omnilateral concatenation* [*allseitige Verkettung*] and which occurs when the $a_1, \ldots, a_m$ are compatible but any two of them have at least one reference in common that does not fall under any of the remaining ideas. Hence we set

$$Onnicon(\overset{m}{a}) \leftrightarrow \top(\overset{m}{a}) \wedge \bigwedge_{j \neq k \leq m} \exists i (i \mathrel{\mathsf{E}} a_j \wedge i \mathrel{\mathsf{E}} a_k \wedge \bigwedge_{j,k \neq l \leq m} \neg i \mathrel{\mathsf{E}} a_l) \qquad \text{SD2.27}$$

Anticipating the use of concepts and notations that will be introduced in the following sections, we mention that Bolzano emphasizes that it follows from the definition that

$$Onnicon(\overset{m}{a}) \leftrightarrow \bigwedge_{i \leq m}([e](\overset{m}{\bar{a}}) \mathrel{\dot{\sqsubseteq}} [e](\bar{a}_1 + \cdots + \bar{a}_{i-1} + \bar{a}_{i+1} + \cdots + \bar{a}_m))\ ^{44} \qquad \text{ST2.9}$$

Other important special cases of concatenation are connected with those of incompatibility. Let us first define the notion of exclusion.

2.4.7 *Exclusion [Ausschließung]* *(or Omnilateral Incompatibility [allseitige Unverträglichkeit])*

This special case of incompatibility occurs when the ideas $a_1, \ldots, a_m$ are incompatible *in pairs*;[45] writing '$\perp\!\!\!\perp (\overset{m}{a})$' for '$a_1, \ldots, a_m$ *exclude one another*', we set

[43] TS §98, I 328: "5) One soon sees that the relation of overlapping is not confined to individual ideas but can be extended to entire *collections* of them. We shall say that the collection of ideas *A*, *B*, *C*, ... *overlaps* the collection of ideas *M*, *N*, *O*, ... if there are objects that stand under one of the ideas *A*, *B*, *C*, ... as well as under one of the ideas *M*, *N*, *O*, ... provided that there are others that are represented by each of these collections alone."

[44] TS §98, I 328: "4) An even more remarkable relation [...] holds if the ideas *A*, *B*, *C*, *D*, ... overlap in pairs in such a way that no two ideas share all objects, but any two have some common objects. In this case the idea of something that is *A*, as well as *B*, *C* etc., or [something]$(a + b + c \ldots)$ is an objectual idea in which none of the parts *a*, *b*, *c*, *d* ... *is redundant* (§69). This is the case because the ideas that result if we leave out any one of these parts e.g., [something]$(b + c \ldots)$, [something]$(a + c \ldots)$ are all wider than the idea [something]$(a + b + c \ldots)$. This relation could be called *collective overlapping.*"

[45] TS §103, I 343: "1) [...] when the ideas *A*, *B*, *C*, *D*, ... are *incompatible* in such a fashion that no two of them are compatible, i.e., that no object is commonly represented by any two of them, I shall call the relation one of *all-sided incompatibility* or, more briefly, *exclusion.*"

$$\perp\!\!\!\perp (\overset{m}{a}) \leftrightarrow (\bigwedge_{j\leq n} G(a_j)) \wedge (\bigwedge_{j\neq k\leq m} a_j \perp a_k) \qquad \text{SD2.28}$$

Obviously,

$$\perp\!\!\!\perp (a, c) \leftrightarrow a \perp c \qquad \text{ST2.10}$$

Bolzano extends this relation to aggregates and even to *more* aggregates of ideas by establishing that "the aggregates $A, B, C, \ldots; M, N, O, \ldots; R, S, T, \ldots$ etc. are said to *exclude* one another, iff not even one object, which falls under one of them, falls also under a second of them" and remarking that, clearly, if the aggregates in question are only two, their exclusion reduces to incompatibility.[46]

Bolzano also introduces, for pairs of incompatible aggregates, the distinction *contradictory–contrary*. The aggregate of the $a_1, \ldots, a_m$ excluding the aggregate of the $c_1, \ldots, c_n$ *contradicts* the latter iff every object that does not fall under any of the $a_1, \ldots, a_m$ falls under one of the $c_1, \ldots, c_n$. whereas it is simply *contrary* iff this is not the case.[47] Writing '$\overset{m}{a} \mathbin{+\!\!\!\!+} \overset{n}{c}$' and '$\overset{m}{a} \parallel \overset{n}{c}$' for '*the aggregate of the* $a_1, \ldots, a_m$ *contradicts that of the* $c_1, \ldots, c_n$' and '*the aggregate of the* $a_1, \ldots, a_m$ *is contrary to that of the* $c_1, \ldots, c_n$', respectively, we may establish

$$\overset{m}{a} \mathbin{+\!\!\!\!+} \overset{n}{c} \leftrightarrow \overset{m}{a} \perp \overset{n}{c} \wedge \forall i((\bigvee_{j\leq m} i \mathrel{\varepsilon} a_j) \vee (\bigvee_{k\leq n} i \mathrel{\varepsilon} c_k))) \qquad \text{SD2.29}$$

$$\overset{m}{a} \parallel \overset{n}{c} \leftrightarrow \overset{m}{a} \perp \overset{n}{c} \wedge \neg(\overset{m}{a} \mathbin{+\!\!\!\!+} \overset{n}{c}) \qquad \text{SD2.30}$$

2.4.8 *Mediate Concatenation [mittelbar verkettet sein]*

This is the special case of both concatenation and exclusion that happens when the ideas $a_1, \ldots, a_m$ exclude one another but are all compatible with an other idea c;[48] i.e.

$$Concmedi(\overset{m}{a}; c) \leftrightarrow (\perp\!\!\!\perp (\overset{m}{a}) \wedge (\bigwedge_{j\leq m} a_j \top c) \qquad \text{SD2.31}$$

[46] TS §103, I 344: "2) The relation of exclusion can also be extended to entire collections of ideas. The collections $A, B, C, \ldots; M, N, O, \ldots; R, S, T, \ldots$ etc., are said to *exclude* each other when no object that stands under one of the ideas [in a given collection] stands under any of the ideas in the others. Again, if we compare only two collections, it matters not whether we call them incompatible or mutually exclusive."

[47] TS §103, I 345: "5) [...] If two collections of ideas $A, B, C, D, \ldots$ and $M, N, O, \ldots$ exclude each other and are so constituted that every object, whatever it may be, is included either under one of the ideas $A, B, C, D, \ldots$ or else under one of $M, N, O, \ldots$ we may say that these collections stand in a relation of *contradictoriness*. If the latter condition is not met, we say that they are merely *contrary*."

[48] TS §98, I 327–8: "3) The first case, where certain ideas $A, B, C, D, \ldots$ overlap a certain other, M, but exclude one another, is quite important [...]. In such a case, we call the ideas $A, B, C, D, \ldots$ *mediately linked* through M."

2.4.9 *Chain [Kette]*

The ideas $a_1, \dots, a_m$ constitute a *chain* iff each one of them is compatible with the immediately preceding idea, but incompatible with all other preceding ideas (i.e. iff for all $j > 1$, $a_{j\text{-}1} \top a_j$, but for all $k > 1$, $a_j \perp a_{j\text{-}k}$.[49] Hence

$$Chain(\overset{m}{a}) \leftrightarrow (\bigwedge_{1<j\leq m} a_{j\text{-}1} \top a_j) \wedge (\bigwedge_{1<j\leq m} (\bigwedge_{k\text{-}1<j} a_k \perp a_j)) \qquad \text{SD2.32}$$

2.4.10 *Cyclic or Closed Chain [in sich selbst zurückkehrende* or *geschlossene Kette]*

The ideas $a_1, \dots, a_m, c$ constitute a *cyclic* or *closed chain* iff they are a chain and c is compatible with a_1:[50]

$$Cyclchain(\overset{m}{a}) \leftrightarrow Cat(\overset{m-1}{a}) \wedge (\bigwedge_{2\leq j<m-1} a_j \perp a_m) \wedge a_{m-1} \top a_m \wedge a_m \top a_1 \qquad \text{SD2.33}$$

2.4.11 *Coordination*

The ideas $a_1, \dots, a_m$ are *coordinated* [*beigeordnet* or *coordinirt*] *under* the idea c iff they exclude one another and are all included in c, that is to say, iff they exclude one another and *arabiclist* c;[51] i.e.

$$Coor(\overset{m}{a}, c) \leftrightarrow \perp\!\!\!\perp (\overset{m}{a}) \wedge \overset{m}{a} \sqsubseteq c \qquad \text{SD2.34}$$

2.4.12 *Exhaustive Coordination*

The ideas $a_1, \dots, a_m$ *exhaust* [*erschöpfen*] the idea c or are *integrant parts* [*ergänzende Theile* or *Partes integrantes*] of c iff they are coordinated under c and every object falling under c also falls under one and only one of the $a_1, \dots, a_m$;[52] in other words, iff they exclude one another but enumerate c and, on its side, c distributes among

[49] TS §98, I 328: "3) [...] If each of the ideas *A, B, C, D, ...* is linked to the preceding one, but is incompatible with all earlier ones, then this series of ideas will be called a *chain*."

[50] TS §98, I 328: "3) [...] If all remains as above, except that the last idea is again linked with the first, then the collection of these ideas will be called a *closed chain*."

[51] TS §104, I 347: "1) [...] if the ideas *A, B, C, D, ...* are all subordinate to the idea *X* and if they exclude each other, they are called ideas *coordinated under the idea X* or, in short, *disjoint* ideas."

[52] TS §104, I 347: "3) [...] the extensions of ideas *A, B, C, D, ... M exhaust* the extension of *X* if every object that stands under one of the ideas *A, B, ... M* also stands under *X* and every object that stands under *X* stands under one, and only one, of the ideas *A, B, ... M*."

them (today we would say that the $a_1, \ldots, a_m$ '*constitute a partition of c*'.):

$$Part(\overset{m}{a}, c) \leftrightarrow Coor(\overset{m}{a}, c) \wedge c \sqsubseteq \overset{m}{a} \qquad \text{SD2.35}$$

$$Part(\overset{m}{a}, c) \leftrightarrow \perp\!\!\!\perp(\overset{m}{a}) \wedge \overset{m}{a} \equiv c \qquad \text{ST2.11}$$

2.5 The Idea of Something

Moving on to the consideration of the ideas and the operations on ideas that are *present* in the Bolzanian universe, the first thing to emphasize is that Bolzano firmly believed that there is the idea *something* [*Etwas*] or *object in general* [*Gegenstand überhaupt*][53] that *refers to all and only all the objects*[54] and is *simple*.[55]

We may account for this Bolzanian firm belief by assuming the presence of the (object) idea *something* [*Etwas*] or *object in general,* denoted by '*e*', which refers to any object and is mereologically simple; that is, by assuming the comprehension axiom

$$\exists a(\forall x(x \mathrel{\text{E}} a \leftrightarrow O(x)) \wedge \neg\exists l(l < a)) \qquad \text{CV1}$$

and defining

$$e = \varepsilon a(\forall x(x \mathrel{\text{E}} a \leftrightarrow O(x)) \wedge \neg\exists l(l < a)) \qquad \text{D2.36}$$

so that, by the first epsilon, we get

$$V(e) \wedge \forall x(x \mathrel{\text{E}} e \leftrightarrow O(x)) \wedge \neg\exists l(l < e) \qquad \text{T2.25}$$

and therefore

$$\begin{aligned} &(1)\ V(e) \\ &(2)\ x \mathrel{\text{E}} e \leftrightarrow O(x) \\ &(3)\ \neg\exists l(l < e) \end{aligned} \qquad \text{T2.26}$$

Immediate corollaries are

$$\exists a \forall i(i \mathrel{\text{E}} a) \qquad \text{T2.27}$$

$$i \mathrel{\text{E}} e \qquad \text{T2.28}$$

[53] TS §60, I 188: "[. . .] the most general of all [*sc.* idea], that of *something* or of an object in general. [. . .]."

[54] TS §99, I 331: "2) [. . .] it is impossible that the extension of an idea should be any more extended than that which encompasses every object that there is, and this extension belongs to the idea 'object' or 'something in general'."

[55] TS §89, I 298–9: "2) [. . .] the idea of something in general, which certainly [. . .] as completely simple [. . .]."

$$G(e) \leftrightarrow \exists x O(x) \qquad \text{T2.29}$$

Obviously we also get the two theorems

$$G(a) \rightarrow e \top a \qquad \text{T2.30}$$ [56]

$$G(a) \rightarrow a \sqsubseteq e \qquad \text{T2.31}$$ [57]

Let us emphasize that *e* refers to *every object,* not *to every thing*—just to all and only all those things that are objects.

2.6 Concretization

In his discussion of the possible *composition modes of ideas* and of the relationships of such modes and the *linguistic expressions* [*sprachliche Ausdrücke*] used to denote [*bezeichnen*] them, Bolzano (§58, n.6) calls attention to the relative pronouns, to which he attributes the role of denoting some modes of composition of an idea starting from a *principal idea* [*Hauptvorstellung*] and a *proposition.* The following are some of the examples he gives: "a space *that* is empty; a triangle the sides *of which* are equal to one another; a man *in whom* nobody believes".[58]

To account for this idea and its fundamental role in the Bolzanian system, we will begin by assuming the presence of the lectological operation of *relative concretization,* which to every object idea a and every quality idea $\bar{c}$ associates the object idea *(an) a that has* $\bar{c}$ [*(ein) a welches* $\bar{c}$ *hat*], called *concretization of* $\bar{c}$ *with respect to the (principal) idea a,* and which is such that it

(1) refers to all those objects referred to by a to which at least one of qualities referred to by $\bar{c}$ belongs;
(2) has as parts a and $\bar{c}$, as well as the parts of these ideas;
(3) is denoted by '$[a](\bar{c})$'.

Hence we first set

$$\forall a \forall \bar{c} \exists b (\forall i (i \mathrel{\text{E}} b \leftrightarrow i \mathrel{\text{E}} a \wedge \exists r (i \in r \wedge r \mathrel{\text{E}} \bar{c})) \wedge \forall l (l < b \leftrightarrow l \leq a \vee l \leq \bar{c})) \qquad \text{CV2}$$

[56] TS §105, I 358: "13) [. . .] The idea of an object in general [. . .] since it is superordinate to all other ideas which have extensions, and hence compatible with them."

[57] TS §105, I 358: "13) [. . .] The idea of an object in general [. . .] is superordinate to all other ideas which have extensions [. . .]."

[58] "Ein Raum, der leer ist; Ein Dreieck, dessen Seiten einander gleich sind; Ein Mann, dem Niemand traut".

and then, proceeding in strict analogy with what has been done in the case of e, we define

$$[a](\bar{c}) = \varepsilon b(\forall i(i \mathrel{\mathsf{E}} b \leftrightarrow i \mathrel{\mathsf{E}} a \wedge \exists r(i \in r \wedge r \mathrel{\mathsf{E}} \bar{c})) \wedge \forall l(l < b \leftrightarrow l \leq a \vee l \leq \bar{c})) \qquad \text{D2.37}$$

getting

$$\begin{aligned} &(1)\ V([a](\bar{c})) && \text{T2.32} \\ &(2)\ \forall i(i \mathrel{\mathsf{E}} [a](\bar{c}) \leftrightarrow i \mathrel{\mathsf{E}} a \wedge \exists r(i \in r \wedge r \mathrel{\mathsf{E}} \bar{c}) \\ &(3)\ \forall l(l < [a](\bar{c}) \leftrightarrow l \leq a \vee l \leq \bar{c}) \end{aligned}$$

Regarding the objectuality of this type of ideas, we have, of course,

$$G([a](\bar{c})) \leftrightarrow \exists i(i \mathrel{\mathsf{E}} a \wedge \exists r(i \in r \wedge r \mathrel{\mathsf{E}} \bar{c})) \qquad \text{T2.33}$$

2.7 Concrete and Abstract Ideas

Bolzano considers two fundamental specializations of the concretization. The first is the case in which the principal *idea* is the idea e. He calls an idea of this kind (i.e. an idea of the form $[e](\bar{a})$ [*something which has* $\bar{a}$]) a *concrete idea* or simply a *concretum*; he calls an *idea abstracted from this concretum*, or even simply its *abstractum*, the quality idea occurring in this *concretum*.[59] We introduce the predicate 'K' [*to be a concrete idea*] by setting

$$K(c) \leftrightarrow \exists \bar{a}(c = [e](\bar{a})) \qquad \text{D2.38}$$

and this yields

$$K(c) \leftrightarrow \exists \bar{a} \underset{L(x)}{\forall x}(x \leq c \leftrightarrow x \leq [e](\bar{a})) \qquad \text{T2.34}$$

Bolzano used 'A' to denote the concretization of a (i.e. [*Etwas*]a), but employed 'a' as the *abstractum* of A *only* when he knew that A was concrete and had the form [*Etwas*]a.[60].

[59] TS §60, I 188–9: "The kind of ideas we considered in §58 includes those of the form 'Something, which has (the attribute) *b*', where the place of the main idea is occupied by the most general of all, that of something or of an object in general. I call such things concrete ideas, or simply concreta. The idea *b* that occurs here, which as the idea of a mere attribute appropriately bears the name of an *attribute-idea* can, with regard to its occurrence as a constituent in this concretum, be called the corresponding *abstractum*, or the idea *abstracted* from it."

[60] TS §60, I 189: "Evidently, for any attribute-idea *b*, one can find a concretum containing it, of which it is the corresponding abstractum, namely, the idea 'something, which has (the attribute) *b*'. For the sake of brevity, I will symbolise the latter as *B*."

By T46, we have $G([e](\overline{a})) \leftrightarrow \exists i(i \sqsubset e \wedge \exists r(i \in r \wedge r \sqsubset \overline{a}))$, but, as every i falls under e and there are no empty qualities, this equivalence may be reduced to $G([e](\overline{a})) \leftrightarrow \exists r(r \sqsubset \overline{a})$, i.e.

$$G([e](\overline{a})) \leftrightarrow G(\overline{a}) \qquad \text{T2.35}$$

An important point is that naturally we have

$$\overline{a}\,\mathsf{T}\,\overline{c} \rightarrow [e](\overline{a})\,\mathsf{T}\,[e](\overline{c}) \qquad \text{T2.36}$$

but the converse does not hold. This was indicated by Bolzano in the following passage:

> "If two quality ideas *a* and *b* are compatible, then their concretizations *A* and *B* are compatible, too. However, we may not at once inversely conclude from the compatibility of two concreta *A* and *B* that their abstracta *a* and *b* are also compatible. Indeed, if *a* and *b* agree, there must be a quality *x* which is both *a* and *b* and therefore an object having the quality *x* is both an *A* and a *B*. On the contrary, the simple fact that *A* and *B* agree, does not allow us to conclude that also *a* and *b* agree. Indeed, although the object which is both *A* and *B* must unite in itself both the qualities *a* and *b*, it does not follow from this that, for this object, the quality falling under *a* is the same as the quality falling under *b*. Thus the abstracta—intelligence and prudence—are compatible and so are, therefore, their concreta—intelligent and prudent. On the other hand, the concreta—pious and erudite—are compatible without their abstracta—piety and erudition—being so; no kind of erudition, indeed, can be said to be a kind of erudition."[61]

It is also worth pointing out that the Bolzanian argument confirms two theses that have given rise to controversies.

The first is that there are no *empty qualities*. Indeed, from the existence of a quality underlying both ideas, the existence is inferred of an object to which this quality belongs. If there were an empty quality, however, both ideas could refer to it and such an inference would not be possible.

The second thesis is that *quality ideas may be common*. Indeed, if they were singular, sharing a common reference, this should be the same reference and so the last part of the argument would be not comprehensible.

Moreover, we have

$$G(\overline{a}) \rightarrow (\overline{a} \sqsubseteq \overline{c} \rightarrow [e](\overline{a}) \sqsubseteq [e](\overline{c})) \qquad \text{T2.37}$$

[61] Cf. TS §94, I 317.

but its inverse does not hold. In fact, an i may fall both under $[e](\bar{a})$ (since it has an $r \mathrm{E} \bar{a}$) and under $[e](\bar{c})$ (since it has $s \mathrm{E} \bar{c}$), but this s may be different from r and not fall under $\bar{a}$.

Bolzano emphasizes the fact that not all object ideas are concrete (such is, for instance, the idea e);[62] nevertheless, he believes that at least objectual ideas are always equivalent to concrete ones. Indeed, he accepts the fundamental principle according to which *falling under an idea is a quality* and that *the concretization of the idea of this quality is equivalent to the original idea.*[63] We can account for this conviction as follows.

First of all, one accepts that *for every objectual idea a there is at least one 'peculiar' quality of the a's* (i.e. *a quality that belongs to all and only to the objects falling under a*). In other words, one accepts the ontological comprehension axiom

$$G(a) \rightarrow \exists r \forall i (i \in r \leftrightarrow i \mathrm{E} a) \qquad \text{CQ1}$$

Note that the following developments do not require the possibility of referring to a well-specified one among the qualities peculiar to the a's; in other words, they do not require any ε-resolution of the comprehension axiom.

Second, one accepts that there is a quality idea having as parts a and the parts of a and, for objectual a, refers to the above qualities whereas, for objectless a, it is itself objectless; i.e.

$$\forall a \exists \bar{c} \begin{pmatrix} (G(a) \rightarrow \forall r (r \mathrm{E} \bar{c} \leftrightarrow \forall i (i \in r \leftrightarrow i \mathrm{E} a))) \\ \wedge (\neg G(a) \rightarrow \neg G(\bar{c})) \wedge \forall l (l < \bar{c} \leftrightarrow l \leq a) \end{pmatrix} \qquad \text{CB1}$$

and, third, one defines

$$\widehat{a} = \varepsilon \bar{c} \begin{pmatrix} (G(a) \rightarrow \forall r (r \mathrm{E} \bar{c} \leftrightarrow \forall i (i \in r \leftrightarrow i \mathrm{E} a))) \\ \wedge (\neg G(a) \rightarrow \neg G(\bar{c})) \wedge \forall l (l < \bar{c} \leftrightarrow l \leq a) \end{pmatrix} \qquad \text{D2.39}$$

so that

$$(1)\ B(\widehat{a}); \qquad \text{T2.38}$$

[62] TS §60, I 189: "Clearly, there are ideas that are neither abstracta nor concreta. For example, the idea 'something' by itself cannot be called a concrete idea, since it is not of the form 'Something, which has (the attribute) *b*'. Nor can it be called abstract, as it is not an attribute-idea. Similar remarks apply to the ideas 'nothing', 'this *A*', 'Socrates', and many others, which neither represent attributes nor are of the form of concrete ideas that I just stipulated. Abstract ideas are sometimes simple, while concrete ideas are always complex, since they consist of the idea 'something', and of the proposition 'which has *b*'."

[63] TS §225, II 286: "2) It is obvious that every true proposition remains true if its subject idea *A* is replaced by an equivalent idea, e.g., by the following: 'anything having the attribute of standing under idea *A*'. Thus it follows that the generality of the investigation is not limited if we replace the form '*A* has *b*' by the form 'Anything having *a* has *b*' or 'Whatever has *a* has *b*'."

$$(2)\ G(a) \to \forall r(r \mathrel{\mathsf{E}} \widehat{a} \leftrightarrow \forall i(i \in r \leftrightarrow i \mathrel{\mathsf{E}} a))$$
$$(3)\ \neg G(a) \to \neg G(\widehat{a})$$
$$(4)\ \forall l(l < \widehat{a} \leftrightarrow l \leq a)$$

Thus, the operation $\widehat{\ }$—called *abstraction*—associates with every object idea a the quality idea '*a-ity*', which, if a is objectual, refers to the quality of falling under a (of '*to be an a*'), but if a is objectless, does not refer to anything.

In this way, we get the fundamental Bolzanian theorem stating, as already hinted, that '*underlying the (objectual) idea a*' is equivalent to falling under the idea '*being something with the quality of underlying a*'. In short, '*being a*' is equivalent to '*being something that has a-ity*':

$$G(a) \to a \equiv [e](\widehat{a}) \qquad \text{T2.39}$$

Proof If $G(a)$, then, by CQ1 and T51,

$$\text{(a) } \exists r \forall i(i \in r \leftrightarrow i \mathrel{\mathsf{E}} a) \text{ and (b) } \forall r(r \mathrel{\mathsf{E}} \widehat{a} \leftrightarrow \forall i(i \in r \leftrightarrow i \mathrel{\mathsf{E}} a))$$

It follows from (a) that, for some r', (i) $\forall i(i \in r' \leftrightarrow i \mathrel{\mathsf{E}} a)$, and hence it follows from (b) that, in particular, we have (ii) $r' \mathrel{\mathsf{E}} \widehat{a} \leftrightarrow \forall i(i \in r' \leftrightarrow i \mathrel{\mathsf{E}} a)$ and therefore (iii) $r' \mathrel{\mathsf{E}} \widehat{a}$. Let $j \mathrel{\mathsf{E}} a$. Then, by (i), $j \in r'$; it follows from this and (iii) that $j \in r' \wedge r' \mathrel{\mathsf{E}} \widehat{a}$ and hence $\exists r(j \in r \wedge r \mathrel{\mathsf{E}} \widehat{a})$, i.e. $j \mathrel{\mathsf{E}} [e](\widehat{a})$. Thus $a \sqsubseteq [e](\widehat{a})$. Vice versa, let $j \mathrel{\mathsf{E}} [e](\widehat{a})$. Then, by definition, there is an r'' such that (iv) $j \in r''$ and (v) $r'' \mathrel{\mathsf{E}} \widehat{a}$. However, by (b), we have, in particular, $r'' \mathrel{\mathsf{E}} \widehat{a} \leftrightarrow \forall i(i \in r'' \leftrightarrow i \mathrel{\mathsf{E}} a))$, and hence, by (v), $\forall i(i \in r'' \leftrightarrow i \mathrel{\mathsf{E}} a)$. It follows from this and (iv) that $j \mathrel{\mathsf{E}} a$. Therefore, also $[e](\widehat{a}) \sqsubseteq a$.

An immediate corollary is that *each (objectual) object idea is equivalent to a concretum*

$$\underset{G(a)K(c)}{\forall a \exists c}(a \equiv c) \qquad \text{T2.40}$$

Now note that, under the hypothesis $G(\overline{a})$, the idea $\widehat{[e](\overline{a})}$ refers exclusively to the quality of falling under $[e](\overline{a})$ (i.e. to the quality *of having some quality falling under* $\overline{a}$). It is clear that the quality falling under $\widehat{[e](\overline{a})}$ is not only distinct from each quality falling under $\overline{a}$ but, in general, not even (ontologically) equivalent to any of them; indeed, it would be equivalent to the 'union' of them, but, in general, such a quality does not fall under $\overline{a}$. Hence the two ideas $\widehat{[e](\overline{a})}$ and $\overline{a}$ are very different from each other. Nevertheless, an object has the quality underlying $\widehat{[e](\overline{a})}$ iff it has at least one of the qualities underlying $\overline{a}$; that is to say, the respective concretizations are equivalent. Formally, let us take as a in T52 a any concrete idea; we have

$$G([e](\overline{a})) \to [e](\overline{a}) \equiv [e](\widehat{[e](\overline{a})}) \qquad \text{T2.41}$$

and, by T35,

$$G(\overline{a}) \rightarrow [e](\overline{a}) \equiv [e](\widehat{[e](\overline{a})}) \qquad \text{T2.42}$$

Thus, *although not every (objectual) quality idea is equivalent to the abstraction of an object idea, nevertheless its concretization is always equivalent to the concretization of an abstraction of that type*; i.e.

$$G(\overline{a}) \rightarrow \exists c([e](\overline{a}) \equiv [e](\widehat{c})) \qquad \text{T2.43}$$

2.8 Intuitions

The second type of special case of concretization is introduced as follows:

> "A particularly remarkable type of ideas is that which we express through the form This (or That). I believe, however, that we understand this expression in two different senses. In the one that I call the most exact This means more o less: This which is an *A*."[64]

Subsequently, Bolzano delves deeper into the question and considers such expressions as denotation of *intuitions* [*Anschauungen*], which he defines as those ideas that are both *simple* and *singular* [*einfache Einzelvorstellungen*] (i.e. without (proper) parts and referring to one and only one object). As we will see subsequently, in the analysis of the Bolzanian theory of knowledge, objects of intuitions are in fact qualities or, more exactly, *adherences* [*Adhärenzen*] (i.e. existent qualities of *substances* [*Substanzen*]). Hence the exhaustive treatment of intuitions presupposes as a logical level at least the 'higher level' (i.e. the level characterized by the axiom $Q(x) \rightarrow O(x)$). Nevertheless, if we set aside the ontological and metaphysical aspects of the issue, we can develop an elementary treatment of the intuitions, based on these three properties:

(1) Intuitions are simple and singular ideas.[65]
(2) Every object has at most one intuition of it.[66]
(3) There are infinitely many intuitions.[67]

[64] TS §59, I 187.

[65] TS §72, I 236: "[...] the ideas which occur in subject position and which we designate by the word this are certainly genuine singular ideas (§68). [...] Moreover, it is no less certain that all these ideas are also *simple*."

[66] Bolzano even thought that all simple ideas had this property. Cf. TS §96, I 320: "3) If, on the contrary, one were to ask whether two simple ideas could be interchangeable, the answer would have to be no. For [...]."

[67] Objects of intuitions are the instantaneous perceptions of the variations produced in us by some (internal or external) agent.

Thus, to simplify our task, we introduce a new predicate '*An*(*x*) [*x is an intuition*]', which is characterized by the defining axiom corresponding to (1):

$$An(x) \leftrightarrow V(x) \wedge \underset{O(u)O(z)}{\exists u \forall z}(z \mathrel{\text{E}} x \leftrightarrow z = u) \wedge \neg \underset{L(u)}{\exists u}(u < x) \qquad \text{D2.40}$$

We also assume, corresponding to (2),

$$An(x) \wedge An(y) \rightarrow (z \mathrel{\text{E}} x \wedge z \mathrel{\text{E}} y \rightarrow x = y) \qquad \text{GA18}$$

Finally, corresponding to (3), we introduce, for every natural number *n*, a constant idea d_n and assume

$$An(d_n) \qquad \text{GA19}$$

$$\begin{aligned} &(1)\ V(d_n) && \text{ST2.12} \\ &(2)\ \exists i \forall j (j \mathrel{\text{E}} d_n \leftrightarrow j = i) \\ &(3)\ \neg \exists a (a < d_n) \\ &(4)\ i \mathrel{\text{E}} d_m \wedge i \mathrel{\text{E}} d_n \rightarrow d_m = d_n \end{aligned}$$

Proof *Ad* (1)–(3) By substitution in the definition, one has

$$An(d_n) \leftrightarrow V(d_n) \wedge \exists i \forall j (j \mathrel{\text{E}} d_n \leftrightarrow j = i) \wedge \neg \exists a (a < d_n)$$

and so, by separating with the second axiom,

$$V(d_n) \wedge \exists i \forall j (j \mathrel{\text{E}} d_n \leftrightarrow j = i) \wedge \neg \exists a (a < d_n)$$

Ad (4) By substitution in the first axiom, one has

$$An(d_m) \wedge An(d_n) \rightarrow (i \mathrel{\text{E}} d_m \wedge i \mathrel{\text{E}} d_n \rightarrow d_m = d_n)$$

and so, by separating with the second axiom, $i \mathrel{\text{E}} d_m \wedge i \mathrel{\text{E}} d_n \rightarrow d_m = d_n$.

Very important for the Bolzanian theory of knowledge is that if '*this that has a-ty*' is objectual, then it is equivalent to '*this*', i.e.

$$G([d_n](\bar{a})) \rightarrow [d_n](\bar{a}) \equiv d_n \qquad \text{ST2.13}$$

Proof [$\rightarrow$] If $i \mathrel{\text{E}} [d_n](\bar{a})$, then $\exists r(i \mathrel{\text{E}} d_n \wedge i \in r \wedge r \mathrel{\text{E}} \bar{a})$ and so, in particular, $i \mathrel{\text{E}} d_n$.

[$\leftarrow$] If $i \sqsubset d_n$ and $G([d_n](\bar{a}))$, then there are j and r such that $j \sqsubset d_n \wedge j \in r \wedge r \sqsubset \bar{a}$; but, by (2) of ST12, from $i \sqsubset d_n$ and $j \sqsubset d_n$ it follows that $i = j$; therefore $i \sqsubset d_n \wedge i \in r \wedge r \sqsubset \bar{a}$ and hence $i \sqsubset [d_n](\bar{a})$.[68]

It may be useful to remark that, formally speaking, intuitions, as simple and singular ideas, are close to the symbols usually referred to as *individual constants.*

Intuitions defined as such constitute the tool by means of which Bolzano deals with the issue of the relationship *knowledge–experience.* In particular, he calls [§ 73] *pure intuition* [*reine Anschauung*] every intuition, as well as every non-simple idea in which all occurring ideas are intuitions and *pure concept* [*reiner Begriff*] every simple or composite idea in which no intuitions occur; i.e.

$$Bg(x) \leftrightarrow V(x) \wedge \forall y(y \leq x \rightarrow \neg An(y)) \qquad \text{D2.41}$$

Any other idea is said to be *mixed* [*gemischte Vorstellung*] and, according to the preeminence in it [*der vorzügliche Bestandtheil*] of the intuitive part or of the conceptual part, *mixed intuition* or *mixed concept,* respectively. Propositions in which no intuition occur are called [§133] *propositions of pure concepts* or *conceptual propositions* [*Sätze aus reinen Begriffen, Begriffssätze*] and, if true, *conceptual truths* [*Begriffswahrheiten*]; any other proposition (i.e. any proposition in which one or more intuitions occur) is called *intuitive* or *perceptive proposition* [*Anschauung-* or *Wahrnehmungssatz*] or also *empirical proposition* [*empirischer Satz*]. Without embarking on a discussion of this aspect, we may remark that the truths of Bolzanian '*speculative physics*' are always conceptual. This, however, does not hinder their *knowledge* from being acquired through experience (i.e. it may be drawn from empirical truths).

2.9 Sum of Quality Ideas

Bolzano conceives of two fundamental operations on quality ideas, the first of which he calls *sum* [*Summe*] and denotes as '+'. This operation is never properly defined; nevertheless, its very frequent use allows us to isolate its essential characteristics. The definition that we propose here satisfies almost all statements about the sum of quality ideas we have encountered in the Bolzanian text, although it probably does not correspond exactly to what Bolzano thought.

[68] TS §69, I 224–5: "3) I must emphasise the ideas of the form 'This *A*' in the strict sense, which I discussed above (§. 59. 68.) as an especially noteworthy example of redundant ideas. Every such idea, provided that it has an object, is redundant, namely, with respect to the constituent 'which is an *A*' which, according to the analysis developed in §59, they all contain. For the mere this, which occurs as the main part of such ideas, only applies to a single object, and indeed an object that has the attribute *a*, so that it also stands under the idea *A*. The addition 'which is an *A*' can thus be omitted without changing the extension."

Thus, we begin by assuming that if two qualities 'intersect' (i.e. if there is an object to which both belong), then they have an 'intersection quality' (i.e. a quality that belongs to all and only those objects to which both qualities belong; i.e.

$$\exists i(i \in r \wedge i \in s) \rightarrow \exists t \forall i(i \in t \leftrightarrow i \in r \wedge i \in s) \qquad \text{CQ2}$$

It is worth remarking that just as in the case of axiom CQ1, it is not necessary to isolate, among all these 'intersections', a particular one.

Now, given two quality ideas $\bar{a}$ and $\bar{c}$, their concretizations may be compatible, that is to say, there is an i such that $i \mathrel{\text{E}} [e](\bar{a}) \wedge i \mathrel{\text{E}} [e](\bar{c})$. This, however, simply means that under $\bar{a}$ there is a quality r that belongs to i and under $\bar{c}$ there is a quality s that also belongs to i. The compatibility of the concretizations of $\bar{a}$ and $\bar{c}$ is thus equivalent to the existence under $\bar{a}$ of a quality that 'intersects' a quality under $\bar{c}$, whereas their incompatibility is equivalent to the non-existence of such a pair of qualities. In the first case, we say that the two ideas *interfere* with each other (and we write '$\bar{a} \mathrel{\text{TT}} \bar{c}$') and in the second case, of course, that they do not. We thus define the predicate $\mathrel{\text{TT}}$ by establishing

$$\bar{a} \mathrel{\text{TT}} \bar{c} \leftrightarrow [e](\bar{a}) \mathrel{\text{T}} [e](\bar{c}) \qquad \text{D2.42}$$

But if $\bar{a}$ and $\bar{c}$ interfere with each other because a certain r under $\bar{a}$ and a certain s under $\bar{c}$ simultaneously belong to an object, then, by the preceding ontological axiom, there is also at least one quality that belongs to all and only all those objects to which both r and s belong. Clearly, under $\bar{a}$ and $\bar{c}$, there may be more pairs of 'intersecting' qualities; hence, corresponding to each of these pairs, there also is a quality belonging to all and only all objects to which both qualities of the pair belong. We assume that there always is a quality idea, the *sum* $\bar{a} + \bar{c}$ *of* $\bar{a}$ *and* $\bar{c}$, which, if the two ideas interfere with each other, refers to all and only all the qualities existing in correspondence to each pair of 'intersecting' qualities, whereas it is objectless if the two ideas do not interfere (i.e. if their concretizations are incompatible). We thus assume

$$\forall \bar{a}\,\bar{c} \exists \bar{b} \begin{pmatrix} (\bar{a} \mathrel{\text{TT}} \bar{c} \rightarrow \forall t(t \mathrel{\text{E}} \bar{b} \leftrightarrow \exists r \exists s(r \mathrel{\text{E}} \bar{a} \wedge s \mathrel{\text{E}} \bar{c} \wedge \forall i(i \in t \leftrightarrow i \in r \wedge i \in s))) \\ \wedge \quad (\neg(\bar{a} \mathrel{\text{TT}} \bar{c}) \rightarrow \neg G(\bar{b})) \quad \wedge \quad \forall l(l < \bar{b} \leftrightarrow l \leq \bar{a} \vee l \leq \bar{c}) \end{pmatrix} \qquad \text{CB2}$$

and define

$$\bar{a} + \bar{c} = \varepsilon \bar{b} \begin{pmatrix} (\bar{a} \mathrel{\text{TT}} \bar{c} \rightarrow \forall t(t \mathrel{\text{E}} \bar{b} \leftrightarrow \exists r \exists s(r \mathrel{\text{E}} \bar{a} \wedge s \mathrel{\text{E}} \bar{c} \wedge \forall i(i \in t \leftrightarrow i \in r \wedge i \in s))) \\ \wedge \quad (\neg(\bar{a} \mathrel{\text{TT}} \bar{c}) \rightarrow \neg G(\bar{b})) \quad \wedge \quad \forall l(l < \bar{b} \leftrightarrow l \leq \bar{a} \vee l \leq \bar{c}) \end{pmatrix} \qquad \text{D2.43}$$

so that, as usual, we obtain

$$(1)\ B(\bar{a}+\bar{c}) \qquad \text{T2.44}$$
$$(2)\ \bar{a} \mathrel{\top\!\!\top} \bar{c} \rightarrow \forall t(t \mathrel{\mathrm{E}} \bar{a}+\bar{c} \leftrightarrow \exists r \exists s(r \mathrel{\mathrm{E}} \bar{a} \wedge s \mathrel{\mathrm{E}} \bar{c} \wedge \forall i(i \in t \leftrightarrow i \in r \wedge i \in s)))$$
$$(3)\ \neg \bar{a} \mathrel{\top\!\!\top} \bar{c} \rightarrow \neg G(\bar{a}+\bar{c})$$
$$(4)\ \forall l(l < \bar{a}+\bar{c} \leftrightarrow l \leq \bar{a} \vee l \leq \bar{c})$$

The following two theorems give us the basic properties of the *sum of quality ideas*:[69]

$$\bar{a} \mathrel{\top\!\!\top} \bar{c} \rightarrow (i \mathrel{\mathrm{E}} [e](\bar{a}+\bar{c}) \leftrightarrow i \mathrel{\mathrm{E}} [e](\bar{a}) \wedge i \mathrel{\mathrm{E}} [e](\bar{c})) \qquad \text{T2.45}$$

Proof Let $\bar{a} \mathrel{\top\!\!\top} \bar{c}$.

[$\rightarrow$] Let $i \mathrel{\mathrm{E}} [e](\bar{a}+\bar{c})$; then, by definition, there is a $t \mathrel{\mathrm{E}} (\bar{a}+\bar{c})$ such that $i \in t$. But $\bar{a} \mathrel{\top\!\!\top} \bar{c}$ and $t \mathrel{\mathrm{E}} (\bar{a}+\bar{c})$ imply, by T44, that there are $r \mathrel{\mathrm{E}} \bar{a}$ and $s \mathrel{\mathrm{E}} \bar{c}$ such that $\forall i(i \in t \leftrightarrow i \in r \wedge i \in s)$. It follows from this and $i \in t$ that $i \in r$ and $i \in s$ and, holding $r \mathrel{\mathrm{E}} \bar{a}$ and $s \mathrel{\mathrm{E}} \bar{c}$, we have $i \mathrel{\mathrm{E}} [e](\bar{a})$ and $i \mathrel{\mathrm{E}} [e](\bar{c})$.

[$\leftarrow$] Vice versa, let $i \mathrel{\mathrm{E}} [e](\bar{a})$ and $i \mathrel{\mathrm{E}} [e](\bar{c})$; then, by definition, there are $r \mathrel{\mathrm{E}} \bar{a}$ and $s \mathrel{\mathrm{E}} \bar{c}$ such that $i \in r$ and $i \in s$; by CQ2, there is then also a t such that $\forall i(i \in t \leftrightarrow i \in r \wedge i \in s)$ and therefore, by T44, $t \mathrel{\mathrm{E}} \bar{a}+\bar{c}$. But $i \in t$. Hence $i \mathrel{\mathrm{E}} [e](\bar{a}+\bar{c})$.

$$\bar{a} \mathrel{\top\!\!\top} \bar{c} \rightarrow (i \mathrel{\mathrm{E}} [e](\bar{a}+\bar{c}) \leftrightarrow i \mathrel{\mathrm{E}} [[e](\bar{a})](\bar{c})) \qquad \text{T2.46}$$

Proof Let $\bar{a} \mathrel{\top\!\!\top} \bar{c}$.

[$\rightarrow$] Let $i \mathrel{\mathrm{E}} [e](\bar{a}+\bar{c})$; therefore, by definition, there is a $t \mathrel{\mathrm{E}} (\bar{a}+\bar{c})$ such that $i \in t$. $\bar{a} \mathrel{\top\!\!\top} \bar{c}$ and $t \mathrel{\mathrm{E}} (\bar{a}+\bar{c})$ imply, by T44, that there are $r \mathrel{\mathrm{E}} \bar{a}$ and $s \mathrel{\mathrm{E}} \bar{c}$ such that $\forall i(i \in t \leftrightarrow i \in r \wedge i \in s)$. It follows from this and $i \in t$ that $i \in r$ and $i \in s$ and, holding $r \mathrel{\mathrm{E}} \bar{a}$, it follows from $i \in r$ that $i \mathrel{\mathrm{E}} [e](\bar{a})$; it follows from this and $i \in s \mathrel{\mathrm{E}} \bar{c}$ that $i \mathrel{\mathrm{E}} [[e](\bar{a})](\bar{c})$. [$\leftarrow$] Vice versa, let $i \mathrel{\mathrm{E}} [[e](\bar{a})](\bar{c})$; then, by definition, there in an s such that $i \mathrel{\mathrm{E}} [e](\bar{a})$ and $i \in s$ and $s \mathrel{\mathrm{E}} \bar{c}$. But $i \mathrel{\mathrm{E}} [e](\bar{a})$ implies that there is an r such that $i \in r$ and $r \mathrel{\mathrm{E}} \bar{a}$. Holding $i \in r \mathrel{\mathrm{E}} \bar{a}$ and $i \in s \mathrel{\mathrm{E}} \bar{c}$, r and s interfere so, by the ontological axiom, they have an 'intersection' t and this, by T44, underlies $\bar{a}+\bar{c}$. Hence $i \mathrel{\mathrm{E}} [e](\bar{a}+\bar{c})$.

[69] TS §96, I 322: "7) Let us denote the idea 'something that has (the attributes) $a, b, c, d, \ldots$' by [something]$(a+b+c+d+\cdots)$. Let us furthermore denote (cf. §60) the idea 'something that has (the attribute) x' by X. Then the following symbols represent equivalent ideas provided they have objects: [something]$(a+b)$, $[A]b$, $[B]a$; so too: [something]$(a+b+c+d+\cdots)$, $[A](b+c+\cdots)$, $[B](a+c+\cdots)$, $[C](a+b+\cdots)$."

TS §225, II 287: "8) Since ideas of the form $[A]\,b$ are always equivalent to those of the form $[B]\,a$, or [something]$(a+b)$, we may derive from the two conclusions we just mentioned four others which are equivalent to them." In TS §233, [Etwas]$(a+b+c+d+\cdots)$, $[A](b+c+d+\cdots)$, $[B](a+c+d+\cdots)$, $[[A]b](c+d+\cdots)$, $[[A]c](b+d+\cdots)$, etc. are declared to be all equivalent.

We must point out that the strictly intensional perspective of the Bolzanian theory naturally excludes that $\overline{a}+\overline{a}=\overline{a}$, $\overline{a}+\overline{c}=\overline{c}+\overline{a}$ or $(\overline{a}+\overline{b})+\overline{c}=\overline{a}+(\overline{b}+\overline{c})$. The polyvocity of quality ideas, however, also implies the fact that, for instance, not even $\overline{a}+\overline{a}\equiv\overline{a}$ or $\overline{a}+(\overline{b}+\overline{c})\equiv(\overline{a}+\overline{b})+\overline{c}$ generally hold. Indeed, under $\overline{a}+\overline{a}$ fall all qualities falling under $\overline{a}$, but, so do others; precisely all 'intersections' of qualities that are different, but nevertheless interfere. Analogously, one can see that the qualities falling under $\overline{a}+(\overline{b}+\overline{c})$ are not, in general, the same as the qualities falling under $(\overline{a}+\overline{b})+\overline{c}$. As the following theorems show, however, *such limitations do not apply to concretizations of the sum.*

$$\overline{a} \mathbin{\top\!\!\top} \overline{c} \rightarrow [e](\overline{a}+\overline{a}) \equiv [e](\overline{a}) \qquad \text{T2.47}$$

Proof $[\rightarrow]$ Let $i \mathrel{\mathrm{E}} [e](\overline{a}+\overline{a})$; then there is a $t \mathrel{\mathrm{E}} (\overline{a}+\overline{a})$ such that $i \in t$. Now $t \mathrel{\mathrm{E}} (\overline{a}+\overline{a})$ implies that there are 'intersecting' $r \mathrel{\mathrm{E}} \overline{a}$ and $s \mathrel{\mathrm{E}} \overline{a}$ such that $i \in r$ and $i \in s$; it follows from $i \in r \mathrel{\mathrm{E}} \overline{a}$ that $i \mathrel{\mathrm{E}} [e](\overline{a})$.

$[\leftarrow]$ Vice versa, let $i \mathrel{\mathrm{E}} [e](\overline{a})$; then there is an r such that $i \in r \mathrel{\mathrm{E}} \overline{a}$; since, trivially, r 'intersects' itself and also trivially $\forall i(i \in r \leftrightarrow i \in r \wedge i \in r)$, r falls under $\overline{a}+\overline{a}$; hence $i \mathrel{\mathrm{E}} [e](\overline{a}+\overline{a})$.

$$\overline{a} \mathbin{\top\!\!\top} \overline{c} \rightarrow [e](\overline{a}+\overline{c}) \equiv [e](\overline{c}+\overline{a}) \qquad \text{T2.48}$$

Proof This is obvious.

$$\overline{a} \mathbin{\top\!\!\top} \overline{c} \rightarrow [e](\overline{a}+(\overline{b}+\overline{c})) \equiv [e]((\overline{a}+\overline{b})+\overline{c}) \qquad \text{T2.49}$$

Proof $[\rightarrow]$ Let $i \mathrel{\mathrm{E}} [e](\overline{a}+(\overline{b}+\overline{c}))$; then, by definition, there is a $t' \mathrel{\mathrm{E}} (\overline{a}+(\overline{b}+\overline{c}))$ for which $i \in t'$. Now $t' \mathrel{\mathrm{E}} (\overline{a}+(\overline{b}+\overline{c}))$ implies, by T44, that there are 'intersecting' $r \mathrel{\mathrm{E}} \overline{a}$ and $t'' \mathrel{\mathrm{E}} (\overline{b}+\overline{c})$ such that for every object, and hence in particular for i, $i \in t \leftrightarrow i \in r \wedge i \in t''$; so $i \in r$ and $i \in t''$. Always, by T44, it follows from $t'' \mathrel{\mathrm{E}} (\overline{b}+\overline{c})$ that there are 'intersecting' $s \mathrel{\mathrm{E}} \overline{b}$ and $t \mathrel{\mathrm{E}} \overline{c}$ such that for every object, and so in particular for i, $i \in t \leftrightarrow i \in s \wedge i \in t$; hence $i \in s$ and $i \in t$. Therefore, altogether, (i) $i \in r \mathrel{\mathrm{E}} \overline{a}$, (ii) $i \in s \mathrel{\mathrm{E}} \overline{b}$, and (iii) $i \in t \mathrel{\mathrm{E}} \overline{c}$. It follows from (i) and (ii), by the ontological axiom, that there is an r' such that for every object, and hence in particular for i, $i \in r' \leftrightarrow i \in r \wedge i \in s$, and, by T44, $r' \mathrel{\mathrm{E}} (\overline{a}+\overline{b})$. It follows from this and from (iii), again by the ontological axiom, that there is an r'' such that for every object, and hence in particular for i, $i \in r'' \leftrightarrow i \in r' \wedge i \in t$ and therefore also, by T44, $r'' \mathrel{\mathrm{E}} ((\overline{a}+\overline{b})+\overline{c}$. Hence $i \mathrel{\mathrm{E}} [e]((\overline{a}+\overline{b})+\overline{c})$.

$[\leftarrow]$ This is analogous to the proof of $[\rightarrow]$.

It is worth pointing out that the objectuality of $\overline{a}+\overline{c}$, i.e. the interference between $\overline{a}$ and $\overline{c}$, is a weaker property than the compatibility of $\overline{a}$ and $\overline{c}$; the latter, indeed, requires the existence under $\overline{a}$ e $\overline{c}$ of a *common quality*, whereas the former simply requires the existence of a quality under the one and of a quality under the other having a *common object*.

This discussion allows us to introduce an '*intersection*' between the object ideas a and c (to be denoted by '$a \sqcap c$') or the idea under which all and only all objects fall that simultaneously fall under a and c (i.e. objects that have both a-ity and c-ity). This can be done by setting

$$a \sqcap c = [e](\widehat{a} + \widehat{c}) \qquad \text{D2.44}$$

It must be emphasized that although without introducing a particular name or a notation for it, Bolzano uses this 'intersection'; indeed, he makes, for example, use of the theorem[70]

$$[e](\overline{a}) \sqcap [e](\overline{c}) \equiv [e](\overline{a} + \overline{c}) \qquad \text{T2.50}$$

Clearly, one has the following:

$$i \,\text{E}\, a \sqcap c \leftrightarrow i \,\text{E}\, a \wedge i \,\text{E}\, c \qquad \text{T2.51}$$

$$\begin{aligned} &(1)\ a \sqcap a \equiv a \\ &(2)\ a \sqcap c \equiv c \sqcap a \\ &(3)\ a \sqcap (b \sqcap c) \equiv (a \sqcap b) \sqcap c \\ &(4)\ a \sqcap e \equiv a \end{aligned} \qquad \text{T2.52}$$

$$G(a \sqcap c) \rightarrow [e](\widehat{a \sqcap c}) \equiv [e](\widehat{a} + \widehat{c}) \qquad \text{T2.53}$$

Proof $G(a \sqcap c) \rightarrow [e](\widehat{a \sqcap c}) \equiv a \sqcap c = [e](\widehat{a} + \widehat{c})$

Moreover, note that, being defined, the parts of $a \sqcap c$ are those of $[e](\widehat{a} + \widehat{c})$ and not simply those of a and c.

A better evaluation of the proposed interpretation of + may be obtained by employing the following set-theoretical consideration. If we treat both relations $\in$ and E as set-theoretical memberships and thus consider the objects as individuals, the qualities as sets of individuals, and the quality ideas as families of sets of individuals, then $[e](\overline{a})$ becomes the union $\bigcup \overline{a}$ over the family of sets $\overline{a}$ (i.e. $i \in \bigcup \overline{a} \leftrightarrow \exists r(i \in r \in \overline{a})$) and, admitting also empty intersections for simplicity,

[70] TS §98, I 328: "4)[. . .] the idea of something that is *A*, as well as *B*, *C*, etc., or the idea [something]($a + b + c$.....) [. . .]."

$\bar{a} + \bar{c}$ becomes the so-called 'complex or direct intersection' of the families $\bar{a}$ and $\bar{c}$, i.e.

$$t \in (\bar{a} + \bar{c}) \leftrightarrow \exists rs(t = r \cap s \wedge r \in \bar{a} \wedge s \in \bar{c}), \text{ i.e. } (\bar{a} + \bar{c}) = [r \cap s | r \in \bar{a} \wedge s \in \bar{c}]$$

The equivalence $[e](\bar{a}) \sqcap [e](\bar{c}) \equiv [e](\bar{a} + \bar{c})$ is thus the set-theoretical equation

$$\bigcup \bar{a} \cap \bigcup \bar{c} = \bigcup (\bar{a} + \bar{c})$$

We close this section by remarking that, as can clearly be seen from the preceding quotations, Bolzano, just like his contemporaries, uses notations of the type '$a + b + c + \ldots$' without really thinking them as iterations of an associative binary operation. Thus, strictly speaking, it would be more correct to introduce, for each $n > 1$, an operation of '*n-ary sum*'. The arrangement of our definitions from this point of view does not present any real difficulties, but only issues related to notational burdening. As the same issue will continue to appear, it seems opportune at this point to develop it. It is, however, understood that hereinafter this type of consideration will always be omitted.

We begin with the assumption, for every $n > 1$, of the comprehension for the 'intersection' among n qualities

$$\exists i(\bigwedge_{k \leq n} i \in r_k) \rightarrow \exists r \forall i(i \in r \leftrightarrow \bigwedge_{k \leq n} i \in r_k) \qquad \text{SCQ2}$$

We thus introduce the predicate of interference among n quality ideas

$$\mathrm{TT}^n(\overset{n}{\bar{a}}) \leftrightarrow \overset{n}{\exists} \overset{n}{r}(\bigwedge_{k \leq n}(r_k \mathsf{E} \bar{a}_k) \wedge \exists i(\bigwedge_{k \leq n} i \in r_k)) \qquad \text{SD2.45}$$

Finally, we assume the comprehension and its ε-resolution for the n-ary *sum of quality ideas*:

$$\forall \overset{n}{\bar{a}} \exists \bar{c} \left(\begin{array}{l} (\mathrm{TT}^n(\overset{n}{\bar{a}}) \rightarrow \forall r(r \mathsf{E} \bar{c} \leftrightarrow \exists \overset{n}{s}(\bigwedge_{k \leq n}(s_k \mathsf{E} \bar{a}_k) \wedge \forall i(i \in r \leftrightarrow \bigwedge_{k \leq n}(i \mathsf{E} s_k))))) \\ \wedge(\neg \mathrm{TT}^n(\overset{n}{\bar{a}}) \rightarrow \neg G(\bar{c})) \wedge \forall l(l < \bar{c} \leftrightarrow \bigvee_{k \leq n}(l \leq \bar{a}_k))) \end{array} \right) \qquad \text{SCB1}$$

$$\sum_{k \leq n}(\bar{a}_k) = \varepsilon \bar{c} \left(\begin{array}{l} (\mathrm{TT}^n(\overset{n}{\bar{a}}) \rightarrow \forall r(r \mathsf{E} \bar{c} \leftrightarrow \exists \overset{n}{s}(\bigwedge_{k \leq n}(s_k \mathsf{E} \bar{a}_k) \wedge \forall i(i \in r \leftrightarrow \bigwedge_{k \leq n}(i \mathsf{E} s_k))))) \\ \wedge(\neg \mathrm{TT}^n(\overset{n}{\bar{a}}) \rightarrow \neg G(\bar{c})) \wedge \forall l(l < \bar{c} \leftrightarrow \bigvee_{k \leq n}(l \leq \bar{a}_k))) \end{array} \right) \qquad \text{SD2.46}$$

From these, we obtain

$$(1)\ B(\sum_{k\leq n}(\overline{a}_k)) \qquad \text{ST2.14}$$

$$(2)\ \overset{n}{\Pi}(\overline{a}) \rightarrow \forall r(r \mathrel{\varepsilon} \sum_{k\leq n}(\overline{a}_k) \leftrightarrow \overset{n}{\exists} \overline{s}(\bigwedge_{k\leq n}(s_k \mathrel{\varepsilon} \overline{a}_k) \wedge \forall i(i \in r \leftrightarrow \bigwedge_{k\leq n}(i \mathrel{\varepsilon} s_k))))$$

$$(3)\ \neg\overset{n}{\Pi}(\overline{a}) \rightarrow \neg G(\sum_{k\leq n}(\overline{a}_k))$$

$$(4)\ \forall l(l < \sum_{k\leq n}\overline{a}_k \leftrightarrow \bigvee_{k\leq n}(l \leq \overline{a}_k))$$

2.10 Another Lectological Operation on Quality Ideas

In a purely systematic perspective, we may also introduce the 'dual' operation of a sum, which can be denoted by '$\overline{a}\cdot\overline{c}$'. We assume

$$\forall\overline{a}\,\overline{c}\exists\overline{b}\begin{pmatrix}(G(\overline{a}) \vee G(\overline{c}) \rightarrow \forall r(r \mathrel{\varepsilon} \overline{b} \leftrightarrow r \mathrel{\varepsilon} \overline{a} \vee r \mathrel{\varepsilon} \overline{c}) \wedge \\ (\neg(G(\overline{a}) \vee G(\overline{c})) \rightarrow \neg G(\overline{b})) \wedge \forall l(l < \overline{b} \leftrightarrow l \leq \overline{a} \vee l \leq \overline{c})\end{pmatrix} \qquad \text{CB3}$$

and define

$$\overline{a}\cdot\overline{c} = \varepsilon\overline{b}\begin{pmatrix}(G(\overline{a}) \vee G(\overline{c}) \rightarrow \forall r(r \mathrel{\varepsilon} \overline{b} \leftrightarrow r \mathrel{\varepsilon} \overline{a} \vee r \mathrel{\varepsilon} \overline{c}) \wedge \\ (\neg(G(\overline{a}) \vee G(\overline{c})) \rightarrow \neg G(\overline{b})) \wedge \forall l(l < \overline{b} \leftrightarrow l \leq \overline{a} \vee l \leq \overline{c})\end{pmatrix} \qquad \text{D2.47}$$

so that

$$(1)\ B(\overline{a}\cdot\overline{c}) \qquad \text{T2.54}$$

$$(2)\ G(\overline{a}) \vee G(\overline{c}) \rightarrow (r \mathrel{\varepsilon} (\overline{a}\cdot\overline{c}) \leftrightarrow r \mathrel{\varepsilon} \overline{a} \vee r \mathrel{\varepsilon} \overline{c})$$

$$(3)\ \neg G(\overline{a}) \wedge \neg G(\overline{c}) \rightarrow \neg G(\overline{a}\cdot\overline{c})$$

$$(4)\ \forall l(l < \overline{a}\cdot\overline{c} \leftrightarrow l \leq \overline{a} \vee l \leq \overline{c})$$

The relevant property of this operation is formulated in the following theorem:

$$G(\overline{a}) \vee G(\overline{c}) \rightarrow (i \mathrel{\varepsilon} [e](\overline{a}) \vee i \mathrel{\varepsilon} [e](\overline{c}) \leftrightarrow i \mathrel{\varepsilon} [e](\overline{a}\cdot\overline{c})) \qquad \text{T2.55}$$

Proof Let $G(\overline{a}) \vee G(\overline{c})$. Therefore $i \mathrel{\varepsilon} [e](\overline{a}) \vee i \mathrel{\varepsilon} [e](\overline{c}) \leftrightarrow \exists r(i \in r \wedge r \mathrel{\varepsilon} \overline{a}) \vee \exists r(i \in r \wedge r \mathrel{\varepsilon} \overline{c}) \leftrightarrow \exists r((i \in r \wedge r \mathrel{\varepsilon} \overline{a}) \vee (i \in r \wedge r \mathrel{\varepsilon} \overline{c})) \leftrightarrow \exists r(i \in r \wedge (r \mathrel{\varepsilon} \overline{a} \vee r \mathrel{\varepsilon} \overline{c})) \leftrightarrow \exists r(i \in r \wedge r \mathrel{\varepsilon} \overline{a}\cdot\overline{c}) \leftrightarrow i \mathrel{\varepsilon} [e](\overline{a}\cdot\overline{c})$.

As already done in the case of + and $\sqcap$, it is possible to introduce an operation $\sqcup$ by setting

$$a \sqcup c := [e](\widehat{a}\cdot\widehat{c}) \qquad \text{D2.48}$$

from which it is easy to get the following:

$$[e](\overline{a}) \sqcup [e](\overline{c}) \equiv [e](\overline{a} \cdot \overline{c}) \qquad \text{T2.56}$$

$$i \sqsubset a \sqcup c \leftrightarrow i \sqsubset a \vee i \sqsubset c \qquad \text{T2.57}$$

$$a \sqcup a \equiv a \qquad \text{T2.58}$$

$$a \sqcup c \equiv c \sqcup a \qquad \text{T2.59}$$

$$a \sqcup (b \sqcup c) \equiv (a \sqcup b) \sqcup c \qquad \text{T2.60}$$

$$a \sqcup e \equiv e \qquad \text{T2.61}$$

$$G(a \sqcup c) \rightarrow [e](\widehat{a \sqcup c}) \equiv [e](\widehat{a} \cdot \widehat{c}) \qquad \text{T2.62}$$

Proof $G(a \sqcup c) \rightarrow [e](\widehat{a \sqcup c}) \equiv a \sqcup c = [e](\widehat{a} \cdot \widehat{c})$.

It should be noted that Bolzano, even if without an explicit introduction, also uses this operation when he refers, as he sometimes does, to the "*sum of the fields of the ideas* $a_1, \ldots, a_n$ *[Summe der Gebiete der Vorstellungen]*".[71]

2.11 Negation

Bolzano devotes two large sections to pursue the clarification of the notion of negation. The first (§89) is in the part dedicated to ideas and just entitled *Affirmative [bejahende]* and *negative [verneinende] ideas,* while the second (§136) comes under the title *Propositions with negative ideas,* in the part dedicated to propositions, on the whole about 20 pages. In this exposition, he is even more careful than usual in distinguishing between lectological things and the *verbal expressions* used *to denote* them;[72] here, indeed,

> "the usual verbal expressions of an idea are rarely sufficiently determinate to allow an immediate judgment, from them alone, on whether this idea is negative, and of what type of negative idea it is; on the contrary, this must mainly be obtained from a careful decomposition of it and from its use."[73]

Let us begin by pointing out that Bolzano identified and, by means of a suitable symbolization, obtained a systematic distinction of three kinds of negation: (1) *the negation of an object idea* (or, better, *the negation of the subject of a proposition*), for which he first used the notation 'Nicht *A*' and later turned to '**Non** *A*'; (2) *the negation of a quality idea* (or, better, *the negation of the predicate of a proposition*), for

[71] See e.g. TS §236.11, II 446.

[72] The precise technical meaning of these locutions will be discussed in Chapter 9.

[73] TS §89, I 299.

which he used the notation '**non** *a*' and later also '**n** *a*'; and finally (3) *the negation of a proposition*, for which he used the notation '**Neg.** *M*'. The most important among these is the second, because the other two are expressible by means of it. Indeed, **Non** *A* is identified with '*something that has* **non** *a*' and **Neg.** *M* is identified with '*M has* **non** *truth*'. In the reconstruction, however, it is more convenient for us to begin with the consideration of the negation of an object idea.

For Bolzano, an idea is *negative* (*in the wider sense*) when the negation occurs at least once among its components. Among them, however, he immediately distinguishes the ideas that are *negative in the strict sense* (i.e. those ideas in which the negations cannot be completely omitted, because this omission would generate ideas that are not equivalent to the originals) from those in which this is not the case, for example those ideas that contain a sequence of an even number of negations and in which pairs of negations can be omitted without altering the extension.[74]

As far as the negative ideas in the strict sense are concerned, he introduces a distinction between *purely* [*rein*] or *absolutely* [*durchaus*] *negative ideas* and only *partially* [*theilweise*] *negative ideas.*

Absolutely negative are

> "those ideas which, contained under the form Non *A*, do not set absolutely anything else than the negation of a certain idea *A*, without requiring that, instead of the negated idea, an other idea be thought, not even the idea of something in general. Just for these reasons I call them purely or absolutely negative and I have no doubt that there are some of them."[75]

Bolzano provides the following examples: 'not blue', 'not round', and, in particular, 'nothing'. Later on, he returned to this type of idea, confirming their presence in the universe so that their representations really matter and their verbal expressions are not devoid of meaning, but also emphasizing that such ideas are always objectless.[76] Consequently, he excluded, on the basis of his notion of truth that we will examine later, that such ideas may ever act as subjects in a true proposition.[77]

[74] In the following, Bolzano also decides to reserve the qualification of 'negative' for the strictly negative ideas and to refer to the remaining ones as 'affirmative'.

[75] TS §89, I 299.

[76] TS §89, I 300: "It is true, admittedly, that a purely negative idea not only fails to apply to anything existing, but in fact applies to no object at all, [. . .]."

TS §89, I 302: "Against this, I wish to object that ideas that are absolutely negative actually have no object at all."

[77] TS §136, II 31: "1) First, concerning *absolutely* negative ideas: I believe that these represent no objects whatsoever, and hence no attributes (for attributes are also objects), and thus can neither occur as the subject, nor as

With respect to the only partially negative ideas, he calls attention to those

> "of the form '*A* which is not *B*'. The concept of curve line has this form; indeed, this is the concept of a line that is not straight. One immediately imagines that there are many varied types of such only partially negative ideas. As a particularly remarkable type of them, which is closest to the type of absolutely negative ideas and is therefore often confused with it, must be regarded the type of ideas of the form 'something which is not *B*'. Such an idea differs from the absolutely negative idea 'Not *A*' only in its really setting something, even if without determining it better than simply not being *B*. Such an idea is the idea denoted by the words 'what has no angles' in the proposition 'what has no angles is not even a triangle'. As, indeed, in each proposition it must necessarily be spoken of an object (something) and in the preceding proposition no other words capable to denote its object are visible except for 'what has no angles', the idea expressed by these words must refer to some object for which, however, it only establishes that it has no angles."[78]

Given that '*an A that is not B*' and '*something that is not B*' are just modes for expressing '*an A that does not have B-ity*' and, respectively, '*something that does not have B-ity*', the issue of interpreting this type of negation centres on the real value of '*not having b*'. This will be discussed in greater detail in §136.

Bolzano was well aware of the fact that traditionally the idea has often been accepted that the notion of *negation* could enter in composition with the *copula* of the propositions.[79] Moreover, he is conscious that this possibility is supported by "the circumstance that in all languages—known to him—the negation sign composes in the most exact way with the *verb* (i.e. with that word of the proposition which contains the notion of connection). Overall, this sign is located as near as possible to it", if not actually merged in it. Nevertheless, he states "After various fluctuations, the opinion has become stable in me that the notion of negation never belongs to the copula or, from my point of view, to the notion of having, but to the predicate of the proposition."[80]

Moreover, the idea that as the *determinations* [*Bestimmungen*] of probability, necessity, possibility, and contingency, the negation expresses a property that belongs to the whole proposition and simply says that the proposition is false is discussed and rejected. It works when the subject is singular, but not when the subject is plural: '*every A does not have b*' is not the negation of '*every A has b*'. The conclusion is that "all propositions which we usually express or could express in the form

the predicative part, of a proposition which is supposed to lay claim to truth. Such ideas therefore give rise to no noteworthy species of propositions."

[78] TS §89, I 299–300.

[79] TS §136, II 31: "2) If, as many logicians believe, the concept of negation can also accompany the copula of a proposition [. . .]."

[80] TS §136, II 31.

'*A has not b*' do not enunciate anything but the *lack* [*Mangel*] of the quality *b*" and, furthermore, "when its components should come out", we could express them in the form '*A—has—lack of b*'.[81]

The objection according to which in the affirmative propositions a quality is attributed whereas in the negative ones it is detracted is rejected, with the argument that '*detracting* [*absprechen*]' a certain quality is really '*attributing* [*beilegen*]' the quality '*lack of that quality*'.[82] In the uneven discussion that follows, Bolzano strives to clear up what this 'lack of a quality' really is. His thesis is

> "Hence it seems to me very probable that in the idea expressed by the words 'lack of *b*', besides the two ideas already indicated [i.e. *b* and a certain negation], there is a third one, as the idea of a quality. The idea 'lack of *b*'—I mean—is composed of the three subideas *b*, not and quality in the way indicated by the words 'quality not *b*'."[83]

And the conclusion is

> "and so I will now consider as proved that there are propositions which contain the notion of negation in the way which is made most recognizable by the following verbal expression: '*A*—has—the quality non *b*'."[84]

Now, our problem is that of accounting for these, anything but perspicuous, Bolzanian insights.

A first firm point that can be drawn from the former discussion is that the simple *not falling under* $\overline{a}$ (as do, for example, all objects that are not also qualities) cannot be regarded as a sufficient condition for the *falling under the negation of* $\overline{a}$, but must (at least) be associated with the being quality. Hence we agree that the negation of the quality idea $\overline{a}$, for which we will use the notation '$\sim \overline{a}$', must be in any case a quality idea.

Now the question is: which are the qualities under $\sim \overline{a}$?

It is clear that if all quality ideas were singular, it would be rather natural to think that $\sim \overline{a}$ refers to the 'complementary quality' (if any) of the quality referred to by $\overline{a}$. However, as we have seen, quality ideas are, in general, not singular. This seems at first to suggest an extension of this to the general case and to accept that (setting aside the limiting cases) the qualities under $\sim \overline{a}$ are precisely the 'complementary qualities' (if any) of the qualities under $\overline{a}$.

[81] TS §136, II 33.

[82] WL §136, II 33: "[...] My reply is that withholding differs from ascribing only in the fact that it is an ascription of an attribute which represents the lack of some other attribute. Clearly, the lack of an attribute is itself an attribute, and the lack of the lack makes the original attribute appear once again."

[83] TS §136, II 34.

[84] TS §136, II 35.

As we can certainly freely assume that by taking the complements of the complements of qualities we get the original qualities, we see that a first immediate positive property of this proposal is that $\overline{a}$ *and* $\sim\sim\overline{a}$ *are equivalent* (in a sense that is obvious, but is just the sense in which Bolzano also used the term). Another positive property of this interpretation of $\sim\overline{a}$ is that, given any object i and a proper quality idea $\overline{a}$, we have *either* i *is 'something that has* $\overline{a}$*'* (i.e. i falls under the idea $[e](\overline{a})$) *or* i *is 'something that does not have* $\overline{a}$*'* (i.e. i falls under the idea $[e](\sim\overline{a})$). Indeed, if i does not fall under $[e](\overline{a})$, then no quality under $\overline{a}$ belongs to it, and as, by the being proper of $\overline{a}$, at least one such quality exists, its complementary quality belongs to i, which hence falls under the idea $[e](\sim\overline{a})$. Unfortunately, however, this interpretation of negation of a quality idea is deeply unsatisfactory because to the mentioned commendable property it adjoins an other, far less recommendable property: there are objectual quality ideas $\overline{a}$ and objects i such that the following holds: i *is 'something that has both* $\overline{a}$ *and* $\sim\overline{a}$*'*, i.e. $i \mathrel{\text{E}} [e](\overline{a}+\sim\overline{a})$. Indeed, if r and s are two qualities under $\overline{a}$ such that $i \in r$ but $i \notin s$, then, if s' is the 'complementary quality' of s, $i \in s'$ and hence, on the one hand, $i \in r \mathrel{\text{E}} \overline{a}$ and, on the other hand, $i \in s' \mathrel{\text{E}} \sim\overline{a}$; thus, the t such that $\forall i(i \in t \leftrightarrow i \in r \wedge i \in s')$ falls under $(\overline{a}+\sim\overline{a})$ and so $i \mathrel{\text{E}} [e](\overline{a}+\sim\overline{a})$.

It is clear that the elimination of this unpleasant possibility requires that under $\sim\overline{a}$ *only* qualities that are 'disjoint' from those occurring under $\overline{a}$ may occur. A simple (and possibly also redundant) way to reach this result is to think that $\sim\overline{a}$ refers to *all and only all* the qualities that are disjoint from each quality under $\overline{a}$. This is the route upon which we will embark, obtaining a theory of the negation of quality ideas that, avoiding the preceding incongruences, posits that for every i and $\overline{a}$, i falls under one and only one of the two ideas $[e](\overline{a})$ and $[e](\sim\overline{a})$, but in which one is limited to the fact that each quality under $\overline{a}$ also falls under $\sim\sim\overline{a}$ but not, in general, vice versa.

The systematic reconstruction requires us to accept the ontological axiom that ensures the presence, for every object, of a 'peculiar' quality (i.e. a quality that belongs to it and exclusively to it):

$$\forall i\exists r\forall j(j \in r \leftrightarrow j = i) \qquad \text{CQ3}$$

It follows, in particular, that if an object i does not have a certain quality r, then there still is at least one quality (one that belongs to it and only to it) that belongs to i and is disjoint from r; i.e.

$$\forall ir(i \notin r \rightarrow \exists s(s \perp r \wedge i \in s)) \qquad \text{T2.63}$$

We now assume that for each quality idea $\overline{a}$ there is an other quality idea, denoted by '$\sim\overline{a}$', such that (1) if $\overline{a}$ is objectual and there are objects to which none of the qualities under $\overline{a}$ belong, then it refers to all and only all qualities that are disjoint from each quality under $\overline{a}$; (2) if, on the contrary, $\overline{a}$ is objectual but there is no object to which none of the qualities under $\overline{a}$ belong, then it is objectless; (3) finally, if $\overline{a}$ is objectless, then it refers, not necessarily to *all qualities*, but at least to a system of qualities such that every object has one of them; i.e.

$$\forall\overline{a}\exists\overline{c}\left(\left(G(\overline{a})\rightarrow\begin{pmatrix}\exists i(\neg i\,\mathsf{E}\,[e](\overline{a}))\rightarrow(\forall r(r\,\mathsf{E}\,\overline{c}\leftrightarrow\forall s(s\,\mathsf{E}\,\overline{a}\rightarrow s\perp r)))\\ \wedge\,\forall i(i\,\mathsf{E}\,[e](\overline{a}))\rightarrow\neg G(\overline{c})\end{pmatrix}\right)\right.\\ \left.\wedge\,(\neg G(\overline{a})\rightarrow\forall i(i\,\mathsf{E}\,[e](\overline{c})))\ \wedge\ \forall l(l<\overline{c}\leftrightarrow l\leq\overline{a})\right)\qquad\text{CB4}$$

Thus, defining

$$\sim\overline{a}=\varepsilon\,\overline{c}\left(\left(G(\overline{a})\rightarrow\begin{pmatrix}\exists i(\neg i\,\mathsf{E}\,[e](\overline{a}))\rightarrow(\forall r(r\,\mathsf{E}\,\overline{c}\leftrightarrow\forall s(s\,\mathsf{E}\,\overline{a}\rightarrow s\perp r)))\\ \wedge\,\forall i(i\,\mathsf{E}\,[e](\overline{a}))\rightarrow\neg G(\overline{c})\end{pmatrix}\right)\right.\\ \left.\wedge\,(\neg G(\overline{a})\rightarrow\forall i(i\,\mathsf{E}\,[e](\overline{c})))\ \wedge\ \forall l(l<\overline{c}\leftrightarrow l\leq\overline{a})\right)\qquad\text{D2.49}$$

we obtain

(1) $B(\sim\overline{a})$ T2.64
(2) $G(\overline{a})\wedge\exists i(\neg i\,\mathsf{E}\,[e](\overline{a}))\rightarrow(\forall r(r\,\mathsf{E}\,\sim\overline{c}\leftrightarrow\forall s(s\,\mathsf{E}\,\overline{a}\rightarrow s\perp r)))$
(3) $\forall i(i\,\mathsf{E}\,[e](\overline{a}))\rightarrow\neg G(\sim\overline{a})$
(4) $\neg G(\overline{a})\rightarrow\forall i(i\,\mathsf{E}\,[e](\sim\overline{a}))$
(5) $\forall l(l<\sim\overline{a}\leftrightarrow l\leq\overline{a})$

We now have some important theorems on $\sim$.

$$\overline{a}\sqsubseteq\overline{c}\rightarrow\sim\overline{c}\sqsubseteq\sim\overline{a}\qquad\text{T2.65}$$

Proof Let $\overline{a}\sqsubseteq\overline{c}$; then, in any case, $G(\overline{a})$ as well as $G(\overline{c})$. Now let $r\,\mathsf{E}\sim\overline{c}$; therefore also $\sim\overline{c}$ is objectual and so, by (3), there is at least an i such that $\neg i\,\mathsf{E}\,[e](\overline{c})$ and hence, by (2), this r is disjoint from every quality under $\overline{c}$. However, as $\overline{a}\sqsubseteq\overline{c}$, this r is also disjoint from every quality under $\overline{a}$ and, moreover, that i is certainly not under $[e](\overline{a})$. Therefore, by (2), $r\,\mathsf{E}\sim\overline{a}$.

$$\overline{a}\equiv\overline{c}\rightarrow\sim\overline{a}\equiv\sim\overline{c}\qquad\text{T2.66}$$

$$\neg G(\overline{a}+\sim\overline{a})\qquad\text{T2.67}$$

Proof If $\overline{a}$ is objectless, then it does not interfere with any other quality idea and hence, in particular, with $\sim\overline{a}$; hence $\overline{a}+\sim\overline{a}$ in objectless. If, on the contrary, $\overline{a}$ is objectual, and every object falls under $[e](\overline{a})$, then, by (3), $\sim\overline{a}$ is objectless and hence, as already seen, also $\overline{a} + \sim \overline{a}$ is objectless. Finally, if $\overline{a}$ is objectual and not every object falls under $[e](\overline{a})$, then, by (2), under $\sim \overline{a}$ fall only qualities disjoint from those falling under $\overline{a}$, and therefore there are not $r \mathrel{\varepsilon} \overline{a}$ and $s \mathrel{\varepsilon} \sim \overline{a}$ such that they interfere each other and hence, a fortiori, there is not even a t such that $t \mathrel{\varepsilon} (\overline{a}+ \sim \overline{a})$.

$$\neg G([e](\overline{a}) \sqcap [e](\sim \overline{a})) \tag{T2.68}$$

That is,

$$\neg i \mathrel{\varepsilon} [e](\overline{a}) \vee \neg i \mathrel{\varepsilon} [e](\sim \overline{a}) \tag{T2.69}$$

$$[e](\overline{a}) \perp [e](\sim \overline{a}) \tag{T2.70}$$

$$\neg \forall i(i \mathrel{\varepsilon} [e](\overline{a})) \wedge \neg \forall i(i \mathrel{\varepsilon} [e](\sim \overline{a})) \rightarrow \overline{a} \sqsubseteq \sim\sim \overline{a} \tag{T2.71}$$

Proof Indeed, let $r \mathrel{\varepsilon} \overline{a}$. Then $G(\overline{a})$. It follows from this and from the hypothesis $\neg \forall i(i \mathrel{\varepsilon} [e](\overline{a}))$, by (2), that every quality under $\sim \overline{a}$ is disjoint from every quality under $\overline{a}$ and hence, in particular, from r; but, by the hypothesis $\neg \forall i(i \mathrel{\varepsilon} [e](\sim \overline{a}))$, every quality disjoint from every quality under $\sim \overline{a}$ (and such is r) falls under $\sim\sim \overline{a}$.

It should be noted that under these hypotheses, $\sim\sim \overline{a} \sqsubseteq \overline{a}$ does not hold, in general. Indeed, if $\overline{a}$ is, for instance, a singular quality idea that refers to a quality r belonging exclusively to the distinct objects i and j, then no quality under $\sim \overline{a}$, being disjoint from r, belongs to i; by the ontological axiom, however, there is at least one quality that belongs exclusively to i, and clearly every such quality is disjoint from every quality under $\sim \overline{a}$ and hence falls under $\sim\sim \overline{a}$, although not under $\overline{a}$, which, by hypothesis, refers uniquely to r.

Of course, an arbitrary quality r does not necessarily fall either under $\overline{a}$ or under $\sim \overline{a}$. Nevertheless, it is an important fact that every object is either something that has $\overline{a}$ or something that has $\sim \overline{a}$:

$$i \mathrel{\varepsilon} [e](\overline{a}) \vee i \mathrel{\varepsilon} [e](\sim \overline{a}) \tag{T2.72}$$

Proof Let us suppose that the object i does not fall under $[e](\overline{a})$, but that $\overline{a}$, and hence also $[e](\overline{a})$, are objectual. By the ontological axiom, there is still a quality that belongs exclusively to i; naturally, such a quality is disjoint from every quality under $\overline{a}$ and so, by (2), it falls under $\sim \overline{a}$. Therefore, as there is a quality that belongs to i and falls under $\sim \overline{a}$, it is $i \mathrel{\varepsilon} [e](\sim \overline{a})$. Suppose, on the contrary, that the object i does not fall under $[e](\overline{a})$, but that $\overline{a}$ is objectless; then, by (4), every object and so, in particular, our i, falls under $[e](\sim \overline{a})$.

This theorem may be, of course, rewritten as $O(x) \rightarrow x \mathrel{\varepsilon} ([e](\overline{a}) \sqcup [e](\sim \overline{a}))$ and, on the other hand, by GA12, its converse also holds. We thus conclude that a thing is an object iff given any quality idea, *it falls under the concretization of this idea or under the concretization of its negation*; i.e.

$$O(x) \leftrightarrow x \,\text{E}\, ([e](\bar{a}) \sqcup [e](\sim \bar{a})) \qquad \text{T2.73}$$

Composing the two theorems T69 and T72, we obtain

$$i \,\text{E}\, [e](\bar{a}) \leftrightarrow \neg(i \,\text{E}\, [e](\sim \bar{a})) \qquad \text{T2.74}$$

Also, any object is something that has $\bar{a}$ iff it is something that has $\sim\sim \bar{a}$; i.e.

$$[e](\bar{a}) \equiv [e](\sim\sim \bar{a}) \qquad \text{T2.75}$$

Proof $i \,\text{E}\, [e](\bar{a}) \leftrightarrow \neg(i \,\text{E}\, [e](\sim \bar{a})) \leftrightarrow \neg(\neg i \,\text{E}\, [e](\sim\sim \bar{a})) \leftrightarrow i \,\text{E}\, [e](\sim\sim \bar{a})$.

As an aside, it should be noted that another advantage of this interpretation is that it *renders superfluous the postulation of the existence, for every quality, or at least for every proper quality, of a complementary quality.*

We are now able to introduce the operation of *negation of an object idea a* (i.e. "the idea the field of which includes all that does not fall under *a*",[85] by setting

$$-a = [e](\sim \widehat{a}) \qquad \text{D2.50}$$

Therefore

$$i \,\text{E}\, -a \leftrightarrow \neg i \,\text{E}\, a \qquad \text{T2.76}$$

$$[[e](\bar{a})](\sim \bar{c}) \equiv [e](\bar{a} + \sim \bar{c}) \equiv [e](\sim \bar{c} + \bar{a}) \equiv [[e](\sim \bar{c})](\bar{a}) \equiv [-[e](\bar{c})](\bar{a}) \qquad \text{T2.77}$$

In particular, $[[e](\bar{a})](\sim \bar{c}) \equiv [-[e](\bar{c})](\bar{a}) \equiv [e](\bar{a}+ \sim \bar{c})$.[86]

$$G(-a) \leftrightarrow \neg a \equiv e \qquad \text{T2.78}$$

$$G(-a) \rightarrow [e](\widehat{-a}) \equiv [e](\sim \widehat{a}) \qquad \text{T2.79}$$

Proof $G(-a) \rightarrow [e](\widehat{-a}) \equiv -a = [e](\sim \widehat{a})$.

[85] TS §103, I 344: "3) For two ideas to exclude one another all that is required is that the extension of one of them be composed only of objects not belonging to the extension of the other; it is by no means required that all the objects that are not found in the extension of one must occur in the extension of the other. However, the latter case is possible. For if the idea *A* does not already represent all the objects there are (i.e., is not equivalent to the idea 'object in general') the idea 'something (any thing you like) that is not *A*' will have an extension that includes everything that does not stand under the idea *A*."

[86] TS §230, II 304: "4) It is obvious that we may replace the idea [*A*]**n** *b* in the above conclusion with the equivalent ideas [**Non** *B*]*a* or [something] (*a*+**n** *b*)."

As announced in §17, we now list the remaining 11 theorems of WL §105:

$a \dot{\sqsubseteq} b \rightarrow -b \dot{\sqsubseteq} -a$ [87] T2.80

$a \dot{\sqsubseteq} b \rightarrow -a \top b$ [88] T2.81

$\neg a \sqsubseteq -b \rightarrow a \top b$ [89] T2.82

$a \top b \rightarrow \neg a \sqsubseteq -b \wedge \neg -a \sqsubseteq b$ [90] T2.83

$a \sqsubseteq -b \rightarrow a \perp b$ [91] T2.84

$\neg a \sqsubseteq b \wedge \neg b \sqsubseteq a \rightarrow a \top -b \wedge -a \top b$ [92] T2.85

$a \top -b \wedge -a \top b \rightarrow a \dot{\top} b \wedge a \perp b$ [93] T2.86

$a \dot{\perp} b \rightarrow \neg a \equiv -b \wedge \neg -a \equiv b \wedge -a \top -b$ [94] T2.87

$\neg b \equiv e \rightarrow (a \dot{\sqsubseteq} b \rightarrow \neg -a \dot{\sqsubseteq} b)$ [95] T2.88

$\exists ab(a \dot{\top} b \wedge -a \dot{\sqsubseteq} b)$ [96] T2.89

$a \dot{\top} b \wedge -a \dot{\top} -b \rightarrow a \sqcup b \equiv e$ [97] T2.90

Subsequently, it will be useful to know that the material relations between two ideas considered in this section may be expressed, by means of $\sqcap$ and $-$, in terms of objectuality. In fact, we have

[87] TS §105, I 357: "9) If an idea *A* is higher than another *B*, then one of the ideas contradictory to *A* (which for the sake of brevity I shall express as non-*A*) is lower than non-*B*. For [. . .]."

[88] TS §105, I 358: "14) If the idea *B* is lower than *A*, then non-*B* is compatible with *A*. For [. . .]."

[89] TS §105, I 358: "15) If an idea *A* is neither equivalent nor subordinate to the idea non-*B*, then it must be compatible with *B*. For [. . .]."

[90] TS §105, I 358: "16) If *A* and *B* are compatible, then of the two ideas *A* and non-*B* and of the two ideas *B* and non-*A*neither includes the other. For [. . .]."

[91] TS §105, I 358: "17) Conversely, if inclusion holds between *A* and non-*B*, then *A* and *B* are incompatible."

[92] TS §105, I 358: "18) If *A* and *B* both have extensions, but neither idea includes the other, then *A* and non-*B*, *B* and non-*A* are compatible, and neither of non-*A* and non-*B* is subordinate to the other. For [. . .]."

[93] TS §105, I 359: "19) If *A* and non-*B* and *B* and non-*A* are compatible, then *A* and *B* are either incompatible, or they overlap. For [. . .]."

[94] TS §105, I 359: "20) If *A* and *B* are merely contrary, then *A* and non-*B*, and *B* and non-*A* are not equivalent, while non-*A* and non-*B* are compatible. For [. . .]."

[95] TS §105, I 359: "21) If the idea *B* is lower than *A*, then non-*B* cannot be lower than *A*, unless *A* is an idea of greatest width. However, if *B* is lower than *A*, a contrary of *B* can be lower than *A*. Since [. . .]."

[96] TS §105, I 360: "22) Two overlapping ideas *A* and *B* can be so constituted that the contradictory of one of them, e.g., non-*A* is completely included in the other *B*. For [. . .]."

[97] TS §105, I 360: "23) Only in the case where a pair of overlapping ideas do not collectively exhaust the extension of the widest idea of something in general are non-*A* and non-*B* also overlapping. For [. . .]."

Compatibility : $a \top c \leftrightarrow G(a \sqcap c)$ T2.91

Inclusion : $a \sqsubseteq c \leftrightarrow G(a \sqcap c) \wedge \neg G(a \sqcap -c)$

Disparateness : $a \dot{\top} c \leftrightarrow G(a \sqcap c) \wedge G(a \sqcap -c) \wedge G(-a \sqcap c)$

Equivalence : $a \equiv c \leftrightarrow G(a \sqcap c) \wedge \neg G(a \sqcap -c) \wedge \neg G(-a \sqcap c)$

Subordination : $a \dot{\sqsubseteq} c \leftrightarrow G(a \sqcap c) \wedge \neg G(a \sqcap -c) \wedge G(-a \sqcap c)$

Incompatibility : $a \perp c \leftrightarrow \neg G(a \sqcap c)$

Contradictoriness : $a \,||\, c \leftrightarrow \neg G(a \sqcap c) \wedge \neg G(e \sqcap -a \sqcap -c)$

Contrariety : $a \dot{\perp} c \leftrightarrow \neg G(a \sqcap c) \wedge G(e \sqcap -a \sqcap -c)$

3

Elementary Level (B): Propositions

3.1 Atomic Propositions

According to Bolzano, the simplest type of propositions (but in his opinion, as we will see later, also the unique type) has the form '*a has* $\bar{c}$' (which we will denote by '$h(a, \bar{c})$'), where

(1) h is the *copula* [*Bindetheil, Copula, Copel*];
(2) a is the *subject idea* [*Subjectvorstellung*] or, simply, the *subject* [*Unterlage*];
(3) $\bar{c}$ is the *predicate idea* [*Prädicatvorstellung*] or the *predicate* [*Aussagetheil*].

The proposition *a has* $\bar{c}$:

(1) *handles* [*handelt*] the objects referred to by a;
(2) *attributes* [*legt bei*] to them the qualities referred to by $\bar{c}$.[1]

Bolzano believed the copula was an idea, too. Nevertheless, we prefer, as is done for example in the case of the operations on ideas, to treat this type of components somewhat differently. As a matter of fact, Bolzano was not unwilling to consider

[1] TS §127, II 5: "In all propositions, first of all, there occurs the concept '*to have*' or, more precisely, the concept expressed by the word '*has*'. Along with this constituent every proposition contains two others, that 'has' connects in the way indicated by the expression '*A has b*'. The first of these constituents, that which is expressed by *A*, occurs as if it represented the *object* the proposition concerns, and the other, *b*, as if it represented the *attribute* that the proposition ascribes to this object. Accordingly, I allow myself to call the part *A*, however it may be constituted, the *basis* or *subject-idea*, and *b* the *attributive part* or *predicate-idea*."

In TS §66,1 I 214–15, Bolzano gives the following as synonyms of '*to attribute* [*beilegen*]': '*to ascribe* [*zuschreiben*]', '*to predicate* [*aussagen*]', and '*to relate* [*beziehen*]'; with regard to the last, he emphasizes the necessity to clearly distinguish it from '*to refer* [*sich beziehen*]'; a proposition *relates* a quality to an object (i.e. it *sets in relation* a quality with an object); an idea *refers* to an object.

propositions (even if false) also constructions of the form $h(a, c)$ (i.e. constructions in which the 'predicate' is an object idea).[2]

In our systematization, we think of $h(a, \bar{c})$ as the privileged solution of a lectological comprehension axiom that ensures the existence, for every object idea and every quality idea, of a proposition having such ideas and their parts as its only lectological parts; i.e.

$$\forall a \forall \bar{c} \exists p \forall l (l \leq p \leftrightarrow l \leq a \vee l \leq \bar{c}) \tag{CS1}$$

$$h(a, \bar{c}) = \varepsilon p \forall l (l \leq p \leftrightarrow l \leq a \vee l \leq \bar{c}) \tag{D3.1}$$

$$S(h(a, \bar{c})) \tag{T3.1}$$

$$\forall l (l \leq h(a, \bar{c}) \leftrightarrow l \leq a \vee l \leq \bar{c}) \tag{T3.2}$$

By means of a defining axiom, we can then introduce the monadic predicate '*being an atomic proposition*' $[S^{\circ}]$:

$$S^{\circ}(p) \leftrightarrow \exists a \exists \bar{c}(p = h(a, \bar{c})) \tag{D3.2}$$

The terminology that has been introduced for the semantic properties of the ideas can now be transferred in a natural way to the (atomic) propositions by establishing that such a proposition is *objectual, objectless, singular* or *universal* iff such is its subject.[3]

One of the most significant novelties of Bolzano's logical work is his definition of truth for propositions of the form $h(a, \bar{c})$:

$h(a, \bar{c})$ *is true* iff:

(1) it handles something (i.e. its subject is objectual);

(2) to every object referred to by a belongs at least one of the qualities referred to by $\bar{c}$.[4]

[2] Cf. TS § 127.7 II.

[3] TS §146, II 53–4: "If I am right to suppose that there are also propositions whose subject represents no object, then perhaps I shall be permitted to call such proposition *objectless*. Other propositions, those for which one can indicate an object they deal with, can be called *objectual*, or propositions concerning an *extension*. If there is only one object, we can call them *singular* propositions and, if more than one, perhaps even infinitely many, we can call them *general* propositions."

[4] TS §131, II 18: "The proposition '*A* has *b*' has no other sense than that every object standing under *A* has one of the attributes that stand under *b*; and if there are several of the latter, it remains undetermined which of them belongs to each *A*."

The first, quite provisional and preliminary, formulation of this definition, entitled *Probable components of this notion* (i.e. truth of a proposition in itself)[5] does not point out that the predicate may refer to more qualities, and this fact has generated some misunderstanding. We must, however, remark that Bolzano was well aware of the novelty of his definition. Indeed, in *Bolzano's Wissenschaftslehre und Religionswissenschaft in einer beurtheilenden Uebersicht,* he wrote

> "Before we leave this first chapter, we think we must still present to our readers the definition of the notion of truth (in itself) attempted by the author; at least, in order that they may more easily recognize the incorrectness of another definition of this notion that has produced so many errors. So, Bolzano says that only propositions and not purely ideas, can be partitioned in true and false. Furthermore, he remarks that each proposition has an object which it handles and to which it attributes a quality (represented by its predicate idea). Now, this proposition is said to be true if it attributes to the object a quality that belongs to it and false, if it does not do this. It seems to us that this definition is at least understandable and also that everyone should agree that the notion to be defined does not occur unanalyzed in any of its parts."[6]

Moving on to the systematization of this conviction, we may introduce, again by a defining axiom, the truth predicate for atomic propositions $[W(h(a,\bar{c}))]$, establishing

$$W(h(a,\bar{c})) \leftrightarrow G(a) \wedge \forall i(i \sqsubset a \rightarrow \exists r(i \in r \wedge r \sqsubset \bar{c})) \qquad \text{D3.3}$$

Immediate corollaries are

$$W(h(a,\bar{c})) \rightarrow G(\bar{c}) \qquad \text{T3.3}$$

$$W(h(A,\bar{c})) \leftrightarrow G(a) \wedge G(\bar{c}) \wedge \forall i(i \sqsubset a \rightarrow \exists r(i \in r \wedge r \sqsubset \bar{c})) \qquad \text{T3.4}$$

BGA IIA 12/2, p.61: "Was heißt nun also die Behauptung: *der Satz A hat b ist wahr?*—Nichts Anders als daß unter der Vorstellung von *A* irgend ein Gegenstand (Einer oder mehrere) stehe, u. daß dieser Gegenstand eine Beschaffenheit habe, welche unter der Vorstellung *b* steht."

BGA IIA 12/2, p.105 "*Definition der Wahrheit.* Man kann sie auch so ausdrücken: Ein Satz ist wahr, wenn jeder Gegenstand, der dem Subjecte des Satzes untersteht, eine Beschaffenheit hat, die dem Prädicate untersteht."

[5] Cf. TS I §28.

[6] WR, pp. 30–1: "Bevor wir dieß erste Hauptstück verlassen, glauben wir unseren Lesern noch die Erklärung, die der Vf. von dem Begriffe der Wahrheit (an sich) versucht, vorlegen zu sollen; wäre es auch nur, damit sie um so leichter die Unrichtigkeit einer anderen Erklärung dieses Begriffes, die viele Irrungen veranlaßt hat, erkennen. B. sagt also daß nur Sätze, nicht aber bloße *Vorstellungen* in wahre und falsche eingetheilt werden können; er bemerkt ferner, daß jeder Satz einen *Gegenstand,* von dem er handelt, habe, und daß er diesem Gegenstande eine *Beschaffenheit* (vorgestellt in seiner Prädicatvorstellung) *beilege. Wahr* heißt nun dieser Satz, wenn er dem Gegenstande eine Beschaffenheit *beilegt,* die diesem *zukommt;* und *falsch,* wenn er dieß nicht thut. Diese Erklärung ist (daücht uns) wenigstens sehr verständlich; auch dürfte Jeder zugeben, daß der zu erklärende Begriff in keinem ihrer Bestandtheile noch unzerlegt stecke."

$$W(h(a,\overline{c})) \leftrightarrow a \sqsubseteq [e](\overline{c})^{7} \qquad \text{T3.5}$$

Among the theorems provable for W, we mention the following:

$$W(h(a,\overline{b}+\overline{c})) \leftrightarrow W(h(a,\overline{b})) \wedge W(h(a,\overline{c})) \qquad \text{T3.6}$$

Proof Let $W(h(a,\overline{b}+\overline{c}))$; then $G(a)$ and $\forall i(i \,\textsc{e}\, a \rightarrow \exists t(i \in t \wedge t \,\textsc{e}\, \overline{b}+\overline{c}))$, i.e. for every $i \,\textsc{e}\, a$, there are $r \,\textsc{e}\, \overline{b}$ and $s \,\textsc{e}\, \overline{c}$, which interfere with each other and are such that their intersection belongs to i; hence $G(a) \wedge \forall i(i \,\textsc{e}\, a \rightarrow \exists r(i \in r \wedge r \,\textsc{e}\, \overline{b}))$ and $G(a) \wedge \forall i(i \,\textsc{e}\, a \rightarrow \exists s(i \in s \wedge s \,\textsc{e}\, \overline{c}))$, i.e. $W(h(a,\overline{b}))$ e $W(h(a,\overline{c}))$. Vice versa, let $W(h(a,\overline{b}))$ and $W(h(a,\overline{c}))$. Then $G(a)$, as well as both $\forall i(i \,\textsc{e}\, a \rightarrow \exists r(i \in r \wedge r \,\textsc{e}\, \overline{b}))$ and $\forall i(i \,\textsc{e}\, a \rightarrow \exists s(i \in s \wedge s \,\textsc{e}\, \overline{c}))$. Let $i \,\textsc{e}\, a$; then there are $r \,\textsc{e}\, \overline{b}$ and $s \,\textsc{e}\, \overline{c}$ both belonging to i. Belonging to i, such r and s interfere with each other, and hence their intersection, which belongs to i, falls under $\overline{b}+\overline{c}$. Therefore $W(h(a,\overline{b}+\overline{c}))$.

$$W(h(a,\overline{b}+\overline{c})) \leftrightarrow a \sqsubseteq [e](\overline{b}) \sqcap [e](\overline{c}) \leftrightarrow a \sqsubseteq [e](\overline{b}) \wedge a \sqsubseteq [e](\overline{c}) \qquad \text{T3.7}$$

Bolzano also affirms that $W(h(a \sqcup b,\overline{c})) \leftrightarrow W(h(a,\overline{c})) \wedge W(h(b,\overline{c}))$, but it must be remarked that in order to prove $\rightarrow$, one must postulate $G(a) \wedge G(b)$.

Without difficulty, we can prove

$$W(h(a,\overline{b})) \vee W(h(a,\overline{c})) \rightarrow W(h(a,\overline{b}\cdot\overline{c})) \qquad \text{T3.8}$$

$$W(h(a,\overline{b}\cdot\overline{c})) \leftrightarrow a \sqsubseteq [e](\overline{b}) \sqcup [e](\overline{c}) \qquad \text{T3.9}$$

$$a \sqsubseteq [e](\overline{b}) \vee a \sqsubseteq [e](\overline{c}) \rightarrow a \sqsubseteq [e](\overline{b}) \sqcup [e](\overline{c}) \qquad \text{T3.10}$$

Support for the formerly proposed interpretation of '*lack of* $\overline{c}$' comes from the proof of the fact that, by objectual a, the proposition '*a has lack of* $\overline{c}$' is true iff no object under a has any quality under $\overline{c}$.

$$W(h(a,\sim\overline{c})) \leftrightarrow G(a) \wedge \forall ir(i \,\textsc{e}\, a \wedge r \,\textsc{e}\, \overline{c} \rightarrow \neg i \in r) \qquad \text{T3.11}$$

Proof By D3, $W(h(a,\sim\overline{c})) \leftrightarrow G(a) \wedge \forall i(i \,\textsc{e}\, a \rightarrow \exists r(i \in r \wedge r \,\textsc{e}\, \sim\overline{c}))$; so it will be sufficient to prove $\forall ir(i \,\textsc{e}\, a \wedge r \,\textsc{e}\, \overline{c} \rightarrow \neg i \in r) \leftrightarrow \forall i(i \,\textsc{e}\, a \rightarrow \exists r(i \in r \wedge r \,\textsc{e}\, \sim\overline{c}))$. Let $\forall ir(i \,\textsc{e}\, a \wedge r \,\textsc{e}\, \overline{c} \rightarrow \neg i \in r)$ and $i \,\textsc{e}\, a$. Let us consider $\{i\}$; as, by hypothesis,

[7] TS §196, II 237: "4) It is similarly obvious [. . .] that the concretum which corresponds to the predicate-idea of a true proposition must always be an objectual idea which stands to the subject in the relation of *inclusion* (§95), and which therefore must be either equivalent to or higher than it. For, if the proposition '*A has b*' is to be true, then the idea *B*, or 'something which has *b*', must include every object that stands under the idea *A*." Cf. also TS §246, II 345: "Here we shall simply point out that a proposition stating that an individual idea *B* is included in another individual idea *A* is equivalent to the proposition '*A has b*'."

i has no quality under $\bar{c}$, $\{i\}$ is disjoint from any $r \varepsilon \bar{c}$ and hence $\{i\} \varepsilon \sim \bar{c}$; so, for each $i \varepsilon a$, $i \in \{i\} \varepsilon \sim \bar{c}$. Vice versa, let $\forall i(i \varepsilon a \rightarrow \exists r(i \in r \wedge r \varepsilon \sim \bar{c}))$ and $i \varepsilon a$; as, by hypothesis, i has at least one quality, say r, that is disjoint from all qualities under $\bar{c}$, i may not have any quality under $\bar{c}$, otherwise r would not be disjoint from such a quality. Hence, for every $i \varepsilon a$ and $r \varepsilon \bar{c}$, we have $\neg i \in r$.

$$W(h(a, \sim \bar{c})) \leftrightarrow a \sqsubseteq -[e](\bar{c}) \qquad \text{T3.12}$$

$$a \equiv c \rightarrow (W(h(a, \bar{c}) \leftrightarrow W(h(c, \bar{b})) \qquad \text{T3.13}$$

$$\bar{b} \equiv \bar{c} \rightarrow (W(h(a, \bar{c}) \leftrightarrow W(h(a, \bar{b})) \qquad \text{T3.14}$$

From these, we have the corollaries

$$W(h(a, \sim \widehat{c})) \leftrightarrow W(h(a, \widehat{-c})) \qquad \text{T3.15}$$

$$W(h(b, \widehat{a} + \widehat{c})) \leftrightarrow W(h(b, \widehat{a \sqcap c})) \qquad \text{T3.16}$$

$$W(h(b, \widehat{a} \cdot \widehat{c})) \leftrightarrow W(h(b, \widehat{a \sqcup c})) \qquad \text{T3.17}$$

3.2 Aristotelian Universal Propositions with Existential Import

We are able to identify the traditional universal propositions with existential import as certain atomic propositions of our universe without employing any 'symbolic' hypothesis.

We define

$$a\mathbb{A}^{+}c = h(a, \widehat{c}) \qquad \text{D3.4}$$

$$a\mathbb{E}^{+}c = h(a, \sim \widehat{c}) \qquad \text{D3.5}$$

The suitability of these definitions results from the following two theorems, which say that $a\mathbb{A}^{+}c$ and $a\mathbb{E}^{+}c$ satisfy exactly the truth conditions that are traditionally imposed on the *universal categorical affirmative* (respectively *negative*) *with existential import*:

$$W(a\mathbb{A}^{+}c) \leftrightarrow \exists i(i \varepsilon a) \wedge \forall i(i \varepsilon a \rightarrow i \varepsilon c) \qquad \text{T3.18}$$

Proof $W(a\mathbb{A}^{+}c) \leftrightarrow W(h(a, \widehat{c})) \leftrightarrow a \sqsubseteq [e](\widehat{c}) \leftrightarrow a \sqsubseteq c$
$\leftrightarrow \exists i(i \varepsilon a) \wedge \forall i(i \varepsilon a \rightarrow i \varepsilon c)$.

$$W(a\mathbb{E}^{+}c) \leftrightarrow \exists i(i \varepsilon a) \wedge \forall i(i \varepsilon a \rightarrow \neg i \varepsilon c) \qquad \text{T3.19}$$

Proof $W(a\mathbb{E}^{+}c) \leftrightarrow W(h(a, \sim \widehat{c}) \leftrightarrow a \sqsubseteq [e](\sim \widehat{c}) \leftrightarrow a \sqsubseteq -[e](\widehat{c}) \leftrightarrow a \sqsubseteq -c$.

3.3 Material Proprieties and Relations of Propositions

Starting with the property of truth (which, however, is defined only for atomic propositions!), Bolzano singles out and defines a set of proprieties and relations of propositions.

The first material property of propositions considered by Bolzano is *falsity* [*Falschheit*]; it is defined as the negation of its truth; i.e.

$$F(p) \leftrightarrow \neg W(p) \qquad \text{D3.6}$$

Bolzano then turns to what he calls [§160] *relations among the propositions arising from the considerations on how many truth or falsities are in an aggregate.* We will treat these issues in strict analogy with what was done with the case of relations among aggregates of ideas. In particular, we will substitute the defining schema of an infinity $P^2, P^3, \ldots$ of 2-adic, 3-adic, ... predicates for the notion 'relation P^* subsistent among the propositions of an aggregate'. Moreover, using the same notational devices introduced for that case, we can write '$\overset{n}{p}$' for '$p_1, \ldots, p_n$' and omit the indication of the adicity of the predicate, when superfluous.

Bolzano explicitly considers five or, better, taking into account also duals, ten relations that may subsist among the n propositions $p_1, \ldots, p_n$, depending on their being true or false:

(1) the propositions are all true,[8] respectively all false;
(2) at least one of them is true, respectively false.;
(3) exactly one of them is true, respectively false;
(4) more than one is true, respectively false;
(5) exactly k of them are true, respectively false.[9]

In fact, however, he also considers other relations:[10]

(6) at least k of them are true, respectively false;
(7) at most k of them are true, respectively false;

[8] TS §160, II 117: "1) First of all, it may happen that all of the propositions in a given collection are true."

[9] TS §252, II 362: "There are five (or ten) forms of such propositions: the propositions $M, N, O, \ldots$ are all true (all false); Among the propositions $M, N, O, \ldots$ there are one or more true (false) ones; There is only one true (false) proposition among; Among the propositions several are true (false); Among the propositions $M, N, O, \ldots$ n are true (false)."

[10] Cf. TS §160 3c,d,e,h 5c, etc.

as well as their composition

(8) at least k, but not more than h, of them are true, respectively false.

For some of these relations, Bolzano introduces special names. In the case in which at least one proposition is true, he speaks of *material integration* or *supplement* (or *help*) [*materiale Ergänzung (Aushülfe)*];[11] in the case in which one and only one is true, he speaks of *material disjunction* or *singular material integration* [*materiale Disjunction, eingliedrige Ergänzung*)];[12] in the case in which more than one is true, he speaks of *material plural integration* or *material redundant integration* [*mehrtheilig* or *überfüllt*]; finally, in the case in which exactly k are true, he speaks of *material k-redundant integration* [*k-überfüllt*].

If we are also allowed to use, as is now obvious, the improper order relation, then it turns out, without any particular duress, that all the cases considered by Bolzano can be obtained starting from cases (6) and (7).

Let us begin by remarking that, given n propositions $p_1, \ldots, p_n$, the fact that *al least the first k* of them are true, respectively false, means that $W(p_1) \wedge \ldots \wedge W(p_k)$, respectively $[F(p_1) \wedge \ldots \wedge F(p_k)]$, holds (i.e. in more compact notation, $\bigwedge_{i \leq k} W(p_i)$ $[\bigwedge_{i \leq k} F(p_i)]$). Hence that *at least k*, not necessarily the *first k* of the $p_1, \ldots, p_n$, are true [false] will mean that for at least one of the permutations π of the set $\Pi(n)$ of all permutations of $\{1, \ldots, n\}$, $\bigwedge_{i \leq k} W(p_{\pi(i)})$ $[\bigwedge_{i \leq k} F(p_{\pi(i)})]$ holds, which means that the disjunction holds, the members of which are constituted by all the $\bigwedge_{i \leq k} W(p_{\pi(i)})$ $[\bigwedge_{i \leq k} F(p_{\pi(i)}))]$ obtained as π varies on $\Pi(n)$.

Writing '$_kW^n(p_1, \ldots, p_n)$' ['$_kF^n(p_1, \ldots, p_n)$'] for '*at least k of the propositions $p_1, \ldots, p_n$ are true* [*false*]', or, using the notational conventions introduced earlier, more simply '$_kW(\overset{n}{p})$' [$_kF(\overset{n}{p})$], we may introduce, for all n and $k \leq n$, the definitions

$$_kW(\overset{n}{p}) \leftrightarrow \bigvee_{\pi \in \Pi(n)} (\bigwedge_{i \leq k} W(p_{\pi(i)})) \qquad \text{SD3.1}$$

[11] TS §160, II 117: "2) It may also occur that we only know of a given collection of propositions that not all of them are false, or true, without knowing [...] that either one, two, three, or perhaps all, are true [...] It is important to recognise this relation when it obtains. [...] I know of no term that has been used to designate this relation, so I will permit myself to call them *complementary* or *supporting* propositions. [...] This relation of complementation can hold for particular propositions $M, N, O, \ldots$ which are given; in this case it may be called material complementation or support."

[12] TS §160, II 117-8: "3) A different relation obtains when we learn that there is only one true (or false) proposition in a given collection of propositions $M, N, O, \ldots$ [...] This relation may be called *single-termed* complementation. It is customarily called *disjunction*. If the given propositions stand in the relation as they are, it is *material* disjunction [...]."

$$ {}_kF(\overset{n}{p}) \leftrightarrow \bigvee_{\pi\in\Pi(n)} (\bigwedge_{i\leq k} F(p_{\pi(i)})) \qquad \text{SD3.2} $$

To say that *at most the first k* among the $p_1, \dots, p_n$ are true [false] will of course mean that if the first k of them are true [false], then all the remaining are false [true], that is to say that $\bigwedge_{i\leq k} W(p_i) \rightarrow \bigwedge_{k<j\leq n} F(p_j))$ $[\bigwedge_{i\leq k} F(p_i) \rightarrow \bigwedge_{k<j\leq n} W(p_j)]$ holds. Hence, to say that *at most k,* not necessarily the *first k,* among the $p_1, \dots, p_n$, are true [false], will mean that, for at least one permutation π of $\Pi(n)$, $\bigwedge_{i\leq k} W(p_{\pi(i)}) \rightarrow \bigwedge_{k<j\leq n} F(p_{\pi(j)}))$ $[\bigwedge_{i\leq k} F(p_{\pi(i)}) \rightarrow \bigwedge_{k<j\leq n} W(p_{\pi(j)})]$ holds, that is to say that the disjunction holds, the members of which are constituted by all the $\bigwedge_{i\leq k} W(p_{\pi(i)}) \rightarrow \bigwedge_{k<j\leq n} F(p_{\pi(j)}))$ $[\bigwedge_{i\leq k} F(p_{\pi(i)}) \rightarrow \bigwedge_{k<j\leq n} W(p_{\pi(j)})]$ obtained as π varies on $\Pi(n)$. So writing '${}^kW(\overset{n}{p})$' ['${}^kF(\overset{n}{p})$'], for '*at most k among the propositions* $p_1, \dots, p_n$ *are true [false]*', may introduce, for all n and $k \leq n$, the definitions

$$ {}^kW(\overset{n}{p}) \leftrightarrow \bigvee_{\pi\in\Pi(n)} (\bigwedge_{i\leq k} (W(p_{\pi(i)}) \rightarrow \bigwedge_{k<j\leq n} F(p_{\pi(j)}))) \qquad \text{SD3.3} $$

$$ {}^kF(\overset{n}{p}) \leftrightarrow \bigvee_{\pi\in\Pi(n)} (\bigwedge_{i\leq k} (F(p_{\pi(i)}) \rightarrow \bigwedge_{k<j\leq n} F(p_{\pi(j)}))) \qquad \text{SD3.4} $$

That *at least k, but not more than h,* among the $p_1, \dots, p_n$ are [false] will now be understood as the simultaneous presence of the two preceding relations, i.e., writing '${}^h_kW(\overset{n}{p})$'['${}^h_kF(\overset{n}{p})$'], we may set, for all n and $k \leq h \leq n$,

$$ {}^h_kW(\overset{n}{p}) \leftrightarrow {}_kW(\overset{n}{p}) \wedge {}^hW(\overset{n}{p}) \qquad \text{SD3.5} $$

$$ {}^h_kF(\overset{n}{p}) \leftrightarrow {}_kF(\overset{n}{p}) \wedge {}^hF(\overset{n}{p}) \qquad \text{SD3.6} $$

Now that *exactly k* among the propositions $p_1, \dots, p_n$ are true [false] or that *at least one* among them is true [false], or that *at most one* of them is true [false], or that *exactly one* among them is true [false], or that *more than one* among them is true [false], will be of course expressible by '${}^k_kW(\overset{n}{p})$' ['${}^k_kF(\overset{n}{p})$'], '${}_1W(\overset{n}{p})$' ['${}_1F(\overset{n}{p})$'], '${}^1W(\overset{n}{p})$' ['${}^1F(\overset{n}{p})$'], '${}^1_1W(\overset{n}{p})$' ['${}^1_1F(\overset{n}{p})$'] and '${}_2W(\overset{n}{p})$' ['${}_2F(\overset{n}{p})$'], respectively.

The notation introduced up to this point, turns out, in some cases, to be somewhat cumbersome. So it will be advantageous to adopt a few simplifying conventions. In particular, let us establish that (1) the relations subsisting among the $p_1, \dots, p_n$ when they are *all true* or *all false,* officially expressed by '${}_nW(\overset{n}{p})$' and

'$_nF(\overset{n}{p})$', will be denoted by '$W(\overset{n}{p})$' and '$F(\overset{n}{p})$', respectively; (2) the relation subsisting among the $p_1, \ldots, p_n$ when *at least one among them* is true, officially '$_1W(\overset{n}{p})$', will be denoted by '$E(\overset{n}{p})$'; (3) the relation subsisting among the $p_1, \ldots, p_n$ when *one and only one among them is true*, officially '$^1_1W(\overset{n}{p})$', will be denoted by '$D(\overset{n}{p})$'; i.e.,

$$W(\overset{n}{p}) := {}_nW(\overset{n}{p}) \qquad \text{MD3.7}$$

$$F(\overset{n}{p}) := {}_nF(\overset{n}{p}) \qquad \text{MD3.8}$$

$$E(\overset{n}{p}) := {}_1W(\overset{n}{p}) \qquad \text{MD3.9}$$

$$D(\overset{n}{p}) := {}^1_1W(\overset{n}{p}) \qquad \text{MD3.10}$$

Obviously, we have

$$W(\overset{n}{p}) \leftrightarrow \bigwedge_{i \leq k} W(p_i) \qquad \text{ST3.1}$$

$$F(\overset{n}{p}) \leftrightarrow \bigwedge_{i \leq k} F(p_i) \qquad \text{ST3.2}$$

$$E(\overset{n}{p}) \leftrightarrow \bigvee_{i \leq k} W(p_i) \qquad \text{ST3.3}$$

3.4 Complex Propositions

Let us now remember that our *propositions* handle *objects* and attribute *qualities* to them. At the elementary level, however, the lectological entities (ideas and propositions) do *not* appear among the objects and so to their properties and to the relations among them there do not correspond qualities present in the universe. This is in particular true for 'semantic' properties and relations (being objectual, objectless, including, etc.; being true, false, implying, etc.). It follows that the attribution to ideas or propositions of properties or relations of this kind cannot be expressed by a *proposition* of our universe. In other words, at the elementary level, we cannot, for example, formulate a *proposition* enunciating that a certain other proposition is false. This does not prevent us, however, from accepting that, for example, for any given proposition (of the universe) p there is a proposition (of the universe) that behaves *as it would enunciate* the falsity of p, in the sense that it is true iff p is false. We may therefore *temporarily* accept (at the symbolic level, we will discuss the Bolzanian conviction of their superfluity) *operations* that, starting from one or more propositions, generate a new proposition that behaves *as it would express* the subsistence among them of one of the relations (or property) of the mentioned type, just in the sense that it turns out to be true iff that relation subsists among (belongs to) those propositions.

We begin with the property of falsity of a proposition. We may reach our desired result by requiring the presence for every p of a proposition (to be denoted by '$\boxed{\neg}(p)$', or, in brief, by '$\boxed{\neg}p$'), that is true iff p is false and that has as its parts p and its parts. Thus, we assume

$$\forall p \exists q((W(q) \leftrightarrow F(p)) \wedge \forall l(l < q \leftrightarrow l \leq p)) \qquad \text{CS2}$$

and define

$$\boxed{\neg}p = \varepsilon q((W(q) \leftrightarrow F(p)) \wedge \forall l(l < q \leftrightarrow l \leq p)) \qquad \text{D3.11}$$

so that

$$\begin{aligned} &(1)\ \ S(\boxed{\neg}p) \\ &(2)\ \ W(\boxed{\neg}p) \leftrightarrow F(p)) \\ &(3)\ \ w < \boxed{\neg}p \leftrightarrow w \leq p) \end{aligned} \qquad \text{T3.20}$$

As Bolzano explicitly recalls, the equivalence $W(\boxed{\neg}p) \leftrightarrow F(p))$ has among its consequences the fact that the preceding 10 relations considered by him (and, of course, also the remaining ones) can be halved by simply replacing their negations with the propositions involved.[13]

Moreover, we may introduce the n-ary operations that, for given propositions $p_1, \ldots, p_n$, give us the proposition that is respectively true iff among those propositions the relations of simultaneous truth, material integration, or material disjunction hold. That is, we assume

$$\forall \overset{n}{p} \exists q((W(q) \leftrightarrow W(\overset{n}{p})) \wedge \forall l(l < q \leftrightarrow l \leq \overset{n}{p})) \qquad \text{SCS1}$$

$$\forall \overset{n}{p} \exists q((W(q) \leftrightarrow E(\overset{n}{p})) \wedge \forall l(l < q \leftrightarrow l \leq \overset{n}{p})) \qquad \text{SCS2}$$

$$\forall \overset{n}{p} \exists q((W(q) \leftrightarrow D(\overset{n}{p})) \wedge \forall l(l < q \leftrightarrow l \leq \overset{n}{p})) \qquad \text{SCS3}$$

and we define

$$\boxed{\wedge^n}(\overset{n}{p}) = \varepsilon q((W(q) \leftrightarrow W(\overset{n}{p})) \wedge \forall l(l < q \leftrightarrow l \leq \overset{n}{p})) \qquad \text{SD3.7}$$

[13] TS §252 II 367: "Since it is evident, in particular, that the ten cases [...] could without much difficulty be reduced by half (namely, by replacing the given propositions $M, N, O, \ldots$ with their negations), we may, without loss of generality, consider only the five that I did not put in parentheses."

$$\boxed{\vee^n}(\overset{n}{p}) = \varepsilon q((W(q) \leftrightarrow E(\overset{n}{p})) \wedge \forall l(l < q \leftrightarrow l \leq \overset{n}{p})) \qquad \text{SD3.8}$$

$$\boxed{\text{Ш}^n}(\overset{n}{p}) = \varepsilon q((W(q) \leftrightarrow D(\overset{n}{p})) \wedge \forall l(l < q \leftrightarrow l \leq \overset{n}{p})) \qquad \text{SD3.9}$$

Thus we have the following:

$$(1)\ S(\boxed{\wedge^n}(\overset{n}{p})) \qquad \text{ST3.4}$$
$$(2)\ W(\boxed{\wedge^n}(\overset{n}{p})) \leftrightarrow W(\overset{n}{p})$$
$$(3)\ \forall l(l < \boxed{\wedge^n}(\overset{n}{p}) \leftrightarrow l \leq \overset{n}{p})$$
$$(1)\ S(\boxed{\vee^n}(\overset{n}{p})) \qquad \text{ST3.5}$$
$$(2)\ W(\boxed{\vee^n}(\overset{n}{p})) \leftrightarrow E(\overset{n}{p});$$
$$(3)\ \forall l(l < \boxed{\vee^n}(\overset{n}{p}) \leftrightarrow l \leq \overset{n}{p})$$
$$(1)\ S(\boxed{\text{Ш}^n}(\overset{n}{p})) \qquad \text{ST3.6}$$
$$(2)\ W(\boxed{\text{Ш}^n}(\overset{n}{p})) \leftrightarrow D(\overset{n}{p});$$
$$(3)\ \forall l(l < \boxed{\text{Ш}^n}(\overset{n}{p}) \leftrightarrow l < \overset{n}{p})$$

In a natural way, instead of '$\boxed{\wedge^2}(p, q)$', '$\boxed{\vee^2}(p, q)$', and '$\boxed{\text{Ш}^2}(p, q)$', we will write '$p\boxed{\wedge}q$', '$p\boxed{\vee}q$', and '$p\boxed{\text{Ш}}q$', respectively; i.e., we will set

$$p\boxed{\wedge}q := \boxed{\wedge^2}(p, q) \qquad \text{MD3.12}$$
$$p\boxed{\vee}q := \boxed{\vee^2}(p, q) \qquad \text{MD3.13}$$
$$p\boxed{\text{Ш}}q := \boxed{\text{Ш}^2}(p, q) \qquad \text{MD3.14}$$

Later on, it will turn out to be useful to have at our disposal an $m + n$-ary operation that, for given aggregates $p_1, \ldots, p_m$ and $q_1, \ldots, q_n$ of propositions, yields a proposition that is true iff, in the case that all the $p_1, \ldots, p_m$ are true, so also are all the $q_1, \ldots, q_n$; that is to say, the proposition that is true iff some one of the $p_1, \ldots, p_m$ is false or all the $q_1, \ldots, q_n$ are true. Hence

$$\exists r((W(r) \leftrightarrow (W(\overset{m}{p}) \rightarrow W(\overset{n}{q})) \wedge \forall l(l < r \leftrightarrow l \leq \overset{m}{p}, \overset{n}{q}))) \qquad \text{SCS4}$$

$$\boxed{\rightarrow^{mn}}(\overset{m}{p}; \overset{n}{q}) = \varepsilon r((W(r) \leftrightarrow (W(\overset{m}{p}) \rightarrow W(\overset{n}{q})) \wedge \forall l(l < r \leftrightarrow l \leq \overset{m}{p}, \overset{n}{q}))) \qquad \text{D3.15}$$

$$(1)\ S(\boxed{\rightarrow^{mn}}(\overset{m}{p}; \overset{n}{q})) \qquad \text{ST3.7}$$

$$(2)\ W(\boxed{\to_{mn}}(\overset{m}{p};\overset{n}{q})) \leftrightarrow (W(\overset{m}{p}) \to W(\overset{n}{q}))$$

$$(3)\ \forall l(l < \boxed{\to_{mn}}(\overset{m}{p};\overset{n}{q}) \leftrightarrow l \leq \overset{m}{p},\overset{n}{q})$$

Instead of '$\boxed{\to_{mn}}(\overset{m}{p};\overset{n}{q})$', we will write '$\overset{m}{p}\boxed{\to}\overset{n}{q}$'; i.e. we set

$$\overset{m}{p}\boxed{\to}\overset{n}{q} := \boxed{\to_{mn}}(\overset{m}{p};\overset{n}{q}) \qquad \text{MD3.16}$$

We have the important

$$W(\overset{m}{p}\boxed{\to}\overset{n}{q}) \leftrightarrow W(\boxed{\neg}((\boxed{\wedge^m}\overset{m}{p})\boxed{\wedge}(\boxed{\vee^n}\boxed{\neg}\overset{n}{q}))) \qquad \text{ST3.8}$$

Proof

$$W(\overset{m}{p}\boxed{\to}\overset{n}{q}) \leftrightarrow (W(\overset{m}{p}) \to W(\overset{n}{q})) \leftrightarrow \neg(W(\overset{m}{p}) \wedge \neg W(\overset{n}{q})) \leftrightarrow \neg(W(\overset{m}{p}) \wedge \neg \bigwedge_{j\leq n} W(q_j))$$

$$\leftrightarrow \neg(W(\overset{m}{p}) \wedge \bigvee_{j\leq n} \neg W(q_j)) \leftrightarrow \neg(W(\overset{m}{p}) \wedge \bigvee_{j\leq n} W(\boxed{\neg}q_j)) \leftrightarrow \neg(W(\overset{m}{p}) \wedge W(\boxed{\vee^n}\boxed{\neg}\overset{n}{q}))$$

$$\leftrightarrow \neg(W(\boxed{\wedge^m}\overset{m}{p}) \wedge W(\boxed{\vee^n}\boxed{\neg}\overset{n}{q})) \leftrightarrow \neg(W((\boxed{\wedge^m}\overset{m}{p})\boxed{\wedge}(\boxed{\vee^n}\boxed{\neg}\overset{n}{q}))) \leftrightarrow W(\boxed{\neg}((\boxed{\wedge^m}\overset{m}{p})\boxed{\wedge}(\boxed{\vee^n}\boxed{\neg}\overset{n}{q})))$$

As a particular case of this theorem, we have

$$W(p\boxed{\to}q) \leftrightarrow W(\boxed{\neg}(p\boxed{\wedge}\boxed{\neg}q) \qquad \text{T3.21}$$

which says that $p\boxed{\to}q$ corresponds to the traditional 'material implication'. It is worth noting that Bolzano also knew very well this interpretation of the 'if... then', which he, however, did not consider adequate and to which he preferred the one we will discuss later in the context of formal relations among propositions. Indeed, he said

> "We however often use the hypothetical form 'If *x* is, then *y* is' in an improper sense; since we will simply express that the proposition *x* and the negation of the proposition *y* (i.e. the proposition *Neg.y*) are not both true."[14]

Of course, we also have

$$W(\overset{m}{p}\boxed{\to}\overset{n}{q}) \leftrightarrow W((\boxed{\vee^m}\boxed{\neg}\overset{m}{p})\boxed{\vee}(\boxed{\wedge^n}\overset{n}{q})) \qquad \text{ST3.9}$$

with the particular case

$$W(p\boxed{\to}q) \leftrightarrow W(\boxed{\neg}p\boxed{\vee}q) \qquad \text{T3.22}$$

[14] TS §252 II 373.

3.5 Variation

The most famous Bolzanian doctrine is certainly his 'variation theory'. Although the best known and most important applications of this theory are to be found in the realm of propositions, it is worth remembering that this doctrine was introduced in the context of the discussion of ideas, as a tool for the extension to objectless ideas of notions such as those introduced (and originally only introduced) for objectual ones. In particular, this occurs with two issues: the first more properly concerns the content [*Inhalt*] of ideas, whereas the second is more directly connected with their extension [*Umfang*].

The first application concerns the notion of *redundancy* of an idea: an objectual idea is said to be *redundant* or *overabundant* [*überfüllt* or *überfließend*] when some of its parts—henceforth called *superfluous* [*überflüssig*]—may be omitted without modifying its reference (i.e., to be exact, neither its object, if it is singular, nor its extension, if it is plural).[15] Writing '*Red*' for 'being redundant', we may thus set

$$\underset{(V(x)\vee B(x))\wedge G(x)}{\forall x} (Red(x) \leftrightarrow \underset{V(z)\vee B(z)}{\exists z} (x < z \wedge x \equiv z)) \qquad \text{D3.17}$$

from which we obtain

$$G(a) \rightarrow (Red(a) \leftrightarrow \exists c(c < a \wedge a \equiv c)) \qquad \text{T3.23}$$

$$G(\overline{a}) \rightarrow (Red(\overline{a}) \leftrightarrow \exists \overline{c}(\overline{c} < \overline{a} \wedge \overline{a} \equiv \overline{c})) \qquad \text{T3.24}$$

It is clear that this criterion is completely ineffective in the case of objectless ideas; nevertheless, Bolzano thinks that, for instance, in the idea of a *round and angular quadrangle*, it is reasonable to regard as superfluous the idea of *angular* and, as a possible technically correct formulation of this conviction, he points to the idea *round* as variable and suggests substituting it for any other idea that makes the first one objectual. Clearly, in any such 'objectualizing' substitution, the idea *angular* will be superfluous in terms of the preceding definition. Thus, in general, his proposal is

> "Also an idea, which in the form in which it presents itself is objectless, will be called redundant, and precisely in the parts $\alpha, \beta, \gamma, \ldots$, in so far as, regarding as *variables* [*veränderlich*] certain other of its parts $i, j, \ldots$, we ascertain that in every case in which we exchange these ideas with other ones in such a way that the resulting idea is objectual, then this idea is redundant in the parts $\alpha, \beta, \gamma, \ldots$ in the sense of n.1."[16]

[15] TS §69, I 224: "1) With ideas composed of several parts it may happen that one combines more parts than are necessary to represent the objects the idea presents; in such cases, one can indicate an idea which has fewer parts than the given idea—some of its parts having been omitted—and yet still represents the same objects. [...] I shall call them *redundant* ideas, and the parts which occur in them that can be omitted without altering the extension I shall call *superfluous*."

[16] TS §69, I 225.

We now turn to the second application of the variation method, which concerns the possibility of extending to objectless ideas the various relations among ideas originally introduced with reference to the respective extensions: *equivalence, inclusion, exclusion, etc.* The solution consists in considering, as in the case of redundancy,

> "as variables some of their components. One must then only pay attention to the infinity of new ideas resulting from the given ones when whichever ideas take the place of the variables parts *i*, *j*, ... At this point we will also attribute to the given objectless ideas the same relationship which these new ideas reciprocally observe *whenever they are objectual*; only *conditionally*, of course, so far as the parts considered as variables are just the parts *i*, *j*, ..."[17]

Bolzano is very careful, already in the first presentation of this method, to underline that it concerns the lectological sphere and, in order to avoid misunderstandings, he 'deprives' it of any material value (strictly speaking, indeed, only material things can vary, or change, or be transformed, etc.), as well as of any possible 'linguistic' reading.[18]

The concepts discussed in Section 1.5 are key to following the subsequent exposition.

[17] TS §108, I 371–2: "*Ways in which the relations considered in §§93 ff. could be extended to objectless ideas.* According to their definitions, the relations that we have considered in §93 ff. only hold between ideas that *have objects*. It is nevertheless certain that we apply, even in common language, several of those relations to ideas which represent no object, or even to those which *cannot* represent any object, since they assign certain contrary attributes to it. [...] In §69, we discussed the concept of redundancy [...] (and) we discovered a means for a suitable expansion of it by considering certain constituents of the given idea *i*, *j*, ... to be variable. This method can be applied in the present case too. We can at once extend all of the definitions mentioned in §§93–107 to objectless ideas, if only we are permitted to envisage certain of their parts as variable. For then it would only be necessary to consider the infinitely many new ideas that are generated from the given ones if the variable parts *i*, *j*, ... are replaced by any other ideas. The same relation that holds between these new ideas, *whenever they have an object*, we also attribute to the given objectless ideas. It is understood that this is the case only relative to the parts *i*, *j*, ... which are taken to be variables. *Note.* Mathematicians make the most important use of this extension of these relations, especially in the theory of equations. According to their original concept, equations are nothing but statements of the equivalence of two ideas. Thus to say that $4 + 5 = 11 - 2$ is to say that the idea $4 + 5$ has the same objects as the idea $11 - 2$. But if we retain this concept then we can never establish equations such as $2 - 2 = 0$, $\frac{1}{\sqrt{-1}} = -\sqrt{-1}$ etc., whose meaning can be easily explained according to the above."

[18] TS §69, I 227: "*Note* 2. In no. 4, I suggested that we should look upon certain parts of an idea as variable, and examine what ideas result if we substitute any other ideas for these variable parts. In the sequel I shall frequently have to make use of this method. I hope that no one will object to the expression 'variable' and its cognates and that it will not be assumed that what I said contradicts the assertion that ideas are not something that exists. From the latter it indeed follows that they cannot undergo genuine variation. If I say that in a given idea, e.g., 'a wise man', a constituent, e.g. 'man' is to be envisaged as variable and may be replaced by any other idea, this merely means that we are to survey all ideas that have the same constituents and order of constituents as 'a wise man', except that they have other ideas in the place of 'man'."

3.6 Formal Properties and Relations of Ideas

Let us first point out that, in a lectological context, the Bolzanian theory of variation concerns ideas in general (i.e. both object and quality ideas). As at the elementary level, we distinguish these two types of idea, our reconstruction should take into account the fact that the aggregate of ideas to be varied is, in general, of the form $a_1, \ldots, a_s, \bar{b}_1, \ldots, \bar{b}_r$. This, however, would complicate—without presenting any real advantage—a notational situation that is in itself already complicated enough. Hence, in the following sections, we renounce our excess of exactitude, understanding, however, that the limitation to object ideas is everywhere simply notational. To emphasize this, we will simply use special variables for object ideas, in this context, but we will speak in general of *ideas* (without any specification).

Building on what we analysed in the previous section, we now extend the property of redundancy to objectless ideas, by introducing, for each $s > 0$, the $s + 1$-ary relation '*the objectless idea a is redundant with respect to the ideas* $a_1, \ldots, a_s$', to be denoted by 'Red^{s+1}', which subsists between a and the $a_1, \ldots, a_s$ iff every objectual idea resulting from a by any substitution in it for the $a_1, \ldots, a_s$ is redundant; i.e.

$$\neg G(a) \rightarrow (Red(a, \overset{s}{a}) \leftrightarrow \forall \overset{s}{c}(G(a[\overset{s}{a}/\overset{s}{c}]) \rightarrow Red(a[\overset{s}{a}/\overset{s}{c}]))) \qquad \text{SD3.10}$$

Concerning the second application of the variation method that was considered above (i.e. for the extension to objectless ideas of the relations considered in Section 2.2), we limit ourselves to an example—the equivalence—because all remaining cases are treated in totally similar way. Thus we introduce, for every $s > 0$, the $s + 2$-ary relation of *equivalence of the two objectless ideas a and c with respect to the ideas* $a_1, \ldots, a_s$, to be denoted by '$a \underset{\overset{s}{a}}{\equiv} c$', which subsists among a, c and $a_1, \ldots, a_s$ iff any couple of objectual ideas resulting from a and c by any substitution in them for the $a_1, \ldots, a_s$ are equivalent; i.e.

$$\neg G(a) \wedge \neg G(c) \rightarrow (a \underset{\overset{s}{a}}{\equiv} c \leftrightarrow \forall \overset{s}{c}(G(a[\overset{s}{a}/\overset{s}{c}]) \wedge G(c[\overset{s}{a}/\overset{s}{c}]) \rightarrow a[\overset{s}{a}/\overset{s}{c}] \equiv c[\overset{s}{a}/\overset{s}{c}])) \qquad \text{SD3.11}$$

3.7 Validity of Propositions

A general problem that arises with respect to the variation method, even if we limit ourselves to the two cases addressed by Bolzano—substitutions for ideas in ideas or in propositions—is clearly the problem concerning the admissibility of a given substitution. An obvious condition, implicit in all applications, is that the substitutions preserve the 'nature' of the thing to which they are applied (i.e. they should transform ideas into ideas and propositions into propositions). This purely formal requirement is quite far from resolving the issue, when we shift over from the formal to the material consideration. For example, which are the substitutions admitted in the idea *father of Charles*, if *Charles* is the variable idea? Certainly the idea *Archie* or that of any other man. How about the idea *Taffy* or that of any other animal? The idea *country*? The idea *triangle*? The idea *to eat*? And what if, on the contrary, we consider *father as* the variable idea? Certainly the idea *uncle* or *teacher* or *friend*. What about *dog*? *watch*? *to sleep*? Bolzano never directly addresses this question; in particular, he never clearly distinguishes the objectlessness of an idea resulting from a substitution and the possible nonsense of the resulting construct. Nevertheless, he somehow conveys that there are certain natural partitions of the ideas in classes of reciprocal intersubstitutability preserving the property of making sense. By the way, it seems unquestionable to us that the theory of *semantic categories*, which Husserl developed in his *Fourth Logical Investigation*, has just found in Bolzano's theory of variation its problematic stimulus as well as, in Brentano's and Stumpf's ideas on the dependence of the parts in a whole, the first hints at the tools necessary for its treatment.

The particular applications of the variation method discussed in the preceding sections, however, make this issue harmless, at least from the formal point of view. As objectual ideas are certainly also provided with sense, the requirement that the resulting idea be objectual make superfluous the worry of distinguishing, in the realm of 'useless' substitutions, between those that are such because they generate objectless ideas and those that are such because they generate senseless constructs.

If substitutions are applied to propositions, the issue is more complex. In this case, a restriction to the objectuality of the resultant ideas is useless; nonetheless, in the applications we consider, its place may be taken by the restriction to the *objectuality of the resulting proposition* or by the requirement of objectuality of their subject-ideas. In the cases that will be considered, doubt remains about whether the requirement of *objectuality* should or should not also regard also the propositions to which the substitution is applied. It seems to us that, at least for the applications that will be considered, such a restriction is often essential. Hence, the tacitly assumed general hypothesis will be that the examined substitutions are substitutions that apply to objectual propositions and generate objectual propositions.

Naturally, these considerations on objectuality make sense (at least up to this point) only with respect to atomic propositions (i.e. to propositions of the form $h(a, \bar{c})$). We could certainly free them from such a restriction; however, the complications related to this operation would be totally out of proportion to the result obtained, which is not, in itself, of great interest. Hence, in the following cases, the propositional variables 'p', 'q', ... are always intended as varying over atomic propositions.

The first application to propositions of the variation method concerns single propositions and yields the notion of the *validity of a proposition.*

Given an objectual proposition p and certain ideas $a_1, \ldots, a_s$ occurring in it, Bolzano considers the collection, which we will denote by '$\mathcal{S}(p, \overset{s}{a})$' of all objectual propositions we obtain from p by substituting in it any $c_1, \ldots, c_s$ for the $a_1, \ldots, a_s$. In general, $\mathcal{S}(p, \overset{s}{a})$ splits into two classes, both non-empty: the class of the true propositions (here denoted by '$\mathcal{W}(p, \overset{s}{a})$'), and the class of false propositions (here denoted by '$\mathcal{F}(p, \overset{s}{a})$').[19]

Bolzano calls (*degree of*) *validity of* p the 'ratio' between $\mathcal{W}(p, \overset{s}{a})$ and $\mathcal{S}(p, \overset{s}{a})$.[20]

Clearly, except for the two limiting cases—when $\mathcal{W}(p, \overset{s}{a})$ and $\mathcal{S}(p, \overset{s}{a})$ coincide (where it is trivially possible to assign to this ratio the value 1) and when, on the contrary, $\mathcal{F}(p, \overset{s}{a})$ and $\mathcal{S}(p, \overset{s}{a})$ coincide (where it is also possible to assign to this ratio the value 0)—the possibility of assigning to this ratio a definite numerical value appears to be connected with the finiteness of $\mathcal{S}(p, \overset{s}{a})$. Well aware of this difficulty, Bolzano points out the possibility of achieving, at least in some cases, a satisfying result by operating with certain restrictions [*nach einer gewissen Regel*], which save, of course, the sense of the whole. With this aim, besides requiring what we have already accepted (namely the objectuality of the result), he proposes to consider the substitutions up to equivalence of the involved ideas. Indeed, having already proved in §96 (and consolidated in §108) that every idea has infinitely many equivalents ($a \equiv --a \equiv ----a \equiv \cdots$; also $a \equiv [e](\widehat{a}) \equiv [e](\widehat{[e](\widehat{a})}) \equiv \cdots$), he knows perfectly well that $\mathcal{S}(p, \overset{s}{a})$ is, *in any case*, infinite; knowing, however, that, at least in the majority of cases, substitutions in propositions of equivalent ideas preserve the truth status of the propositions involved, his proposal is to take, instead of $\mathcal{S}(p, \overset{s}{a})$, $\mathcal{W}(p, \overset{s}{a})$, and $\mathcal{F}(p, \overset{s}{a})$, their subclasses $\mathcal{S}^\circ(p, \overset{s}{a})$, $\mathcal{W}^\circ(p, \overset{s}{a})$, and $\mathcal{F}^\circ(p, \overset{s}{a})$ in which 'repetitions' are avoided, and his conviction is that in this way it will be possible to treat the 'degree of validity' as a true fraction in many interesting cases.[21]

[19] TS §147, II 55: "Most propositions by far are such that the propositions which evolve from them are neither all true nor all false, but some of them are true and others are false. Here the question arises *how many* are true and how many false, or what the ratio is of one multitude to the other (or the total)."

[20] TS §147, II 56–7: "I therefore want to give a special name to the concept of the ratio of all true propositions to the total of all propositions which can be generated by treating certain ideas in a proposition as variables and replacing them with others according to a certain rule. I wish to call it the *validity* [*Gültigkeit*] of the proposition."

[21] TS §147, II 55–6: "If we allow a replacement of the ideas which we consider variable by any other idea whatever, then the totality of the true as well as false propositions which can be generated from a given proposition will be infinitely large. [...] Thus if we do not somehow limit the kinds of ideas which may be put into the

We will return to these issues in the final sections of this chapter, when discussing probability.

3.8 Analytic and Synthetic Truth and Falsity

The notion of the degree of validity of a proposition also provides Bolzano with the basis for an explanation of the notions of *analytic* (or *synthetic*) *truth* (or *falsity*). We saw earlier that the possibility of assigning a numerical value to the ratio between $\mathcal{W}(p, \overset{s}{a})$ and $\mathcal{S}(p, \overset{s}{a})$ is always given in the two extreme cases: (1) when all admissible substitutions for the $\overset{s}{a}$ generate true propositions; (2) when all admissible substitutions for the $\overset{s}{a}$ generate false propositions.

In the first case, Bolzano speaks of *universally* or *completely valid* propositions [*allgemein*—or *vollgültig*] or also of propositions *true by their type* [*ihrer ganzen Art nach wahr*] or *formally true* [*ihrer Form nach wahr*]; in the second case, he speaks of *universally* or *absolutely invalid* propositions [*allgemein oder durchaus ungültig*] or also *false by their type* [*ihrer ganzen Art nach*] or *formally false* [*ihrer Form nach falsch*]. All, of course, are relativized to the $\overset{s}{a}$.[22]

Hence, writing '$\mathbb{W}^{s+1}$' ('$\mathbb{F}^{s+1}$') for the $s+1$-ary relation that subsists between the proposition p and the ideas $a_1, \ldots, a_s$ (occurring in it) iff all substitutions for the $a_1, \ldots, a_s$ which serve the objectuality of p make it true (false), we set

$$\mathbb{W}_{\overset{s}{a}}(p) \leftrightarrow G(p) \wedge \forall \overset{s}{c}(G(p[\overset{s}{a}/\overset{s}{c}]) \rightarrow W(p[\overset{s}{a}/\overset{s}{c}])) \qquad \text{SD3.12}$$

$$\mathbb{F}_{\overset{s}{a}}(p) \leftrightarrow G(p) \wedge \forall \overset{s}{c}(G(p[\overset{s}{a}/\overset{s}{c}]) \rightarrow F(p[\overset{s}{a}/\overset{s}{c}])) \qquad \text{SD3.13}$$

place of a variable one, we will rarely be in a position to determine the ratio of the number of true or false propositions to the number of all propositions which can be generated in this way. On the other hand, if we restrict the choice, and stipulate that no two equivalent ideas may replace the ideas *i, j, . . .* which are treated as variable (equivalent in the sense of §96 or in the wider sense of §108), if we furthermore require that only those ideas be chosen which generate objectual propositions, then the number of ideas which are still permitted, hence also the number of propositions which can thus be generated, will be significantly reduced. Thus, under these restrictions, we will more often succeed in assigning numerical values to the ratio of the multitude of true or false propositions to the multitude of all propositions which can be generated."

[22] TS §147, II 57: "Let the proposition *A* be such that all the propositions which can be generated from it are true, if the ideas *i, j, . . .* alone are considered variable, and if only objectual propositions may be formed. Then the degree of validity of *A* with respect to *i, j, . . .* is the largest possible, i.e., it equals 1, and we can call the proposition *universally* or *fully valid.* If, on the other hand, all propositions developed from *A* are false, then the degree of validity of *A* is the smallest possible, i.e., it equals 0. We can therefore say that it is a *universally* or *absolutely invalid* proposition. Universally valid propositions could also be said to be *true after their kind or form,* universally invalid propositions *false after their kind or form,* where by kind is meant the collection of all propositions, which differ from *A* only in the ideas *i, j, . . .*"

It is obviously possible to omit the condition $\overset{s}{a} < p$; the trivial consequence is that every true (false) is to be considered universally valid (invalid) with respect to any variable not occurring in it. As alluded to above, in order to avoid trivial situations, the condition $G(p)$ appears to be unavoidable, at least in the case of invalidity.

A total variation freedom allows the transformation of any atomic proposition into any other one $(h(a, \overline{b})[a, b/c, \overline{d}] = h(c, \overline{d}))$ and so the notions of universal validity or invalidity lose any interest. A fundamental distinction among the propositions arises, on the contrary, when one focuses one's attention on single ideas occurring in a proposition. Two cases then become possible: either the proposition is, with respect to at least one of its ideas, universally true (universally false) or there is in it no idea with respect to which it has such a property (i.e., given any of its ideas, there are always substitutions for it that will generate a true proposition and substitutions that will generate a false proposition). Propositions of the first type (i.e. propositions containing at least one idea with respect to which they are universally true (false)) are said to be *analytic*; all others are *synthetic*.[23]

Hence, writing '*AW*' ('*AF*') for 'analytically true (false)' and '*SW*' ('*SF*') for 'synthetically true (false)':

$$AW(p) \leftrightarrow \exists a \underset{a}{\mathbb{W}}(p) \qquad \text{D3.18}$$

$$AF(p) \leftrightarrow \exists a \underset{a}{\mathbb{F}}(p) \qquad \text{D3.19}$$

$$SW(p) \leftrightarrow W(p) \wedge \neg AW(p) \qquad \text{D3.1}$$

$$SF(p) \leftrightarrow F(p) \wedge \neg AF(p) \qquad \text{D3.2}$$

Although difficult to regulate, an important, distinction that arises when one begins to examine propositions with respect to their possible analyticity is that which, in epistemological terms, may be described referring to the type of knowledge necessary in order to recognize this property. There are propositions, for example *a is a* $[h(a, \widehat{a})]$ or *an a that as* $\overline{c}$*, has* $\overline{c}$ $[h([a](\overline{c}), \overline{c})]$, the analyticity of which does not require, in order to be recognized, any scientific knowledge. On the contrary, to recognize the analyticity of *the double of the natural number* 5 *is even*, $[h(d(5), \overline{p})]$, i.e. to recognize that in this proposition there is an idea (namely the idea of 5) such that on every exchange of it with any other idea that renders objectual the subject-idea

[23] TS §148, II 58: "1) [...] But if a proposition contained even a *single idea* which could be arbitrarily changed without altering the truth or falsity of the proposition; i.e., if the propositions which could be obtained from it through the arbitrary alteration of this one idea would either all be true or all false, provided only that they have objectuality, then this property of the proposition would be noteworthy enough to distinguish it from those that do not have it. Borrowing an expression from KANT, I allow myself to call propositions of this kind *analytic*."

(i.e. with the idea of any natural number), the proposition remains true, requires arithmetical knowledge. Propositions of the first type are said to be *logically analytic* [*logisch analytische*] or *analytic in a strict sense* [*analytische in der engeren Bedeutung*]; propositions of the second type are said to be *analytic in a broad sense* [*analytische in der weitern Bedeutung*]. Bolzano emphasizes, however, that the demarcation line between logical and non-logical ideas is anything but clearly traceable and that consequently not even the preceding bipartition of analytic propositions is well defined.[24]

The preceding properties may be extended to finite systems of propositions. Hence the propositions $p_1, \ldots, p_n$ are called *formally true (false)* [*ihrer Form nach wahr (falsch)*], of course with respect to certain ideas $a_1, \ldots, a_s$, iff they result true (false) in every substitution for the $a_1, \ldots, a_s$ that renders them objectual; i.e.

$$\underset{\overset{s}{a}}{\mathbb{W}}(\overset{n}{p}) \leftrightarrow \bigwedge_{i \leq n} G(p_i) \wedge \forall \overset{s}{c}(\bigwedge_{i \leq n} G(p_i[\overset{s}{a}/\overset{s}{c}]) \rightarrow \bigwedge_{i \leq n} W(p_i[\overset{s}{a}/\overset{s}{c}]))) \qquad \text{SD3.3}$$

$$\underset{\overset{s}{a}}{\mathbb{F}}(\overset{n}{p}) \leftrightarrow \bigwedge_{i \leq n} G(p_i) \wedge \forall \overset{s}{c}(\bigwedge_{i \leq n} G(p_i[\overset{s}{a}/\overset{s}{c}]) \rightarrow \bigwedge_{i \leq n} F(p_i[\overset{s}{a}/\overset{s}{c}]))) \qquad \text{SD3.4}$$

3.9 Formal Relations Among Propositions

We now turn to the second and even more significant application of the variation method to propositions, the one that, for Bolzano, succeeds in disclosing the most important relations among propositions.[25]

The following analysis focuses on aggregates of propositions and hence we will use, with the necessary adjustments, the notational techniques and conventions

[24] TS §148, II 58–9: "2) The following are some very general examples of analytic propositions which are also true: '*A* is *A*', '*A* which is *B* is *A*', '*A* which is *B* is *B*', 'Every object is either *B* or non-*B*', etc. Propositions of the first kind, i.e., propositions cast in the form '*A* is *A*' or '*A* has (the attribute) *a*' are commonly called *identical* or *tautological* propositions. 3) The difference between the last mentioned analytic propositions and those under no. 1 lies in the following: In order to appraise the analytic nature of the propositions under no. 2, nothing but logical knowledge is needed, since the concepts which form the invariable part of these propositions all belong to logic. On the other hand, for the appraisal of the truth and falsity of propositions like those given in no. 1 a wholly different kind of knowledge is required, since concepts alien to logic intrude. This distinction, I admit, is rather unstable, as the whole domain of concepts belonging to logic is not circumscribed so sharply that controversies could not arise at times. Nevertheless, it might be profitable to keep this distinction in mind. Hence propositions like those in no. 2 may be called logically analytic, or analytic in the narrower sense; those of no. 1, analytic in the broader sense."

[25] TS §154, II 70: "1) The most important relations among propositions come to light when we follow the method introduced in §147, and envisage certain ideas contained in them as variable; we then consider the new propositions which are generated, if these ideas are replaced by any other ideas whatever, and observe how they fare with respect to their truth or falsity."

that we have been using so far. It is important, however, to understand that, in general, Bolzano, when considering substitutions for certain $a_1, \dots, a_s$ in certain $p_1, \dots, p_n$, not only did not require, but *explicitly excluded the requirement* that all the $a_1, \dots, a_s$ be present in each one of the $p_1, \dots, p_n$. Moreover, he does not seem to require (S. 154.6) that each one of the $a_1, \dots, a_s$ be present in at least one of the $p_1, \dots, p_n$, even though, of course, he does not express any interest in trivial situations. Thus, in our exposition, unless we state otherwise, we will also adopt this liberal attitude.

Thus, given certain $p_1, \dots, p_n$, certain $a_1, \dots, a_s$, and certain $c_1, \dots, c_s$, let us consider the n propositions $p_1[\overset{s}{a}/\overset{s}{c}], \dots, p_n[\overset{s}{a}/\overset{s}{c}]$ obtained from the $p_1, \dots, p_n$ by uniformly substituting $c_1, \dots, c_s$ for the $a_1, \dots, a_s$.

We will also use the following mereological definitions:

$$a < \overset{s}{a} \leftrightarrow a = a_1 \vee \dots \vee a = a_s \qquad \text{SD3.5}$$

$$a < \overset{n}{p} \leftrightarrow a < p_1 \vee \dots \vee a < p_n \qquad \text{SD3.6}$$

$$p < \overset{n}{p} \leftrightarrow p = p_1 \vee \dots \vee p = p_n \qquad \text{SD3.7}$$

3.9.1 *Compatibility, Integrability, and their Complements*

Bolzano focuses on two cases, for the given propositions $p_1, \dots, p_n$: the case in which *there are* ideas $c_1, \dots, c_s$ such that one has $W(\overset{n}{p}[\overset{s}{a}/\overset{s}{c}])$ [i.e. $\bigwedge_{i \leq n} W(p_i[\overset{s}{a}/\overset{s}{c}])$] and the case in which for all ideas $c_1, \dots, c_s$ one has $E(\overset{n}{p}[\overset{s}{a}/\overset{s}{c}])$ [i.e. $\bigvee_{i \leq n} W(p_i[\overset{s}{a}/\overset{s}{c}])$]. In the first case, Bolzano states that among the propositions $p_1, \dots, p_n$ there subsists, *with respect to the ideas* $a_1, \dots, a_s$, a relation of *compatibility* [*Verträglichkeit*] or *agreement* [*Einstimmung*] or *consonance* [*Einhelligkeit*],[26] while, in the second case, he states that among the propositions $p_1, \dots, p_n$ there subsists, *with respect to the ideas* $a_1, \dots, a_s$, a relation of *(formal) integration* [*(formale) Ergänzung*] or *(formal) help* [*(formale) Aushilfe*].[27] Writing '$\curlywedge^{n,s}$' for the $n + s$-ary relation of compatibility

[26] TS §154, II 70: "2) We know already that almost any proposition in which we replace certain ideas by different, arbitrarily chosen, ideas will sometimes turn into a true, and sometimes into a false, proposition. Let us compare several propositions *A, B, C, D,...* and consider as variable certain ideas *i, j,...*, which they have in common (each containing one or the other of them). The question arises whether there are any ideas which can be put into the places of *i, j,...* and which are of such a nature that they make all of the above propositions true at the same time. If this question must be answered in the affirmative, then I wish to call the relation between propositions *A, B, C, D,...* a relation of *compatibility* or *concordance*, and *A, B, C, D,...* will be called *compatible* or *concordant* propositions."

[27] TS §160, II 117: "2) It may also occur that we only know of a given collection of propositions that not all of them are false, or true, without knowing of any particular proposition that is true or false. [...] It is important to recognise this relation when it obtains. [...] I know of no term that has been used to designate this relation, so I

('$\underset{\overset{s}{a}}{\Lambda\!\!\!\Lambda}(\overset{n}{p})$' is thus '*the propositions* $p_1, \ldots, p_n$ *are compatible with respect to the idea* $a_1, \ldots, a_s$') and '$\mathbb{W}^{n,s}$' for the $n + s$-ary relation of integration ('$\underset{\overset{s}{a}}{\mathbb{W}}(\overset{n}{p})$' is thus '*the propositions* $p_1, \ldots, p_n$ *integrate each other with respect to the ideas* $a_1, \ldots, a_s$), we have the two basic Bolzanian definitions:

$$\underset{\overset{s}{a}}{\Lambda\!\!\!\Lambda}(\overset{n}{p}) \leftrightarrow \exists\overset{s}{c}W(\overset{n}{p}[\overset{s}{a}/\overset{s}{c}]) \qquad \text{SD3.8}$$

$$\underset{\overset{s}{a}}{\mathbb{W}}(\overset{n}{p}) \leftrightarrow \forall\overset{s}{c}E(\overset{n}{p}[\overset{s}{a}/\overset{s}{c}]) \qquad \text{SD3.9}$$

It must be emphasized that the notions of compatibility and integrability among propositions are always relative to a particular system of ideas;[28] nevertheless, it is also possible to give 'absolutized' versions of them: the $p_1, \ldots, p_n$ *are compatible* (*integrate each other*) iff there are $a_1, \ldots, a_s$ such that the $p_1, \ldots, p_n$ are compatible (integrate each other) with respect to them;[29] i.e.

$$\Lambda\!\!\!\Lambda(\overset{n}{p}) \leftrightarrow \exists\overset{s}{a}\underset{\overset{s}{a}}{\Lambda\!\!\!\Lambda}(\overset{n}{p}) \qquad \text{SD3.10}$$

$$\mathbb{W}(\overset{n}{p}) \leftrightarrow \exists\overset{s}{a}\underset{\overset{s}{a}}{\mathbb{W}}(\overset{n}{p}) \qquad \text{SD3.11}$$

Indeed—and this remark should be kept in mind in all the analogous cases of 'absolutization' that follow—strictly speaking, the preceding definitions of $\Lambda\!\!\!\Lambda(\overset{n}{p})$ and $\mathbb{W}(\overset{n}{p})$ are not completely satisfying, because the formulas $\exists\overset{s}{a}\underset{\overset{s}{a}}{\Lambda\!\!\!\Lambda}(\overset{n}{p})$ and $\exists\overset{s}{a}\underset{\overset{s}{a}}{\mathbb{W}}(\overset{n}{p})$ *depend also on s*. As the direct elimination of this dependence, by the assumptions $\Lambda\!\!\!\Lambda(\overset{n}{p}) \leftrightarrow \exists s\exists\overset{s}{a}\underset{\overset{s}{a}}{\Lambda\!\!\!\Lambda}(\overset{n}{p})$ and $\mathbb{W}(\overset{n}{p}) \leftrightarrow \exists s\exists\overset{s}{a}\underset{\overset{s}{a}}{\mathbb{W}}(\overset{n}{p})$, would drag us out of the logical framework in which we have decided to move, we could think to define, instead of the predicates $\Lambda\!\!\!\Lambda$ and $\mathbb{W}$, for each $s > 0$, the predicates $\Lambda\!\!\!\Lambda_s$ and $\mathbb{W}_s$, by

$$\Lambda\!\!\!\Lambda_s(\overset{n}{p}) \leftrightarrow \exists\overset{s}{a}\underset{\overset{s}{a}}{\Lambda\!\!\!\Lambda}(\overset{n}{p}) \qquad \text{and} \qquad \mathbb{W}_s(\overset{n}{p}) \leftrightarrow \exists\overset{s}{a}\underset{\overset{s}{a}}{\mathbb{W}}(\overset{n}{p})$$

will permit myself to call them *complementary* or *supporting* propositions. [...] This relation of complementation can hold for particular propositions $M, N, O, \ldots$ which are given; in this case it may be called *material* complementation or support. But the relation may continue to hold, no matter what ideas are put in the place of certain variable ideas $i, j, \ldots$ [...]; in this case we speak of *formal* complementation or support."

[28] TS §154, II 72: "5) Thus, if we say of a given class of propositions $A, B, C, D, \ldots$ that they are compatible or incompatible, we must indicate, for the sake of precision, *in what respect*, i.e., in relation to what variable ideas $i, j, \ldots$ this is meant."

[29] TS §164, II 141: "If we claim that certain propositions $A, B, C, \ldots$ are compatible without caring to determine the ideas with respect to which they are compatible, we actually only claim that there are certain ideas in the propositions which, if considered variable, render them compatible."

Nonetheless, the question is somewhat more complicated, because, if we want to respect the principle that the ideas subjected to variation should, in any case, occur in at least one of the propositions of the aggregate, the index s cannot be greater than the total number $s(\overset{n}{p})$ of ideas occurring in the $\overset{n}{p}$. Hence, strictly speaking, the preceding schemata are to be intended as a summary presentation, for any given n, of

$$\begin{cases} \mathbb{A}_1(\overset{n}{p}) \leftrightarrow \exists a \underset{a}{\mathbb{A}}(\overset{n}{p}) \\ \vdots \\ \mathbb{A}_{s(\overset{n}{p})}(\overset{n}{p}) \leftrightarrow \exists \overset{s(\overset{n}{p})}{a} \underset{\overset{s(\overset{n}{p})}{a}}{\mathbb{A}}(\overset{n}{p}) \end{cases}$$

and of

$$\begin{cases} \mathbb{W}_1(\overset{n}{p}) \leftrightarrow \exists a \underset{a}{\mathbb{W}}(\overset{n}{p}) \\ \vdots \\ \mathbb{W}_{s(\overset{n}{p})}(\overset{n}{p}) \leftrightarrow \exists \overset{s(\overset{n}{p})}{a} \underset{\overset{s(\overset{n}{p})}{a}}{\mathbb{W}}(\overset{n}{p}) \end{cases}$$

Obviously, both the relation of compatibility and the relation of integrability have a complement: the relation of *incompatibility* [*Unverträglichkeit*] or *dissonance* [*Mißhelligkeit*] of the propositions $p_1, \ldots, p_n$ with respect to the $a_1, \ldots, a_s$ (i.e. the relation that subsists between those propositions and those ideas iff there is no substitution for the latter that simultaneously makes the former true[30] and the relation of *formal* or *intrinsic* [*ihrer ganzen Art nach*] falsity[31] of the propositions $p_1, \ldots, p_n$ with respect to the $a_1, \ldots, a_s$ (i.e. the relation that subsists between those propositions and those ideas iff there is no substitution for the latter that makes at least one of the former true. That is,

$$\neg \underset{\overset{s}{a}}{\mathbb{A}}(\overset{n}{p}) \leftrightarrow \forall \overset{s}{c} \bigvee_{i \leq n} F(p_i[\overset{s}{a}/\overset{s}{c}]) \leftrightarrow \forall \overset{s}{c} \bigvee_{i \leq n} W(\boxminus p_i[\overset{s}{a}/\overset{s}{c}]) \qquad \text{ST3.10}$$

30 TS §154, II 71: "2) [. . .] if there are no ideas whose substitution for *i, j, . . .* will make all of the propositions *A, B, C, D, . . .* true, I say that they stand in a relation of *incompatibility*, and the propositions themselves I call *incompatible* or *discordant.*"

31 TS §160, II 117: "1) First of all, it may happen that all of the propositions in a given collection are false—[. . .] in the forms they assume when certain of the ideas in them are changed [. . .], with the condition that their bases continue to be genuine object-ideas. Propositions which stand in such a relation [. . .], we call [. . .] *formally* [. . .] *false* (§147)."

$$\neg \underset{\overset{s}{a}}{\mathbb{W}}(\overset{n}{p}) \leftrightarrow \forall \overset{s}{c} \bigwedge_{i \leq n} F(p_i[\overset{s}{a}/\overset{s}{c}]) \leftrightarrow \forall \overset{s}{c} \bigwedge_{i \leq n} W(\boxminus p_i[\overset{s}{a}/\overset{s}{c}]) \qquad \text{ST3.11}$$

Bolzano devotes §§ 154–9 to the examination of the notion of compatibility, of its complement, and of their respective specializations, and §160 to integration and its specializations, a total of 71 pages. It is worth noting that in this examination Bolzano is very careful to point out not only the 'positive' situations (i.e. those situations in which certain hypotheses imply certain consequences), but also the 'negative' ones, those in which certain consequences do not follow from certain hypotheses. With the exception of the most trivial cases, Bolzano provides proofs (and often examples) or an exhibition of a counterexample.

3.9.2 *Compatibility and Incompatibility*

One can immediately see that

$$\text{If } \pi \text{ is a permutation of } \{1, \ldots, n\}, \text{ then } \exists \overset{s}{c} \bigwedge_{i \leq n} W(p_i[\overset{s}{a}/\overset{s}{c}]) \leftrightarrow \exists \overset{s}{c} \bigwedge_{i \leq n} W(p_{\pi(i)}[\overset{s}{a}/\overset{s}{c}]) \qquad \text{ST3.12}$$

therefore, in particular,[32]

$$\underset{\overset{s}{a}}{\mathbb{A}}(\overset{m}{p}, \overset{n}{q}) \leftrightarrow \underset{\overset{s}{a}}{\mathbb{A}}(\overset{n}{q}, \overset{m}{p}) \qquad \text{ST3.13}$$

Besides the relation of compatibility, with respect to certain ideas, of *one* aggregate of propositions, it is sometimes useful to consider the relation of compatibility, with respect to certain ideas, between *two* aggregates of propositions; such a relation will be defined in a natural way as compatibility of the aggregate built up from the union of the two;[33] i.e.

$$\overset{m}{p} \underset{\overset{s}{a}}{\mathbb{A}} \overset{n}{q} \leftrightarrow \underset{\overset{s}{a}}{\mathbb{A}}(\overset{m}{p}, \overset{n}{q}) \qquad \text{SD3.12}$$

All truths are compatible with each other with respect to any system of ideas:[34]

$$W(\overset{n}{p}) \rightarrow \underset{\overset{s}{a}}{\mathbb{A}}(\overset{n}{p}) \qquad \text{ST3.14}$$

[32] TS §154, II 71: "3) It follows from the above definition that the relations of compatibility and incompatibility are symmetrical."

[33] GL p. 62: "Wenn ein oder mehrere Sätze *A*, *B*, *C*, ... mit einem oder mehreren anderen *M*, *N*, ... verträglich heiß en sollen, und dieß zwar hinsichtlich auf die Bestandtheile *i*, *j*, ... ; so muß nach dem Gesagten wenigstens einige an die Stelle der von *i*, *j*, ... zu setzende <Bestandheile>geben, die nicht nur die Sätze *A*, *B*, *C*, ... sondern auch die *M*, *N*, ... wahr machen."

[34] TS §154, II 72: "6) All truths are compatible with each other no matter which of their ideas are considered variable: they are already true by virtue of the ideas of which they originally consist."

Consequently, among incompatible propositions, at least one must be false:[35]

$$\neg \underset{\overset{s}{a}}{\bigwedge\!\!\!\!\bigwedge}(\overset{n}{p}) \rightarrow \bigvee_{i \leq n} F(p_i) \qquad \text{ST3.15}$$

Among compatible propositions one, more than one, even all, may be false:[36]

$$\underset{\overset{s}{a}}{\bigwedge\!\!\!\!\bigwedge}(\overset{n}{p}) \not\rightarrow W(\overset{n}{p}) \qquad \text{ST3.16}$$

If the $\overset{n}{q}$ are only some of the $\overset{m}{p}$ and are incompatible with respect to certain ideas, then the $\overset{m}{p}$ are also incompatible with respect to the same ideas:[37]

$$\text{For } \{q_1, \ldots, q_n\} \subset \{p_1, \ldots, p_m\} : \neg \underset{\overset{s}{a}}{\bigwedge\!\!\!\!\bigwedge}(\overset{n}{q}) \rightarrow \neg \underset{\overset{s}{a}}{\bigwedge\!\!\!\!\bigwedge}(\overset{m}{p}) \qquad \text{ST3.17}$$

The converse, however, does not hold:[38]

$$\text{For } \{q_1, \ldots, q_n\} \subset \{p_1, \ldots, p_m\} : \neg \underset{\overset{s}{a}}{\bigwedge\!\!\!\!\bigwedge}(\overset{m}{p}) \not\rightarrow \neg \underset{\overset{s}{a}}{\bigwedge\!\!\!\!\bigwedge}(\overset{n}{q}) \qquad \text{ST3.18}$$

If the $\overset{r}{c}$ are only some of the $\overset{s}{a}$, then compatibility with respect to the former implies compatibility with respect to the latter, and incompatibility with respect to the latter implies incompatibility with respect to the former; not, however, vice versa:[39]

$$\text{For } \{c_1, \ldots, c_r\} \subseteq \{a_1, \ldots, a_s\} : \underset{\overset{r}{c}}{\bigwedge\!\!\!\!\bigwedge}(\overset{n}{p}) \rightarrow \underset{\overset{s}{a}}{\bigwedge\!\!\!\!\bigwedge}(\overset{n}{p}) \qquad \text{ST3.19}$$

[35] TS §154, II 72: "7) Hence of any given set of incompatible propositions at least one must be false, but several, even all of them, may be false."

[36] TS §154, II 72: "8) In a set of compatible propositions, there may also be false ones, indeed all of them may be false. The fact that certain propositions are false, given the ideas of which they originally consist, does not prevent them from becoming true if certain other ideas are introduced. It is obviously necessary, however, that each of these false propositions should contain at least one of the ideas which are considered variable, since otherwise it would not change and hence not turn into a truth."

[37] TS §154, II 72: "10) If a part of the propositions *A, B, C, D,* . . . , for example, *A, B,* . . . are not compatible with each other with respect to certain ideas *i, j,* . . . , then the total collection is also not compatible with respect to the same ideas. For [. . .]."

[38] TS §154, II 72: "9) Even if the *n* propositions *A, B, C, D,* . . . are incompatible with respect to ideas *i, j,* . . . , there may still be a relation of compatibility with respect to the same ideas *i, j,* . . . among any smaller number of these propositions e.g., between any $(n-1)$, $(n-2)$, . . . of them. For [. . .]."

[39] TS §154, II 73: "11) If certain propositions are compatible with each other with respect to the ideas *i, j,* . . . , then they are also compatible with respect to the more inclusive set *i, j, k, l,* . . . which contains the former. If certain propositions are incompatible with respect to the larger set of ideas *i, j, k, l,* . . . , then they are also incompatible with respect to the smaller set *i, j,* . . . . Conversely, it does not follow that, if a number of propositions are incompatible with respect to ideas *i, j,* . . . they are also incompatible with respect to *i, j, k, l,* . . . ; and if they are compatible with respect to ideas *i, j, k, l,* . . . , it does not follow that they are also compatible with respect to *i, j,* . . . "

$$\text{For } \{c_1, \ldots, c_r\} \subseteq \{a_1, \ldots, a_s\} : \neg \underset{\substack{s\\a}}{\wedge\!\!\!\wedge}(\overset{n}{p}) \rightarrow \neg \underset{\substack{r\\c}}{\wedge\!\!\!\wedge}(\overset{n}{p}) \qquad \text{ST3.20}$$

$$\text{For } \{c_1, \ldots, c_r\} \subseteq \{a_1, \ldots, a_s\} : \neg \underset{\substack{r\\c}}{\wedge\!\!\!\wedge}(\overset{n}{p}) \not\rightarrow \neg \underset{\substack{s\\a}}{\wedge\!\!\!\wedge}(\overset{n}{p}) \qquad \text{ST3.21}$$

$$\text{For } \{c_1, \ldots, c_r\} \subseteq \{a_1, \ldots, a_s\} : \underset{\substack{s\\a}}{\wedge\!\!\!\wedge}(\overset{n}{p}) \not\rightarrow \underset{\substack{r\\c}}{\wedge\!\!\!\wedge}(\overset{n}{p}) \qquad \text{ST3.22}$$

The compatibility with respect to certain ideas is not 'euclidean' and therefore, taking into account its symmetry, it is not even transitive;[40] i.e.

$$\overset{m}{p} \underset{\substack{s\\a}}{\wedge\!\!\!\wedge} \overset{n}{q} \wedge \overset{s}{r} \underset{\substack{s\\a}}{\wedge\!\!\!\wedge} \overset{n}{q} \not\rightarrow \overset{m}{p} \underset{\substack{s\\a}}{\wedge\!\!\!\wedge} \overset{s}{r} \qquad \text{ST3.23}$$

$$\overset{m}{p} \underset{\substack{s\\a}}{\wedge\!\!\!\wedge} \overset{n}{q} \wedge \overset{n}{q} \underset{\substack{s\\a}}{\wedge\!\!\!\wedge} \overset{s}{r} \not\rightarrow \overset{m}{p} \underset{\substack{s\\a}}{\wedge\!\!\!\wedge} \overset{s}{r} \qquad \text{ST3.24}$$

The same holds for incompatibility:[41]

$$\neg(\overset{m}{p} \underset{\substack{s\\a}}{\wedge\!\!\!\wedge} \overset{n}{q}) \wedge \neg(\overset{s}{r} \underset{\substack{s\\a}}{\wedge\!\!\!\wedge} \overset{n}{q}) \not\rightarrow \neg(\overset{m}{p} \underset{\substack{s\\a}}{\wedge\!\!\!\wedge} \overset{s}{r}) \qquad \text{ST3.25}$$

Nor does compatibility among certain propositions imply compatibility among their negations:[42]

$$\underset{\substack{s\\a}}{\wedge\!\!\!\wedge}(\overset{n}{p}) \not\rightarrow \underset{\substack{s\\a}}{\wedge\!\!\!\wedge}(\boxed{\neg}\overset{n}{p}) \qquad \text{ST3.26}$$

The fact that the negation of *each one of* the $p_1, \ldots, p_n$ is compatible (with respect to certain ideas) with all remaining ones, does not imply that the negation of *two* of them is also compatible (with respect to the same ideas) with the remaining ones:[43]

$$p_i \neq p_j \wedge \begin{pmatrix} \underset{\substack{s\\a}}{\wedge\!\!\!\wedge}(\boxed{\neg}p_i, p_1, \ldots, p_{i-1}, p_{i+1}, \ldots, p_n) \\ \wedge \underset{\substack{s\\a}}{\wedge\!\!\!\wedge}(\boxed{\neg}p_j, p_1, \ldots, p_{j-1}, p_{j+1}, \ldots, p_n) \end{pmatrix} \qquad \text{ST3.27}$$
$$\not\rightarrow \underset{\substack{s\\a}}{\wedge\!\!\!\wedge}(\boxed{\neg}p_i, \boxed{\neg}p_j, p_1, \ldots, p_{i-1}, p_{i+1}, \ldots, p_{j-1}, p_{j+1}, \ldots, p_n)$$

[40] TS §154, II 73: "12) From the fact that propositions $A, B, C, D, \ldots$ as well as propositions $G, H, I, K, \ldots$ are compatible with propositions $M, N, O, \ldots$ with respect to ideas $i, j, \ldots$ it does not follow that propositions $A, B, C, D, \ldots$ and $G, H, I, K, \ldots$ are compatible *with each other* with respect to these same ideas. For [...]."

[41] TS §154, II 73: "13) Similarly, it does not follow from the fact that propositions $A, B, C, D, \ldots$ as well as $G, H, I, K, \ldots$ are incompatible with propositions $M, N, O, \ldots$ with respect to certain ideas $i, j, \ldots,$ that $A, B, C, D, \ldots$ and $G, H, I, K, \ldots$ are incompatible *with each other* with respect to the same ideas. For [...]."

[42] TS §154, II 73-4: "14) From the fact that certain propositions $A, B, C, D, \ldots$ are compatible with respect to $i, j, \ldots$ it does not follow that their negations, i.e., the propositions Neg.*A*,Neg.*B*,Neg.*C*,Neg.*D*, ..., (§141) are compatible with respect to the same ideas. For [...]."

[43] TS §154, II 74: "15) From the fact that the negation of any one of the propositions $A, B, C, D, \ldots$ is compatible with the remaining propositions with respect to ideas $i, j, \ldots$ it does not follow that the negations of any two or more of these propositions are compatible with the remainder with respect to the same ideas."

A false proposition that does not contain any one of the ideas considered as variable in the $p_1, \ldots, p_n$ is always incompatible with them (with respect to those variables):[44]

$$F(q) \land \bigwedge_{i \leq s} \neg(a_i < q) \rightarrow \neg(\overset{n}{p} \underset{\overset{s}{a}}{\curlywedge} q) \qquad \text{ST3.28}$$

Hence, conversely, a proposition that is compatible with certain others with respect to certain ideas, which do not occur in it, is true:[45]

$$\overset{n}{p} \underset{\overset{s}{a}}{\curlywedge} q \land \bigwedge_{i \leq s} \neg(a_i < q) \rightarrow W(q) \qquad \text{ST3.29}$$

Finally, the compatibility of certain propositions with respect to certain ideas is equivalent to the compatibility, with respect to certain other ideas, of the propositions obtained from the first by replacing their ideas with these other ideas:[46]

$$\underset{\overset{s}{a}}{\curlywedge}(\overset{n}{p}) \leftrightarrow \underset{\overset{s}{c}}{\curlywedge}(\overset{n}{p}[\overset{s}{a}/\overset{s}{c}]) \qquad \text{ST3.30}$$

Bolzano devotes great attention to the consideration of particular cases of compatibility and of incompatibility.

3.9.2.1 Derivability

Certainly the most famous application of the Bolzanian theory of variation is found in the special case of compatibility that Bolzano called *derivability* [*Ableitbarkeit*] but for which, according to the current terminology, the term 'consequence' may be more suitable. It is in fact the relation that subsists between two aggregates of propositions and an aggregate of ideas when the first aggregate is compatible with respect to those ideas and, moreover, each substitution (for those ideas) that makes true all the propositions of the first aggregate also makes true all the propositions of the second aggregate. We can reformulate all of this as follows.

Let us denote by '$\Rightarrow^{m,n,s}$' the $m + n + s$-ary relation that subsists among the propositions $p_1, \ldots, p_m$, the propositions $q_1, \ldots, q_n$, and the ideas $a_1, \ldots, a_s$

[44] TS §154, II 76: "20) If a proposition is false and does not contain the ideas $i, j, \ldots$, which we consider variable in a certain collection of propositions, it is incompatible with these propositions, no matter how they may be constituted. Since [...]."

[45] TS §154, II 76: "21) Hence, if a certain proposition F is compatible with certain others $A, B, C, D, \ldots$ even though it does not contain a single one of the variable ideas $i, j, \ldots$ then it must be a true proposition."

[46] TS §154, II 76: "22) Whatever the relation between propositions $A, B, C, D, \ldots$ with respect to ideas $i, j, \ldots$, that same relation must obtain between propositions $A', B', C', D', \ldots$ with respect to ideas $i', j', \ldots$, where the propositions $A', B', C', D', \ldots$ have resulted from propositions $A, B, C, D, \ldots$ merely by replacing ideas $i, j, \ldots$ with $i', j', \ldots$. For [...]."

iff the $p_1, \ldots, p_m$ are compatible with the $q_1, \ldots, q_n$ with respect to the $a_1, \ldots, a_s$ and, moreover, for every aggregate $c_1, \ldots, c_s$ of ideas substituted in the $p_1, \ldots, p_m, q_1, \ldots, q_n$ for the $a_1, \ldots, a_s$, we have the following: if all the $p_1, \ldots, p_m$ become true, then also all the $q_1, \ldots, q_n$ become true. Furthermore, given certain propositions $p_1, \ldots, p_m, q_1, \ldots, q_n$, we say that *the* $q_1, \ldots, q_n$ *are derivable from the* $p_1, \ldots, p_m$, and denote by '$\Rightarrow^{m,n}$' this $m + n$-ary relation, iff this happens with respect to some aggregate of ideas.[47] So we set

$$\overset{m}{p} \underset{\overset{s}{a}}{\Rightarrow} \overset{n}{q} \leftrightarrow \overset{m}{p} \underset{\overset{s}{a}}{\mathbin{\wedge\!\!\!\wedge}} \overset{n}{q} \wedge \forall \overset{s}{c}(W(\overset{m}{p}[\overset{s}{a}/\overset{s}{c}]) \rightarrow W(\overset{n}{q}[\overset{s}{a}/\overset{s}{c}])) \qquad \text{SD3.13}$$

$$\overset{m}{p} \Rightarrow \overset{n}{q} \leftrightarrow \exists \overset{s}{a}(\overset{m}{p} \underset{\overset{s}{a}}{\Rightarrow} \overset{n}{q}) \qquad \text{SD3.14}$$

Clearly, in the presence of $\forall \overset{s}{c}(W(\overset{m}{p}[\overset{s}{a}/\overset{s}{c}]) \rightarrow W(\overset{n}{q}[\overset{s}{a}/\overset{s}{c}]))$, the compatibility of the $p_1, \ldots, p_m$ with respect to the $a_1, \ldots, a_s$ also implies the compatibility of the $p_1, \ldots, p_m$ with the $q_1, \ldots, q_n$ with respect to the same ideas and in particular that of the $q_1, \ldots, q_n$. Therefore

$$\overset{m}{p} \underset{\overset{s}{a}}{\Rightarrow} \overset{n}{q} \leftrightarrow \underset{\overset{s}{a}}{\mathbin{\wedge\!\!\!\wedge}}(\overset{m}{p}) \wedge \forall \overset{s}{c}(W(\overset{m}{p}[\overset{s}{a}/\overset{s}{c}]) \rightarrow W(\overset{n}{q}[\overset{s}{a}/\overset{s}{c}])) \qquad \text{ST3.31}$$

$$\overset{m}{p} \Rightarrow \overset{n}{q} \rightarrow \exists \overset{s}{a}(\underset{\overset{s}{a}}{\mathbin{\wedge\!\!\!\wedge}}(\overset{m}{p}) \wedge \forall \overset{s}{c}(W(\overset{m}{p}[\overset{s}{a}/\overset{s}{c}]) \rightarrow W(\overset{n}{q}[\overset{s}{a}/\overset{s}{c}]))) \qquad \text{ST3.32}$$

It is worth remarking that this relation *does not admit general thinning*; if q is derivable from the $p_1, \ldots, p_m$ (with respect to the $a_1, \ldots, a_s$), then q is derivable from $p_1, \ldots, p_m, p_{m+1}$ *only if* p_{m+1} *is compatible (with respect to the* $a_1, \ldots, a_s$*)* with the $p_1, \ldots, p_m$.

Clearly, derivability is not, in general, symmetrical:[48]

$$\overset{m}{p} \underset{\overset{s}{a}}{\Rightarrow} \overset{n}{q} \not\rightarrow \overset{n}{q} \underset{\overset{s}{a}}{\Rightarrow} \overset{m}{p} \qquad \text{ST3.33}$$

[47] TS §155, II 80: "Let us consider, first of all, the case where among the compatible propositions *A, B, C, D, . . . , M, N, O, . . .* the following relation obtains: all ideas whose substitution for the variable ideas *i, j, . . .* turns a certain part of these propositions, namely, *A, B, C, D, . . .* into truths, are also so constituted as to make a certain other part of these propositions, namely, *M, N, O, . . .* true. [. . .] I wish to give the name of *deducibility* [*Ableitbarkeit*] to this relation between propositions *A, B, C, D, . . .* on one hand and *M, N, O, . . .* on the other. Hence I say that propositions *M, N, O, . . .* are *deducible* from propositions *A, B, C, D, . . .* with respect to variable parts *i, j, . . .* if every collection of ideas whose substitution for *i, j, . . .* makes all of *A, B, C, D, . . .* true, also makes all of *M, N, O, . . .* true."

[48] TS §155, II 81: "3) [. . .] the relation of deducibility is not necessarily mutual."

A premise that does not contain any of the ideas to be varied is superfluous:[49]

$$\overset{m}{p} \underset{\overset{s}{a}}{\Rightarrow} \overset{n}{q} \wedge \neg(\overset{s}{a} < p_i) \rightarrow (p_1, \ldots, p_{i-1}, p_{i+1}, \ldots, p_n \underset{\overset{s}{a}}{\Rightarrow} \overset{n}{q}) \tag{ST3.34}$$

Proof As it does not contain the $\overset{s}{a}$, p_i remains unaltered by all substitutions and so, being compatible with the remaining propositions, it must be true; but then, every substitution for the $\overset{s}{a}$ that makes all those remaining true also makes true all the $\overset{n}{q}$, which, hence, by definition, are derivable from the remaining propositions.

If one of the conclusions is false, then so is one of the premises:[50]

$$\overset{m}{p} \underset{\overset{s}{a}}{\Rightarrow} \overset{n}{q} \rightarrow (\bigvee_{i \leq n} F(q_i) \rightarrow \bigvee_{j \leq m} F(p_j)) \tag{ST3.35}$$

If every conclusion of certain premises is true, then so are all the premises:[51]

$$\forall q(\overset{m}{p} \underset{\overset{s}{a}}{\Rightarrow} q \rightarrow W(q)) \rightarrow W(\overset{m}{p}) \tag{ST3.36}$$

From no proposition, for no system of variables, is derivable the negation of this proposition,[52]

$$\neg(p \Rightarrow \boxed{\neg}p) \tag{ST3.37}$$

Derivability is substitutive with respect to compatibility:[53]

$$\overset{m}{p} \underset{\overset{s}{a}}{\Rightarrow} \overset{n}{q} \wedge \overset{m}{p} \underset{\overset{s}{a}}{\mathbin{\wedge\!\!\!\wedge}} \overset{u}{r} \rightarrow \overset{n}{q} \underset{\overset{s}{a}}{\mathbin{\wedge\!\!\!\wedge}} \overset{u}{r} \tag{ST3.38}$$

As far as the possibility of compatibility among premises and conclusions of distinct derivability relations is concerned, Bolzano establishes[54]

$$\neg(\overset{n}{q} \underset{\overset{s}{a}}{\mathbin{\wedge\!\!\!\wedge}} \overset{n'}{s}) \rightarrow \neg(\overset{m}{p} \underset{\overset{s}{a}}{\Rightarrow} \overset{n}{q} \wedge \overset{m'}{r} \underset{\overset{s}{a}}{\Rightarrow} \overset{n'}{s} \wedge \overset{m}{p} \underset{\overset{s}{a}}{\mathbin{\wedge\!\!\!\wedge}} \overset{m'}{r}) \tag{ST3.39}$$

[49] TS §155, II 81: "4) Assume that propositions *M*, *N*, *O*, ... are deducible with respect *i*, *j*, ... from propositions *A*, *B*, *C*, *D*, ..., and that proposition *A*does not contain any of these ideas. In this case *A* may be omitted [...]."

[50] TS §155, II 81: "5) If certain propositions *M*, *N*, *O*, ... are deducible from certain others *A*, *B*, *C*, *D*, ... and one of the former is false, then there must also be a false one among the latter. For [...]."

[51] TS §155, II 81: "6) If all propositions which are deducible from *A*, *B*, *C*, *D*, ... with respect to certain ideas *i*, *j*, ... are true, then *A*, *B*, *C*, *D*, ... must themselves be true. For [...]."

[52] TS §155, II 82: "7) The negation of a proposition *A*, Neg.*A*, i.e., the proposition '*A* is false', is never deducible from *A*, no matter what ideas *i*, *j*, ..., which occur only in A, are envisaged as variable. For [...]."

[53] TS §155, II 82: "8) All conclusions *M*, *N*, *O*, ... which follow from certain propositions *A*, *B*, *C*, *D*, ... with respect to ideas *i*, *j*, ... are compatible with respect to *i*, *j*, ... with all propositions, with which the premises *A*, *B*, *C*, *D*, ... are compatible with respect to *i*, *j*, ... For [...]."

[54] TS §155, II 82: "9) Propositions which are not compatible are never conclusions from propositions which are compatible with respect to the same variable ideas. For [...]."

that is to say,[55]

$$\overset{m}{p} \underset{\substack{s\\a}}{\Rightarrow} \overset{n}{q} \wedge \overset{m'}{r} \underset{\substack{s\\a}}{\Rightarrow} \overset{n'}{s} \wedge \neg(\overset{n}{q} \underset{\substack{s\\a}}{\not\curlywedge} \overset{n'}{s}) \rightarrow \neg(\overset{m}{p} \underset{\substack{s\\a}}{\not\curlywedge} \overset{m'}{r}) \qquad \text{ST3.40}$$

the converse, however, does not hold (i.e. from incompatible propositions compatible propositions may well be derived):[56]

$$\overset{m}{p} \underset{\substack{s\\a}}{\Rightarrow} \overset{n}{q} \wedge \overset{m'}{r} \underset{\substack{s\\a}}{\Rightarrow} \overset{n'}{s} \wedge \neg(\overset{m}{p} \underset{\substack{s\\a}}{\not\curlywedge} \overset{m'}{r}) \not\rightarrow \neg(\overset{n}{q} \underset{\substack{s\\a}}{\not\curlywedge} \overset{n'}{s}) \qquad \text{ST3.41}$$

A proposition that is not always true can never be derivable both from a proposition and from its negation:[57]

$$\neg \underset{\substack{s\\a}}{\mathbb{W}}(q) \rightarrow \neg(p \underset{\substack{s\\a}}{\Rightarrow} q \wedge \boxminus p \underset{\substack{s\\a}}{\Rightarrow} q) \qquad \text{ST3.42}$$

When, however, there is more than one premise, nothing prevents the same proposition from also being derivable from the negation of one or more, even all, of the premises:[58]

$$\text{If } 1 < m = r + s, \text{ then } \overset{m}{p} \underset{\substack{s\\a}}{\Rightarrow} q \not\rightarrow \neg(\boxminus \overset{r}{p}, \overset{s}{p} \underset{\substack{s\\a}}{\Rightarrow} q) \qquad \text{ST3.43}$$

The negation of a proposition that is compatible with certain others is never derivable from them:[59]

$$\overset{m}{p} \underset{\substack{s\\a}}{\curlywedge} q \rightarrow \neg(\overset{m}{p} \underset{\substack{s\\a}}{\Rightarrow} \boxminus q) \qquad \text{ST3.44}$$

If, on the contrary, a proposition is incompatible with certain other propositions, then its negation is derivable from them:[60]

$$\underset{\substack{s\\a}}{\curlywedge}(\overset{m}{p}) \wedge \neg(\overset{m}{p} \underset{\substack{s\\a}}{\curlywedge} q) \rightarrow \overset{m}{p} \underset{\substack{s\\a}}{\Rightarrow} \boxminus q \qquad \text{ST3.45}$$

[55] TS §155, II 82: "10) If certain conclusions are incompatible, then their premises must also be incompatible with respect to the same variable ideas. For [. . .]."

[56] TS §155, II 82: "11) But it is quite possible that conclusions are compatible when their respective premises are incompatible with respect to the same variable parts. [. . .]."

[57] TS §155, II 83: "12) No proposition which is not true after its form with respect to ideas *i*, *j*, . . . (§147) is deducible from both the individual proposition *A*as well as its negation Neg.*A*. For [. . .]."

[58] TS §155, II 83: "13) But if there are *several* premises from which a certain proposition *M* is to be deducible, e.g., *A*, *B*, *C*, *D*, . . . , then it is quite possible that *M* is also deducible from the negation of some or even all of them. For [. . .]."

[59] TS §155, II 83: "14) If *M* is compatible with propositions *A*, *B*, *C*, *D*, . . . with respect to ideas *i*, *j*, . . . , then its negation Neg.*M* is not deducible from these propositions with respect to the same ideas. For [. . .]."

[60] TS §155, II 83: "15) If propositions *A*, *B*, *C*, *D*, . . . are compatible with each other with respect to ideas *i*, *j*, . . . but incompatible with *M*, then the proposition Neg.*M* is deducible from them with respect to the same ideas. For [. . .]."

Bolzano states the principle of *contraposition*, with the usual compatibility restrictions:[61]

$$\overset{m}{p}, p \underset{\substack{s\\a}}{\Rightarrow} q \wedge \overset{m}{p} \underset{\substack{s\\a}}{\mathrel{\wedge\!\!\wedge}} \boxed{\neg}\, q \rightarrow \overset{m}{p}, \boxed{\neg}\, q \underset{\substack{s\\a}}{\Rightarrow} \boxed{\neg}\, p \qquad \text{ST3.46}$$

He then shows the superfluity of a premise that may be replaced by its negation:[62]

$$\overset{m}{p}, p \underset{\substack{s\\a}}{\Rightarrow} \overset{n}{q} \wedge \overset{m}{p}, \boxed{\neg}\, p \underset{\substack{s\\a}}{\Rightarrow} \overset{n}{q} \rightarrow \overset{m}{p} \underset{\substack{s\\a}}{\Rightarrow} \overset{n}{q} \qquad \text{ST3.47}$$

but emphasizes that this does not hold for two (or more) premises:[63]

$$\overset{m}{p}, p, r \underset{\substack{s\\a}}{\Rightarrow} \overset{n}{q} \wedge \overset{m}{p}, \boxed{\neg}\, p, \boxed{\neg}\, r \underset{\substack{s\\a}}{\Rightarrow} \overset{n}{q} \not\rightarrow \overset{m}{p} \underset{\substack{s\\a}}{\Rightarrow} \overset{n}{q} \qquad \text{ST3.48}$$

Subsequently, Bolzano turns to the derivability relation with respect to more or fewer ideas:[64]

$$\text{For } \{c_1, \ldots, c_r\} \subseteq \{a_1, \ldots, a_s\} : \overset{m}{p} \underset{\substack{s\\a}}{\Rightarrow} \overset{n}{q} \wedge \underset{\substack{r\\c}}{\mathrel{\wedge\!\!\wedge}} (\overset{m}{p}) \rightarrow \overset{m}{p} \underset{\substack{r\\c}}{\Rightarrow} \overset{n}{q} \qquad \text{ST3.49}$$

$$\text{For } \{c_1, \ldots, c_r\} \subseteq \{a_1, \ldots, a_s\} : \wedge \overset{m}{p} \underset{\substack{r\\c}}{\Rightarrow} \overset{n}{q} \wedge \underset{\substack{r\\c}}{\mathrel{\wedge\!\!\wedge}} (\overset{m}{p}) \not\rightarrow \overset{m}{p} \underset{\substack{s\\a}}{\Rightarrow} q \qquad \text{ST3.50}$$

At this point, Bolzano establishes that the generality introduced by the arbitrariness of the choice of the ideas to be varied does not trivialize the notion of derivability. Indeed:

> "Not every proposition M and even less so aggregate of propositions $M, N, O, \ldots$ can be set in a relation of derivability with any single proposition A or even also with any aggregate of more propositions $A, B, C, \ldots$, simply on account of the fact that

[61] TS §155, II 84: "16) If we delete any one of the premises $A, B, C, D, \ldots$ of a conclusion M, for example A, and replace it by the negation of M, Neg.M, then we can deduce the negation of the missing proposition, i.e., Neg.A, from the collection of propositions $B, C, D, \ldots$, Neg.M, provided only that Neg.M is compatible with the remaining premises. For [...]."

[62] TS §155, II 84: "17) If propositions $M, N, O, \ldots$ are deducible with respect to the same ideas $i, j, \ldots$ from the collection of propositions $A, B, C, D, \ldots$ and X as well as from the collection of propositions $A, B, C, D, \ldots$ and Neg.X, then they are also deducible from propositions $A, B, C, D, \ldots$ alone with respect to the same ideas. For [...]."

[63] TS §155, II 84: "18) The matter would be quite different if one of the above collections contained the negation of more than one proposition from the other collection. For, if M is deducible from the collection of propositions $A, B, C, D, \ldots, X, Y$, as well as from the collection of propositions $A, B, C, D, \ldots$, Neg.X,Neg.Y, it does not follow that M is deducible from $A, B, C, D, \ldots$ alone. For [...]."

[64] TS §155, II 84: "19) If propositions $M, N, O, \ldots$ are deducible from $A, B, C, D, \ldots$ with respect to ideas $i, j, k, \ldots$ then they are also deducible from the same propositions with respect to the fewer ideas $j, k, \ldots$ (which are part of the former), provided that propositions $A, B, C, D, \ldots$ are compatible with each other with respect to the fewer ideas $j, k, \ldots$. For [...]."

we may assume at will which and how many ideas *i*, *j*, ... in such propositions are to be regarded as variables'.'[65]

With the usual restriction of compatibility, a principle of *additionability* is proved:[66]

$$\overset{m}{p} \underset{\substack{s\\a}}{\Rightarrow} \overset{n}{q} \wedge \overset{m'}{r} \underset{\substack{s\\a}}{\Rightarrow} \overset{n'}{s} \wedge \overset{m}{p} \underset{\substack{s\\a}}{\mathrel{\wedge\!\!\wedge}} \overset{m'}{r} \rightarrow \overset{m}{p}, \overset{m'}{r} \underset{\substack{s\\a}}{\Rightarrow} \overset{n}{q}, \overset{n'}{s} \qquad \text{ST3.51}$$

as is its generalization:[67]

$$\overset{m}{p} \underset{\substack{s\\a}}{\Rightarrow} \overset{n}{q} \wedge \overset{m'}{r} \underset{\substack{r\\c}}{\Rightarrow} \overset{n'}{s} \wedge \overset{r}{c} \setminus \overset{s}{a} \not\prec \overset{m}{p} \wedge \overset{s}{a} \setminus \overset{r}{c} \not\prec \overset{m'}{r} \wedge \overset{m}{p}, \overset{m'}{r} \underset{\substack{s\,r\\a,c}}{\mathrel{\wedge\!\!\wedge}} \overset{s}{a}, \overset{i}{c} \rightarrow \overset{m}{p}, \overset{m'}{r} \underset{\substack{s\,r\\a,c}}{\Rightarrow} \overset{n}{q}, \overset{n'}{s} \qquad \text{ST3.52}$$

We then have the *cut* principle[68]

$$\overset{m}{p} \underset{\substack{s\\a}}{\Rightarrow} \overset{n}{q} \wedge \overset{n}{q}, \overset{m'}{r} \underset{\substack{s\\a}}{\Rightarrow} \overset{n'}{s} \rightarrow \overset{m}{p}, \overset{m'}{r} \underset{\substack{s\\a}}{\Rightarrow} \overset{n'}{s} \qquad \text{ST3.53}$$

and its generalization[69]

$$\overset{m}{p} \underset{\substack{s\\a}}{\Rightarrow} \overset{n}{q} \wedge \overset{n}{q}, \overset{m'}{r} \underset{\substack{r\\c}}{\Rightarrow} \overset{n'}{s} \wedge \overset{i}{c} \setminus \overset{s}{a} \not\prec \overset{m}{p} \wedge \overset{s}{a} \setminus \overset{i}{c} \not\prec \overset{n}{q}, \overset{m'}{r} \rightarrow \overset{m}{p}, \overset{m'}{r} \underset{\substack{s\,r\\a,c}}{\Rightarrow} \overset{n'}{s} \qquad \text{ST3.54}$$

[65] Cf. TS §155, II 85 (21).

[66] TS §155, II 85: "22) If propositions *M*, *N*, *O*, ... are deducible from propositions *A*, *B*, *C*, *D*, ... with respect to ideas *i*, *j*, ..., and propositions *P*, *Q*, *R*, ... deducible from propositions *F*, *G*, *H*, ... with respect to the same ideas; and if *A*, *B*, *C*, *D*, ... are compatible with *F*, *G*, *H*, ... (also with respect to the same ideas), then the collection of propositions *M*, *N*, *O*, ..., *P*, *Q*, *R*, ... is deducible from the collection of propositions *A*, *B*, *C*, *D*, ..., *F*, *G*, *H*, ... For [...]."

[67] TS §155, II 85–6: "23) Even in the case where propositions *M*, *N*, *O*, ... are deducible from propositions *A*, *B*, *C*, *D*, ... with respect to certain ideas *i*, *j*, ..., and propositions *P*, *Q*, *R*, ... are deducible from propositions *F*, *G*, *H*, ... with respect to certain other ideas *k*, *l*, ..., which are either altogether, or at least in part, different from *i*, *j*, ..., the collection of propositions *M*, *N*, *O*, ..., *P*, *Q*, *R*, ... is deducible from the collection of propositions *A*, *B*, *C*, *D*, ..., *F*, *G*, *H*, ... with respect to the collection of ideas *i*, *j*, ... *k*, *l*, ..., provided that none of the ideas of *k*, *l*, ..., which are not also contained in *i*, *j*, ... appear in propositions *A*, *B*, *C*, *D*, ..., that none of the ideas of *i*, *j*, ... which are not also contained in *k*, *l*, ... occur in propositions *F*, *G*, *H*, ..., and also that propositions *A*, *B*, *C*, *D*, ..., *F*, *G*, *H*, ... are compatible with respect to ideas *i*, *j*, ... *k*, *l*, ... For [...]."

[68] TS §155, II 86: "24) If propositions *M*, *N*, *O*, ... are deducible from propositions *A*, *B*, *C*, *D*, ... with respect to certain ideas *i*, *j*, ..., and propositions *X*, *Y*, *Z*, ... are deducible from propositions *M*, *N*, *O*, ... and *R*, *S*, *T*, ... with respect to the same ideas, then propositions *X*, *Y*, *Z*, ... are also deducible from propositions *A*, *B*, *C*, *D*, ... *R*, *S*, *T*, ... with respect to the same ideas. For [...]."

[69] TS §155, II 86–7: "25) Even if propositions *M*, *N*, *O*, ... are deducible from *A*, *B*, *C*, *D*, ... with respect to ideas *i*, *j*, ..., while propositions *X*, *Y*, *Z*, ... are deducible from *M*, *N*, *O*, ..., *R*, *S*, *T*, ... with respect to ideas *k*, *l*, ..., which are either altogether or in part different from *i*, *j*, ..., propositions *X*, *Y*, *Z*, ... are still deducible from propositions *A*, *B*, *C*, *D*, ..., *R*, *S*, *T*, ... with respect to the collection of ideas *i*, *j*, ..., *k*, *l*, ..., provided only that none of the ideas among *k*, *l*, ... that differ from *i*, *j*, ... occurs in propositions *A*, *B*, *C*, *D*, ... and none of the *i*, *j*, ... that differ from *k*, *l*, ... occurs in *M*, *N*, *O*, ..., *R*, *S*, *T*, ... For [...]."

Moving in a perspective that we might refer to today as 'of relevance', Bolzano introduces a distinction between an *exact* [*genaue*] or *exactly measured* [*genau bemessene*] or *adequate* [*adäquate*] derivability of a proposition q from certain premises $p_1, \ldots, p_m$ and a *redundant* [*überfüllte*] derivability of q from the $p_1, \ldots, p_m$. The first relation subsists when, by omitting even only one of the $p_1, \ldots, p_m$, or, rather, even just one of the ideas occurring in them, the q is no longer derivable from the $p_1, \ldots, p_m$; the second relation subsists in the opposite case;[70] i.e.

$$\overset{n}{p} \underset{\substack{s\\a}}{\overset{=}{\Rightarrow}} q \leftrightarrow \overset{n}{p} \underset{\substack{s\\a}}{\Rightarrow} q \wedge \forall p(p < \overset{m}{p} \rightarrow \neg(\overset{m}{p} \setminus p \underset{\substack{s\\a}}{\Rightarrow} q)) \wedge$$
$$\forall a(a < \overset{m}{p} \wedge a \not\leq b \rightarrow \neg(\overset{n}{p}[a/b] \underset{\substack{s\\a}}{\Rightarrow} q)) \qquad \text{SD3.15}$$

$$\overset{n}{p} \underset{\substack{s\\a}}{\overset{+}{\Rightarrow}} q \leftrightarrow \overset{n}{p} \underset{\substack{s\\a}}{\Rightarrow} q \wedge \neg(\overset{n}{p} \underset{\substack{s\\a}}{\overset{=}{\Rightarrow}} q) \qquad \text{SD3.16}$$

In an exact derivability, propositions that are always true cannot occur, either as premises or as conclusions:[71]

$$\overset{n}{p} \underset{\substack{s\\a}}{\overset{=}{\Rightarrow}} q \rightarrow \bigwedge_{i \leq n} \neg \mathbb{W}(p_i) \wedge \neg \mathbb{W}(q) \qquad \text{ST3.55}$$

If the negation of a conclusion of certain premises is incompatible with some of the premises, then the negation of the negation of that conclusion is already derivable from those premises and so is a conclusion itself; hence, in such a case, the derivability would not be exact; therefore[72]

$$\overset{n}{p} \underset{\substack{s\\a}}{\overset{=}{\Rightarrow}} q \wedge \{q_1, \ldots, q_m\} \subset \{p_1, \ldots, p_n\} \rightarrow \overset{m}{q} \underset{\substack{s\\a}}{\wedge\!\!\!\wedge} \boxed{\neg} q \qquad \text{ST3.56}$$

In an exact derivability, no premise is derivable from the remaining ones:[73]

$$\overset{n}{p} \underset{\substack{s\\a}}{\overset{=}{\Rightarrow}} q \rightarrow \neg(\overset{n}{p} \setminus p_i \underset{\substack{s\\a}}{\Rightarrow} p_i) \qquad \text{ST3.57}$$

[70] TS §155, II 87: "26) Let proposition *M* be deducible from premises *A, B, C, D,...* with respect to ideas *i, j,....* If *A, B, C, D,...* are such that none of them, nor even any of their parts, may be omitted, with *M* still deducible from the remainder with respect to the same ideas *i, j,...,* I call the relation of deducibility of proposition *M* from *A, B, C, D,... exact, precise,* or *adequate.* A relation of deducibility that is not exact is said to be *redundant.*"

[71] TS §155, II 87: "27) If a relation of exact deducibility holds between propositions, then neither the conclusion nor any of the premises may be a formally true proposition. [...]."

[72] TS §155, II 88: "28) If the relation of deducibility with respect to ideas *i, j,...* between premises *A, B, C, D,...* and the conclusion *M* is exact, then the negation of this conclusion, Neg.*M*, must be compatible with any part of the premises with respect to the same ideas *i, j,....* For [...]."

[73] TS §155, II 88: "29) If a relation of deducibility is exact with respect to ideas *i, j,...,* then no premise must be deducible from the remaining premises with respect to the same ideas. For [...]."

moreover, the negation of each premise must be compatible not only with all the other premises, but also with the negation of the conclusion:[74]

$$\overset{n}{p} \underset{\substack{s\\a}}{\overset{=}{\Rightarrow}} q \rightarrow \boxed{\neg} p_i \underset{\substack{s\\a}}{\mathbin{\wedge\!\!\wedge}} (\overset{n}{p} \setminus p_i, \boxed{\neg} q) \qquad \text{ST3.58}$$

An exact derivability, however, does not exclude that the negations of two or more premises could be incompatible with the remaining ones:[75]

$$\overset{n}{p} \underset{\substack{s\\a}}{\overset{=}{\Rightarrow}} q \not\rightarrow \neg(\boxed{\neg} p_i, \boxed{\neg} p_j) \underset{\substack{s\\a}}{\mathbin{\wedge\!\!\wedge}} (\overset{n}{p} \setminus (p_i, p_j))) \qquad \text{ST3.59}$$

A particularly interesting question concerns the relationships between exact derivability and cut. In general, a relation of derivability that is obtained by cut from two relations of exact derivability is not itself exact:[76]

$$\overset{n}{p} \underset{\substack{s\\a}}{\overset{=}{\Rightarrow}} q \wedge \overset{n}{q}, q \underset{\substack{s\\a}}{\overset{=}{\Rightarrow}} r \not\rightarrow \overset{m}{p}, \overset{n}{q} \underset{\substack{s\\a}}{\overset{=}{\Rightarrow}} r \qquad \text{ST3.60}$$

on the other hand, however, it is not excluded that it could be exact:[77]

$$\overset{n}{p} \underset{\substack{s\\a}}{\overset{=}{\Rightarrow}} q \wedge \overset{n}{q}, q \underset{\substack{s\\a}}{\overset{=}{\Rightarrow}} r \not\rightarrow \neg(\overset{m}{p}, \overset{n}{q} \underset{\substack{s\\a}}{\overset{=}{\Rightarrow}} r) \qquad \text{ST3.61}$$

In the last case, when a relation of exact derivability is obtainable from a cut of two relations of exact derivability, Bolzano speaks of *composed* [*zusammengesetzt*] (exact) derivability, whereas in the opposite case, when an exact derivability cannot be obtained by a cut of two relations of exact derivability, he speaks of *simple* [*einfach*] (exact) derivability. Writing '$\overset{=}{\Rightarrow}^{+}$' and '$\overset{=}{\Rightarrow}^{-}$' for the $m + n + 1 + s$-ary relation of composed, respectively simple (exact), derivability, we then have

$$\overset{m}{p}, \overset{n}{q} \underset{\substack{s\\a}}{\overset{=}{\Rightarrow}}{}^{+} r \leftrightarrow \exists q(\overset{m}{p} \underset{\substack{s\\a}}{\overset{=}{\Rightarrow}} q \wedge \overset{n}{q}, q \underset{\substack{s\\a}}{\overset{=}{\Rightarrow}} r \wedge \overset{m}{p}, \overset{n}{q} \underset{\substack{s\\a}}{\overset{=}{\Rightarrow}} r) \qquad \text{SD3.17}$$

[74] TS §155, II 88: "30) If a relation of deducibility is exact, then the negation of any single premise must be compatible not only with all other premises, but also with the added negation of the conclusion *M*, given that the same ideas are variable throughout. For [. . .]."

[75] TS §155, II 88: "31) It also holds for exact deducibility that the negation of two or more of the premises can be incompatible with the remainder, given the same variable ideas throughout. [. . .]."

[76] TS §155, II 89: "32) Even if the relation of deducibility between premises *A, B, C, D*, . . . Even if the relation of deducibility between premises *M*, as well as the relation between premises *M, R, S, T*, . . . and the conclusion *X*, is exact with respect to the same ideas *i, j*, . . . , it does not follow that the relation of deducibility which holds between premises *A, B, C, D*, . . . , *R, S, T*, . . . and the conclusion *X* (cf. no. 24) must also be exact [. . .]."

[77] TS §155, II 89: "33) But there can be no doubt that *sometimes* when the conclusion *M* is exactly deducible from premises *A, B, C, D*, . . . and the conclusion *X* from premises *M, R, S, T*, . . . , the relation of deducibility between conclusion *X* and premises *A, B, C, D*, . . . , *R, S, T*, . . . is also exact [. . .]."

$$\overset{m}{p}, \overset{n}{q} \underset{\overset{s}{a}}{\overset{=}{\Rightarrow}} {}^{-}r \leftrightarrow \overset{m}{p}, \overset{n}{q} \underset{\overset{s}{a}}{\overset{=}{\Rightarrow}} r \wedge \neg(\overset{m}{p}, \overset{n}{q} \underset{\overset{s}{a}}{\overset{=}{\Rightarrow}} {}^{+}r) \qquad \text{SD3.18}$$

As an example of simple derivability, Bolzano gives[78] $a\mathbb{A}^{+}b, b\mathbb{A}^{+}c \Rightarrow a\mathbb{A}^{+}c$.

Subsequently, Bolzano turns to derivability between atomic propositions having the same predicate and, respectively, the same subject and, taking this common element as sole variable, establishes the two equivalencies

$$h(a, \bar{c}) \underset{\bar{c}}{\Rightarrow} h(b, \bar{c}) \leftrightarrow b \sqsubseteq a \text{ [79]} \qquad \text{T3.25}$$

$$h(a, \bar{b}) \underset{a}{\Rightarrow} h(a, \bar{c}) \leftrightarrow [e](\bar{c}) \sqsubseteq [e](\bar{b}) \text{ [80]} \qquad \text{T3.26}$$

3.9.2.2 Equivalence, Subordination, and Independence

In the following sections, Bolzano considers the two special cases: 'bilateral' and 'unilateral' derivability.

In the first case, he speaks of *equivalence* [*Gleichgültigkeit*].[81] Writing '$\Leftrightarrow^{m,n,s}$' and '$\Leftrightarrow^{m,n}$', for this $m + n + s$-ary relation and for its $m + n$-ary absolutization, respectively, we set

$$\overset{m}{p} \underset{\overset{s}{a}}{\Leftrightarrow} \overset{n}{q} \leftrightarrow \overset{m}{p} \underset{\overset{s}{a}}{\Rightarrow} \overset{n}{q} \wedge \overset{n}{q} \underset{\overset{s}{a}}{\Rightarrow} \overset{m}{p} \qquad \text{SD3.19}$$

$$\overset{m}{p} \Leftrightarrow \overset{n}{q} \leftrightarrow \exists \overset{s}{a}(\overset{m}{p} \underset{\overset{s}{a}}{\Leftrightarrow} \overset{n}{q}) \qquad \text{SD3.20}$$

from which

$$\overset{m}{p} \Leftrightarrow \overset{n}{q} \rightarrow \overset{m}{p} \Rightarrow \overset{n}{q} \wedge \overset{n}{q} \Rightarrow \overset{m}{p} \qquad \text{ST3.62}$$

[78] TS §155, II 89: "34) [. . .]."

[79] TS §155, II 90: "35) If two propositions '*A has x*' and '*B has x*' have the same predicate, which is also the only variable idea in them, then the second is deducible from the first if the subject-idea of the first, *A*, stands in the relation of inclusion to the subject-idea of the second, *B* (§95). If this is not the case, then they do not stand in this relation. For [. . .]."

[80] TS §155, II 90: "36) If two propositions '*X has a*' and '*X has b*' have the same subject, which is also the only variable idea in them, then the second proposition is deducible from the first if the idea *B* (the concretum belonging to *b*) includes idea *A*; failing this, the relation of deducibility does not hold. For [. . .]."

[81] TS §156, II 95: "1) If the relation of deducibility holds reciprocally between propositions *A, B, C, D, . . .* and *M, N, O, . . .* with respect to the same ideas *i, j, . . .* , i.e., if every collection of ideas whose substitution for *i, j. . .* makes all of *A, B, C, D, . . .* true also makes all of *M, N, O, . . .* true and, conversely, when every collection of ideas whose substitution for *i, j, . . .* makes all of *M, N, O, . . .* true also makes all of *A, B, C, D, . . .* true, then I say that propositions *A, B, C, D, . . .* and *M, N, O, . . .* stand in the relation of *equivalence* with respect to the ideas *i, j, ...*" In the *Größenlehre* (p.64), Bolzano also uses the term 'equipollence [*Aequipollenz*]'.

Bolzano emphasizes that equivalence among aggregates of propositions is a quite different matter from equivalence among propositions.[82]

Propositions contained in equivalent aggregates behave analogously with respect to truth and falsity;[83] more precisely,

$$\overset{m}{p} \underset{\overset{s}{a}}{\Leftrightarrow} \overset{n}{q} \rightarrow \forall \overset{s}{c}(W(\overset{m}{p}[\overset{s}{a}/\overset{s}{c}]) \leftrightarrow W(\overset{n}{q}[\overset{s}{a}/\overset{s}{c}])) \qquad \text{ST3.63}$$

$$\overset{m}{p} \Leftrightarrow \overset{n}{q} \rightarrow (W(\overset{m}{p}) \leftrightarrow W(\overset{n}{q})) \qquad \text{ST3.64}$$

Subsequently, Bolzano presents the laws of substitutivity of the equivalence with respect to the relations considered up to this point (compatibility and incompatibility, derivability in both directions, equivalence):[84]

$$\overset{m}{p} \underset{\overset{s}{a}}{\Leftrightarrow} \overset{n}{q} \rightarrow \begin{pmatrix} (\overset{m}{p} \underset{\overset{s}{a}}{\mathrel{\wedge\!\!\wedge}} \overset{u}{r} \leftrightarrow \overset{n}{q} \underset{\overset{s}{a}}{\mathrel{\wedge\!\!\wedge}} \overset{u}{r}) \wedge (\overset{m}{p} \underset{\overset{s}{a}}{\Rightarrow} \overset{u}{r} \leftrightarrow \overset{n}{q} \underset{\overset{s}{a}}{\Rightarrow} \overset{u}{r}) \wedge \\ (\overset{u}{r} \underset{\overset{s}{a}}{\Rightarrow} \overset{m}{p} \leftrightarrow \overset{u}{r} \underset{\overset{s}{a}}{\Rightarrow} \overset{n}{q}) \wedge (\overset{m}{p} \underset{\overset{s}{a}}{\Leftrightarrow} \overset{u}{r} \leftrightarrow \overset{n}{q} \underset{\overset{s}{a}}{\Leftrightarrow} \overset{u}{r}) \end{pmatrix} \qquad \text{ST3.65}$$

Moreover, he also proves for the equivalence, under the usual condition of compatibility, the principle of *additionability*:[85]

$$\overset{m}{p} \underset{\overset{s}{a}}{\Leftrightarrow} \overset{n}{q} \wedge \overset{u}{r} \underset{\overset{s}{a}}{\Leftrightarrow} \overset{v}{s} \wedge (\overset{m}{p} \underset{\overset{s}{a}}{\mathrel{\wedge\!\!\wedge}} \overset{u}{r}) \rightarrow \overset{m}{p}, \overset{u}{r} \underset{\overset{s}{a}}{\Leftrightarrow} \overset{n}{q}, \overset{v}{s} \qquad \text{ST3.66}$$

[82] TS §156, II 95: "2) If propositions $A, B, C, D, \dots$ are together equivalent to propositions $M, N, O, \dots$, also taken together, then according to this definition it is not required that for each individual proposition in $A, B, C, D, \dots$ there should be an equivalent proposition in $M, N, O, \dots$ It is not even necessary that the number of propositions $A, B, C, D, \dots$ should be the same as the number of propositions $M, N, O, \dots$"

[83] TS §156, II 95: "3) If propositions $A, B, C, D, \dots$ are all true, then the equivalent propositions $M, N, O, \dots$ must also be all true. Conversely, if the former are not all true, then the latter are also not all true."

[84] TS §156, II 95: "4) If propositions $A, B, C, D, \dots$ taken together, are equivalent to propositions $A', B', C', D', \dots$, also taken together, with respect to ideas $i, j, \dots$ and if propositions $A, B, C, D, \dots$ stand in any one of the relations so far considered to certain other propositions $M, N, O, \dots$, then the equivalent propositions $A', B', C', D', \dots$ stand in the same relation to $M, N, O, \dots$ with respect to the same ideas. This holds with respect to the relations of compatibility, incompatibility, equivalence, and deducibility, no matter whether $M, N, O, \dots$ are deducible from $A, B, C, D, \dots$, or $A, B, C, D, \dots$ from $M, N, O, \dots$ [...]."

[85] TS §156, II 95: "5) If propositions $A, B, C, D, \dots$ are equivalent to $A', B', C', D', \dots$ and propositions $D, E, F, \dots$ equivalent to $D', E', F', \dots$ with respect to the same ideas $i, j, \dots$, and if propositions $A, B, C, \dots$ are compatible with propositions $D, E, F, \dots$ with respect to the same ideas, then propositions $A, B, C, \dots, D, E, F, \dots$, taken together, are equivalent to propositions $A', B', C', \dots, D', E', F', \dots$ also taken together, and again with respect to the same ideas. For [...]."

Propositions equivalent to given other propositions are also equivalent to the union of these other propositions with their consequences:[86]

$$\overset{m}{p} \underset{\overset{s}{a}}{\Leftrightarrow} \overset{n}{q} \wedge \overset{m}{p} \underset{\overset{s}{a}}{\Rightarrow} \overset{u}{r} \rightarrow \overset{m}{p}, \overset{u}{r} \underset{\overset{s}{a}}{\Leftrightarrow} \overset{n}{q} \qquad \text{ST3.67}$$

In general, however, consequences of equivalent premises are not themselves equivalent:[87]

$$\overset{m}{p} \underset{\overset{s}{a}}{\Leftrightarrow} \overset{n}{q} \wedge \overset{m}{p} \underset{\overset{s}{a}}{\Rightarrow} \overset{u}{r} \wedge \overset{n}{q} \underset{\overset{s}{a}}{\Rightarrow} \overset{v}{s} \nrightarrow \overset{u}{r} \underset{\overset{s}{a}}{\Leftrightarrow} \overset{v}{s} \qquad \text{ST3.68}$$

nor from the equivalence of two aggregates may the equivalence of subaggregates of them be inferred, not even when this holds for a part of them:[88]

$$\overset{m}{p}, \overset{u}{r} \underset{\overset{s}{a}}{\Leftrightarrow} \overset{n}{q}, \overset{v}{s} \wedge \overset{m}{p} \underset{\overset{s}{a}}{\Leftrightarrow} \overset{n}{q} \nrightarrow \overset{u}{r} \underset{\overset{s}{a}}{\Leftrightarrow} \overset{v}{s} \qquad \text{ST3.69}$$

The negations of two equivalent (but not always true) propositions are in turn equivalent:[89]

$$p \underset{\overset{s}{a}}{\Leftrightarrow} q \wedge \neg \mathbb{W}(p; \overset{s}{a}) \wedge \neg \mathbb{W}(q; \overset{s}{a}) \rightarrow \boxed{\neg} p \underset{\overset{s}{a}}{\Leftrightarrow} \boxed{\neg} q \qquad \text{ST3.70}$$

This, however, does not extend to equivalent aggregates of propositions:[90]

$$\overset{m}{p} \underset{\overset{s}{a}}{\Leftrightarrow} \overset{n}{q} \nrightarrow\Leftrightarrow \boxed{\neg} \overset{m}{p} \underset{\overset{s}{a}}{\Leftrightarrow} \boxed{\neg} \overset{n}{q} \qquad \text{ST3.71}$$

In the second case, Bolzano speaks of *subordination* [*Unterordnung*] or *unilateral derivability* [*einseitige Ableitbarkeit*].[91] Writing '$\overset{m,n,s}{\Rightarrow}$' for this $m + n + s$-ary relation, we set

[86] TS §156, II 96: "6) If propositions $A, B, C, \ldots$ are equivalent to propositions $A', B', C', \ldots$ with respect to ideas $i, j, \ldots$ and if propositions $A, B, C, \ldots$ are deducible from propositions $M, N, O, \ldots$ with respect to the same ideas, then $A, B, C, \ldots, M, N, O, \ldots$ taken together stand in the relation of equivalence to propositions $A', B', C', \ldots$ with respect to the same ideas. For [...]."

[87] TS §156, II 96: "7) Merely from the fact that propositions $A, B, C, \ldots$ are equivalent to propositions $A', B', C', \ldots$ with respect to ideas $i, j, \ldots$ it does not follow that propositions $M, N, O, \ldots$ and $M', N', O', \ldots$, which are deducible one set from the first, the other from the second of these two collections, also with respect to $i, j, \ldots$ are also equivalent with respect to the same ideas. For [...]."

[88] TS §156, II 97: "8) Assume that the sum of propositions $A, B, C, D, \ldots$ is equivalent to the sum of propositions $A', B', C', D', \ldots$ with respect to ideas $i, j, \ldots$ and that a part of the first sum, e.g., $A, B, \ldots$ is itself equivalent to a part of the second sum, e.g., $A', B', \ldots$ with respect to the same ideas. It does then not necessarily follow that the remaining parts $C, D, \ldots$ and $C', D', \ldots$ are equivalent with respect to the same ideas. For [...]."

[89] TS §156, II 97: "9) If two propositions A and A' are equivalent with respect to ideas $i, j, \ldots$ and if neither of them is formally true, then propositions Neg.A and Neg.A' are also equivalent with respect to the same ideas. For [...]."

[90] TS §156, II 97: "10) The same does not hold for whole collections of propositions. [...]."

[91] TS §157, II 101: "1) Let us assume that a relation of deducibility between propositions $A, B, C, D, \ldots$ and $M, N, O, \ldots$ is not reciprocal [...] but holds only in one direction [...] In such a case, we call the relation [...] a relation of *subordination*."

$$\overset{m}{p} \underset{\overset{s}{a}}{\dot{\Rightarrow}} \overset{n}{q} \leftrightarrow \overset{m}{p} \underset{\overset{s}{a}}{\Rightarrow} \overset{n}{q} \land \neg(\overset{n}{q} \underset{\overset{s}{a}}{\Rightarrow} \overset{m}{p}) \qquad \text{SD3.21}$$

$$\overset{m}{p} \dot{\Rightarrow} \overset{n}{q} \leftrightarrow \exists \overset{s}{a}(\overset{m}{p} \underset{\overset{s}{a}}{\dot{\Rightarrow}} \overset{n}{q}) \qquad \text{SD3.22}$$

Subordination always implies the existence of propositions that are only compatible with the premises:[92]

$$\overset{m}{p} \underset{\overset{s}{a}}{\dot{\Rightarrow}} \overset{n}{q} \rightarrow \exists r(\overset{m}{p} \underset{\overset{s}{a}}{\mathrel{\wedge\!\!\wedge}} r) \land \neg(\overset{n}{q} \underset{\overset{s}{a}}{\mathrel{\wedge\!\!\wedge}} r)) \qquad \text{ST3.72}$$

The relation of subordination is transitive:[93]

$$\overset{m}{p} \underset{\overset{s}{a}}{\dot{\Rightarrow}} \overset{n}{q} \land \overset{n}{q} \underset{\overset{s}{a}}{\dot{\Rightarrow}} \overset{u}{r} \rightarrow \overset{m}{p} \underset{\overset{s}{a}}{\dot{\Rightarrow}} \overset{u}{r} \qquad \text{ST3.73}$$

As far as atomic propositions having the same predicate or the same subject are concerned, Bolzano establishes the theorems corresponding to those already proved for derivability in general:[94]

$$h(a, \bar{c}) \underset{\bar{c}}{\dot{\Rightarrow}} h(b, \bar{c}) \leftrightarrow b \mathrel{\dot{\sqsubseteq}} a \qquad \text{T3.27}$$

$$h(a, \bar{b}) \underset{a}{\dot{\Rightarrow}} h(a, \bar{c}) \leftrightarrow [e](\bar{c}) \mathrel{\dot{\sqsubseteq}} [e](\bar{b}) \qquad \text{T3.28}$$

He also considers the relation of *concatenation* [*Verkettung*] or *entanglement* [*Verschlungenheit*] or *independence* [*Unabhängigkeit*] which subsists when compatibility is accompanied by no derivability.[95] Writing '$\mathbin{\top\!\!\top}^{m,n,s}$' for this $m + n + s$-ary relation, we set

[92] TS §157, II 101-2: "2) If propositions $M, N, O, \ldots$ are unilaterally deducible from propositions $A, B, C, \ldots$ with respect to ideas $i, j, \ldots$, then there are always certain propositions which are compatible with $M, N, O, \ldots$ without being compatible with $A, B, C, \ldots$ with respect to the same ideas $i, j, \ldots$. Since [...]."

[93] TS §157, II 102: "3) If propositions $M, N, O, \ldots$ are unilaterally deducible from propositions $A, B, C, D, \ldots$, and propositions $R, S, T, \ldots$ are also unilaterally deducible from propositions $M, N, O, \ldots$ with respect to the same ideas $i, j, \ldots$, then propositions $R, S, T, \ldots$ are unilaterally deducible from propositions $A, B, C, D, \ldots$ with respect to the same ideas. [...]."

[94] TS §157, II 102: "If two propositions '*A has x*' and '*B has x*' have one and the same predicate, which is to be considered variable, then the second is unilaterally deducible from the first if, and only if, its subject-idea, *B*, is subordinate to the subject-idea of the first, *A*. For, [...] 5) If two propositions '*X has a*' and '*X has b*' have the same subject, which is to be the only variable part of them, then the second is unilaterally deducible from the first if, and only if, the idea *A* is subordinate to the idea *B*. [...]."

[95] TS §158, II 102-3: "1) It remains to consider the case where there is a relation of compatibility between propositions $A, B, C, \ldots$ and $M, N, O, \ldots$, but in such a way that neither propositions $M, N, O, \ldots$ are deducible from $A, B, C, \ldots$ nor, conversely, $A, B, C, \ldots$ from $M, N, O, \ldots$ with respect to the same ideas $i, j, \ldots$. In other words, where there are ideas whose substitution $i, j, \ldots$ makes all of $A, B, C, \ldots$ and $M, N, O, \ldots$ true, but others which make only $A, B, C, \ldots$ but not all of $M, N, O, \ldots$ true, and still others which make only $M, N, O, \ldots$ but not all of $A, B, C, \ldots$ true. This relation bears a great similarity to the relation which I have called, in the case of ideas, *overlapping* or *linking* (§98). Hence it is reasonable to give it the same name in the case of propositions, unless the designation *independence* is preferred."

$$\overset{m}{p} \underset{\overset{s}{a}}{\top\!\!\top} \overset{n}{q} \leftrightarrow \overset{m}{p} \underset{\overset{s}{a}}{\wedge\!\!\!\wedge} \overset{n}{q} \wedge \neg(\overset{m}{p} \underset{\overset{s}{a}}{\Rightarrow} \overset{n}{q}) \wedge \neg(\overset{n}{q} \underset{\overset{s}{a}}{\Rightarrow} \overset{m}{p}) \qquad \text{SD3.23}$$

The negations of two concatenated propositions are either incompatible or in turn concatenated:[96]

$$p \underset{\overset{s}{a}}{\top\!\!\top} q \rightarrow (\boxed{\neg}\, p \underset{\overset{s}{a}}{\wedge\!\!\!\wedge} \boxed{\neg}\, q \rightarrow \boxed{\neg}\, p \underset{\overset{s}{a}}{\top\!\!\top} \boxed{\neg}\, q) \qquad \text{ST3.74}$$

It is worth emphasizing that Bolzano is always careful about the substitutivity of $\Leftrightarrow$; every time he introduces a new relation, he carefully establishes the substitutivity of $\Leftrightarrow$ with respect to it. Thus, he establishes that equivalence is substitutive also with respect to concatenation:[97]

$$\overset{m}{p} \underset{\overset{s}{a}}{\Leftrightarrow} \overset{n}{q} \rightarrow (\overset{m}{p} \underset{\overset{s}{a}}{\top\!\!\top} r \leftrightarrow \overset{n}{q} \underset{\overset{s}{a}}{\top\!\!\top} r) \qquad \text{ST3.75}$$

The whole of § 159, which takes up pp.146–62 of WL II, is devoted to the study of various particular types of incompatibility: *exclusion* [*Ausschließung*], *reciprocal exclusion* [*wechselseitige Ausschließung*], *contradictoriness* [*Widerspruch*], and *contrariety* [*Widerstreit*].

3.9.2.3 Exclusion and Reciprocal Exclusion

The first particular type of incompatibility is constituted by *exclusion*. This is the relation that subsists, with respect to the $a_1, \ldots, a_s$, between given $p_1, \ldots, p_m$ (*the excluding agent* [*die ausschließenden*]) and certain $q_1, \ldots, q_n$ (the *excluded* agent[die *ausgeschloßenen*]) if the first propositions are compatible with respect to those ideas, but each substitution that makes all of them true makes all the second propositions false.[98] Denoting this $m + n + s$-ary relation by '$\perp^{m,n,s}$', we then set

$$\overset{m}{p} \underset{\overset{s}{a}}{\perp} \overset{n}{q} \leftrightarrow \underset{\overset{s}{a}}{\wedge\!\!\!\wedge}(\overset{n}{p}) \wedge \forall \overset{s}{c}(W(\overset{m}{p}[\overset{s}{a}/\overset{s}{c}]) \rightarrow F(\overset{n}{q}[\overset{s}{a}/\overset{s}{c}])) \qquad \text{SD3.24}$$

$$\overset{m}{p} \perp \overset{n}{q} \leftrightarrow \exists \overset{s}{a}(\overset{m}{p} \underset{\overset{s}{a}}{\perp} \overset{n}{q}) \qquad \text{SD3.25}$$

[96] TS §158, II 104: "6) If a pair of propositions *A* and *B* are independent, then their negations, the propositions Neg.*A* and Neg.*B*, are either independent or incompatible, all with respect to the same ideas. For [...]."

[97] TS §158, II 104: "7) Propositions which are equivalent to independent propositions are themselves independent, all with respect to the same ideas. [...]."

[98] TS II §159, II 105: "1) [...] I shall say that one or several propositions *M*, *N*, *O*,... are *excluded* by certain others *A*, *B*, *C*,... with respect to variable ideas *i*, *j*,... if every collection of ideas whose substitution for *i*, *j*,... makes all of *A*, *B*, *C*,... true, makes all of *M*, *N*, *O*,... false."

It should be noted that the Bolzanian definition does not explicitly mention the condition $\underset{\overset{s}{a}}{\bigwedge\!\!\!\bigwedge}(\overset{m}{p})$, but its tacit assumption is confirmed by the immediately following theorem:[99]

$$\overset{m}{p} \underset{\overset{s}{a}}{\perp} \overset{n}{q} \leftrightarrow \overset{m}{p} \underset{\overset{s}{a}}{\Rightarrow} \boxminus \overset{n}{q} \qquad \text{ST3.76}$$

i.e.

$$\overset{m}{p} \underset{\overset{s}{a}}{\perp} \overset{n}{q} \leftrightarrow \underset{\overset{s}{a}}{\bigwedge\!\!\!\bigwedge}(\overset{n}{p}) \wedge \forall \overset{s}{c}(W(\overset{m}{p}[\overset{s}{a}/\overset{s}{c}]) \rightarrow W(\boxminus \overset{n}{q}[\overset{s}{a}/\overset{s}{c}])$$

If $m, n > 1$, the exclusion, with respect to the $a_1, \ldots, a_s$, of certain $p_1, \ldots, p_m$ versus certain $q_1, \ldots, q_n$, is not necessarily symmetric and Bolzano explicitly distinguishes as a special case of the incompatibility that of the *reciprocal* [*wechselseitig*] incompatibility.[100] Writing '$\perp\!\!\!\perp^{m,n,s}$' for this $m + n + s$-ary relation, we set

$$\overset{m}{p} \underset{\overset{s}{a}}{\perp\!\!\!\perp} \overset{n}{q} \leftrightarrow \overset{m}{p} \underset{\overset{s}{a}}{\perp} \overset{n}{q} \wedge \overset{n}{q} \underset{\overset{s}{a}}{\perp} \overset{m}{p} \qquad \text{SD3.26}$$

$$\overset{m}{p} \perp\!\!\!\perp \overset{n}{q} \leftrightarrow \exists \overset{s}{a}(\overset{m}{p} \underset{\overset{s}{a}}{\perp\!\!\!\perp} \overset{n}{q}) \qquad \text{SD3.27}$$

The fact that the preceding statement is an equivalence and not simply an implication, as Bolzano's verbal formulation seems to suggest, is confirmed by the following statement:[101]

$$\overset{m}{p} \underset{\overset{s}{a}}{\perp\!\!\!\perp} \overset{n}{q} \leftrightarrow \overset{m}{p} \underset{\overset{s}{a}}{\Rightarrow} \boxminus \overset{n}{q} \wedge \overset{n}{q} \underset{\overset{s}{a}}{\Rightarrow} \boxminus \overset{m}{p} \qquad \text{ST3.77}$$

3.9.2.4 Contrariness and Contradictoriness

The two limiting cases of the reciprocal exclusion are those in which both relations of derivability are relations of equivalence, respectively of subordination. In the first

[99] TS §159, II 105-6: "2) If propositions $A, B, C, \ldots$ exclude certain others $M, N, O, \ldots$ with respect to certain ideas $i, j, \ldots$, then the negations of the latter, i.e., the propositions Neg.M,Neg.N,Neg.$O, \ldots$ must be deducible from the collection of propositions $A, B, C, \ldots$ with respect to the same ideas. For [...]."

[100] TS §159, II 106: "3) It can also be the case that the relation of exclusion holds mutually and with respect to the same ideas $i, j, \ldots$ [...] between propositions $A, B, C, \ldots$ and $M, N, O, \ldots$ [...] We can properly call this relation between propositions $A, B, C, \ldots$ and $M, N, O, \ldots$ a relation of *mutual exclusion.*"

[101] TS §159, II 106: "4) When we say that propositions $A, B, C, \ldots$ and $M, N, O, \ldots$ are mutually exclusive with respect to ideas $i, j, \ldots$, then all that is claimed is that as soon as all of $A, B, C, \ldots$ are true, all of $M, N, O, \ldots$ are false, and as soon as all of $M, N, O, \ldots$ are true, all of $A, B, C, \ldots$ are false; in other words that propositions Neg.M,Neg.N, Neg.$O, \ldots$ are deducible from propositions $A, B, C, \ldots$, and Neg.A,Neg.B,Neg.$C, \ldots$ are deducible from propositions $M, N, O, \ldots$."

case, Bolzano speaks of *contradictoriness*, in the second of *contrariness*.[102] Writing '$\#^{m,n,s}$' and respectively '$\parallel^{m,n,s}$' for these two $m + n + s$-ary relations we then have:[103]

$$\overset{m}{p} \underset{\overset{s}{a}}{\#} \overset{n}{q} \leftrightarrow \overset{m}{p} \underset{\overset{s}{a}}{\dot{\Leftrightarrow}} \boxed{\neg}\, \overset{n}{q} \wedge \overset{n}{q} \underset{\overset{s}{a}}{\dot{\Leftrightarrow}} \boxed{\neg}\, \overset{m}{p} \qquad \text{SD3.28}$$

$$\overset{m}{p} \# \overset{n}{q} \leftrightarrow \exists \overset{s}{a}(\overset{m}{p} \underset{\overset{s}{a}}{\#} \overset{n}{q} \qquad \text{SD3.29}$$

$$\overset{m}{p} \underset{\overset{s}{a}}{\parallel} \overset{n}{q} \leftrightarrow \overset{m}{p} \underset{\overset{s}{a}}{\dot{\Rightarrow}} \boxed{\neg}\, \overset{n}{q} \wedge \overset{n}{q} \underset{\overset{s}{a}}{\dot{\Rightarrow}} \boxed{\neg}\, \overset{m}{p} \qquad \text{SD3.30}$$

$$\overset{m}{p} \parallel \overset{n}{q} \leftrightarrow \exists \overset{s}{a}(\overset{m}{p} \underset{\overset{s}{a}}{\parallel} \overset{n}{q}) \qquad \text{SD3.31}$$

i.e.

$$\overset{m}{p} \underset{\overset{s}{a}}{\#} \overset{n}{q} \leftrightarrow \overset{m}{p} \underset{\overset{s}{a}}{\Rightarrow} \boxed{\neg}\, \overset{n}{q} \wedge \boxed{\neg}\, \overset{n}{q} \underset{\overset{s}{a}}{\Rightarrow} \overset{m}{p} \wedge \overset{n}{q} \underset{\overset{s}{a}}{\Rightarrow} \boxed{\neg}\, \overset{m}{p} \wedge \boxed{\neg}\, \overset{m}{p} \underset{\overset{s}{a}}{\Rightarrow} \overset{n}{q} \qquad \text{ST3.78}$$

$$\overset{m}{p} \underset{\overset{s}{a}}{\parallel} \overset{n}{q} \leftrightarrow \overset{m}{p} \underset{\overset{s}{a}}{\Rightarrow} \boxed{\neg}\, \overset{n}{q} \wedge \neg(\boxed{\neg}\, \overset{n}{q} \underset{\overset{s}{a}}{\Rightarrow} \overset{m}{p}) \wedge \overset{n}{q} \underset{\overset{s}{a}}{\Rightarrow} \boxed{\neg}\, \overset{m}{p} \wedge \neg(\boxed{\neg}\, \overset{m}{p} \underset{\overset{s}{a}}{\Rightarrow} \overset{n}{q}) \qquad \text{ST3.79}$$

Always true or always false propositions cannot occur in a relation of *contradictoriness*:[104]

$$\overset{m}{p} \underset{\overset{s}{a}}{\#} \overset{n}{q} \rightarrow \bigwedge_{i \leq m}(\neg \underset{\overset{s}{a}}{\mathbb{W}}(p_i) \wedge \neg \underset{\overset{s}{a}}{\mathbb{F}}(p_i)) \wedge \bigwedge_{j \leq n}(\neg \underset{\overset{s}{a}}{\mathbb{W}}(q_j) \wedge \neg \underset{\overset{s}{a}}{\mathbb{F}}(q_j)) \qquad \text{ST3.80}$$

Every proposition that is neither always true nor always false is always in contradiction with its negation:[105]

$$\neg \underset{\overset{s}{a}}{\mathbb{W}}(p) \wedge \neg \underset{\overset{s}{a}}{\mathbb{F}}(p) \rightarrow p \underset{\overset{s}{a}}{\#} \boxed{\neg}\, p \qquad \text{ST3.81}$$

[102] TS §159, II 106: "4) [...] let us assume that the relation of deducibility between propositions *A, B, C,...* and Neg.*M*,Neg.*N*,Neg.*O*,... and between the propositions *M, N, O,...* and Neg.*A*,Neg.*B*,Neg.*C*,... is mutual, i.e., that every collection of ideas whose substitution for *i, j,...* makes all of *A, B, C,...* true or false will also make all of *M, N, O,...* false or true respectively. In this case, we say that propositions *A, B, C,...* on the one hand, and *M, N, O,...* on the other stand in the relation of *contradiction*, or that they *contradict* each other. We also say that propositions *A, B, C,...* taken together are the contradictory of *M, N, O,...*, and that the latter are the contradictory of the former. But if the relation of deducibility holds only in one direction, then we say that propositions *A, B, C,...* on the one hand and *M, N, O,...* on the other, merely stand in the relation of *contrariety*."

[103] WL §159, II 108: "5) It follows directly from the definition of no. 4 that if propositions *A, B, C,...* stand in the relation of contradiction to propositions *M, N, O,...*, then propositions *A, B, C,...* must be equivalent to Neg.*M*,Neg.*N*,Neg.*O*,..., and propositions *M, N, O,...* equivalent to Neg.*A*, Neg.*B*,Neg.*C*,..., all with respect to the same variable ideas. Similarly, if propositions *A, B, C,...* are all true or all false, then propositions *M, N, O,...* are all false or all true."

[104] TS §159, II 108: "6) Among propositions which stand in the relation of contradiction to each other there can be no proposition which is formally true or formally false with respect to the same variable ideas. For [...]."

[105] TS §159, II 108: "7) Any arbitrary proposition *A* and its negation Neg.*A* stand in the relation of contradiction to each other with respect to any ideas whatsoever, provided only that is neither formally true nor formally false. For [...]."

The negations of contradictory are, in turn, contradictory:[106]

$$\overset{m}{p} \mathrel{\underset{\substack{s\\a}}{+\!\!\!\!+}} \overset{n}{q} \rightarrow \boxed{\neg}\overset{m}{p} \mathrel{\underset{\substack{s\\a}}{+\!\!\!\!+}} \boxed{\neg}\,\overset{n}{q} \qquad \text{ST3.82}$$

Once again, the equivalence is substitutive with respect to contradictoriness:[107]

$$\overset{m}{p} \mathrel{\underset{\substack{s\\a}}{\Leftrightarrow}} \overset{n}{q} \wedge \overset{u}{r} \mathrel{\underset{\substack{s\\a}}{\Leftrightarrow}} \overset{v}{s} \rightarrow (\overset{m}{p} \mathrel{\underset{\substack{s\\a}}{+\!\!\!\!+}} \overset{u}{r} \leftrightarrow \overset{n}{q} \mathrel{\underset{\substack{s\\a}}{+\!\!\!\!+}} \overset{v}{s}) \qquad \text{ST3.83}$$

Contradictories of contradictories are equivalent:[108]

$$\overset{m}{p} \mathrel{\underset{\substack{s\\a}}{+\!\!\!\!+}} \overset{n}{q} \wedge \overset{n}{q} \mathrel{\underset{\substack{s\\a}}{+\!\!\!\!+}} \overset{u}{r} \rightarrow \overset{m}{p} \mathrel{\underset{\substack{s\\a}}{\Leftrightarrow}} \overset{u}{r} \qquad \text{ST3.84}$$

For the additionability of contradictoriness, Bolzano proves[109]

$$\overset{m}{p} \mathrel{\underset{\substack{s\\a}}{+\!\!\!\!+}} \overset{n}{q} \wedge \overset{u}{r} \mathrel{\underset{\substack{s\\a}}{+\!\!\!\!+}} \overset{v}{s} \wedge \overset{m}{p} \mathrel{\underset{\substack{s\\a}}{\wedge\!\!\!\wedge}} \overset{u}{r} \wedge \boxed{\neg}\,\overset{m}{p} \mathrel{\underset{\substack{s\\a}}{\wedge\!\!\!\wedge}} \boxed{\neg}\,\overset{u}{r} \rightarrow \overset{m}{p}, \overset{u}{r} \mathrel{\underset{\substack{s\\a}}{+\!\!\!\!+}} \overset{n}{q}, \overset{v}{s} \qquad \text{ST3.85}$$

emphasizing, however, the indispensability of the premises[110] $\overset{m}{p} \mathrel{\underset{\substack{s\\a}}{\wedge\!\!\!\wedge}} \overset{u}{r} \wedge \boxed{\neg}\,\overset{m}{p} \mathrel{\underset{\substack{s\\a}}{\wedge\!\!\!\wedge}} \boxed{\neg}\,s$ and also showing that this is not convertible:[111]

$$\overset{m}{p}, \overset{u}{r} \mathrel{\underset{\substack{s\\a}}{+\!\!\!\!+}} \overset{n}{q}, \overset{v}{s} \wedge \overset{m}{p} \mathrel{\underset{\substack{s\\a}}{+\!\!\!\!+}} \overset{n}{q} \nrightarrow \overset{u}{r} \mathrel{\underset{\substack{s\\a}}{+\!\!\!\!+}} \overset{v}{s} \qquad \text{ST3.86}$$

106 TS §159, II 108: "8) If propositions *A, B, C,...* stand in the relation of contradiction to propositions *M, N, O,...*, then propositions Neg.*A*,Neg.*B*,Neg.*C*,... stand in the same relation to propositions Neg.*M*, Neg.*N*,Neg.*O*,... all with respect to the same ideas."

107 TS §159, II 108: "9) If some propositions stand in a relation of contradiction to each other, then all propositions equivalent to them stand in the same relation to each other, with respect to the same ideas."

108 TS §159, II 108–9: "10) If propositions *A, B, C, D,...* are contradictory to *M, N, O,...* and *M, N, O,...* contradictory to *R, S, T,...*, all with respect to the same ideas, then propositions *A, B, C,...* and propositions *R, S, T,...* are equivalent to each other with respect to the same ideas. For the same ideas which make all of *A, B, C,...* true or false make all of *M, N, O,...* false or true and consequently all of *R, S, T,...* true or false. [...]."

109 TS §159, II 109: "11) Let propositions *A, B, C,...* be contradictory to propositions *M, N, O,...* and propositions *E, F, G,...* contradictory to *P, Q, R,...* with respect to the same ideas; in addition, let propositions *A, B, C, D,..., E, F, G,...*, as well as Neg.*A*,Neg.*B*,Neg.*C*,...,Neg.*E*,Neg.*F*,Neg.*G*,... be compatible with each other with respect to the same ideas. In this case, the totality of propositions *A, B, C,..., E, F, G,...* stands in the relation of contradiction to the totality of propositions *M, N, O,..., P, Q, R,...*, again with respect to the same ideas. For [...]."

110 TS §159, II 109: "12) The twofold condition which was just made, namely, that propositions *A, B, C, D,..., E, F, G,...*, as well as their negations, Neg.*A*,Neg.*B*,Neg.*C*,...,Neg.*E*, Neg.*F*, Neg.*G*,... should be compatible with each other, is by no means superfluous. For [...]."

111 TS §159, II 109–10: "13) Given that the totality of propositions *A, B, C,..., E, F, G,...* stands in relation of contradiction to the totality of propositions *M, N, O,..., P, Q, R,...*, and that part of the first, *A, B, C,...* stands in the relation of contradiction to part of the second, *M, N, O,...* with respect to the same ideas, it does not follow that the remaining propositions, *E, F, G,...*, also stand in the relation of contradiction to the remainder, *P, Q, R,...*. For [...]."

If more propositions are in contradiction with a single one, then they are all equivalent:[112]

$$\overset{m}{p} \underset{\substack{s\\a}}{+\!\!\!\!+} q \rightarrow \Big(\bigwedge_{i\neq j\leq m} p_i \underset{\substack{s\\a}}{\Leftrightarrow} p_j\Big) \qquad \text{ST3.87}$$

If propositions that are not all equivalent to each other are in contradiction with other propositions, then there must be at least one proposition among the former that is compatible with one of the latter:[113]

$$\overset{m}{p} \underset{\substack{s\\a}}{+\!\!\!\!+} \overset{n}{q} \wedge \neg\Big(\bigwedge_{i\neq j\leq m} p_i \underset{\substack{s\\a}}{\Leftrightarrow} p_j\Big) \rightarrow \Big(\bigvee_{i\leq m, j\leq n} p_i \underset{\substack{s\\a}}{\curlywedge} q_j\Big) \qquad \text{ST3.88}$$

Two equivalent, but not always true, propositions are each in contradiction with the negation of the other[114]: "16) If two propositions *A* and *M* are not formally true with respect to certain ideas *i*, *j*, ... and are equivalent to each other with respect to the same ideas, then each of them stands in the relation of contradiction to the negation of the other, *A* to Neg.*M*, and *M* to Neg.*A*, with respect to the same ideas. For [...]."

$$\neg \underset{\substack{s\\a}}{\mathbb{W}}(p) \wedge \neg \underset{\substack{s\\a}}{\mathbb{W}}(q) \wedge p \underset{\substack{s\\a}}{\Leftrightarrow} q \rightarrow p \underset{\substack{s\\a}}{+\!\!\!\!+} \boxed{\neg} q \wedge q \underset{\substack{s\\a}}{+\!\!\!\!+} \boxed{\neg} p \qquad \text{ST3.89}$$

This, however, does not hold when more then two propositions are involved:[115]

$$\neg \underset{\substack{s\\a}}{\mathbb{W}}(\overset{m}{p}) \wedge \neg \underset{\substack{s\\a}}{\mathbb{W}}(\overset{n}{q}) \wedge \overset{m}{p} \underset{\substack{s\\a}}{\Leftrightarrow} \overset{n}{q} \not\rightarrow \overset{m}{p} \underset{\substack{s\\a}}{+\!\!\!\!+} \boxed{\neg} \overset{n}{q} \qquad \text{ST3.90}$$

Analogously,[116]

$$p \underset{\substack{s\\a}}{\Rightarrow} q \wedge \boxed{\neg} p \underset{\substack{s\\a}}{\Rightarrow} \boxed{\neg} q \rightarrow p \underset{\substack{s\\a}}{+\!\!\!\!+} \boxed{\neg} q \wedge q \underset{\substack{s\\a}}{+\!\!\!\!+} \boxed{\neg} p \qquad \text{ST3.91}$$

[112] TS §159, II 110: "14) [...] several propositions *A*, *B*, *C*, *D*, ... can stand in the relation of contradiction to an individual proposition only if every substitution for ideas *i*, *j*, ... will make either all of them true or all of them false, i.e., if they are equivalent to each other. [...]."

[113] TS §159, II 110: "15) If several propositions *A*, *B*, *C*, ..., which are not equivalent to each other, stand in the relation of contradiction to several others *M*, *N*, *O*, ..., then there must be some among them, as well as among *M*, *N*, *O*, ..., which are compatible with each other, all with respect to the same variable ideas. For [...]."

[114] TS §159, II 110

[115] TS §159, II 111: "17) But if a whole collection of propositions *A*, *B*, *C*, ... is equivalent to a collection of others *M*, *N*, *O*, ... then we cannot conclude that a relation of contradiction holds between propositions *A*, *B*, *C*, ... on the one hand and Neg.*M*,Neg.*N*,Neg.*O*, ... on the other. For [...]."

[116] TS §159, II 111: "18) If an individual proposition *M* is deducible from the individual proposition *A* with respect to certain ideas, and if the proposition Neg.*M* is deducible from the individual proposition Neg.*A* with respect to the same ideas, then proposition *A* contradicts Neg.*M*, and Neg.*A* contradicts *M* with respect to the same ideas. For [...]."

However, when no more than two propositions are involved,[117]

$$\overset{m}{p} \underset{\substack{s\\a}}{\Rightarrow} \overset{n}{q} \wedge \boxminus \overset{m}{p} \underset{\substack{s\\a}}{\Rightarrow} \boxminus \overset{n}{q} \nrightarrow \overset{m}{p} \underset{\substack{s\\a}}{\mathrel{\text{⫲}}} \boxminus \overset{n}{q} \wedge \overset{n}{q} \underset{\substack{s\\a}}{\mathrel{\text{⫲}}} \boxminus \overset{m}{p} \qquad \text{ST3.92}$$

Unlike what happens when a single proposition contradicts one or more propositions, from the contradictoriness of two aggregates of propositions, it cannot be inferred that those of one of them are all true and those of the other are all false;[118] i.e.

$$p \mathrel{\text{⫲}} \overset{n}{q} \rightarrow (W(p) \wedge F(\overset{n}{q})) \vee (F(p) \wedge W(\overset{n}{q})) \qquad \text{ST3.93}$$

$$\overset{m}{p} \mathrel{\text{⫲}} \overset{n}{q} \nrightarrow (W(\overset{m}{p}) \wedge F(\overset{n}{q})) \vee (F(\overset{m}{p}) \wedge W(\overset{n}{q})) \qquad \text{ST3.94}$$

The substitutivity of equivalence also holds with respect to contrariness:[119]

$$\overset{m}{p} \underset{\substack{s\\a}}{\Leftrightarrow} \overset{n}{q} \wedge \overset{u}{r} \underset{\substack{s\\a}}{\Leftrightarrow} \overset{v}{s} \rightarrow \overset{m}{p} \underset{\substack{s\\a}}{\parallel} \overset{u}{r} \leftrightarrow \overset{n}{q} \underset{\substack{s\\a}}{\parallel} \overset{v}{s} \qquad \text{ST3.95}$$

Contrary propositions may all be false:[120]

$$\overset{m}{p} \parallel \overset{n}{q} \nrightarrow E(\overset{m}{p}) \vee E(\overset{n}{q}) \qquad \text{ST3.96}$$

A proposition that is not trivially excluded from certain other propositions is in contradiction with some proposition derivable from them:[121]

$$\overset{m}{p} \underset{\substack{s\\a}}{\perp} q \wedge \neg \underset{\substack{s\\a}}{\mathbb{F}}(q) \rightarrow \exists r(\overset{m}{p} \underset{\substack{s\\a}}{\Rightarrow} r \wedge r \underset{\substack{s\\a}}{\mathrel{\text{⫲}}} q) \qquad \text{ST3.97}$$

As a generalization of the contraposition principle, we have that from a proposition that contradicts non-trivial consequences of a given proposition, the negation

[117] TS §159, II 111: "19) But the same does not hold for several propositions, [...]."

[118] TS §159, II 111–12: "20) [...] if a whole collection of propositions *A, B, C,*... contradict an individual proposition *M*, then either all of *A, B, C,*... sind wahr, wenn nämlich *M* falsch ist; oder *A, B, C,*... are true (if *M* is false) or all of are false (if *M* is true). But if several propositions *A, B, C,*... contradict several others *M, N, O,*..., we may not conclude that the one collection consists only of true propositions, and the other only of false ones; rather, it may be the case that both collections contain some true, and some false, propositions."

[119] TS §159, II 112: "21) Propositions which are equivalent to contrary propositions are themselves contrary [...]."

[120] TS §159, II 112: "22) Of a pair of contraries, whether they are individual propositions, such as *A* and *M*, or collections of propositions, such as *A, B, C,*... and *M, N, O,*..., all may be false."

[121] TS §159, II 122: "23) From any individual proposition , as well as from any collection of propositions *A, B, C,*... which are incompatible with an individual proposition *M*, we can deduce propositions which are contradictory to the latter, provided only that is not formally false. (All with respect to the same ideas.) For [...]."

of this given proposition is derivable:[122]

$$p \underset{\substack{s\\a}}{\Rightarrow} \overset{n}{q} \wedge \neg \underset{\substack{s\\a}}{\mathbb{W}}(\overset{n}{q}) \wedge r \underset{\substack{s\\a}}{\boldsymbol{+}} \overset{n}{q} \rightarrow r \underset{\substack{s\\a}}{\Rightarrow} \boxed{\neg} p \qquad \text{ST3.98}$$

Two contradictory propositions cannot both be incompatible with one and the same (not always false) proposition:[123]

$$p \underset{\substack{s\\a}}{\boldsymbol{+}} q \wedge \neg \underset{\substack{s\\a}}{\mathbb{F}}(r) \rightarrow p \underset{\substack{s\\a}}{/\!\!\!\wedge\!\!\wedge} r \vee q \underset{\substack{s\\a}}{/\!\!\!\wedge\!\!\wedge} r \qquad \text{ST3.99}$$

However, this does not hold if the contradictoriness subsists between two aggregates of propositions.

Propositions contradicting other propositions are not derivable from propositions compatible with those they contradict; moreover, if they are not derivable from certain propositions, then these latter are compatible with at least one of those they contradict:[124]

$$\overset{m}{p} \underset{\substack{s\\a}}{+\!\!\!\!+} \overset{n}{q} \rightarrow (\overset{m}{p} \underset{\substack{s\\a}}{/\!\!\!\wedge\!\!\wedge} \overset{u}{r} \rightarrow \neg(\overset{u}{r} \underset{\substack{s\\a}}{\Rightarrow} \overset{n}{q})) \wedge (\neg(\overset{u}{r} \underset{\substack{s\\a}}{\Rightarrow} \overset{n}{q}) \rightarrow (\bigvee_{i \leq m} p_i \underset{\substack{s\\a}}{/\!\!\!\wedge\!\!\wedge} \overset{u}{r})) \qquad \text{ST3.100}$$

3.9.3 *Absolute and Conditioned Integrations*

Returning to the relations of integration, we must first remark that, as already seen in the 'material' case, also in the formal case a rigorous distinction subdivides *absolute* [*unbedingt*] and *conditioned* [*bedingt*] integration.

Insofar as absolute integrations are concerned, Bolzano considers, as he has already done with derivability, the 'adequate' case, in which trivial situations are eliminated, and he says that the integration among the $p_1, \ldots, p_n$, with respect to the $a_1, \ldots, a_s$, is *adequate* or *minimal* [*von der geringsten Anzahl*] if none of the $p_1, \ldots, p_n$ is always true or always false, with respect to the $a_1, \ldots, a_s$; i.e., writing '$\overset{+}{\mathbb{V}}^{n,s}$' for this $n+s$-ary relation,

$$\underset{\substack{s\\a}}{\overset{+}{\mathbb{V}}}(\overset{n}{p}) \leftrightarrow \underset{\substack{s\\a}}{\mathbb{V}}(\overset{n}{p}) \wedge \bigwedge_{i \leq n} \neg(\underset{\substack{s\\a}}{\mathbb{W}}(p_i) \vee \underset{\substack{s\\a}}{\mathbb{F}}(p_i)) \qquad \text{SD3.32}$$

[122] TS §159, II 112: "24) If certain propositions $M, N, O, \ldots$ which are not formally true are deducible from an individual proposition A, then the contradictory of A must be deducible from the contradictory of the propositions $M, N, O, \ldots$ all with respect to the same ideas. For [...]."

[123] TS §159, II 112: "25) No proposition which is not formally false is *incompatible* with both of a pair of contradictory propositions. [...]."

[124] TS §159, II 113: "27) If propositions $A, B, C, \ldots$ and $M, N, O, \ldots$ are contradictories of each other, and $A, B, C, \ldots$ are compatible with certain propositions $X, Y, Z, \ldots$, then $M, N, O, \ldots$ are not deducible from the latter, and if $M, N, O, \ldots$ are not deducible from them, then at least one of the propositions $A, B, C, \ldots$ is compatible with them, all with respect to the same variable parts. For [...]."

Bolzano then turns to the problem of the number of propositions that in a given integration become true when the substitutions vary. He makes an initial distinction between those in which that number is *always equal to* 1 and those in which that number is *always greater than* 1. He calls the integrations of the first type *(formal) one-member* [*eingliedrig*] *integrations* or *(formal) disjunctions*;[125] those of the second type, he calls *(formal) multipartite* [*mehrtheilig*] or *redundant* [*überfüllt*] integrations.[126] He then takes into consideration the further refinement of multipartiteness in the case in which the number of true propositions remains, under all variations, constant, and if this number is *n*, then he speaks of *(formal) aggregates of n true propositions* [*(formale Inbegriffe) von n wahren Sätzen*].

In strict analogy with what has already been done in the material cases, we may systematize these various relations through the following definition schemata:

(1) For *each* substitution for the $\overset{s}{a}$, *at least k* of the $\overset{n}{p}$ become *true* (*false*):

$$\underset{\overset{s}{a}}{{}_k\mathbb{W}}(\overset{n}{p}) \leftrightarrow \forall \overset{s}{c}\, {}_kW(\overset{n}{p}[\overset{s}{a}/\overset{s}{c}]) \qquad \text{SD3.33}$$

$$\underset{\overset{s}{a}}{{}_k\mathbb{F}}(\overset{n}{p}) \leftrightarrow \forall \overset{s}{c}\, {}_kF(\overset{n}{p}[\overset{s}{a}/\overset{s}{c}]) \qquad \text{SD3.34}$$

(2) For *each* substitution for the $\overset{s}{a}$, *at most k* of the $\overset{n}{p}$ become *true* (*false*):

$$\underset{\overset{s}{a}}{{}^k\mathbb{W}}(\overset{n}{p}) \leftrightarrow \forall \overset{s}{c}\, {}^kW(\overset{n}{p}[\overset{s}{a}/\overset{s}{c}]) \qquad \text{SD3.35}$$

$$\underset{\overset{s}{a}}{{}^k\mathbb{F}}(\overset{n}{p}) \leftrightarrow \forall \overset{s}{c}\, {}^kF(\overset{n}{p}[\overset{s}{a}/\overset{s}{c}]) \qquad \text{SD3.36}$$

(3) For *each* substitution for the $\overset{s}{a}$, *at least k*, but *not more* than *h*, of the $\overset{n}{p}$ become *true* (*false*):

$$\underset{\overset{s}{a}}{{}_k^h\mathbb{W}}(\overset{n}{p}) \leftrightarrow \forall \overset{s}{c}\, {}_k^hW(\overset{n}{p}[\overset{s}{a}/\overset{s}{c}]) \qquad \text{SD3.37}$$

$$\underset{\overset{s}{a}}{{}_k^h\mathbb{F}}(\overset{n}{p}) \leftrightarrow \forall \overset{s}{c}\, {}_k^hF(\overset{n}{p}[\overset{s}{a}/\overset{s}{c}]) \qquad \text{SD3.38}$$

[125] TS §160, II 117–18: "3) A different relation obtains when we learn that there is only one true (or false) proposition in a given collection of propositions *M, N, O,* ... [...] This relation may be called *single-termed complementation.* It is customarily called *disjunction.* If the given propositions stand in the relation as they are, it is *material disjunction* [...] It is *formal disjunction* if the relation obtains no matter what ideas are put in the place of those considered arbitrarily variable [...]

[126] TS §160, II 118–19: "5) Another special case arises when we learn of the propositions *M, N, O,* ... that more than one of them is true, without knowing how man [...] This relation is called *multi-termed* or *redundant complementation.*"

Obviously, we have

$$\underset{\overset{s}{a}}{\mathbb{W}}(\overset{n}{p}) \leftrightarrow {}_n\underset{\overset{s}{a}}{\mathbb{W}}(\overset{n}{p}) \qquad \text{ST3.101}$$

$$\underset{\overset{s}{a}}{\mathbb{F}}(\overset{n}{p}) \leftrightarrow {}_n\underset{\overset{s}{a}}{\mathbb{F}}(\overset{n}{p}) \qquad \text{ST3.102}$$

Moreover, we fix, for convenience, the two abbreviations

$$\underset{\overset{s}{a}}{E}(\overset{n}{p}) := {}_1\underset{\overset{s}{a}}{\mathbb{W}}(\overset{n}{p}) \qquad \text{MD3.39}$$

$$\underset{\overset{s}{a}}{D}(\overset{n}{p}) := {}^1_1\underset{\overset{s}{a}}{\mathbb{W}}(\overset{n}{p}) \qquad \text{MD3.40}$$

We remark that, in particular,

$$\underset{\overset{s}{a}}{D}(\overset{n}{p}) \leftrightarrow \forall \overset{s}{c} \bigvee_{i\leq n} (W(p_i[\overset{s}{a}/\overset{s}{c}]) \wedge \bigwedge_{j\neq i\leq n} F(p_j[\overset{s}{a}/\overset{s}{c}])) \qquad \text{ST3.103}$$

and that Bolzano emphasizes that if this relation subsists among the $p_1, \ldots, p_n$, then each one of them is incompatible with all the remaining ones;[127] i.e.

$$\underset{\overset{s}{a}}{D}(\overset{n}{p}) \rightarrow \bigwedge_{i\leq n} (\neg \underset{\overset{s}{a}}{\mathbb{F}}(p_i) \rightarrow p_i \underset{\overset{s}{a}}{\perp} \overset{n}{p} \setminus p_i) \qquad \text{ST3.104}$$

Bolzano also emphasizes that ${}^k_k\underset{\overset{s}{a}}{\mathbb{W}}(\overset{n}{p})$ subsists among the $\overset{n}{p}$ iff for each one (not always false) of them there are other $k-1$ that are compatible with it and, together with it, exclude;[128] i.e.

$$ {}^k_k\underset{\overset{s}{a}}{\mathbb{W}}(\overset{n}{p}) \leftrightarrow \forall p \left(p < \overset{n}{p} \wedge \neg \underset{\overset{s}{a}}{\mathbb{F}}(p) \rightarrow \overset{k-1}{\exists!!} q \left(\begin{array}{l} q < \overset{n}{p} \wedge q \neq p \wedge \wedge p \underset{\overset{s}{a}}{\wedge\!\!\wedge} \overset{k-1}{q} \wedge \\ \forall r (r \leq \overset{n}{p} \wedge r \neq p \wedge r \neq q \rightarrow p, \overset{k-1}{q} \underset{\overset{s}{a}}{\perp} r) \end{array} \right) \right) \qquad \text{ST3.105}$$

[127] TS §160, II 118: "3) [. . .] we can also say that each of the propositions $M, N, O, \ldots$ stands in the relation of exclusion to all the others, so long as it is not formally false."

[128] TS §160, II 119: "6) More noteworthy is the relation between propositions $M, N, O, \ldots$ when the number of true propositions among them is known, and when this number remains the same when one arbitrarily alters certain ideas $i, j, \ldots$ in these propositions. [. . .] If, in the given collection of propositions $M, N, O, \ldots, n$ are always true, but none is formally false, then it must be that each one of them is compatible with $(n-1)$ others, and that these, taken together, constitute a collection of n propositions that stand in a relation of exclusion with respect to the others. For if the former were not the case, the number of true propositions would have to be less than n; and if the latter were not the case, it would have to be greater than n."

As a special and particularly interesting case of absolute generic (i.e. not of the type $\overset{m}{D}$) integrations, Bolzano points out one in which the integrating propositions are also compatible with each other.[129] Thus, we can define this $n + s$-ary relation by setting

$$\underset{\overset{s}{a}}{\mathbb{X}}(\overset{n}{p}) \leftrightarrow \underset{\overset{s}{a}}{\mathbb{W}}(\overset{n}{p}) \wedge \underset{\overset{s}{a}}{\mathbb{M}}(\overset{n}{p}) \qquad \text{SD3.41}$$

thus obtaining

$$\underset{\overset{s}{a}}{\mathbb{X}}(\overset{n}{p}) \leftrightarrow \forall\overset{s}{c}(\bigvee_{i\leq n} W(p_i[\overset{s}{a}/\overset{s}{c}])) \wedge \exists\overset{s}{c}(\bigwedge_{i\leq n} W(p_i[\overset{s}{a}/\overset{s}{c}])) \qquad \text{ST3.106}$$

With regard to the conditioned integrations, we can quite generally remark that a *formal conditioned integration* is a type of integration relationship in which substitutions are not completely arbitrary, but dependent to one or more conditions. A particularly important case is that of an imposed condition to make the propositions of a certain other aggregate of propositions true.[130]

For given propositions $p_1, \ldots, p_m, q_1, \ldots, q_n$ and ideas $a_1, \ldots, a_s$, we say that the $q_1, \ldots, q_n$ *integrate one another with respect to the* $a_1, \ldots, a_s$, *under conditions* $p_1, \ldots, p_m$ [*einander aushelfen unter der Bedingung der Sätze*], when for each aggregate of ideas $c_1, \ldots, c_s$ that take the place of the $a_1, \ldots, a_s$ in the $p_1, \ldots, p_m$ and the $q_1, \ldots, q_n$ we have that if all the $p_1, \ldots, p_m$ become true, then at least one of the $q_1, \ldots, q_n$ also becomes true.[131] Writing '$\overset{\vee}{\Rightarrow}^{m,n,s}$' for this very important $m + n + s$-ary relation, the first consideration of which is usually attributed to Gerhard Gentzen, we thus set

$$\overset{m}{p} \underset{\overset{s}{a}}{\overset{\vee}{\Rightarrow}} \overset{n}{q} \leftrightarrow \forall\overset{s}{c}(W(\overset{m}{p}[\overset{s}{a}/\overset{s}{c}]) \rightarrow E(\overset{n}{q}[\overset{s}{a}/\overset{s}{c}])) \qquad \text{SD3.42}$$

$$\overset{m}{p} \overset{\vee}{\Rightarrow} \overset{n}{q} \leftrightarrow \exists\overset{s}{a}(\overset{m}{p} \underset{\overset{s}{a}}{\overset{\vee}{\Rightarrow}} \overset{n}{q}) \qquad \text{SD3.43}$$

Bolzano also considers the special cases of this relation in which, for each system of ideas that makes the $p_1, \ldots, p_m$ true, one and only one, more than one, or exactly $k\,(1 \leq k \leq n)$ of the $q_1, \ldots, q_n$ become true:

129 TS §160, II 119: "7) Among collections where the number of true propositions varies, the most noteworthy are those for which there are ideas that, when put in the place of the variable ones, make all of them true simultaneously. [...] Accordingly, they may be called collections of *complementary* and *compatible* propositions."

130 TS §160, II 120: "8) [...] An especially noteworthy case, rightly placed in the theory of relations between propositions, is that where the variation of the ideas *i*, *j*, ... is restricted by means of certain other propositions, namely, in that it is stipulated that only such ideas are to be chosen as make certain *other propositions A*, *B*, *C*, ... true."

131 TS §160, II 120: "8) [...] Hence if every class of ideas whose substitution for *i*, *j*, ... makes all of *A*, *B*, *C*, ... true also produces one or several truths in the class of propositions *M*, *N*, *O*, ..., then I shall call propositions *M*, *N*, *O*, ... *complementary under the condition that A*, *B*, *C*, ..."

$$\overset{n}{p} \overset{\vee 1}{\underset{\overset{s}{a}}{\Rightarrow}} q \leftrightarrow \forall \overset{s}{c}(W(\overset{n}{p}[\overset{s}{a}/\overset{s}{c}]) \rightarrow D(\overset{m}{q}[\overset{s}{a}/\overset{s}{c}])) \qquad \text{SD3.44}$$

$$\overset{n}{p} \overset{\vee +}{\underset{\overset{s}{a}}{\Rightarrow}} q \leftrightarrow \forall \overset{s}{c}(W(\overset{n}{p}[\overset{s}{a}/\overset{s}{c}]) \rightarrow {}_2W(\overset{m}{q}[\overset{s}{a}/\overset{s}{c}])) \qquad \text{SD3.45}$$

$$\overset{n}{p} \overset{\vee k}{\underset{\overset{s}{a}}{\Rightarrow}} q \leftrightarrow \forall \overset{s}{c}(W(\overset{n}{p}[\overset{s}{a}/\overset{s}{c}]) \rightarrow {}^k_kW(\overset{m}{q}[\overset{s}{a}/\overset{s}{c}])) \qquad \text{SD3.46}$$

He also explicitly considers the case in which *all* the $q_1, \ldots, q_n$ become true. In this case, he says that $q_1, \ldots, q_n$ constitute an aggregate of propositions, *which, under the conditions* $p_1, \ldots, p_m$, *integrate each other and are compatible.*[132] It is, however, somewhat surprising that he mentions neither that this relation is the formal version of what in the material case he called the "improper interpretation of the hypothetical proposition",[133] nor that the relation of derivability is just this relation strengthened by the requirement of the compatibility of the $p_1, \ldots, p_m$.

3.10 Formal Connectives

As in the material case, we may introduce, at the elementary level, operations on propositions that generate new propositions behaving as if they were enunciating the subsistence of a formal relation among the propositions to which they are applied. We will limit ourselves to the introduction of the connectives that correspond to formal truth, integration, derivability, and conditioned integration. Moreover, as all these connectives will later turn out to be superfluous, we will omit, for the sake of simplicity, the requirements relative to their mereological structure.

$$\forall \overset{m}{p} \forall \overset{s}{a} \exists q(W(q) \leftrightarrow \forall \overset{s}{c} W(\overset{m}{p}[\overset{s}{a}/\overset{s}{c}])) \qquad \text{SCS5}$$

$$\underset{\overset{s}{a}}{\boxed{\wedge}}(\overset{m}{p}) = \varepsilon q(W(q) \leftrightarrow \forall \overset{s}{c} W(\overset{m}{p}[\overset{s}{a}/\overset{s}{c}])) \qquad \text{SD3.47}$$

$$W(\underset{\overset{s}{a}}{\boxed{\wedge}}(\overset{m}{p})) \leftrightarrow \forall \overset{s}{c} W(\overset{n}{p}[\overset{s}{a}/\overset{s}{c}]) \leftrightarrow \underset{\overset{s}{a}}{\mathbb{W}}(\overset{n}{p}) \qquad \text{ST3.107}$$

$$\forall \overset{m}{p} \forall \overset{s}{a} \exists q(W(q) \leftrightarrow \forall \overset{s}{c} E(\overset{m}{p}[\overset{s}{a}/\overset{s}{c}])) \qquad \text{SCS6}$$

$$\underset{\overset{s}{a}}{\boxed{\vee}}(\overset{m}{p}) = \varepsilon q(W(q) \leftrightarrow \forall \overset{s}{c} E(\overset{m}{p}[\overset{s}{a}/\overset{s}{c}])) \qquad \text{SD3.48}$$

[132] TS §160, II 120: "8) [...] When, finally, there are ideas which, when put in the place of *i, j,* ... make all of *A, B, C,* ... as well as all of *M, N, O,* ... true, I say that *M, N, O,* ... are *compatible* and *complementary* under the conditions *A, B, C,* ...."

[133] See Note 13.

$$W(\underset{\overset{s}{a}}{\boxed{\vee}}(\overset{m}{p})) \leftrightarrow \forall \overset{s}{c} E(\overset{n}{p}[\overset{s}{a}/\overset{s}{c}]) \leftrightarrow \underset{\overset{s}{a}}{\mathbb{W}}(\overset{n}{p}) \tag{ST3.108}$$

$$\forall \overset{m}{p} \forall \overset{n}{q} \forall \overset{s}{a} \exists q'(W(q') \leftrightarrow \overset{m}{p} \underset{\overset{s}{a}}{\Rightarrow} \overset{n}{q}) \tag{SCS7}$$

$$\overset{m}{x} \underset{\overset{s}{v}}{\boxed{\rightarrow}} \overset{n}{z} = \varepsilon q'(W(q') \leftrightarrow \overset{m}{p} \underset{\overset{s}{a}}{\Rightarrow} \overset{n}{q}) \tag{SD3.49}$$

$$W(\overset{m}{p} \underset{\overset{s}{a}}{\boxed{\rightarrow}} \overset{n}{q}) \leftrightarrow \overset{m}{p} \underset{\overset{s}{a}}{\Rightarrow} \overset{n}{q} \leftrightarrow \overset{m}{p} \underset{\overset{s}{a}}{\mathbin{\wedge\!\!\!\wedge}} \overset{n}{q} \wedge \forall \overset{s}{c}(W(\overset{m}{p}[\overset{s}{a}/\overset{s}{c}]) \rightarrow W(\overset{n}{q}[\overset{s}{a}/\overset{s}{c}])) \tag{ST3.109}$$

$$\forall \overset{m}{p} \forall \overset{n}{q} \forall \overset{s}{a} \exists q'(W(q') \leftrightarrow \overset{m}{p} \underset{\overset{s}{v}}{\overset{\vee}{\Rightarrow}} \overset{n}{q}) \tag{SCS8}$$

$$\overset{m}{x} \underset{\overset{s}{a}}{\boxed{\overset{\vee}{\rightarrow}}} \overset{n}{z} = \varepsilon q'(W(q') \leftrightarrow \overset{m}{p} \underset{\overset{s}{v}}{\overset{\vee}{\Rightarrow}} \overset{n}{q}) \tag{ST3.110}$$

$$W(\overset{m}{p} \underset{\overset{s}{a}}{\boxed{\overset{\vee}{\rightarrow}}} \overset{n}{q}) \leftrightarrow \overset{m}{p} \underset{\overset{s}{a}}{\overset{\vee}{\Rightarrow}} \overset{n}{q} \leftrightarrow \forall \overset{s}{c}(W(\overset{m}{p}[\overset{s}{a}/\overset{s}{c}]) \rightarrow E(\overset{n}{q}[\overset{s}{a}/\overset{s}{c}])) \tag{T3.29}$$

3.11 Probability

In the 21 pages of §161, under the title *Relation of the comparative validity or probability of a proposition with respect to other propositions,* Bolzano simultaneously generalizes the notion of validity of a proposition and the notion of derivability of a proposition from certain other propositions into the notion of *comparative validity or probability* [*vergleichungsweise Gültigkeit* or *Wahrscheinlichkeit*] *of a proposition relatively to other propositions,* and on this basis he develops, often even in mathematical detail, a theory of probability of a type, which in the twentieth century will come to be called 'logical' or 'logical-inductive' and will characterize the attempts of J. M. Keynes, R. Carnap, H. Jeffreys, and others to give a satisfying interpretation of probability.

Given the propositions $p_1, \ldots, p_n, p$ and the ideas $a_1, \ldots, a_s$, we generalize the notation '$\mathcal{W}(p, \overset{s}{a})$' to the notations '$\mathcal{W}(\overset{n}{p}; \overset{s}{a})$' and '$\mathcal{W}(\overset{n}{p}; p; \overset{s}{a})$', which must denote the collections of true propositions obtained by executing all possible substitutions for the $a_1, \ldots, a_s$ in the $p_1, \ldots, p_n$, and, respectively, in the $p_1, \ldots, p_n, p$. Let us suppose that $\mathcal{W}(\overset{n}{p}; \overset{s}{a})$ is not empty (i.e. that the $p_1, \ldots, p_n$ are compatible with respect to the $a_1, \ldots, a_s$); then the *comparative validity or probability of p relative to the $p_1, \ldots, p_n$, with respect to the $a_1, \ldots, a_s$,* [$\mathcal{P}(\overset{n}{p}; p; \overset{s}{a})$] is the 'ratio' between $\mathcal{W}(\overset{n}{p}; q; \overset{s}{a})$

and $\mathcal{W}(\overset{n}{p}; \overset{s}{a})$,[134] a ratio that we will write as a fraction; i.e. we set

$$\mathcal{P}(\overset{n}{p}; p; \overset{s}{a}) = \frac{\mathcal{W}(\overset{n}{p}; p; \overset{s}{a})}{\mathcal{W}(\overset{n}{p}; \overset{s}{a})} \qquad \text{SD3.50}$$

The problems related to reading this ratio as a true numerical ratio are obviously the same as those discussed in the case of validity.[135]

In any case, it is clear that in the first of the two limiting cases—that in which $\mathcal{W}(\overset{n}{p}; p; \overset{s}{a})$ and $\mathcal{W}(\overset{n}{p}; \overset{s}{a})$ coincide—the value 1 may be assigned to this ratio, and it is no less clear that in this case, the probability of p relative to the $p_1, \ldots, p_n$ (with respect to the $a_1, \ldots, a_s$) is nothing other than the derivability of p from the $p_1, \ldots, p_n$ (with respect to the $a_1, \ldots, a_s$). Hence *derivability is simply a limit case of relative probability*. In the other limiting case, in which $\mathcal{W}(\overset{n}{p}; p; \overset{s}{a})$ is empty, that probability will obviously be 0 and the relation in question reduces to the *incompatibility* (with respect to the $a_1, \ldots, a_s$) of q with the $p_1, \ldots, p_n$.[136]

Hence, for the moment, we have

$$\mathcal{P}(\overset{n}{p}; p; \overset{s}{a}) = 1 \leftrightarrow \overset{n}{p} \underset{\overset{s}{a}}{\Rightarrow} p \qquad \text{ST3.111}$$

$$\mathcal{P}(\overset{n}{p}; p; \overset{s}{a}) = 0 \leftrightarrow \overset{n}{p} \underset{\overset{s}{a}}{\Rightarrow} \boxed{\neg} p \qquad \text{ST3.112}$$

[134] TS §161, II 122: "1) [...] One of the relations among several propositions is very reminiscent of (the) concept of validity: given that in an individual proposition *A* or several propositions *A, B, C, D,...* certain ideas *i, j,...* are variable, and given that *A, B, C, D,...* are compatible with respect to these ideas, it will often be important to find the relation between the multitude of cases in which *A, B, C, D,...* all become true and those in which all of these along with a certain other proposition *M* become true. For, if we accept propositions *A, B, C, D,...*, then this proportion tells us whether we should also accept *M*. If *M* becomes true in more than half of the cases in which *A, B, C, D,...* are true, the truth of *A, B, C, D,...* entitles us to accept *M* as well, otherwise not. I wish to call this relation between the indicated multitudes the *relative validity* of proposition *M* with respect to propositions *A, B, C, D,...* or the *probability* which accrues to *M* from the *assumptions A, B, C, D,...*"

[135] TS §161, II 123: "2) Since it is a relation between two multitudes, the relative validity or probability of a proposition has a certain *magnitude* which, if it is determinable at all, can be represented by a fraction whose denominator and numerator are related as these two multitudes."

TS §161, II 124: "7) If there is any idea which makes both *A, B, C, D,...* and *M* true, then there must also be an infinite number of such ideas, since every equivalent idea has the same result. Therefore the multitudes which were mentioned in no. 1, if they exist at all, must always both be infinite. Consequently, the relation between them can never be found *directly*, i.e., by counting them; hence we must seek to determine it through considerations of another sort."

[136] TS §161, II 123: "3) Since the number of cases in which both *A, B, C, D,...* and *M* become true can never be larger, but often smaller, than the number of cases in which propositions *A, B, C, D,...* alone become true, the degree of probability can never be larger than 1, and it reaches this magnitude only when all ideas which make all of *A, B, C, D,...* true also make *M* true, i.e., when *M* is deducible from *A, B, C, D,...* In this special case we tend to say that proposition *M* is certain with respect to propositions *A, B, C, D,....* On the other hand, if there is not a single collection of ideas whose substitution for *i, j,...* makes both *A, B, C, D,...* and *M* true, i.e., if *M* is incompatible with *A, B, C, D,...* then the degree of probability of *M* relative to *A, B, C, D,...* equals zero."

Naturally, outside these cases, the situation is far more complex and we will discuss it later. For the moment, we will analyse situations in which such a reading is possible. Thus we certainly have

$$\mathcal{P}(\overset{n}{p}; p; \overset{s}{a}) \leq 1 \qquad \text{A1}$$

$$\mathcal{P}(\overset{n}{p}; p; \overset{s}{a}) \geq 0 \qquad \text{A2}$$

Bolzano explicitly states the substitutivity of the equivalence with respect to the operation $\mathcal{P}$;[137] i.e.

$$\overset{m}{p} \underset{\overset{s}{a}}{\Leftrightarrow} \overset{n}{q} \rightarrow \mathcal{P}(\overset{m}{p}; r; \overset{s}{a}) = \mathcal{P}(\overset{n}{q}; r; \overset{s}{a}) \qquad \text{A3}$$

$$q \underset{\overset{s}{a}}{\Leftrightarrow} r \rightarrow \mathcal{P}(\overset{n}{p}; q; \overset{s}{a}) = \mathcal{P}(\overset{n}{p}; r; \overset{s}{a}) \qquad \text{A4}$$

and, with regard to the probabilities with respect to the same propositions, of two propositions between which a subordination holds, he sets[138]

$$p \underset{\overset{s}{a}}{\dot{\Rightarrow}} q \rightarrow \mathcal{P}(\overset{n}{p}; q; \overset{s}{a}) \geq \mathcal{P}(\overset{n}{p}; p; \overset{s}{a}) \qquad \text{A5}$$

as well as[139]

$$p \underset{\overset{s}{a}}{\dot{\Rightarrow}} q \rightarrow \mathcal{P}(\overset{n}{p}; p; \overset{s}{a}) \leq \mathcal{P}(\overset{n}{p}; q; \overset{s}{a}) \qquad \text{A6}$$

With regard to the probability of the negation, it is fixed by setting[140]

$$\mathcal{P}(\boxminus\, p) = 1 - \mathcal{P}(p) \qquad \text{SD3.51}$$

[137] TS §161, II 123: "4) If the assumptions *A*, *B*, *C*,... are equivalent to *A*′, *B*′, *C*′,..., then the probability of *M* relative to *A*, *B*, *C*,... is the same as its probability relative to *A*′, *B*′, *C*′,..., all with respect to the same variable ideas *i*, *j*,.... Furthermore, if proposition *M* is equivalent to *M*′, then the probability of *M* relative to certain premises *A*, *B*, *C*,... is the same as the probability of *M*′ relative to the same premises, all with respect to the same ideas *i*, *j*,.... For the same ideas which make *A*, *B*, *C*,... true, also make *A*′, *B*′, *C*′,... true, and the same ideas which make *M* true also make *M*′ true. For [...]."

[138] TS §161, II 126: "9) Let proposition *M* have degree of probability μ under the assumptions *A*, *B*, *C*,... and with respect to ideas *i*, *j*,..., and let proposition *R* be unilaterally deducible from *M* with respect to the same ideas. In this case, the degree of probability of proposition *R* with respect to the same assumptions *A*, *B*, *C*,... can *never be smaller than* μ. For [...]. It may be greater, however: for [...]."

[139] TS §161, II 127: "10) Let *M* have degree of probability μ under the assumptions *A*, *B*, *C*,... and with respect to ideas *i*, *j*,..., and let *M* be unilaterally deducible with respect to the same ideas from some other proposition *L*. In this case, the probability of the proposition *L* with respect to the same assumptions *A*, *B*, *C*,... can never be greater than μ. For [...]."

[140] TS §161, II 123: "5) If the probability of a proposition *M* relative to *A*, *B*, *C*, *D*,... and with respect to ideas *i*, *j*,... equals μ then the probability of its negation, namely, Neg.*M* equals $1 - \mu$[...]."

Bolzano also agrees that a proposition the probability of which is equal to the probability of its negation (i.e. = $\frac{1}{2}$) is '*doubtful* [*zweifelhaft*]' and a proposition the probability of which is $< \frac{1}{2}$ is '*improbable* [*unwahrscheinlich*].[141]

Bolzano then turns to the previously mentioned problem of how to reduce the determination of the probabilities to a finite situation. He begins by remarking that there are situations in which, even without knowing the degree of probability of two or more propositions with respect to certain premises and certain variables, it is nevertheless possible to ascertain that, with respect to those premises, the propositions in question have the same probability. In particular, this will happen when all the propositions in question are, with those premises, in one and the same relation, in such a way that no other difference occurs among them besides that of deriving from the transformation of the one in the other through the interchange of certain ideas taken as variables. So, for example, if, given the premises 'This urn contains 100 balls of which 90 are white and 10 black' and 'Caius has drawn a ball from this urn', we call for our convenience the balls using the first hundred natural numbers, then clearly the 100 propositions 'Caius has drawn *n*.1', 'Caius has drawn *n*.2', ..., 'Caius has drawn *n*.100' have the same probability.[142]

The general procedure suggested by Bolzano to attempt to dominate 'infinitary' situations follows.

Given $p_1, \ldots, p_n$, q and $a_1, \ldots, a_s$, one must determine whether it is possible to identify a certain finite number of propositions $q_1, \ldots, q_m$ such that:

(1) all of them have, with respect to $p_1, \ldots, p_n$ and $a_1, \ldots, a_s$, the same probability, i.e.

$$\bigwedge_{i,j \leq m} \left(\mathcal{P}(\overset{n}{p}; q_i; \overset{s}{a}) = \mathcal{P}(\overset{n}{p}; q_j; \overset{s}{a}) \right)$$

(2) they constitute, with respect to $a_1, \ldots, a_s$, a *disjunction conditioned by the* $p_1, \ldots, p_n$ (i.e. they are such that every system $c_1, \ldots, c_s$ that, substituted for the $a_1, \ldots, a_s$, makes all the $p_1, \ldots, p_n$ true makes true one and only one of them); i.e.,

$$\overset{n}{p} \underset{\overset{s}{a}}{\overset{\veebar\!\!\!\veebar}{\Rightarrow}} \overset{m}{q}$$

[141] TS §161, II 123–4: "6) If the degree of probability of proposition *M* is the same as the degree of probability of its negation, i.e., if it equals one half, then we say that it is merely *doubtful*. Degrees of probability which are still lower are sometimes called *improbabilities*; we say that an improbability is greater the smaller the fraction in question (cf. no. 2)."

[142] TS §161, II 124.

(3) each of them decides, with respect to the $a_1, \ldots, a_s$, the q, i.e.

$$\bigwedge_{i \leq m} (q_i \underset{\overset{s}{a}}{\Rightarrow} q \vee q_i \underset{\overset{s}{a}}{\Rightarrow} \boxminus q)$$

Now, if k is the number of the q_i such that $q_i \underset{\overset{s}{a}}{\Rightarrow} q$, then

$$\mathcal{P}(\overset{n}{p}; q; \overset{s}{a}) = \frac{m}{k}$$

The argument supporting this procedure is as follows. As, by (2), every substitution that makes true all of the $\overset{n}{p}$ makes true one and only one of the $\overset{m}{q}$, the class of all the substitutions that make true all of the $\overset{n}{p}$ subdivides into exactly m disjoint subclasses: the class of those substitutions that make the q_1 true; . . . ; the class of those substitutions that make the q_m true. Let us now consider an arbitrary substitution, the class of those that make the q_i true; by (3), either the q is derivable from the q_i or its negation is derivable from the q_i, and hence either all the substitutions of that class make the q true or all of them make it false. Let us suppose that the number of the $\overset{m}{q}$ from which q is derivable is k and hence that k is the number of the subclasses of substitutions that, besides making true all of the $\overset{n}{p}$, also make the q true; so the probability of q is the ratio between the m classes of all substitutions that make all of the $\overset{n}{p}$ true and the k classes among them that make the q true (i.e. $\frac{m}{k}$).[143]

On this basis, Bolzano tackles the central question of the theory, namely the determination of the probability of a conjunction as a function of the probabilities of the conjuncts. Clearly, Bolzano ignores the fundamental aspect of the question that we treat in term of '*stochastic (in)dependence*'. Said bluntly, his idea was that *the value of the probability of a conjunction is – always and in any case – the product of the probabilities of the conjuncts* and this both in the 'absolute' case, i.e.

$$\mathcal{P}(p \boxed{\wedge} q; \overset{s}{a})) = \mathcal{P}(p; \overset{s}{a}) \cdot \mathcal{P}(q; \overset{s}{a}) \qquad (*)$$

and in the 'relative' one, i.e.

$$\overset{m}{p} \underset{\overset{s}{a}}{\top} \overset{n}{q} \rightarrow \mathcal{P}(\overset{m}{p}, \overset{n}{q}; p \boxed{\wedge} q; \overset{s}{a}) = \mathcal{P}(\overset{m}{p}; p; \overset{s}{a}) \cdot \mathcal{P}(\overset{n}{q}; q; \overset{s}{a}) \qquad (*)$$

One could of course put forward the hypothesis that—at least in the applications—Bolzano tacitly understands this limitation, and in fact various of the examples he puts forward in the sequel of his treatment satisfy this condition.

[143] TS §161, II 124.

However this is not true of all of them—in particular not the one by means of which he wanted to illustrate this principle. He says:

> "Thus the probability that from 6 balls, 4 of which are black and 5 sweet-smelling, someone draws one which is both black and sweet-smelling is equal to $\frac{4}{6} \cdot \frac{5}{6} = \frac{5}{9}$." [144]

Now, under these hypotheses, there are, in fact, two possible cases. In the first case, all black balls are sweet-smelling and therefore the probability of drawing a black and sweet-smelling ball coincides with the probability of drawing a black ball (i.e. $\frac{4}{6} = \frac{2}{3} \neq \frac{5}{9}$). In the second case, when at least one of the black balls is not sweet-smelling, as there is only one not sweet-smelling ball, then there are exactly 3 black and sweet-smelling balls and hence the probability to draw a black and sweet-smelling ball is $\frac{3}{6} = \frac{1}{2} \neq \frac{5}{9}$.

In reality, however, the Bolzanian presentation of the 'conjunction principle' is much more elaborate and complex and the procedure discussed above is just the beginning.

Bolzano maintains that

if

(1) a proposition p has, with respect to the premises $\overset{m}{p}$ and the ideas $\overset{s}{a}$, a degree of probability μ;

(2) another proposition q has, with respect to the premises $\overset{n}{q}$ and the same ideas $\overset{s}{a}$, a degree of probability ν;

(3) the $\overset{m}{p}$ and the $\overset{n}{q}$ are compatible among them with respect to the $\overset{s}{a}$;

(4) the $\overset{u}{r}$ are the propositions of equal probability obtained from the $\overset{m}{p}$ according to the above procedure;

(5) the $\overset{v}{s}$ are the propositions of equal probability obtained from the $\overset{n}{q}{}^{m}$ via the above procedure;

(6) all pairs of propositions obtained by conjoining each of the $\overset{u}{r}$ with each one of the $\overset{v}{s}$ also all have the same probability with respect to the premises $\overset{m}{p}, \overset{n}{q}$ and the ideas $\overset{s}{a}$;

then

> the probability of the conjunction of p and q, with respect to the premises $\overset{m}{p}, \overset{n}{q}$ and the ideas $\overset{s}{a}$ is equal to $\mu \cdot \nu$.[145]

[144] Cf. TS §161, II 127 (11)."

[145] In fact, Bolzano's formulation is even more general—indeed, it regards not only two but any number of propositions—but Bolzano himself recognizes that the fall of two propositions suffices for the justification. TS

Let us write '$\mathcal{K}(\overset{m}{p}; \overset{u}{r}; \overset{s}{a})$' to denote the relation subsisting among the $\overset{m}{p}$, the $\overset{u}{r}$ and the $\overset{s}{a}$ when they satisfy the conditions hypothesized in the procedure considered above; i.e.

$$\mathcal{K}(\overset{m}{p}; \overset{u}{r}; \overset{s}{a}) \leftrightarrow \bigwedge_{i,j \leq u} (\mathcal{P}(\overset{m}{p}; r_i; \overset{s}{a}) = \mathcal{P}(\overset{m}{p}; r_j; \overset{s}{a})) \wedge \overset{m}{p} \underset{\overset{s}{a}}{\overset{\forall}{\Rightarrow}} \overset{u}{r} \wedge \bigwedge_{i \leq u} (r_i \underset{\overset{s}{a}}{\Rightarrow} p \vee r_i \underset{\overset{s}{a}}{\Rightarrow} \boxed{\neg}\, p) \qquad \text{SD3.52}$$

Then the Bolzanian 'principle of the conjunction' can be presented as

$$\text{A} \quad \left[\begin{array}{c} \overset{m}{p} \underset{\overset{s}{a}}{\top} \overset{n}{q} \wedge K(\overset{m}{p}; \overset{u}{r}; \overset{s}{a}) \wedge K(\overset{n}{q}; \overset{v}{s}; \overset{s}{a}) \wedge \\ \bigwedge_{i,k \leq u} \bigwedge_{i,j \leq m} (\mathcal{P}(\overset{m}{p}, \overset{n}{q}; r_i \boxed{\wedge} s_j; \overset{s}{a}) = \mathcal{P}(\overset{m}{p}, \overset{n}{q}; r_k \boxed{\wedge} s_h; \overset{s}{a})) \end{array} \right]$$
$$\rightarrow \mathcal{P}(\overset{m}{p}, \overset{n}{q}; p \boxed{\wedge} q; \overset{s}{a})) = \mathcal{P}(\overset{m}{p}; p; \overset{s}{a}) \cdot \mathcal{P}(\overset{n}{q}; q; \overset{s}{a})$$

Subsequently, Bolzano introduces six theorems, each endowed with the premise "If all remains as in 11." The sense of these words, however, is not perfectly clear. They might mean that the propositions and the involved probabilities satisfy the same conditions that they satisfy in the preceding theorem (in particular the compatibility of the premises) so that the '(Bolzanian) conjunction principle' is applicable to them .

Inspecting the theorems, the first states that if t is subordinate to p, q, then its probability with respect to the $\overset{m}{p}, \overset{n}{q}$ is greater than or equal to the product of the probability of p with respect to the $\overset{m}{p}$, and the probability of q with respect to the $\overset{n}{q}$;[146] i.e.

$$\overset{m}{p} \underset{\overset{s}{a}}{\top} \overset{n}{q} \rightarrow ((p, q \underset{\overset{s}{a}}{\dot{\Rightarrow}} t) \rightarrow (\mathcal{P}(\overset{m}{p}, \overset{n}{q}; t; \overset{s}{a}) \geq \mathcal{P}(\overset{m}{p}; p; \overset{s}{a}) \cdot \mathcal{P}(\overset{n}{q}; q; \overset{s}{a})) \qquad \text{ST3.113}$$

Bolzano then remarks that, owing to the fact that a product of numbers smaller than 1 is always smaller than each of its factors and decreases with an increase in the

§161, II 127 "11) Let M have degree of probability μ under the assumptions $A, B, C, \ldots$ and with respect to ideas $i, j, \ldots$, and suppose that a second proposition N has the degree of probability ν with respect to the same ideas $i, j, \ldots$ and under assumptions $D, E, F, \ldots$, which are compatible with $A, B, C, \ldots$. Suppose, furthermore, that the assumptions $A, B, C, \ldots; D, E, F, \ldots$ are so constituted that every combination of propositions of equal probability associated with $A, B, C, \ldots$ (namely, $K, K', \ldots$) on the one hand, and with $D, E, F, \ldots$ on the other (namely, $L, L', \ldots$) by means of propositions of the following kind: 'K and L are true', 'K' and L' are true' have mutually equal probabilities. In this case, the degree of probability that M and N are both true relative to the combined assumptions $A, B, C, \ldots; D, E, F, \ldots$ with respect to $i, j, \ldots$ is equal to $\mu \times \nu \times \ldots$. Clearly, it suffices to prove this theorem for the case involving two propositions M and N."

[146] TS §161, II 128: "12) If everything remains as in no. 11, and a proposition R is unilaterally deducible from $M, N, \ldots$ with respect to the same ideas $i, j, \ldots$, then the probability of R with respect to the assumptions $A, B, C, \ldots D, E, F, \ldots$ can *never be smaller* than the product $\mu \times \nu \times \ldots$. For [...]."

number of factors, the probability of a conclusion will always be smaller than the probability of each of its merely 'probable' premises ('certain' premises, i.e. those of probability 1, have no influence of course) and will decrease with an increase in their number.[147]

The next theorem concerns the composition of probabilities, and states that if the probability of p with respect to the $\overset{m}{p}$ is μ and that of the q with respect to the $\overset{n}{q}$ is ν and, moreover, the probability of a certain t with respect to p, q is ρ, then the probability of the same t with respect to the $\overset{m}{p}, \overset{n}{q}$ is greater than or equal to $\mu \cdot \nu \cdot \rho$:[148]

$$\overset{m}{p} \underset{\overset{s}{a}}{\top} \overset{n}{q} \rightarrow \mathcal{P}(\overset{m}{p}, \overset{n}{q}; t; \overset{s}{a}) \geq \mathcal{P}(\overset{m}{p}; p; \overset{s}{a}) \cdot \mathcal{P}(\overset{n}{q}; q; \overset{s}{a}) \cdot \mathcal{P}(p, q; t; \overset{s}{a}) \qquad \text{ST3.114}$$

The subsequent theorem deals with the situation that is, in a certain way, the dual of that discussed in the first of these theorems, i.e. the case in which the t, instead of being subordinated to p and q, is such that p and q are subordinated to it:[149]

$$\overset{m}{p} \underset{\overset{s}{a}}{\top} \overset{n}{q} \rightarrow (t \underset{\overset{s}{a}}{\dot{\Rightarrow}} a, b \rightarrow \mathcal{P}(\overset{m}{p}, \overset{n}{q}; t; \overset{s}{a}) \leq \mathcal{P}(\overset{m}{p}; p; \overset{s}{a}) \cdot \mathcal{P}(\overset{n}{q}; q; \overset{s}{a})) \qquad \text{ST3.115}$$

The next theorem concerns the '(Bolzanian) disjunction principle'

$$\overset{m}{p} \underset{\overset{s}{a}}{\top} \overset{n}{q} \rightarrow \mathcal{P}(\overset{m}{p}, \overset{n}{q}; p \boxed{\vee}\, q; \overset{s}{a}) = 1 - (1 - \mathcal{P}(\overset{m}{p}; p; \overset{s}{a})) \times (1 - \mathcal{P}(\overset{n}{q}; q; \overset{s}{a})) \qquad \text{ST3.116}$$

which is obtained from the conjunction principle, using De Morgan's law and D83:[150]

[147] TS §161, II 128: "13) Since the product $\mu \times \nu \times \ldots$ is smaller than each of its factors $\mu, \nu, \ldots$, and since it decreases as the number of its factors increases, we can see that any lower bound we assign to the probability of a conclusion must be less than that of each individual premise, and that it must decrease if the number of probable premises increases. Premises that are certain have no effect on the probability of the conclusion. For if M is certain, $\mu = 1$."

[148] TS §161, II 128–9: "14) If everything is as in no. 11, and the proposition R has the probability ρ with respect to propositions $M, N, \ldots$ and the same ideas $i, j, \ldots$, then the probability of R with respect to the assumptions $A, B, C, \ldots; D, E, F, \ldots$ and the ideas $i, j, \ldots$ is certainly no smaller than the product $\rho \times \mu \times \nu \times \ldots$. For the number of cases in which the propositions $A, B, C, \ldots; D, E, F, \ldots$ all come out true is to the number of cases in which all of $M, N, \ldots$ come out true as $1 : \mu \times \nu \times \ldots$, and the number of cases in which $M, N, \ldots$ all come out true is to the number of cases in which R comes out true as $1 : \rho$. Hence it is certain that the ratio of the number of cases in which R comes out true to the number of cases in which $A, B, C, \ldots; D, E, F, \ldots$ all come out true is no smaller than $\rho \times \mu \times \nu \times \ldots : 1$."

[149] TS §161, II 129: "15) If everything is as in no.11, and propositions $M, N, \ldots$ are unilaterally deducible from R with respect to the same ideas $i, j, \ldots$, then the probability of R with respect to the assumptions $A, B, C, \ldots; D, E, F, \ldots$ is certainly no larger than the product $\mu \cdot \nu \cdot \ldots$. For R is not true more often than all of $M, N, \ldots$ are."

[150] TS §161, II 129: "16) If everything is as in no. 11, then the probability of the claim that there is a true proposition among $M, N, \ldots$ (or that the idea of a true proposition among $M, N, \ldots$ has objectuality) relative to the

$$\mathcal{P}(\overset{m}{p}, \overset{n}{q}; p \boxed{\vee}\, q; \overset{s}{a}) = \mathcal{P}(\overset{m}{p}, \overset{n}{q}; \boxed{\neg}(\boxed{\neg}p \boxed{\wedge} \boxed{\neg}q); \overset{s}{a}) = 1 - \mathcal{P}(\overset{m}{p}, \overset{n}{q}; \boxed{\neg}p \boxed{\wedge} \boxed{\neg}q; \overset{s}{a})$$
$$= 1 - (\mathcal{P}(\overset{m}{p}, \overset{n}{q}; \boxed{\neg}p; \overset{s}{a})) \cdot \mathcal{P}(\overset{m}{p}, \overset{n}{q}; \boxed{\neg}q; \overset{s}{a}) = 1 - (1 - \mathcal{P}(\overset{m}{p}; p; \overset{s}{a})) \times (1 - \mathcal{P}(\overset{n}{q}; q; \overset{s}{a}))$$

Finally, the case is considered in which the premise also presents the condition that p and q are either both true or both false:[151]

$$\overset{m}{p} \underset{\overset{s}{a}}{\top} \overset{n}{q} \rightarrow \mathcal{P}(\overset{m}{p}, \overset{n}{q}, (p \boxed{\wedge} q) \boxed{\vee} (\boxed{\neg}p \boxed{\wedge} \boxed{\neg}q); p \boxed{\wedge} q; \overset{s}{a}) \qquad \text{ST3.117}$$
$$= \frac{\mathcal{P}(\overset{m}{p}; p; \overset{s}{a}) \cdot \mathcal{P}(\overset{n}{q}; q; \overset{s}{a})}{\mathcal{P}(\overset{m}{p}; p; \overset{s}{a}) \cdot \mathcal{P}(\overset{n}{q}; q; \overset{s}{a}) + (1 - \mathcal{P}(\overset{m}{p}; p; \overset{s}{a})) \cdot (1 - \mathcal{P}(\overset{n}{q}; q; \overset{s}{a}))}$$

The mathematical treatment of the logical probability ends with some other remarks. First, there is the role played by a 'doubtful' ($= \frac{1}{2}$) or even 'improbable' ($< \frac{1}{2}$) conjunct in the determination of the probability of a conjunction: in the first case, it is irrelevant, in the second negative:[152]

$$\mathcal{P}(\overset{m}{p}; p; \overset{s}{a}) = \frac{1}{2} \rightarrow \mathcal{P}(\overset{m}{p}; p \boxed{\wedge} q; \overset{s}{a}) = \mathcal{P}(\overset{m}{p};\ q; \overset{s}{a}) \qquad \text{ST3.118}$$
$$\mathcal{P}(\overset{m}{p}; p; \overset{s}{a}) < \frac{1}{2} \rightarrow \mathcal{P}(\overset{m}{p}; p \boxed{\wedge} q; \overset{s}{a}) < \mathcal{P}(\overset{m}{p};\ q; \overset{s}{a}) \qquad \text{ST3.119}$$

Moreover: if, of two reciprocally incompatible propositions p and q, respectively with probabilities μ and ν with respect to certain premises $\overset{m}{p}$, at least one is true,

assumptions and ideas $i, j, \ldots$ = is $1 - (1 - \mu) \times (1 - \nu) \times \ldots$. For if the probabilities of the propositions M and N are μ and ν, then (no. 5) the probabilities of Neg.M and Neg.N are $= (1 - \mu)$, and $(1 - \nu)$. Thus (according to no. 11) the probability of the claim that Neg.M and Neg.N are both true relative to the assumptions $A, B, C, \ldots; D, E, F, \ldots$ is $= (1 - \mu) \cdot (1 - \nu)$. Hence the probability that not both of Neg.M, Neg.N are true, i.e., the probability that among M, N there is a true proposition is $= 1 - (1 - \mu) \cdot (1 - \nu)$. Thus if there are two urns, where one contains 40 black balls and 50 altogether and the other 45 black out of 60, the probability that someone drawing a ball from each of them will pick a black one is $= 1 - (1 - \frac{40}{50})(1 - \frac{45}{60}) = \frac{19}{20}$."

151 TS §161, II 129–30: "17) If everything is as in no. 11 and if we add the condition that propositions $M, N, \ldots$ must either all be true or all false, then the degree of probability for the claim that they are all true with respect to the combined assumptions $A, B, C, \ldots; D, E, F, \ldots$ and the ideas $i, j, \ldots$ is equal to: $= \frac{\mu \times \nu \ldots}{\mu \times \nu \times \ldots + (1-\mu) \times (1-\nu) \times \ldots}$. It suffices to show that this is so for the case involving two propositions M and N. If we add to the conditions of no. 11 the stipulation that M, N are either both true or both false, we find that only some of the combinations of the propositions of the form 'K and L' (in no. 11) are true, but only some of them, namely, those which either make both of M, N true or make both of them false. Now if $\mu = \frac{m}{k}$, and $\nu = \frac{n}{l}$, then the number of combinations in which both M and N are true is $= m \times n$, while the number under which both are false is $= (k - m)(l - n)$. Accordingly, the relevant number of cases of equal probability here is $mn + (k - m)(l - n)$, and the number of cases in which M and N are both true is $= mn$. Thus the degree of probability of the latter occurring is $= \frac{mn}{mn + (k-m)(l-n)}$ $= \frac{\mu \cdot \nu}{m \cdot \nu + (1-\mu)(1-\nu)}$."

152 TS §161, II 130: "18) If the probability of one of the propositions $M, N, \ldots$ is $= \frac{1}{2}$, then this proposition will neither decrease nor increase the probability of the claim that they are all true, for $\frac{\frac{1}{2} \cdot \nu}{\frac{1}{2} \cdot \nu + (1 - \frac{1}{2})(1-\nu)} = \nu$. Propositions whose probability is smaller decrease the degree of probability which would otherwise obtain."

then the probability that the true proposition is p is equal to $\frac{\mu}{\mu+\nu}$:[153]

$$\neg \underset{\overset{s}{a}}{\wedge\!\!\!\wedge}(p,q) \rightarrow \mathcal{P}(\overset{m}{p}, p \boxed{\vee} q; p; \overset{s}{a}) = \frac{\mathcal{P}(\overset{m}{p}; p; \overset{s}{a})}{\mathcal{P}(\overset{m}{p}; p; \overset{s}{a}) + \mathcal{P}(\overset{m}{p}; q; \overset{s}{a})} \qquad \text{ST3.120}$$

Example:[154] If, in an urn of 1000 coloured balls, 10 are black and 1 is white, then the probability that a black ball is drawn is $\frac{10}{1000} = \frac{1}{100}$ and the probability that a white ball be drawn is $\frac{1}{1000}$. If we now know that a ball has been drawn, which is either black or white, then the probability that it is black is $\frac{10}{11}$ $(= \frac{\frac{1}{100}}{\frac{1}{100}+\frac{1}{1000}})$, whereas the probability that it is white is $\frac{1}{11}(= \frac{\frac{1}{1000}}{\frac{1}{100}+\frac{1}{1000}})$.

Finally, we remark on the extreme 'relativity' of the 'greatness of probability': indeed, as $\frac{\mu}{\mu+\nu+\cdots} = 1 - \frac{\nu+\cdots}{\mu+\nu+\cdots}$, it is clear that

> "also a propositions having a very low degree of probability may obtain a very high degree of probability when circumstances intervene so that one has only to choose between it and other, even more improbable propositions. How little may be μ, indeed, $1 - \frac{\nu+\cdots}{\mu+\nu+\cdots}$ approximates the unity as much as one will, if only $\nu + \cdots$ is still much smaller."[155]

3.12 Probability of Events

As we have seen, the treatment of probability developed by Bolzano in the WL is characterized by the fact that this is conceived as a property of propositions or a relation among *propositions*. A well-known alternative point of view is that which sees in the probability a property of a relation among *events*. Bolzano is very clear about this other possibility:

> "the probability is a property which, strictly speaking, belongs only to propositions and hence also to judgements in so far as they are propositions. Even if we often use

[153] TS §161, II 130: "19) Let μ be the probability of a proposition M with respect to the assumptions $A, B, C, D, \ldots$ and ideas $i, j, \ldots$, and let ν be the probability of a second proposition N with respect to the same assumptions and ideas, etc. If it is the case that these propositions are incompatible with respect to the same ideas, and if one of them is said to be true, then the degree of probability that M is the true proposition is $= \frac{\mu}{\mu+\nu+\ldots}$. For [...]."

[154] TS §161, II 131: "19) [...] If, for example, we know that in an urn there are 1000 balls of varied colours, but only 10 black and one white, the probability that someone drawing a single ball shall choose a black one is $= \frac{10}{1000} = \frac{1}{100}$, and that he shall choose a white one $= \frac{1}{1000}$. If now someone simply states that either a white or a black ball was drawn without specifying which,the probability that it was black is $= \frac{10}{11}$."

[155] TS §161, II 131: "20) Since $\frac{\mu}{\mu+\nu+\cdots} = 1 - \frac{\nu+\cdots}{\mu+\nu+\cdots}$, it follows that even a proposition which has a very low degree of probability can receive a very high one if circumstances arise where we have to choose between it and some other proposition which is even less probable. For, no matter how small μ is, if only $\nu + \ldots$ is still much smaller, $1 - \frac{\nu}{\mu+\nu+\ldots}$ can be brought as close to unity as desired."

expressions such as 'This event is probable' or similar, we certainly only want to state that nothing but the proposition enunciating this event is probable. As much we improperly call an event true (with which one simply wants to state that the proposition enunciating it is true),it can also improperly be called probable."[156]

It is, however, interesting to remark that in the synthetic treatment of probability that Bolzano presents in §15 of the second volume of his *Religionswissenschaft*, the 'eventistic' terminology prevails. This treatment appears in the context of the problem of the credibility of the historical testimonies, in particular those concerning the life and work of Christ:

> "All historical judgements, which are such in strict sense of the word, have, by their proper nature, *mere probability*. Hence this should not be the wrong place to better determine this important notion and its subspecies and to also shortly present some of the most notable theorems; in particular, those the knowledge of which is necessary for a radical confutation of some objections."[157]

In particular, insofar as the 'logical' dimension is concerned, the last subsection, number 15, which is of special interest, is introduced as follows:

> "Who has some knowledge of the literal calculus will be capable of easily understanding the following mathematical propositions, which I want to adjoin here only because they are useful for a radical confutation of those objections to the possibility of the historical crediting of a miracle, which, under the appearance of scholarship, have been put forward also by *mathematicians*, for example, by Joh. Grayg."[158]

This treatment, which is mainly developed not in terms of *propositions* and *premises*, but of *events* [*Erfolge*] and *circumstances* [*Umstände*], really addresses four questions:

(1) The probability of the opposite event:

> "If the degree of probability of the assertion of the real happening of the event M is equal to x, then the degree of probability of the contradictorily

[156] Cf. TS §317, III 184, Note 3.

[157] RW II, §15 45: "1. Alle historischen Urtheile, die es in dieses Wortes engerer Bedeutung sind, haben, wie dieß bereits bemerkt wurde, ihrer Natur nach bloße Wahrscheinlichkeit. Daher dürfte es nicht am unrechten Orte seyn, hier diesen wichtigen Begriff mit seinen Unterarten genauer zu bestimmen; auch einige der merkwürdigsten Lehrsätze über die Wahrscheinlichkeit, besonders solche, deren Kenntniß zu einer gründlichen Widerlegung gewisser Einwürfe nothwendig ist, in Kürze mitzutheilen."

[158] RW II, §15 45: "15. Wer einige Kenntnisse in der Buchstabenrechnung hat, wird auch noch folgende mathematische Sätze leicht zu verstehen vermögen, die ich nur darum hier beifügen will, weil sie zur gründlichen Widerlegung jener Einwürfe dienen, die selbst von Mathematikern, z. B. von Joh. Grayg, gegen die Möglichkeit der historischen Beglaubigung eines Wunders mit einem Anscheine von Gelehrsamkeit vorgebracht worden sind."

opposite event (i.e., the assertion that the event will not happen) is equal to 1 – x":[159]

$$\mathcal{P}(\boxed{\neg}\, p) = 1 - \mathcal{P}(p) \qquad \text{D3.53}$$

(2) The probability of an event under two (or more) circumstances as a function of the probabilities of each one of these circumstances:

> "The probability of an event M, which is produced only by the union of two circumstances A and B, each one of them being only probable, is equal to the product of the probabilities of these two."[160]

So, formulated in terms of propositions, Bolzano assumes

$$p, q \Rightarrow r \rightarrow \mathcal{P}(p \boxed{\wedge} q; r) = \mathcal{P}(p) \cdot \mathcal{P}(q) \qquad \text{A1}$$

which is then generalized as

$$\overset{n}{p} \Rightarrow r \rightarrow \mathcal{P}(\boxed{\wedge^n}(\overset{n}{p}); r) = \prod_{i=1}^{n} \mathcal{P}(p_i) \qquad \text{A2}$$

(3) The probability of an event under two (or more) circumstances as a function of its probability under each separately:

> "If the probability of an event M under a certain respect (i.e., in presence of the circumstance A) is equal to x and under another respect (i.e., in presence of the circumstance B) is equal to y, then the degree of probability of the event M from the union of the two respects (i.e., the simultaneous presence of both circumstances) or the absolute probability resulting from the union of those two relative probabilities is equal to $\frac{xy}{xy+(1-x)(1-y)}$."[161]

[159] RW II, §15 45–6: "15.a) Wenn der Grad der Wahrscheinlichkeit der Behauptung des wirklichen Eintreffens eines Erfolges = x ist: so ist der Grad der Wahrscheinlichkeit des contradictorischen Gegentheils, oder der Behauptung, daß sich der Erfolg nicht zutragen werde, = 1 – x."

[160] RW II, §15 46: "15. b) Die Wahrscheinlichkeit eines Erfolges M, der nur bewirkt wird durch die Vereinigung zweier Umstände A und B, deren jeder nur wahrscheinlich ist, gleicht dem Producte aus den Wahrscheinlichkeiten dieser beiden; d. h. wenn die Wahrscheinlichkeit des Satzes, daß sich der Umstand A einfinden werde, = x, die Wahrscheinlichkeit des Satzes aber, daß sich der Umstand B einfinden werde, = y ist: so ist die Wahrscheinlichkeit des Satzes, daß der Erfolg M, der nur durch die Vereinigung von A und B erzeugt wird, zum Vorschein kommen werde, = xy."

[161] RW II, §15 47: "15. d) Wenn die Wahrscheinlichkeit eines Erfolges M in einer gewissen Rücksicht A (d. h. wegen des Vorhandenseyns des Umstandes A) = x; in einer andern Rücksicht B (d. h. wegen des Vorhandenseyns des Umstandes B) = y ist: so ist der Grad der Wahrscheinlichkeit des Erfolges M aus der Vereinigung von beiden Rücksichten (d. h. weil beide Umstände zugleich vorhanden sind) oder die absolute Wahrscheinlichkeit, die aus Vereinigung jener zwei relativen hervorgehet = $\frac{xy}{xy+(1-x)\cdot(1-y)}$."

Hence, still in terms of propositions,

$$p \Rightarrow r \wedge q \Rightarrow r \rightarrow \mathcal{P}(p \boxed{\wedge} q; r) = \frac{\mathcal{P}(p) \cdot \mathcal{P}(q)}{\mathcal{P}(p) \cdot \mathcal{P}(q) + \mathcal{P}(\boxed{\neg} p) \cdot \mathcal{P}(\boxed{\neg} q)} \qquad \text{A3}$$

which is then generalized as[162]

$$\bigwedge_{1 \leq i \leq n} (p_i \Rightarrow r) \rightarrow \mathcal{P}(\boxed{\wedge^n} (\overset{n}{p}); r) = \frac{\prod_{i=1}^{n} \mathcal{P}(p_i)}{\prod_{i=1}^{n} \mathcal{P}(p_i) + \prod_{i=1}^{n} \mathcal{P}(\boxed{\neg} p_i)} \qquad \text{A4}$$

From A3, Bolzano draws, in particular, the consequence that if the p is *doubtful* ($\mathcal{P}(p) = \frac{1}{2}$), or even *improbable* ($\mathcal{P}(p) < \frac{1}{2}$), then the probability of the r coincides with, or is, respectively, smaller than, the probability of the q; i.e.

$$p \Rightarrow r \wedge q \Rightarrow r \wedge \mathcal{P}(p) = \frac{1}{2} \rightarrow \mathcal{P}(p \boxed{\wedge} q; r) = \mathcal{P}(q) \qquad \text{T3.30}$$

$$p \Rightarrow r \wedge q \Rightarrow r \wedge \mathcal{P}(p) < \frac{1}{2} \rightarrow \mathcal{P}(p \boxed{\wedge} q; r) < \mathcal{P}(q) \qquad \text{T3.31}$$

(4) The probability that an event rather than its opposite takes place as a function of the respective probabilities:

> "If the probability of an event M is equal to x and the probability of another event N, which contradicts it, is equal to y, then the degree of probability with which we can assume that the event M rather than the event N happens is equal to $\frac{x}{x+y}$."[163]

Hence

$$\mathcal{P}(p \boxed{\wedge} q) = 0 \wedge \mathcal{P}(p \boxed{\vee} q) > 0 \rightarrow \mathcal{P}(p \boxed{\vee} q; p) = \frac{\mathcal{P}(p)}{\mathcal{P}(p) + \mathcal{P}(q)} \qquad \text{A5}$$

[162] RW II, §15 48: "15. e) Hieraus erhellet von selbst, wie die Wahrscheinlichkeit einer Sache zu berechnen sey, für die mehr als zwei Umstände sprechen, z. B. die drei Umstände A, B, C, welche derselben jeder für sich die Wahrscheinlichkeiten x, y, z ertheilen. Diese Wahrscheinlichkeit ist nämlich = $\frac{xyz}{xyz+(1-x)(1-y)(1-z)}$; u.s.w."

[163] RW II, §15 48: "15. g) Wenn die Wahrscheinlichkeit eines Erfolges $M = x$, und eines andern N, der jenem widerstreitet, $= y$ ist: so ist der Grad der Wahrscheinlichkeit, mit dem wir annehmen können, daß der Erfolg M eher, als der Erfolg N, Statt finden werde, = $\frac{x}{x+y}$."

4

Symbolic Level

4.1 Preliminary Remark: Properties of Things and Quality of Objects

As we have seen, for example, in the case of object ideas or of propositions, Bolzano gradually introduces a series of predicates expressing properties or relations of things occurring in the universe. A basic fact is that, in some cases, he assumes that qualities of the involved things correspond to these properties; in other words, he assumes that to certain properties $Q(x)$ there correspond things characterized by the same condition. This presupposes, of course, that the things to which that quality is attributed are objects, (i.e. things satisfying $O(x)$). So, for instance, on the level that we have called the symbolic level of the system and that is characterized by the assumption that ideas and propositions are objects, it becomes possible to assume that qualities correspond to certain properties there. As we will see shortly, we may, for instance, assume

$$\exists r(a \in r \leftrightarrow G(a))$$

(i.e. there is a *quality* that belongs to all and only all the object ideas that are objectual), or we may also assume

$$\exists r(p \in r \leftrightarrow W(p))$$

(i.e. there is a *quality* that belongs to all and only all the propositions that are true).

A comparison with set theory may be useful here to clarify this rather delicate issue. In set theory, both natural and ordinal numbers are identified with certain particular sets (which particular sets is irrelevant here). So two predicates $N(x)$ and $\Omega(x)$ are introduced together with two definitory axioms $N(x) \leftrightarrow \alpha[x]$ and $\Omega(x) \leftrightarrow \beta[x]$, where α and β express the 'properties' that are required from a set

Bolzano's Logical System. Ettore Casari.

(i.e. from an 'object' of the theory) in order to be (regarded as) a natural and respectively ordinal number. Hence N and Ω isolate two collections of objects in the universe of sets. The objects of set theory, however, are in themselves collections of sets; setting aside their extensionality and joining our terminology, they are 'qualities' of objects. The question now is whether there are 'qualities' that correspond to the 'properties' of objects expressed by N and Ω; i.e. do the comprehension principles

$$\exists z \forall x (x \in z \leftrightarrow N(x))$$
$$\exists z \forall x (x \in z \leftrightarrow \Omega(x))$$

hold or can they at least be assumed scot-free?

It is a widespread opinion that the first of these may be freely assumed; it is one of the possible forms of the '*axiom of infinity*', which is also usually resolved with 'ω', i.e. by setting

$$\omega = \varepsilon z \forall x (x \in z \leftrightarrow N(x))$$

so as to obtain

$$x \in \omega \leftrightarrow N(x)^{1}$$

On the contrary, it is well known that the assumption of the latter leads to a contradiction, the '*Burali–Forti paradox*'.

In other words, whereas it may have seemed acceptable to assume the presence in the universe of sets of a 'quality' corresponding to the 'property' of being a natural number, it is known for certain that the same cannot be done insofaras the 'property' of being an ordinal number is concerned.

4.2 Bolzano's Conjecture

At a certain point, Bolzano puts forward the idea that every proposition—or at least every proposition having any probability of being true—is of the form $h(a, \bar{c})$. In Bolzano's *Wissenschaftslehre und Religionswissenschaft in einer beurtheilenden Uebersicht*, a sort of self-review, Bolzano summarizes this conviction as follows:

> "Turning now to the second chapter 'On the propositions in themselves' and entering into its first section on the 'general qualities of the propositions', we find one of the

[1] By virtue of 'extensionality', $\exists z \forall x (x \in z \leftrightarrow N(x))$ is in fact $\exists !! z \forall x (x \in z \leftrightarrow N(x))$, and so, more than with a resolution, we have here to do with an 'implicit definition'.

least complete parts of B.'s book. In §127, Bolzano goes so far to posit the opinion that all propositions underlie one and the same form and precisely the following '*A* has (the quality) *b*'. In other words, they always and only have three next components: a substrate or subject idea, an enunciative part or predicate idea and the notion of having, which links the other two and for this reason is often called copula. B. doesn't trust himself to regard this opinion as definitely decided; because he doesn't know how to prove it other than by an induction, which is by its own nature incomplete, so he tries to show, partly here and partly in the following, that whenever he comes across a seemingly deviant form, this expression could also be traced back to the form '*A* has *b*', or better that it should be traced back if what one is here properly saying must be really understandable. Weather, however, readers will always be satisfied with this tracing back we are not able to say; what is certain is that the whole can never be regarded as a complete proof."[2]

In the development of this idea in the *Wissenschaftslehre*, Bolzano affirms:

"If, however, someone should still find it too improbable that the here attempted interpretation of these verbal expressions really gives us the same thought that we connect with them, insofar as they are composed of words that are quite different from those they should be according to this interpretation, I will not discuss the issue further, but I only demand that it be at least conceded to me that the sense connected with these expressions and the sense I have given in my interpretation are equivalent in the sense that whenever one proposition is true then the other one is also true. And it is unlikely that someone will be able to challenge this."[3]

This idea, this 'Bolzano's conjecture', thus has two versions: one, which we will call 'weak', according to which *every proposition is equivalent to a proposition in atomic form*, and which we formulate as

$$\forall p \exists a \exists \bar{c}(W(p) \leftrightarrow W(h(a, \bar{c}))) \tag{BC1}$$

and another, which we will call 'strong', according to which *there are no other propositions beside those of atomic form*, and which we can formulate as

$$\forall p \exists a \exists \bar{c}(p = h(a, \bar{c})) \tag{BC2}$$

[2] WR, 48–9: "Er wagt es, §.127. die Ansicht aufzustellen, daß alle Sätze nur einer und eben derselben Form, nämlich der folgenden: 'A hat (die Beschaffenheit) b', unterständen; daß sie also immer nur drei nächste Bestandtheile hätten: eine Unterlage oder Subiectvorstellung A, einen Aussagentheil oder eine Prädicatvorstellung b, und den Begriff des Habens, der jene beide verbindet, daher auch Copel genannt wird. B. getrauet sich selbst nicht, diese Ansicht als eine ganz entschiedene gelten zu machen; denn er weiß sie nicht anders als durch eine ihrer Natur nach nur unvollständige Induction zu erweisen; indem er theils gleich hier, theils in der Folge überall, wo eine scheinbar abweichende Form von Sätzen ihm vorkommt, zu zeigen sucht, daß man auch diesen Ausdruck auf die Form: 'A hat b', zurückführen könne, ja müsse, wenn recht verständlich werden soll, was man hier eigentlich sage. Ob nun die Leser mit dieser Zurückführung überall zufrieden seyn werden, wissen wir nicht zu sagen: gewiß ist nur, daß das Ganze nie als ein vollständiger Beweis angesehen werden könne."

[3] Cf. TS §137, II 37.

A series of remarks are necessary in order to justify the use of such a respectable name as '*conjecture*' for something that may, at first glance, appear to be nothing more than a naive assumption—and one that is capable of completely trivializing the system.

First of all, we must remark that, as has already been seen—and will be seen still better in the following—nobody before Bolzano ever had such great awareness of the variety of the ways in which a proposition may result from the composition of other propositions. Indeed, we may say that it was his opinion that any relation among any number of propositions univocally determines one of those ways. His conjecture was that *despite* the fact that the *verbal expression* of a composition of this kind is in general certainly not of the form '*A has b*', it is nonetheless thinkable that its sense could always be an *atomic proposition* (in itself) or at least an equivalent. In this formulation, the 'conjecture' appears to be more properly a '*thesis*', an assertion that cannot be proven, but only made plausible and hence, first of all, empirically verified.

To this end, verbal propositions must be isolated insofar as possible and it must be subsequently shown how they should be reformulated, *preserving their sense*, in order to assume the form '*A has b*', so that the proposition expressed by the latter, immediately recognizable as atomic, could be understood as the real *sense* of the former. In other words, those verbal propositions must be *interpreted* [*ausgelegt*] as expressing an atomic proposition (in itself).[4] Clearly each successful test of this kind makes it definitely plausible that, even by ransacking the propositions of that particular form, it will never be possible to find a counterexample to the conjecture.

The plausibility of the thesis, however, is not only supported by presenting and successfully solving an increasing number of particular cases, but also by simultaneously offering a uniform strategy to obtain the desired interpretation.

The conceptual kernel of the Bolzanian strategy may be summarized as *the sense of a composed verbal proposition is a 'semantic' proposition* (i.e. a proposition that attributes semantic qualities to lectological things).

As the two fundamental semantic properties are the *objectuality* of ideas and the *truth* of propositions, it is clear that the attainment of that purpose presupposes at least that

(1) both ideas and propositions are *objects*;
(2) both objectuality and truth are *qualities*;
(3) there are *object ideas* that refer to ideas and to propositions;

[4] Chapter 9 of this book is devoted to the Bolzanian theory of signs and hence also to the explanation of the terms we use here.

(4) there are *quality ideas* that refer to objectuality and to truth.

Indeed, it is only under these conditions that a proposition may attribute objectuality or objectlessness to an idea or, respectively, truth or falsity to a proposition.

In the following sections of this chapter, we will try to account for these fundamental Bolzanian insights. We will begin by only considering ideas (*restricted symbolic level*), subsequently extending our investigation to propositions (*extended symbolic level*).

Let us also preliminarily remark that for a certain class of propositions, the validation of the weak form of the conjecture also allows that of the strong form. Indeed, once it has been established that each proposition of a certain class is equivalent to a proposition of the atomic form, it will then be possible to simply *define* such a proposition as its atomic equivalent.

4.3 Restricted Symbolic Level—Ideas of Ideas

In §90, Bolzano addresses for the first time the fact that ideas are objects, too. He does this in an indirect manner by facing those ideas, in the content of which, and precisely as the principal part, the notion itself of an idea occurs (i.e. ideas having the form '*an idea that has (the quality)* $\overline{b}$'. For lack of a better name, Bolzano proposes they be called *symbolic* [*symbolische*] *ideas* or *ideas of ideas* [*Vorstellungsvorstellung*].[5] This issue reappears in a more determined way in §137.[6]

For a satisfying systematization of this subject, it will be necessary to begin by postulating the following:

(1) a general axiom ensuring that *object ideas are objects*; i.e.

$$\forall x(V(x) \rightarrow O(x)) \qquad \text{GA20}$$

i.e.

$$O(a) \qquad \text{T4.1}$$

[5] TS §90 I 306: "I will mention one more kind of ideas, which have the peculiarity that they contain the concept of an *idea* as their main part (§.58.), and thus are of the general form "an idea, which has (the attribute) *b* ." For want of a better term, I shall call such ideas *symbolic* or *ideas of ideas*."

[6] TS §137 II 36: "A most noteworthy genus of propositions[...] are those which treat of *ideas* or of *entire propositions*. [...] If the object a proposition treats of is a mere idea, then understandably its subject-idea or basis must be an idea of an idea. Now we called such ideas *symbolic* (§.90.). Hence all such propositions have as a common feature that their subject is a *symbolic* idea."

(2) an ontological axiom ensuring that *objectuality is a quality*; i.e.

$$\exists r \forall a(a \in r \leftrightarrow G(a)) \qquad \text{CQ4}$$

(3) a lectological axiom, with its relative resolution, ensuring the presence in the universe of a quality idea $\bar{g}$ *referring to the quality of being objectual*; i.e.

$$\exists \bar{c} \forall r(r \mathrel{\mathrm{E}} \bar{c} \leftrightarrow \forall a(a \in r \rightarrow G(a)) \qquad \text{CB5}$$

$$\bar{g} = \varepsilon \bar{c} \forall r(r \mathrel{\mathrm{E}} \bar{c} \leftrightarrow \forall a(a \in r \rightarrow G(a)) \qquad \text{D4.1}$$

from which, as usual,

$$(1)\ B(\bar{g}) \qquad \text{T4.2}$$
$$(2)\ r \mathrel{\mathrm{E}} \bar{g} \leftrightarrow \forall a(a \in r \rightarrow G(a))$$

On the basis of these assumptions, it is possible to prove that for an idea, *being objectual is equivalent to falling under the idea of something that has objectuality*; i.e.

$$a \mathrel{\mathrm{E}} [e](\bar{g}) \leftrightarrow G(a) \qquad \text{T4.3}$$

Proof If $a \mathrel{\mathrm{E}} [e](\bar{g})$, then there is an r such that $a \in r$ and $r \mathrel{\mathrm{E}} \bar{g}$; therefore, by T2, $\forall a(a \in r \rightarrow G(a))$; thus, because $a \in r$, it is $G(a)$. Vice versa, let $G(a)$; by CQ4, there is an r such that $\forall a(a \in r \leftrightarrow G(a))$; hence (1) $a \in r$; (2) $\forall a(a \in r \rightarrow G(a))$ and so, by T2, $r \mathrel{\mathrm{E}} \bar{g}$. Therefore $a \mathrel{\mathrm{E}} [e](\bar{g})$.

The preceding postulations may be expanded: first by assuming (in line with Bolzano's convictions) that *every finite width is a quality* too, as well as that *having a width at least equal to n* or *having a width at most equal to n are also qualities*; i.e.

$$\exists r \forall a(a \in r \leftrightarrow {}_{n}A(a)) \qquad \text{SCQ3}$$
$$\exists r \forall a(a \in r \leftrightarrow {}^{n}A(a)) \qquad \text{SCQ4}$$
$$\exists r \forall a(a \in r \leftrightarrow {}^{n}_{n}A(a)) \qquad \text{SCQ5}$$

and subsequently by accepting the presence, for every *n*, of a quality idea $\overline{\geq n}$ referring to *having at least width n*, a quality idea $\overline{\leq n}$ referring to *having at most width n*, and a quality idea $\bar{n}$ referring to *having (exactly) width < n*:

$$\exists \bar{c} \forall r(r \mathrel{\mathrm{E}} \bar{c} \leftrightarrow \forall a(a \in r \rightarrow {}_{n}A(u))) \qquad \text{SCB2}$$
$$\exists \bar{c} \forall r(r \mathrel{\mathrm{E}} \bar{c} \leftrightarrow \forall a(a \in r \rightarrow {}^{n}A(u))) \qquad \text{SCB3}$$
$$\exists \bar{c} \forall r(r \mathrel{\mathrm{E}} \bar{c} \leftrightarrow \forall a(a \in r \rightarrow {}^{n}_{n}A(u))) \qquad \text{SCB4}$$

$$\overline{\scriptstyle \geq n} = \varepsilon\bar{c}\forall r(r \mathsf{E} \bar{c} \leftrightarrow \forall a(a \in r \rightarrow {}_{n}A(u))) \qquad \text{SD4.20}$$

$$\overline{\scriptstyle \leq n} = \varepsilon\bar{c}\forall r(r \mathsf{E} \bar{c} \leftrightarrow \forall a(a \in r \rightarrow {}^{n}A(u))) \qquad \text{SD4.21}$$

$$\overline{\scriptstyle n} = \varepsilon\bar{c}\forall r(r \mathsf{E} \bar{c} \leftrightarrow \forall a(a \in r \rightarrow {}^{n}_{n}A(u))) \qquad \text{SD4.22}$$

$$\begin{aligned} &(1)\ B(\overline{\scriptstyle \geq n}) && \text{ST4.4}\\ &(2)\ r \mathsf{E} \overline{\scriptstyle \geq n} \leftrightarrow \forall a(a \in r \rightarrow {}_{n}A(a)) &&\\ &(3)\ a \mathsf{E} [e](\overline{\scriptstyle \geq n}) \leftrightarrow {}_{n}A(a) && \end{aligned}$$

$$\begin{aligned} &(1)\ B(\overline{\scriptstyle \leq n}) && \text{ST4.5}\\ &(2)\ r \mathsf{E} \overline{\scriptstyle \leq n} \leftrightarrow \forall a(a \in r \rightarrow {}^{n}A(a)) &&\\ &(3)\ a \mathsf{E} [e](\overline{\scriptstyle \leq n}) \leftrightarrow {}^{n}A(a) && \end{aligned}$$

$$\begin{aligned} &(1)\ B(\overline{\scriptstyle n}) && \text{ST4.6}\\ &(2)\ r \mathsf{E} \overline{\scriptstyle n} \leftrightarrow \forall a(a \in r \rightarrow {}^{n}_{n}A(a)) &&\\ &(3)\ a \mathsf{E} [e](\overline{\scriptstyle n}) \leftrightarrow {}^{n}_{n}A(a) && \end{aligned}$$

At this point, in order to present the statement that an idea a is objectual or has a determinate finite width as an authentic *proposition*, we still need to ensure the presence of a ('*symbolic*') idea that refers to a and may thus act as the *subject* of such a proposition. Actually, Bolzano was convinced that for every object there was at least an idea referring to it and only to it[7] and so, given the fact that at the symbolic level ideas are objects, this Bolzanian conviction, if transformed into an axiom, would render superfluous a special assumption for that particular type of objects that are the ideas. The question, however, is not so simple. The point is that in the developments of the Bolzanian theory, there is a point in which it becomes unavoidable to *accept that the idea a itself is a part of the symbolic idea referring to it.* Now, such a property cannot belong to the type of idea that is at the centre of the Bolzanian conviction: an elephant, indeed, is certainly not a part of the idea that refers to it. It is therefore convenient to keep the two questions separate and accept what we will call the '*principle of material supposition of an idea*', that is to say, a comprehension axiom ensuring the presence, for each a, *of an idea referring to it and only to it and having a as its next part*, as well as its ε-resolution, to be denoted by '$\ulcorner a \urcorner$'.Thus

$$\forall a \exists c(\forall i(i \mathsf{E} c \leftrightarrow i = a) \wedge \forall l(l < c \leftrightarrow l \leq a)) \qquad \text{CV3}$$

$$\ulcorner a \urcorner = \varepsilon c(\forall i(i \mathsf{E} c \leftrightarrow i = a) \wedge \forall l(l < c \leftrightarrow l \leq a)) \qquad \text{D4.23}$$

[7] TS §101, I 339: "For it is certain that for every individual object, [. . .] there is and must be an idea applying exclusively to it [. . .]."

$$\begin{aligned} &(1)\ V(\ulcorner a \urcorner) \\ &(2)\ i\,\mathrm{E}\,\ulcorner a \urcorner \leftrightarrow i = a \\ &(3)\ l < \ulcorner a \urcorner \leftrightarrow l \leq a \end{aligned} \qquad \text{T4.7}$$

In this way, we know that among the propositions present in our universe, there are those having the form '$h(\ulcorner a \urcorner, \overline{g})$', which Bolzano calls '*enunciations* or *assertions of objectuality* [*Aussagen* or *Behauptungen einer Gegenständlichkeit*]',[8] and those of the form '$h(\ulcorner a \urcorner, \sim \overline{g})$', which Bolzano calls '*negations of objectuality* [*Verneinungen der Gegenständlichkeit*]'.[9]

It is now possible to prove the fundamental fact that *an idea is objectual iff the proposition that enunciates the belonging to it of objectuality is true, whereas it is objectless iff the proposition is true that enunciates the belonging to it of the lack of objectuality*; i.e.

$$\begin{aligned} &(1)\ \ G(a) \leftrightarrow W(h(\ulcorner a \urcorner, \overline{g})) \\ &(2)\ \ \neg G(a) \leftrightarrow W(h(\ulcorner a \urcorner, \sim \overline{g})) \end{aligned} \qquad \text{T4.8}$$

Proof *Ad* (1) $W(h(\ulcorner a \urcorner, \overline{g})) \leftrightarrow \exists i(i\,\mathrm{E}\,\ulcorner a \urcorner) \wedge \forall i(i\,\mathrm{E}\,\ulcorner a \urcorner \rightarrow \exists r(i \in r \wedge r\,\mathrm{E}\,\overline{g})) \leftrightarrow$
$\exists i(i = a) \wedge \forall i(i = a \rightarrow \exists r(i \in r \wedge r\,\mathrm{E}\,\overline{g})) \leftrightarrow \forall i(i = a \rightarrow \exists r(i \in r \wedge r\,\mathrm{E}\,\overline{g})) \leftrightarrow$
$\exists r(a \in r \wedge r\,\mathrm{E}\,\overline{g}) \leftrightarrow a\,\mathrm{E}\,[e](\overline{g}) \leftrightarrow G(a)$.

Ad (2) $W(h(\ulcorner a \urcorner, \sim \overline{g})) \leftrightarrow \exists i(i\,\mathrm{E}\,\ulcorner a \urcorner) \wedge \forall i(i\,\mathrm{E}\,\ulcorner a \urcorner \rightarrow \exists r(i \in r \wedge r\,\mathrm{E} \sim \overline{g})) \leftrightarrow$
$\exists i(i = a) \wedge \forall i(i = a \rightarrow \exists r(i \in r \wedge r\,\mathrm{E} \sim \overline{g})) \leftrightarrow$
$\forall i(i = a \rightarrow \exists r(i \in r \wedge r\,\mathrm{E} \sim \overline{g})) \leftrightarrow$
$\exists r(a \in r \wedge r\,\mathrm{E} \sim \overline{g}) \leftrightarrow a\,\mathrm{E}\,[e](\sim \overline{g}) \leftrightarrow \neg a\,\mathrm{E}\,[e](\overline{g}) \leftrightarrow \neg G(a)$.

This, together with the final remarks of Chapter 2, has as an immediate consequence that the subsistence of the material relations between two ideas considered in that place, may be presented as (conjunctions of) true assertions of objectuality (objectlessness). Indeed:

(1) *Compat.* : $a \top c \leftrightarrow W(h(\ulcorner a \sqcap c \urcorner, \overline{g}))$

(2) *Inclus.* : $a \sqsubseteq c \leftrightarrow W(h(\ulcorner a \sqcap c \urcorner, \overline{g})) \wedge W(h(\ulcorner a \sqcap -c \urcorner, \sim \overline{g}))$

(3) *Dispar.* : $a\, \dot{\top}\, c \leftrightarrow W(h(\ulcorner a \sqcap c \urcorner, \overline{g})) \wedge W(h(\ulcorner a \sqcap -c \urcorner, \overline{g})) \wedge W(h(\ulcorner -a \sqcap c \urcorner, \overline{g}))$

[8] TS §137 II 36–7: "The first question that may occur to us when considering an idea is whether or not it has objects or, to use the corresponding abstractum, whether it has *objectuality*. If we answer in the affirmative, we express a proposition which is contained under the following general form: '*The idea A has objectuality.*' Permit me to call such propositions statements or *claims of objectuality*."

[9] TS §138 II 38: "If we answer in the negative to the question raised in the previous section, [. . .] we form a proposition of the form: '*The idea A-has-no objectuality.*' I allow myself to call such propositions *denials of objectuality*."

$$
\begin{array}{llll}
(4)\ \textit{Equival.} & : a \equiv c & \leftrightarrow & W(h(\ulcorner a \sqcap c \urcorner, \bar{g})) \wedge W(h(\ulcorner a \sqcap -c \urcorner, \sim \bar{g})) \wedge W(h(\ulcorner -a \sqcap c \urcorner, \sim \bar{g})) \\
(5)\ \textit{Subord.} & : a \dot{\sqsubseteq} c & \leftrightarrow & W(h(\ulcorner a \sqcap c \urcorner, \bar{g})) \wedge W(h(\ulcorner a \sqcap -c \urcorner, \sim \bar{g})) \wedge W(h(\ulcorner -a \sqcap c \urcorner, \bar{g})) \\
(6)\ \textit{Incomp.} & : a \perp c & \leftrightarrow & W(h(\ulcorner a \sqcap c \urcorner, \sim \bar{g})) \\
(7)\ \textit{Contrad.} & : a \,||\, c & \leftrightarrow & W(h(\ulcorner a \sqcap c \urcorner, \sim \bar{g})) \wedge W(h(\ulcorner e \sqcap -a \sqcap -c \urcorner, \sim \bar{g})) \\
(8)\ \textit{Contrar.} & : a \dot{\perp} c & \leftrightarrow & W(h(\ulcorner a \sqcap c \urcorner, \sim \bar{g})) \wedge W(h(\ulcorner e \sqcap -a \sqcap -c \urcorner, \bar{g}))
\end{array}
$$

T4.9

Note that for $\sqsubseteq$ and $\equiv$, we also have, more simply,

$$
\begin{array}{l}
(1)\ a \sqsubseteq c \leftrightarrow W(h(a, \widehat{c})) \\
(2)\ a \equiv c \leftrightarrow W(h(a \sqcup c, \widehat{a} + \widehat{c}))
\end{array}
\qquad \text{T4.10}
$$

Besides the assertions of objectuality, those of *singularity, plurality* and *determinate finite width* are now at our disposal: $h(\ulcorner a \urcorner, \bar{1})$, $h(\ulcorner a \urcorner, \bar{+})$, $h(\ulcorner a \urcorner, \bar{n})$, $h(\ulcorner a \urcorner, \overline{\leq n})$, $h(\ulcorner a \urcorner, \overline{\geq n})$.

4.4 The Traditional Categorical Propositions

It is now possible to prove an important fact: *the 'Bolzano's conjecture'* (which will henceforth be called 'Bolzano's thesis') *can be proved for all traditional categorical propositions already at the restricted symbolic level.* Indeed, if we set[10,11]

$$
\begin{array}{llll}
(1) & a\mathbb{A}^*c = h(a, \widehat{c}) & (2) & a\mathbb{E}^*c = h(a, \sim \widehat{c}) \\
(3) & a\mathbb{A}c \ = h(\ulcorner [a](\sim \widehat{c}) \urcorner, \sim \bar{g}) & (4) & a\mathbb{E}c \ = h(\ulcorner [a](\widehat{c}) \urcorner, \sim \bar{g}) \\
(5) & a\mathbb{I}c \ = h(\ulcorner [a](\widehat{c}) \urcorner, \bar{g}) & (6) & a\mathbb{O}c \ = h(\ulcorner [a](\sim \widehat{c}) \urcorner, \bar{g})
\end{array}
\qquad \text{D4.24}
$$

then we can prove without any difficulty that the propositions so defined behave in accordance with the traditional values of $\mathbb{A}^*$, $\mathbb{E}^*$, $\mathbb{A}$, $\mathbb{E}$, $\mathbb{I}$, and $\mathbb{O}$, i.e.

$$
\begin{array}{l}
(1)\ W(a\mathbb{A}^*c) \leftrightarrow \exists i(i \,\mathsf{E}\, a) \wedge \forall i(i \,\mathsf{E}\, a \rightarrow i \,\mathsf{E}\, c) \\
(2)\ W(a\mathbb{E}^*c) \leftrightarrow \exists i(i \,\mathsf{E}\, a) \wedge \forall i(i \,\mathsf{E}\, a \rightarrow \neg i \,\mathsf{E}\, c) \\
(3)\ W(a\mathbb{A}c) \ \leftrightarrow \forall i(i \,\mathsf{E}\, c \rightarrow i \,\mathsf{E}\, c)
\end{array}
\qquad \text{T4.11}
$$

[10] TS §138, II 38: "[...] It may seem doubtful whether propositions whose linguistic expression is contained in the form 'No *A* is *B*' should be reckoned to be denials of objectuality, i.e., interpreted to mean 'The idea of an *A* which is also a *B* (or which has the attribute *b* hat) has no objectuality'; or whether they should be looked upon as merely negative propositions of the form 'Every *A* is a non-*B*', i.e., 'Every *A* has the attribute non-*b*.' If, as is often the case, we take the expression 'No *A* is a *B*' in the sense where we do not presuppose that there is an *A*, then it is not even equivalent to the expression 'Every *A* is a non-*B*.' For [...]."

[11] TS §233, II 312–13: "[...] the meaning of the expression 'Some *A* are (or are not) *B*' as used by logicians is simply that the idea of an *A* that is at the same time *B* (or not *B*)—i.e., has or lacks the attribute *b*—has objectuality. Particular judgements are thus a special case of statements of the objectuality of an idea, in particular about an idea of the form [*A*]*b* or [*A*]*nb*."

$$
\begin{aligned}
&(4)\ W(a\mathbb{E}b) \leftrightarrow \forall i(i \,\text{\textsc{e}}\, a \rightarrow \neg i \,\text{\textsc{e}}\, c)\\
&(5)\ W(a\mathbb{I}c) \ \ \leftrightarrow \exists i(i \,\text{\textsc{e}}\, a \wedge i \,\text{\textsc{e}}\, c)\\
&(6)\ W(a\mathbb{O}c) \leftrightarrow \exists i(i \,\text{\textsc{e}}\, a \wedge \neg i \,\text{\textsc{e}}\, c)
\end{aligned}
$$

To complete the subject, let us remark that, as far as the *Reductiones ad subalternatam* are concerned, we have

$$
\begin{array}{llll}
(1)\ \ W(a\mathbb{A}^*c) \rightarrow W(a\mathbb{I}c) & (2)\ \ W(a\,\text{\textsc{e}}^*c) \rightarrow W(a\mathbb{O}c) & \text{T4.12}\\
(3)\ \ W(a\,\text{\textsc{e}}\,c) \not\rightarrow W(a\mathbb{O}c) & (4)\ \ W(a\,\text{\textsc{e}}c) \not\rightarrow W(a\mathbb{O}c) &
\end{array}
$$

whereas, as far as the *Conversiones simplices* are concerned, we have

$$
\begin{array}{lll}
(1)\ \ W(a\mathbb{E}c) \rightarrow W(c\mathbb{E}a) & (2)\ \ W(a\,\mathbb{I}\,c) \rightarrow W(c\,\mathbb{I}\,a) & \text{T4.13}\\
(3)\ \ W(a\mathbb{E}^*c) \not\rightarrow W(a\mathbb{E}^*c) & [\text{if } G(a) \text{ but } \neg G(c), W(a\mathbb{E}^*c) \text{ but } \neg W(c\mathbb{E}^*a)] &
\end{array}
$$

Finally, as far as the *Conversiones per accidens* are concerned, we have

$$W(a\mathbb{A}^*c) \rightarrow W(a\mathbb{I}c) \qquad \text{T4.14}$$

and, as $\mathbb{E}^*$ does not convert *simpliciter* and $\mathbb{E}$ does not subalternate, neither of them converts *per accidens.*

4.5 Extended Symbolic Level: Ideas of Propositions

Let us now deal with the extension to propositions of the symbolic level,[12] accepting first of all that *propositions are objects.*

We thus assume

$$\forall x(S(x) \rightarrow O(x)) \qquad \text{GA21}$$

i.e.

$$O(p) \qquad \text{T4.15}$$

As in the case of ideas, we have promptly ensured that to the fundamental property of objectuality there corresponds, within the universe of things, at least one quality, so we must now make sure that to the fundamental property that a proposition may enjoy, the *truth,* there corresponds in the same universe (at least) one quality. We thus assume

$$\exists r \forall p(p \in r \leftrightarrow W(p)) \qquad \text{CQ5}$$

[12] TS §141, II 43: "Just as noteworthy as propositions which treat of *ideas* are those which deal with *propositions themselves.* For the present, though, I shall only mention those stating attributes of propositions that these propositions have considered in and of themselves [. . .]."

Also in strict parallelism with the case of objectuality, we must ensure the presence of a quality idea, denoted by '$\overline{w}$', *that refers to being true*. Thus

$$\exists\bar{c}\forall r(r \mathrel{\mathrm{E}} \bar{c} \leftrightarrow \forall p(p \in r \rightarrow W(p))) \qquad \text{CB6}$$

as well as

$$\overline{w} = \varepsilon\bar{c}\forall r(r \mathrel{\mathrm{E}} \bar{c} \leftrightarrow \forall p(p \in r \rightarrow W(p))) \qquad \text{D4.25}$$

so that

$$\begin{aligned} &(1)\ \ B(\overline{w}) && \text{T4.16}\\ &(2)\ \ r \mathrel{\mathrm{E}} \overline{w} \leftrightarrow \forall p(p \in r \rightarrow W(p)) && \end{aligned}$$

We can now prove the fundamental fact that *being true is equivalent to falling under the idea of something that has truth*:

$$p \mathrel{\mathrm{E}} [e](\overline{w}). \leftrightarrow W(p) \qquad \text{T4.17}$$

Proof The proof schema is the same as has already been applied elsewhere. As an example, the following is a purely formal proof in our logical system:

$$\begin{aligned}
&1.\ \ p \in r \wedge \forall q(q \in r \rightarrow W(q)) \rightarrow W(p)\\
&2.\ \ r \mathrel{\mathrm{E}} \overline{w} \leftrightarrow \forall q(q \in r \rightarrow W(q))\\
&3.\ \ p \in r \wedge r \mathrel{\mathrm{E}} \overline{w} \rightarrow W(p)\\
&4.\ \ \exists r(p \in r \wedge r \mathrel{\mathrm{E}} \overline{w}) \rightarrow W(p)\\
&5.\ \ p \mathrel{\mathrm{E}} [e](\overline{w}) \rightarrow W(p)\\
&6.\ \ \forall p(p \in r \leftrightarrow W(p)) \wedge W(p) \rightarrow p \in r\\
&7.\ \ \forall p(p \in r \leftrightarrow W(p)) \rightarrow \forall p(p \in r \rightarrow W(p))\\
&8.\ \ \forall p(p \in r \leftrightarrow W(p)) \rightarrow r \mathrel{\mathrm{E}} \overline{w}\\
&9.\ \ \forall p(p \in r \leftrightarrow W(p)) \wedge W(p) \rightarrow p \in r \wedge r \mathrel{\mathrm{E}} \overline{w}\\
&10.\ \ \forall p(p \in r \leftrightarrow W(p)) \wedge W(p) \rightarrow \exists r(p \in r \wedge r \mathrel{\mathrm{E}} \overline{w})\\
&11.\ \ \forall p(p \in r \leftrightarrow W(p)) \rightarrow (W(p) \rightarrow p \mathrel{\mathrm{E}} [e](\overline{w}))\\
&12.\ \ \exists r\forall p(p \in r \leftrightarrow W(p)) \rightarrow (W(p) \rightarrow p \mathrel{\mathrm{E}} [e](\overline{w}))\\
&13.\ \ \exists r(\forall p(p \in r \leftrightarrow W(p))\\
&14.\ \ W(p) \rightarrow p \mathrel{\mathrm{E}} [e](\overline{w})
\end{aligned}$$

We now proceed to extend the '*Principle of material supposition*' to include propositions too; that is to say, we quite generally assume that for every lectological thing there is an object idea that refers exclusively to it and that has as its parts the thing itself and its parts. Thus

$$\forall l\exists a(\forall i(i \mathrel{\mathrm{E}} a \leftrightarrow i = l) \wedge \forall l'(l' < a \leftrightarrow l' \leq l)) \qquad \text{CV4}$$

$$\lceil l \rceil = \varepsilon a(\forall i(i \mathrel{\text{E}} a \leftrightarrow i = l) \wedge \forall l'(l' < a \leftrightarrow l' \leq l)) \qquad \text{D4.26}$$

from which, in particular, besides what is inferred from the restricted axiom for object ideas, we may also infer the corresponding theorems both for quality ideas (what we leave aside) and for propositions; i.e.

$$\begin{aligned} &(1)\ V(\lceil p \rceil) \\ &(2)\ i \mathrel{\text{E}} \lceil p \rceil \leftrightarrow i = p \\ &(3)\ l < \lceil p \rceil \leftrightarrow l \leq p \end{aligned} \qquad \text{T4.18}$$

At this point, we know that among the propositions in our universe there are not only propositions that say that a certain idea has or does not have objectuality, but also *affirmative propositions* or *affirmations* [*Bejahungssätze* or *Bejahungen*] asserting that a certain proposition is true and which are all and only all those having the form

$$h(\lceil p \rceil, \overline{w})$$

as well as *negative* or *corrective propositions* or *negations* [*Verneinungs-* or *Berichtigungssätze* or *Verneinungen*] asserting that a certain proposition is false and which are those and only those having the form

$$h(\lceil p \rceil, \sim \overline{w})$$ [13]

It is now possible to prove the other fundamental fact that *a proposition is true iff its affirmation (i.e. the proposition enunciating the belonging to it of the truth) is true* and *it is false iff its negation (i.e. the proposition enunciating the belonging to it of the lack of truth) is true*; i.e.

$$\begin{aligned} &(1)\ W(p) \leftrightarrow W(h(\lceil p \rceil, \overline{w})) \\ &(2)\ F(p) \leftrightarrow W(h(\lceil p \rceil, \sim \overline{w})) \end{aligned} \qquad \text{T4.19}$$

Proof *Ad* (1) $W(h(\lceil p \rceil, \overline{w})) \leftrightarrow \exists i(i \mathrel{\text{E}} \lceil p \rceil) \wedge \forall i(i \mathrel{\text{E}} \lceil p \rceil \rightarrow \exists r(i \in r \wedge r \mathrel{\text{E}} \overline{w})) \leftrightarrow$
$\exists i(i = p) \wedge \forall i(i = p \rightarrow \exists r(i \in r \wedge r \mathrel{\text{E}} \overline{w})) \leftrightarrow$
$\forall i(i = p \rightarrow \exists r(i \in r \wedge r \mathrel{\text{E}} \overline{w})) \leftrightarrow$
$\exists r(p \in r \wedge r \mathrel{\text{E}} \overline{w})) \leftrightarrow p \mathrel{\text{E}} [e](\overline{w}) \leftrightarrow W(p)$.

[13] TS §141, II 44: "[. . .] The most important thing we can ask about a given proposition *A* is doubtless whether it is true. If we answer this question in the affirmative, I intend to call the judgement we then make an *affirmation* of *A*, or an *affirmative proposition*, while if we answer in the negative, this may be called a *denial* of *A*, or a *corrective proposition*. The general form of affirmative propositions is then 'The proposition *A*-has-truth'; while the general form of denials is: 'The proposition *A*-has-no truth', expressions from which one readily gathers the constituents of which I consider these sorts of propositions to be composed."

Ad (2) $W(h(\lceil p \rceil, \sim \overline{w})) \leftrightarrow \exists i(i \,\text{E}\, \lceil p \rceil) \wedge \forall i(i \,\text{E}\, \lceil p \rceil \rightarrow \exists r(i \in r \wedge r \,\text{E} \sim \overline{w})) \leftrightarrow$
$\exists i(i = p) \wedge \forall i(i = p \rightarrow \exists r(i \in r \wedge r \,\text{E} \sim \overline{w})) \leftrightarrow$
$\forall i(i = p \rightarrow \exists r(i \in r \wedge r \,\text{E} \sim \overline{w})) \leftrightarrow$
$\exists r(p \in r \wedge r \,\text{E} \sim \overline{w})) \leftrightarrow p \,\text{E}\, [e](\sim \overline{w}) \leftrightarrow \neg p \,\text{E}\, [e](\overline{w}) \leftrightarrow \neg W(p) \leftrightarrow F(p).$

4.6 Material Connectives

What we have learned in the previous sections allows us to take further steps towards understanding *Bolzano's thesis*, establishing that it *holds for propositions obtained by material connection of propositions for which it already holds*. To this end, let us begin by defining, following Bolzano,

$$\boxed{\neg} p = h(\lceil p \rceil, \sim \overline{w})^{14} \qquad \text{D4.27}$$

From the last theorem, we immediately obtain

$$W(\boxed{\neg} p) \leftrightarrow \neg W(p) \leftrightarrow F(p) \qquad \text{T4.20}$$

Let us remark (and this remark will also be relevant in the following sections) that although the connectives (material or formal) introduced at the elementary level behave with respect to truth and falsity just like their homonymous counterparts that we will introduce at the symbolic level, they are not at all the same. The most evident manifestation of this difference lies on the mereological plane: the 'elementary' $\boxed{\neg} p$ has as its proper parts only p and the parts of p; the 'symbolic' $\boxed{\neg} p$, as can be seen, has as its proper parts: $\lceil p \rceil$, p, the parts of p and of $\overline{w}$ (on the mereological structure of the latter, however, no hypothesis has been made).

Writing in short '$\lceil \overset{n}{p} \rceil$' for '$\lceil p_1 \rceil \sqcup \ldots \sqcup \lceil p_n \rceil$', we may now define

$$\boxed{\wedge}(\overset{n}{p}) = h(\lceil \overset{n}{p} \rceil, \overline{w}) \qquad \text{SD4.28}$$

and prove that the conjunction so defined really does satisfy the condition imposed at its time, thus also showing that at the symbolic level the axioms introduced in Section 3.3 become superfluous. Indeed,

$$W(\boxed{\wedge}(\overset{n}{p})) \leftrightarrow W(\overset{n}{p}) \qquad \text{ST4.21}$$

[14] TS §141, II 44: "Where *A* designates a proposition I shall frequently designate its denial, or the proposition that *A* is false, by '*Neg.A*'."

TS §147, II 57: "[. . .] the denial of *A*, or the proposition '*A* is false' [. . .]."

TS §155, II 81: "7) The *negation* of a proposition *A*, *Neg.A*, i.e., the proposition '*A* is false', is never deducible from *A*, no matter what ideas *i*, *j*, . . . , which occur only in *A*, are envisaged as variable."

Proof $W(\boxed{\wedge}(\overset{n}{p})) \leftrightarrow W(h(\ulcorner \overset{n}{p} \urcorner, \overline{w}))$

$$\leftrightarrow \exists i(i \,\mathrm{E} \ulcorner \overset{n}{p} \urcorner) \wedge \forall i(i \,\mathrm{E} \ulcorner \overset{n}{p} \urcorner \rightarrow \exists r(i \in r \wedge r \,\mathrm{E}\, \overline{w}))$$
$$\leftrightarrow \exists i(i \,\mathrm{E} \lceil p_1 \rceil \vee \ldots \vee i \,\mathrm{E} \lceil p_n \rceil) \wedge \forall i(i \,\mathrm{E} \lceil p_1 \rceil \vee \ldots \vee i \,\mathrm{E} \lceil p_n \rceil \rightarrow \exists r(i \in r \wedge r \,\mathrm{E}\, \overline{w}))$$
$$\leftrightarrow \exists i(i = p_1 \vee \ldots \vee i = p_n) \wedge \forall i(i \,\mathrm{E} \lceil p_1 \rceil \vee \ldots \vee i \,\mathrm{E} \lceil p_n \rceil \rightarrow \exists r(i \in r \wedge r \,\mathrm{E}\, \overline{w}))$$

but, by T15, $\exists i(i = p_1 \vee \ldots \vee i = p_n)$ is always true; therefore

$$\leftrightarrow \forall i(i \,\mathrm{E} \lceil p_1 \rceil \vee \ldots \vee i \,\mathrm{E} \lceil p_n \rceil \rightarrow \exists r(i \in r \wedge r \,\mathrm{E}\, \overline{w}))$$
$$\leftrightarrow \forall i((i \,\mathrm{E} \lceil p_1 \rceil \rightarrow \exists r(i \in r \wedge r \,\mathrm{E}\, \overline{w})) \wedge \ldots \wedge (i \,\mathrm{E} \lceil p_n \rceil \rightarrow \exists r(i \in r \wedge r \,\mathrm{E}\, \overline{w})))$$
$$\leftrightarrow \forall i(i \,\mathrm{E} \lceil p_1 \rceil \rightarrow \exists r(i \in r \wedge r \,\mathrm{E}\, \overline{w})) \wedge \ldots \wedge \forall i(i \,\mathrm{E} \lceil p_n \rceil \rightarrow \exists r(i \in r \wedge r \,\mathrm{E}\, \overline{w}))$$
$$\leftrightarrow \forall i(i = p_1 \rightarrow \exists r(i \in r \wedge r \,\mathrm{E}\, \overline{w})) \wedge \ldots \wedge \forall i(i = p_n \rightarrow \exists r(i \in r \wedge r \,\mathrm{E}\, \overline{w}))$$
$$\leftrightarrow \exists r(p_1 \in r \wedge r \,\mathrm{E}\, \overline{w}) \wedge \ldots \wedge \exists r(p_n \in r \wedge r \,\mathrm{E}\, \overline{w})$$
$$\leftrightarrow p_1 \,\mathrm{E}\, [e](\overline{w}) \wedge \ldots \wedge p_n \,\mathrm{E}\, [e](\overline{w})$$
$$\leftrightarrow W(p_1) \wedge \ldots \wedge W(p_n)$$
$$\leftrightarrow W(\overset{n}{p})$$

The 'material integration' of propositions $p_1, \ldots, p_n$, on the other hand, is conceived as the proposition stating that the idea of a true proposition among them has objectuality; i.e.

$$\boxed{\vee}(\overset{n}{p}) = h([\ulcorner \overset{n}{p} \urcorner](\overline{w}), \overline{g})^{15} \qquad \text{SD4.29}$$

As in the case of conjunction, we now prove

$$W(\boxed{\vee}(\overset{n}{p})) \leftrightarrow E(\overset{n}{p}) \qquad \text{ST4.22}$$

Proof $W(\boxed{\vee}(\overset{n}{p})) \leftrightarrow W(h([\ulcorner \overset{n}{p} \urcorner](\overline{w}), \overline{g}))$

$$\leftrightarrow G([\ulcorner \overset{n}{p} \urcorner](\overline{w})) \leftrightarrow \exists i(i \,\mathrm{E} \ulcorner \overset{n}{p} \urcorner \wedge \exists r(i \in r \wedge r \,\mathrm{E}\, \overline{w}))$$
$$\leftrightarrow \exists i((i \,\mathrm{E} \lceil p_1 \rceil \vee \ldots \vee i \,\mathrm{E} \lceil p_n \rceil) \wedge \exists r(i \in r \wedge r \,\mathrm{E}\, \overline{w}))$$
$$\leftrightarrow \exists i((i = p_1 \vee \ldots \vee i = p_n)) \wedge \exists r(i \in r \wedge r \,\mathrm{E}\, \overline{w}))$$
$$\leftrightarrow \exists r((p_1 \in r \vee \ldots \vee p_n \in r) \wedge r \,\mathrm{E}\, \overline{w}))$$
$$\leftrightarrow \exists r((p_1 \in r \wedge r \,\mathrm{E}\, \overline{w}) \vee \ldots \vee (p_n \in r \wedge r \,\mathrm{E}\, \overline{w}))$$
$$\leftrightarrow \exists r(p_1 \in r \wedge r \,\mathrm{E}\, \overline{w}) \vee \ldots \vee \exists r(p_n \in r \wedge r \,\mathrm{E}\, \overline{w}))$$
$$\leftrightarrow p_1 \,\mathrm{E}\, [e](\overline{w}) \vee \ldots \vee p_n \,\mathrm{E}\, [e](\overline{w})$$
$$\leftrightarrow W(p_1) \vee \ldots \vee W(p_n)$$
$$\leftrightarrow E(\overset{n}{p})$$

[15] TS §166, II 145: "1) The relation we considered under the name of *complementation* (§.160) also gives rise to propositions which must be mentioned here. First, if we merely claim that one or more of the propositions $M, N, O, \ldots$ are true, our judgement is (according to §.137): 'The idea of a true proposition among $M, N, O, \ldots$ has objectuality.' A judgement of this sort may be called a statement of material complementation (§.160. n°.2.)."

Let us make the familiar notational conventions

$$p \boxed{\wedge} q := \boxed{\wedge}(p, q) \qquad \text{MD4.30}$$

$$p \boxed{\vee} q := \boxed{\vee}(p, q) \qquad \text{MD4.31}$$

and remark that, as particular cases of the preceding definitions, we have

$$p \boxed{\wedge} q = h(\lceil p \rceil \sqcup \lceil q \rceil, \overline{w}) \qquad \text{T4.23}$$

$$p \boxed{\vee} q = h([\lceil p \rceil \sqcup \lceil q \rceil](\overline{w}), \overline{g}) \qquad \text{T4.24}$$

whereas, as particular cases of the preceding theorems, we have

$$W(p \boxed{\wedge} q) \leftrightarrow W(p) \wedge W(q) \qquad \text{T4.25}$$

$$W(p \boxed{\vee} q) \leftrightarrow W(p) \vee W(q) \qquad \text{T4.26}$$

We also remark that for the so-defined $\boxed{\neg}$, $\boxed{\wedge}$ e $\boxed{\vee}$, the *principle of the excluded middle* holds:

$$W(p \boxed{\vee} \boxed{\neg}\, p) \qquad \text{T4.27}$$

Proof $W(p) \vee \neg W(p) \leftrightarrow W(p) \vee W(\boxed{\neg}\, p) \leftrightarrow W(p \boxed{\vee} \boxed{\neg}\, p)$.

The same holds for the *principle of non-contradiction*:

$$W(\boxed{\neg}(p \boxed{\wedge} \boxed{\neg}\, p)) \qquad \text{T4.28}$$

Proof $\neg(W(p) \wedge \neg W(p)) \leftrightarrow \neg(W(p) \wedge W(\boxed{\neg} p)) \leftrightarrow \neg W(p \boxed{\wedge} \boxed{\neg} p) \leftrightarrow W(\boxed{\neg}(p \boxed{\wedge} \boxed{\neg}\, p))$.

We may also define

$$\overset{m}{p} \boxed{\rightarrow} q = \boxed{\neg}((\boxed{\wedge^m} \overset{m}{p}) \boxed{\wedge} (\boxed{\vee^n} (\boxed{\neg} \overset{n}{q}))) \qquad \text{SD4.32}$$

obtaining

$$\overset{m}{p} \boxed{\rightarrow} q = h(\lceil h(\lceil h(\lceil \overset{m}{p} \rceil, \overline{w}) \rceil \sqcup \lceil h([\lceil h(\lceil q_1 \rceil, \sim \overline{w}) \rceil \sqcup \ldots \sqcup \lceil h(\lceil q_n \rceil, \sim \overline{w}) \rceil] (\overline{w}), \overline{g}) \rceil, \overline{w} \rceil, \sim \overline{w})$$

ST4.29

Proof $\overset{m}{p}\boxed{\rightarrow}q = \boxed{\neg}((\boxed{\wedge^m}\overset{m}{p})\boxed{\wedge}(\boxed{\vee^n}(\boxed{\neg}\overset{n}{q}))) = h(\ulcorner(\boxed{\wedge^m}\overset{m}{p})\boxed{\wedge}(\boxed{\vee^n}(\boxed{\neg}\overset{n}{q}))\urcorner, \sim\overline{w})$; but

$$(\boxed{\wedge^m}\overset{m}{p})\boxed{\wedge}(\boxed{\vee^n}(\boxed{\neg}\overset{n}{q})) = h(\ulcorner\boxed{\wedge^m}\overset{m}{p}\urcorner \sqcup \ulcorner\boxed{\vee^n}\urcorner, \overline{w}),$$

and hence

$$\overset{m}{p}\boxed{\rightarrow}q = h(\ulcorner h(\ulcorner\boxed{\wedge^m}\overset{m}{p}\urcorner \sqcup \ulcorner\boxed{\vee^n}(\boxed{\neg}\overset{n}{q})\urcorner, \overline{w})\urcorner, \sim\overline{w})$$

Now $\boxed{\wedge^m}\overset{m}{p} = h(\ulcorner\overset{m}{p}\urcorner, \overline{w})$, whereas $\boxed{\vee^n}(\boxed{\neg}\overset{n}{q}) = h([\ulcorner\boxed{\neg}\overset{n}{q}\urcorner](\overline{w}), \overline{g})$, and so

$$\ulcorner\boxed{\wedge^m}\overset{m}{p}\urcorner = \ulcorner h(\ulcorner\overset{m}{p}\urcorner,\overline{w})\urcorner)\ \ulcorner\boxed{\vee^n}(\boxed{\neg}\overset{n}{q})\urcorner = \ulcorner h([\ulcorner\boxed{\neg}\overset{n}{q}\urcorner](\overline{w}),\overline{g})\urcorner$$

therefore

$$\overset{m}{p}\boxed{\rightarrow}q = h(\ulcorner h(\ulcorner h(\ulcorner\overset{m}{p}\urcorner,\overline{w})\urcorner \sqcup \ulcorner h([\ulcorner\boxed{\neg}\overset{n}{q}\urcorner](\overline{w}),\overline{g})\urcorner,\overline{w})\urcorner,\sim\overline{w})$$

Finally,

$$\ulcorner\boxed{\neg}\overset{n}{q}\urcorner = \ulcorner\boxed{\neg}q_1\urcorner \sqcup\ldots\sqcup \ulcorner\boxed{\neg}q_n\urcorner = \ulcorner h(\ulcorner q_1\urcorner,\sim\overline{w})\urcorner \sqcup\ldots\sqcup \ulcorner h(\ulcorner q_n\urcorner,\sim\overline{w})\urcorner$$

and so

$$\overset{m}{p}\boxed{\rightarrow}\overset{n}{q} = h(\ulcorner h(\ulcorner h(\ulcorner\overset{m}{p}\urcorner,\overline{w})\urcorner \sqcup \ulcorner h([\ulcorner h(\ulcorner q_1\urcorner,\sim\overline{w})\urcorner \sqcup\ldots\sqcup \ulcorner h(\ulcorner q_n\urcorner,\sim\overline{w})\urcorner] (\overline{w}),\overline{g})\urcorner,\overline{w})\urcorner,\sim\overline{w})$$

From this, we have

$$W(\overset{m}{p}\boxed{\rightarrow}\overset{n}{q}) \leftrightarrow (W(\overset{m}{p}) \rightarrow W(\overset{n}{q})) \qquad \text{ST4.30}$$

Proof $W(\overset{m}{p}\boxed{\rightarrow}\overset{n}{q}) \leftrightarrow W(\boxed{\neg}((\boxed{\wedge^m}\overset{m}{p})\boxed{\wedge}(\boxed{\vee^n}(\boxed{\neg}\overset{n}{q})))) \leftrightarrow$
$\neg W((\boxed{\wedge^m}\overset{m}{p})\boxed{\wedge}(\boxed{\vee^n}(\boxed{\neg}\overset{n}{q})))$
$\leftrightarrow \neg(W(\boxed{\wedge^m}\overset{m}{p}) \wedge W(\boxed{\vee^n}(\boxed{\neg}\overset{n}{q})) \leftrightarrow \neg(W(\overset{m}{p}) \wedge E(\boxed{\neg}\overset{n}{q})) \leftrightarrow$
$\neg(W(\overset{m}{p}) \wedge E(\boxed{\neg}\overset{n}{q}))$
$\leftrightarrow \neg(W(\overset{m}{p}) \wedge \neg W(\overset{n}{q})) \leftrightarrow (W(\overset{m}{p}) \rightarrow W(\overset{n}{q}))$.

As a particular case, we have

$$p\boxed{\rightarrow}q = \boxed{\neg}(p\boxed{\wedge}\boxed{\neg}q) \qquad \text{T4.31}$$

as well as, easily,

$$p\boxed{\rightarrow}q = \boxed{\neg}p\boxed{\vee}q) \qquad \text{T4.32}$$

Applying T23 to T9, we immediately obtain a proof of the validity of the thesis, so far as the material relations between two ideas are concerned:

Validity of Bolzano's thesis for the material relations between two ideas

$$
\begin{array}{llll}
(1)\ \textit{Compat.} & : a \top c & \leftrightarrow & W(h(\ulcorner a \sqcap c \urcorner, \overline{g}))\\
(2)\ \textit{Inclus.} & : a \sqsubseteq c & \leftrightarrow & W(h(\ulcorner h(\ulcorner a \sqcap c \urcorner, \overline{g}) \urcorner \sqcup \ulcorner h(\ulcorner a \sqcap -c \urcorner, \sim \overline{g}) \urcorner, \overline{w}))\\
(3)\ \textit{Dispar.} & : a\ \dot{\top}\ c & \leftrightarrow & W(h(\ulcorner h(\ulcorner a \sqcap c \urcorner, \overline{g}) \urcorner \sqcup \ulcorner h(\ulcorner a \sqcap -c \urcorner, \overline{g}) \urcorner \sqcup \ulcorner h(\ulcorner -a \sqcap c \urcorner, \overline{g}) \urcorner, \overline{w}))\\
(4)\ \textit{Equival.} & : a \equiv c & \leftrightarrow & W(h(\ulcorner h(\ulcorner a \sqcap c \urcorner, \overline{g}) \urcorner \sqcup \ulcorner h(\ulcorner a \sqcap -c \urcorner, \sim \overline{g}) \urcorner \sqcup \ulcorner h(\ulcorner -a \sqcap c \urcorner, \sim \overline{g}) \urcorner, \overline{w}))\\
(5)\ \textit{Subord.} & : a\ \dot{\sqsubseteq}\ c & \leftrightarrow & W(h(\ulcorner h(\ulcorner a \sqcap c \urcorner, \overline{g}) \urcorner \sqcup \ulcorner h(\ulcorner a \sqcap -c \urcorner, \sim \overline{g}) \urcorner \sqcup \ulcorner h(\ulcorner -a \sqcap c \urcorner, \overline{g}) \urcorner, \overline{w}))\\
(6)\ \textit{Incomp.} & : a \perp c & \leftrightarrow & W(h(\ulcorner a \sqcap c \urcorner, \sim \overline{g}))\\
(7)\ \textit{Contrad.} & : a \,||\, c & \leftrightarrow & W(\ulcorner h(\ulcorner a \sqcap c \urcorner, \sim \overline{g}) \urcorner \sqcup \ulcorner h(\ulcorner e \sqcap -a \sqcap -c \urcorner, \sim \overline{g}) \urcorner, \overline{w}))\\
(8)\ \textit{Contrar.} & : a\ \dot{\perp}\ c & \leftrightarrow & W(\ulcorner h(\ulcorner a \sqcap c \urcorner, \sim \overline{g}) \urcorner \sqcup \ulcorner h(\ulcorner e \sqcap -a \sqcap -c \urcorner, \overline{g}) \urcorner, \overline{w}))
\end{array}
$$

T4.33

4.7 Formal Properties and Relations

The symbolic treatment of formal properties and relations among finite aggregate of propositions rests on the availability of two new lectological operations that, although not explicitly isolated by him, were certainly familiar to Bolzano. It is clear, for instance, from the following quotation

> "Every aggregate of ideas, which in the place of the $i, j, \ldots$ in the propositions A, $B, C, \ldots M, N, O, \ldots$, makes true all the $A, B, C, \ldots$—has—the quality of making true also all the $M, N, O, \ldots$"[16]

that Bolzano considered it obvious that the capability of an aggregate of ideas (a fortiori, of a single idea) to make true, when set in the place of a certain other aggregate of ideas (a fortiori, of a certain other idea), every proposition of a certain set of propositions *is a quality* of that aggregate (a fortiori, of that idea) and hence, in particular, he accepted that for any given collection of propositions $\overset{n}{p}$ and idea a occurring in them, there is a quality that belongs to an idea c iff, set in the place of a, it makes all the $\overset{n}{p}$ true. Of course, Bolzano believed that such a quality could exist only if there is at least one idea of this type, that is to say, if the $\overset{n}{p}$ are compatible with respect to a. Thus we can assume

$$\underset{a}{\bigwedge\!\!\!\bigwedge}(\overset{n}{p}) \rightarrow \exists r \forall c (c \in r \leftrightarrow W(\overset{n}{p}\,[a/c])) \qquad \text{SCQ6}$$

[16] Cf. TS §164, II 146 (2).

On the other hand, from the following quotation

> "Every aggregate of ideas, which in the place of the $i, j, \ldots$ makes true all the $A, B, C, \ldots$—has—the quality that also the idea of a true proposition among the $M, N, O, \ldots$ becomes objectuality"[17]

it is equally clear that Bolzano was convinced that the same holds for the capability of making true *at least one* proposition among the $\overset{n}{p}$. Indeed, one should not forget that for Bolzano the truth of the disjunction of a set of propositions is given by the objectuality of the idea of a true proposition among the propositions of the set. Also, in this case, of course, such a quality could exist only if there is at least one idea satisfying that condition, that is to say, if the $\overset{n}{p}$ are not formally false with respect to the idea a. We thus can also accept

$$\neg \underset{a}{\mathbb{F}}(\overset{n}{p}) \rightarrow \exists r \forall c(c \in r \leftrightarrow E(\overset{n}{p}[a/c])) \qquad \text{SCQ7}$$

Obviously, he also accepted that there were ideas referring to such qualities. Therefore, we assume the presence of ideas referring to qualities of the first type, which we denote by '$\overset{n\wedge}{p}_a$', as well as ideas referring to qualities of the second type, which we denote by '$\overset{n\vee}{p}_a$'. We also establish that such ideas have as their next parts the $\overset{n}{p}$ and a. Therefore

$$\forall a \exists \bar{c} \begin{pmatrix} (\underset{a}{\bigwedge\!\!\!\!\bigwedge}(\overset{n}{p}) \rightarrow \forall r(r \mathrm{E} \bar{c} \leftrightarrow \forall b(b \in r \leftrightarrow W(\overset{n}{p}[a/b])))) \wedge \\ (\neg \underset{a}{\bigwedge\!\!\!\!\bigwedge}(\overset{n}{p}) \rightarrow \neg G(\bar{c})) \wedge \forall l(l < \bar{c} \leftrightarrow l \leq \overset{n}{p} \vee l \leq a)) \end{pmatrix} \qquad \text{SCB5}$$

$$\forall a \exists \bar{c} \begin{pmatrix} (\neg \underset{a}{\mathbb{F}}(\overset{n}{p}) \rightarrow \forall r(r \mathrm{E} \bar{c} \leftrightarrow \forall b(b \in r \leftrightarrow E(\overset{n}{p}[a/b])))) \wedge \\ ((\underset{a}{\mathbb{F}}(\overset{n}{p}) \rightarrow \neg G(\bar{c})) \wedge \forall l(l < \bar{c} \leftrightarrow l \leq \overset{n}{p} \vee l \leq a)) \end{pmatrix} \qquad \text{SCB6}$$

$$\overset{n\wedge}{p}_a = \varepsilon \bar{c} \begin{pmatrix} (\underset{a}{\bigwedge\!\!\!\!\bigwedge}(\overset{n}{p}) \rightarrow \forall r(r \mathrm{E} \bar{c} \leftrightarrow \forall b(b \in r \leftrightarrow W(\overset{n}{p}[a/b])))) \wedge \\ (\neg \underset{a}{\bigwedge\!\!\!\!\bigwedge}(\overset{n}{p}) \rightarrow \neg G(\bar{c})) \wedge \forall l(l < \bar{c} \leftrightarrow l \leq \overset{n}{p} \vee l \leq a)) \end{pmatrix} \qquad \text{SD4.33}$$

$$\overset{n\vee}{p}_a = \varepsilon \bar{c} \begin{pmatrix} (\neg \underset{a}{\mathbb{F}}(\overset{n}{p}) \rightarrow \forall r(r \mathrm{E} \bar{c} \leftrightarrow \forall b(b \in r \leftrightarrow E(\overset{n}{p}[a/b])))) \wedge \\ ((\underset{a}{\mathbb{F}}(\overset{n}{p}) \rightarrow \neg G(\bar{c})) \wedge \forall l(l < \bar{c} \leftrightarrow l \leq \overset{n}{p} \vee l \leq a)) \end{pmatrix} \qquad \text{SD4.34}$$

so that

$$(1.1)\ B(\overset{n\wedge}{p}_a) \qquad \text{ST4.34}$$

$$(1.2)\ \underset{a}{\bigwedge\!\!\!\!\bigwedge}(\overset{n}{p}) \rightarrow \forall r(r \mathrm{E} \overset{n\wedge}{p}_a \leftrightarrow \forall c(c \in r \leftrightarrow W(\overset{n}{p}[a/c]))$$

[17] Cf. TS §166, II 146 (5).

$$(1.3)\ \neg\underset{a}{\mathbb{A}}(\overset{n}{p}) \rightarrow \neg G(\overset{n\wedge}{p}_a)$$

$$(1.4)\ l < \overset{n\wedge}{p}_a \leftrightarrow l \leq \overset{n}{p} \vee l \leq a)$$

$$(2.1)\ B(\overset{n\vee}{p}_a)$$

$$(2.2)\ \neg\underset{a}{\mathbb{F}}(\overset{n}{p}) \rightarrow \forall r(r \mathrel{\scriptstyle\mathrm{E}} \overset{n\vee}{p}_a \leftrightarrow \forall c(c \in r \leftrightarrow E(\overset{n}{p}\,[a/c]))$$

$$(2.3)\ \underset{a}{\mathbb{F}}(\overset{n}{p}) \rightarrow \neg G(\overset{n\vee}{p}_a)$$

$$(2.4)\ l < \overset{n\vee}{p}_a \leftrightarrow l \leq \overset{n}{p} \vee l \leq a)$$

The following theorem shows that our operations do exactly what is expected of them: *an object idea is* (1) *something that has* $\overset{n\wedge}{p}_a$ *iff set in the place of* a *in the* $\overset{n}{p}$ *makes all them true;* (2) *something that has* $\overset{n\vee}{p}_a$ *iff set in the place of* a *in the* $\overset{n}{p}$ *makes true at least one of them.*

$$(1)\quad c \mathrel{\scriptstyle\mathrm{E}} [e](\overset{n\wedge}{p}_a) \leftrightarrow W(\overset{n}{p}[a/c]) \qquad \text{ST4.35}$$

$$(2)\quad c \mathrel{\scriptstyle\mathrm{E}} [e](\overset{n\vee}{p}_a) \leftrightarrow E(\overset{n}{p}[a/c])$$

Proof *Ad* (1) Let $c \mathrel{\scriptstyle\mathrm{E}} [e](\overset{n\wedge}{p}_a)$. It follows, on the one hand, that $[e](\overset{n\wedge}{p}_a)$ is objectual and hence that $\overset{n\wedge}{p}_a$ is also objectual and therefore, by ST34.1.3, (i) $\underset{a}{\mathbb{A}}(\overset{n}{p})$; on the other hand, there is an r such that (ii) $c \in r$ and (iii) $r \mathrel{\scriptstyle\mathrm{E}} \overset{n\wedge}{p}_a$. From (i) and (iii), it follows, by ST34.1.2, $\forall c(c \in r \leftrightarrow W(\overset{n}{p}\,[a/c]))$, and from this and (iii), as desired $W(\overset{n}{p}\,[a/c]))$. Vice versa, let $W(\overset{n}{p}\,[a/c]))$. Therefore, a fortiori, $\underset{a}{\mathbb{A}}(\overset{n}{p})$; then, by axiom SCQ5, there is an r such that (iv) $\forall c(c \in r \leftrightarrow W(\overset{n}{p}\,[a/c]))$ and hence, because $W(\overset{n}{p}\,[a/c]))$, (v) $c \in r$. But, by ST34.1.2, for an r such that $\forall c(c \in r \leftrightarrow W(\overset{n}{p}\,[a/c]))$, also (vi) $r \mathrel{\scriptstyle\mathrm{E}} \overset{n\wedge}{p}_a$ holds. From (v) and (vi), $c \mathrel{\scriptstyle\mathrm{E}} [e](\overset{n\wedge}{p}_a)$ follows.

Ad (2) Let $c \mathrel{\scriptstyle\mathrm{E}} [e](\overset{n\vee}{p}_a)$. It follows, on the one hand, that $[e](\overset{n\vee}{p}_a)$ is objectual and so $\overset{n\vee}{p}_a$ is also objectual and therefore, by ST34.2.3, (i) $\neg\underset{a}{\mathbb{F}}(\overset{n}{p})$; on the other hand, there is an r such that (ii) $c \in r$ and (iii) $r \mathrel{\scriptstyle\mathrm{E}} \overset{n\vee}{p}_a$. From (i) and (iii), it follows, by ST34.2.2, $\forall c(c \in r \leftrightarrow W(\overset{n}{p}\,[a/c]))$ and from this and (iii), as desired $E(\overset{n}{p}\,[a/c]))$. Vice versa, let $E(\overset{n}{p}\,[a/c]))$. Therefore, a fortiori, $\neg\underset{a}{\mathbb{F}}(\overset{n}{p})$ and so, by axiom SCQ6, there is an r such that (iv) $\forall c(c \in r \leftrightarrow E(\overset{n}{p}\,[a/c]))$ and hence, because $E(\overset{n}{p}\,[a/c]))$, (v) $c \in r$. But, by ST34.2.2., for an r such that $\forall c(c \in r \leftrightarrow E(\overset{n}{p}\,[a/c]))$, also (vi) $r \mathrel{\scriptstyle\mathrm{E}} \overset{n\vee}{p}_a$ holds. From (v) and (vi), $c \mathrel{\scriptstyle\mathrm{E}} [e](\overset{n\vee}{p}_a)$ follows.

As we know, (T2.74), it holds that $\neg i \sqsubset [e](\overline{a}) \leftrightarrow i \sqsubset [e](\sim \overline{a})$; therefore we have the two corollaries:

$$(1)\quad c \sqsubset [e](\sim \overset{n\wedge}{p_a}) \leftrightarrow \neg W(\overset{n}{p}[a/c]) \qquad \text{ST4.36}$$

$$(2)\quad c \sqsubset [e](\sim \overset{n\vee}{p_a}) \leftrightarrow \neg E(\overset{n}{p}[a/c])$$

4.8 Formal Properties of Collections of Propositions

From the two preceding theorems, by simple application of the definitions, we obtain

$$(1)\quad G([e](\overset{n\wedge}{p_a})) \leftrightarrow \exists c W(\overset{n}{p}[a/c]) \qquad \text{ST4.37}$$

$$(2)\quad G([e](\overset{n\vee}{p_a})) \leftrightarrow \exists c E(\overset{n}{p}[a/c])$$

$$(3)\quad G([e](\sim\overset{n\wedge}{p_a})) \leftrightarrow \exists c \neg W(\overset{n}{p}[a/c])$$

$$(4)\quad G([e](\sim\overset{n\vee}{p_a})) \leftrightarrow \exists c F(\overset{n}{p}[a/c])$$

Hence it is possible to reformulate—in terms of objectuality or objectlessness of the concretizations of the two new ideas or of their lack—the eight fundamental formal properties of a finite set of propositions with respect to an idea:

$$\begin{array}{lll} (1)\ \textit{Compatibility} & : \underset{a}{⩓}(\overset{n}{p}) & \leftrightarrow G([e](\overset{n\wedge}{p_a})) \qquad \text{ST4.38} \\ (2)\ \textit{Integration} & : \underset{a}{⩔}(\overset{n}{p}) & \leftrightarrow \neg G([e](\sim\overset{n\vee}{p_a})) \\ (3)\ \textit{Formal truths} & : \underset{a}{\mathbb{W}}(\overset{n}{p}) & \leftrightarrow \neg G([e](\sim\overset{n\wedge}{p_a})) \\ (4)\ \textit{Formal falsities} & : \underset{a}{\mathbb{F}}(\overset{n}{p}) & \leftrightarrow \neg G([e](\overset{n\vee}{p_a})) \\ (5)\ \textit{Incompatibility} & : \neg\underset{a}{⩓}(\overset{n}{p}) & \leftrightarrow \neg G([e](\overset{n\wedge}{p_a})) \\ (6)\ \textit{Global falsifiability} & : \neg\underset{a}{⩔}(\overset{n}{p}) & \leftrightarrow G([e](\sim\overset{n\vee}{p_a})) \\ (7)\ \textit{Local falsifiability} & : \neg\underset{a}{\mathbb{W}}(\overset{n}{p}) & \leftrightarrow G([e](\sim\overset{n\wedge}{p_a})) \\ (8)\ \textit{Local verifiability} & : \neg\underset{a}{\mathbb{F}}(\overset{n}{p}) & \leftrightarrow G([e](\overset{n\vee}{p_a})). \end{array}$$

From this, by virtue of the two theorems that allow us to express the belonging or not belonging of objectuality to an idea as the truth of propositions of objectuality or objectlessness, we have the following:

Validity of Bolzano's thesis for formal properties of a finite set of propositions with respect to an idea

$$
\begin{array}{lll}
(1)\ \textit{Compatibility} & : \underset{a}{\wedge\!\!\!\wedge}(\overset{n}{p}) & \leftrightarrow W(h(\left\lceil [e](\overset{n\wedge}{p}_a) \right\rceil, \bar{g}))^{18} \\
(2)\ \textit{Integration} & : \underset{a}{\vee\!\!\!\vee}(\overset{n}{p}) & \leftrightarrow W(h(\left\lceil [e](\sim\overset{n\vee}{p}_a) \right\rceil, \sim\bar{g})) \\
(3)\ \textit{Formal truths} & : \underset{a}{\mathbb{W}}(\overset{n}{p}) & \leftrightarrow W(h(\left\lceil [e](\sim\overset{n\wedge}{p}_a) \right\rceil, \sim\bar{g})) \\
(4)\ \textit{Formal falsities} & : \underset{a}{\mathbb{F}}(\overset{n}{p}) & \leftrightarrow W(h(\left\lceil [e](\overset{n\vee}{p}_a) \right\rceil, \sim\bar{g})) \\
(5)\ \textit{Incompatibility} & : \neg\underset{a}{\wedge\!\!\!\wedge}(\overset{n}{p}) & \leftrightarrow W(h(\left\lceil [e](\overset{n\wedge}{p}_a) \right\rceil, \sim\bar{g})) \\
(6)\ \textit{Global falsifiability} & : \neg\underset{a}{\vee\!\!\!\vee}(\overset{n}{p}) & \leftrightarrow W(h(\left\lceil [e](\sim\overset{n\vee}{p}_a) \right\rceil, \bar{g})) \\
(7)\ \textit{Local falsifiability} & : \neg\underset{a}{\mathbb{W}}(\overset{n}{p}) & \leftrightarrow W(h(\left\lceil [e](\sim\overset{n\wedge}{p}_a) \right\rceil, \bar{g})) \\
(8)\ \textit{Local verifiability} & : \neg\underset{a}{\mathbb{F}}(\overset{n}{p}) & \leftrightarrow W(h(\left\lceil [e](\overset{n\vee}{p}_a) \right\rceil, \bar{g}))
\end{array}
\qquad \text{ST4.39}
$$

4.9 Formal Relations Among Propositions

First of all, we reformulate the main formal relations among propositions with respect to an idea in terms of material properties and relations among concretizations of the two new ideas. Notice that from now on, in accordance with Bolzano's ideas emerging from the passage quoted in footnote 52 above, *compatibility* of two lists of propositions with respect to *a* certain idea *a* is to be intended as follows: there is at least one idea *c* such that, when substituted for a, makes simultaneously true at least one proposition in each of the two lists.

$$
\begin{array}{ll}
(1)\ \textit{Compatibility} & \overset{m}{p}\underset{a}{\wedge\!\!\!\wedge}\overset{n}{q} \leftrightarrow G([e](\overset{m\vee}{p}_a + \overset{n\vee}{q}_a)) \\
(2)\ \textit{Derivability} & \overset{m}{p} \underset{a}{\Rightarrow} \overset{n}{q} \leftrightarrow [e](\overset{m\wedge}{p}_a) \sqsubseteq [e](\overset{n\wedge}{q}_a) \\
(3)\ \textit{Equivalence} & \overset{m}{p} \underset{a}{\Leftrightarrow} \overset{n}{q} \leftrightarrow [e](\overset{m\wedge}{p}_a) \equiv [e](\overset{n\wedge}{q}_a) \\
(4)\ \textit{Subordination} & \overset{m}{p} \underset{a}{\dot{\Rightarrow}} \overset{n}{q} \leftrightarrow [e](\overset{m\wedge}{p}_a) \sqsubseteq [e](\overset{n\wedge}{q}_a) \wedge G([e](\overset{n\wedge}{q}_a + \sim\overset{m\wedge}{p}_a)) \\
(5)\ \textit{Concatenation} & \overset{m}{p} \underset{a}{\top\!\!\!\top} \overset{n}{q} \leftrightarrow \left(\begin{array}{l} G([e](\overset{m\vee}{p}_a + \overset{n\vee}{q}_a)) \wedge G([e](\overset{m\wedge}{p}_a + \sim\overset{n\wedge}{q}_a)) \\ \wedge G([e](\overset{n\wedge}{q}_a + \sim\overset{m\wedge}{p}_a)) \end{array} \right) \\
(6)\ \textit{Incompatibility} & \neg(\overset{m}{p}\underset{a}{\wedge\!\!\!\wedge}\overset{n}{q}) \leftrightarrow \neg G([e](\overset{m\vee}{p}_a + \overset{n\vee}{q}_a)) \\
(7)\ \textit{Exclusion} & \overset{m}{p} \underset{a}{\perp} \overset{n}{q} \leftrightarrow [e](\overset{m\wedge}{p}_a) \sqsubseteq [e](\sim\overset{n\wedge}{q}_a)
\end{array}
$$

[18] TS §164, II 141: "1) [. . .] Thus if our proposition is to be expressed in a way which makes its logical constituents sufficiently distinct, it must (according to §137) run roughly as follows: 'The idea of a collection of ideas which, when put in the places of i,j,. . . in the propositions A,B,C,D,. . . , makes all of them true, has objectuality." Usually one simply says that 'the propositions A,B,C,D,. . . are *compatible* or that they can all be true *simultaneously*'."

$$\text{(8) \textit{Reciprocal execlusion}} \quad \overset{m}{p} \underset{a}{\perp\!\!\!\perp} \overset{n}{q} \leftrightarrow [e](\sim \overset{m\wedge}{p}_a) \equiv [e](\overset{n\wedge}{q}_a)$$

$$\text{(9) \textit{Conditionated Integration}} \; \overset{m}{p} \underset{a}{\overset{\vee}{\Rightarrow}} \overset{n}{q} \leftrightarrow [e](\overset{m\wedge}{p}_a) \sqsubseteq [e](\overset{n\vee}{q}_a) \qquad \text{ST4.40}$$

Proof *Ad* (1) $\overset{m}{p} \underset{a}{\wedge\!\!\!\wedge} \overset{n}{q} \exists c(E(\overset{m}{p}[a/c]) \wedge E(\overset{n}{q}[a/c])) \leftrightarrow \exists c(c \in [e](\overset{m\vee}{p}_a) \wedge c \in [e](\overset{n\vee}{q}_a)) \leftrightarrow \exists c(c \in [e](\overset{m\vee}{p}_a + \overset{n\vee}{q}_a))$.

Ad 2. $\overset{m}{p} \underset{a}{\Rightarrow} \overset{n}{q} \leftrightarrow \underset{a}{\wedge\!\!\!\wedge}(\overset{m}{p}) \wedge \forall c(W(\overset{m}{p}[a/c])) \rightarrow W(\overset{n}{q}[a/c]))$

$\leftrightarrow G([e](\overset{m\wedge}{p}_a)) \wedge \forall c(c \in [e](\overset{m\wedge}{p}_a) \rightarrow c \in [e](\overset{n\wedge}{q}_a) \leftrightarrow [e](\overset{m\wedge}{p}_a) \sqsubseteq [e](\overset{n\wedge}{q}_a)$.

Ad (3) This is trivial.

Ad (4) $\overset{m}{p} \underset{a}{\dot{\Rightarrow}} \overset{n}{q} \leftrightarrow \overset{m}{p} \underset{a}{\Rightarrow} \overset{n}{q} \wedge \neg(\overset{m}{p} \underset{a}{\Rightarrow} \overset{n}{q}) \leftrightarrow [e](\overset{m\wedge}{p}_a) \sqsubseteq [e](\overset{n\wedge}{q}_a) \wedge \neg([e](\overset{n\wedge}{q}_a) \sqsubseteq [e](\overset{m\wedge}{p}_a))$

$\leftrightarrow [e](\overset{m\wedge}{p}_a) \sqsubseteq [e](\overset{n\wedge}{q}_a) \wedge \exists c(c \in [e](\overset{n\wedge}{q}_a) \wedge \neg(c \in [e](\overset{m\wedge}{p}_a)))$

$\leftrightarrow [e](\overset{m\wedge}{p}_a) \sqsubseteq [e](\overset{n\wedge}{q}_a) \wedge \exists c(c \in [e](\overset{n\wedge}{q}_a) \wedge c \in [e](\sim \overset{m\wedge}{p}_a)))$

$\leftrightarrow [e](\overset{m\wedge}{p}_a) \sqsubseteq [e](\overset{n\wedge}{q}_a) \wedge \exists c(c \in [e](\overset{n\wedge}{q}_a + \sim \overset{m\wedge}{p}_a))) \leftrightarrow [e](\overset{m\wedge}{p}_a) \sqsubseteq [e](\overset{n\wedge}{q}_a) \wedge G([e](\overset{n\wedge}{q}_a + \sim \overset{m\wedge}{p}_a)))$.

Ad (5) $\overset{m}{p} \underset{a}{\top\!\!\!\top} \overset{n}{q} \leftrightarrow \overset{m}{p} \underset{a}{\wedge\!\!\!\wedge} \overset{n}{q} \wedge \neg(\overset{m}{p} \underset{a}{\Rightarrow} \overset{n}{q}) \wedge \neg(\overset{n}{q} \underset{a}{\Rightarrow} \overset{m}{p})$

$\leftrightarrow G([e](\overset{m\vee}{p}_a + \overset{n\vee}{q}_a)) \wedge \neg([e](\overset{m\wedge}{p}_a) \sqsubseteq [e](\overset{n\wedge}{q}_a)) \wedge \neg([e](\overset{n\wedge}{q}_a) \sqsubseteq [e](\overset{m\wedge}{p}_a))$

$\leftrightarrow G([e](\overset{m\vee}{p}_a + \overset{n\vee}{q}_a)) \wedge G([e](\overset{m\wedge}{p}_a + \sim \overset{n\wedge}{q}_a)) \wedge G([e](\overset{n\wedge}{q}_a + \sim \overset{m\wedge}{p}_a)))$.

Ad (6) This is trivial.

Ad (7) $\overset{m}{p} \underset{a}{\perp} \overset{n}{q} \leftrightarrow \underset{a}{\wedge\!\!\!\wedge}(\overset{m}{p}) \wedge \forall c(W(\overset{m}{p}[a/c]) \rightarrow F(\overset{n}{q}[a/c])))$

$\leftrightarrow G([e](\overset{m\wedge}{p}_a)) \wedge \forall c(c \in [e](\overset{m\wedge}{p}_a) \rightarrow c \in [e](\sim \overset{n\wedge}{q}_a)) \leftrightarrow [e](\overset{m\wedge}{p}_a) \sqsubseteq [e](\sim \overset{n\wedge}{q}_a)$.

Ad (8) $\overset{m}{p} \underset{a}{\perp\!\!\!\perp} \overset{n}{q} \leftrightarrow [e](\overset{m\wedge}{p}_a) \equiv [e](\sim \overset{n\wedge}{q}_a) \leftrightarrow [e](\sim \overset{m\wedge}{p}_a) \equiv [e](\overset{n\wedge}{q}_a)$.

(Note that $[e](\overline{a}) \equiv [e](\sim \overline{c}) \leftrightarrow [e](\sim \overline{a}) \equiv [e](\overline{c})$.)

Ad (9) $\overset{m}{p} \underset{a}{\overset{\vee}{\Rightarrow}} \overset{n}{q} \leftrightarrow \underset{a}{\wedge\!\!\!\wedge}(\overset{m}{p}) \wedge \forall c(W(\overset{m}{p}[a/c]) \rightarrow E(\overset{n}{q}[a/c]))$

$\leftrightarrow \underset{a}{\wedge\!\!\!\wedge}(\overset{m}{p}) \wedge G([e](\overset{m\wedge}{p}_a)) \wedge \forall c(c \in [e](\overset{m\wedge}{p}_a) \rightarrow c \in [e](\overset{n\vee}{q}_a) \leftrightarrow [e](\overset{m\wedge}{p}_a) \sqsubseteq [e](\overset{n\vee}{q}_a)$.

Applying the equivalencies $a \sqsubseteq [e](\overline{c}) \leftrightarrow W(h(a, \overline{c}))$ and $G(a) \leftrightarrow W(h(\ulcorner a \urcorner, \overline{g}))$, we obtain the following reformulation of the relations in question:

$$\text{(1)} \; \overset{m}{p} \underset{a}{\wedge\!\!\!\wedge} \overset{n}{q} \leftrightarrow W(h(\left\lceil [e](\overset{m\vee}{p}_a + \overset{n\vee}{q}_a) \right\rceil, \overline{g})) \qquad \text{ST4.41}$$

$$\text{(2)} \; \overset{m}{p} \underset{a}{\Rightarrow} \overset{n}{q} \leftrightarrow W(h([e](\overset{m\wedge}{p}_a), \overset{n\wedge}{q}_a))$$

$$(3)\ \overset{m}{p} \underset{a}{\Leftrightarrow} \overset{n}{q} \quad \leftrightarrow W(h([e](\overset{m\wedge}{p_a}), \overset{n\wedge}{q_a})) \wedge W(h([e](\overset{n\wedge}{q_a}), \overset{m\wedge}{p_a}))$$

$$(4)\ \overset{m}{p} \underset{a}{\overset{\cdot}{\Rightarrow}} \overset{n}{q} \quad \leftrightarrow W(h([e](\overset{m\wedge}{p_a}), \overset{n\wedge}{q_a})) \wedge \neg W(h([e](\overset{n\wedge}{q_a}), \overset{m\wedge}{p_a}))$$

$$(5)\ \overset{m}{p} \underset{a}{\top\!\!\top} \overset{n}{q} \quad \leftrightarrow \begin{pmatrix} W(h(\left\lceil [e](\overset{m\vee}{p_a} + \overset{n\vee}{q_a}) \right\rceil, \overline{g}) \wedge W(h(\left\lceil [e](\overset{m\vee}{p_a} + \sim \overset{n\vee}{q_a}) \right\rceil, \overline{g}) \\ \wedge W(h(\left\lceil [e](\sim \overset{m\vee}{p_a} + \overset{n\vee}{q_a}) \right\rceil, \overline{g}) \end{pmatrix}$$

$$(6)\ \neg(\overset{m}{p} \underset{a}{\barwedge} \overset{n}{q}) \leftrightarrow W(h(\left\lceil [e](\overset{m\vee}{p_a} + \overset{n\vee}{q_a}) \right\rceil, \sim \overline{g})$$

$$(7)\ \overset{m}{p} \underset{a}{\perp} \overset{n}{q} \quad \leftrightarrow W(h([e](\overset{m\wedge}{p_a}), \sim \overset{n\wedge}{q_a}))$$

$$(8)\ \overset{m}{p} \underset{a}{\perp\!\!\perp} \overset{n}{q} \quad \leftrightarrow W(h([e](\overset{m\wedge}{p_a}), \sim \overset{n\wedge}{q_a})) \wedge W(h([e](\overset{n\wedge}{q_a}), \sim \overset{m\wedge}{p_a}))$$

$$(9)\ \overset{m}{p} \underset{a}{\overset{\vee}{\Rightarrow}} \overset{n}{q} \quad \leftrightarrow W(h([e](\overset{m\wedge}{p_a}), \overset{n\vee}{q_a}))$$

Finally, using the theorem $W(\boxed{\wedge}(\overset{m}{p})) \leftrightarrow W(h(\ulcorner \overset{m}{p} \urcorner, \overline{w}))$, we obtain the confirmation of the *validity of Bolzano's thesis for formal relations among finite sets of propositions with respect to an idea*:

ST4.42

$$(1)\ \overset{m}{p} \underset{a}{\barwedge} \overset{n}{q} \quad \leftrightarrow W(h(\left\lceil [e](\overset{m\vee}{p_a} + \overset{n\vee}{q_a}) \right\rceil, \overline{g}))$$

$$(2)\ \overset{m}{p} \underset{a}{\Rightarrow} \overset{n}{q} \quad \leftrightarrow W(h([e](\overset{m\wedge}{p_a}), \overset{n\wedge}{q_a})){}^{19}$$

$$(3)\ \overset{m}{p} \underset{a}{\Leftrightarrow} \overset{n}{q} \quad \leftrightarrow W(h(\left\lceil h([e](\overset{m\wedge}{p_a}), \overset{n\wedge}{q_a}) \right\rceil \sqcup \left\lceil h([e](\overset{n\wedge}{q_a}), \overset{m\wedge}{p_a}) \right\rceil, \overline{w}))$$

$$(4)\ \overset{m}{p} \underset{a}{\overset{\cdot}{\Rightarrow}} \overset{n}{q} \quad \leftrightarrow W(h(\left\lceil h([e](\overset{m\wedge}{p_a}), \overset{n\wedge}{q_a}) \right\rceil \sqcup \left\lceil h(\left\lceil h([e](\overset{n\wedge}{q_a}), \overset{m\wedge}{p_a}) \right\rceil, \sim \overline{w}) \right\rceil, \overline{w}))$$

$$(5)\ \overset{m}{p} \underset{a}{\top\!\!\top} \overset{n}{q} \quad \leftrightarrow W(h\begin{pmatrix} \left\lceil h(\left\lceil [e](\overset{m\vee}{p_a} + \overset{n\vee}{q_a}) \right\rceil, \overline{g}) \right\rceil \sqcup \left\lceil h(\left\lceil [e](\overset{m\vee}{p_a} + \sim \overset{n\vee}{q_a}) \right\rceil, \overline{g}) \right\rceil \\ \sqcup \left\lceil h(\left\lceil [e](\sim \overset{m\vee}{p_a} + \overset{n\vee}{q_a}) \right\rceil, \overline{g}) \right\rceil \end{pmatrix}, \overline{w}))$$

$$(6)\ \neg(\overset{m}{p} \underset{a}{\barwedge} \overset{n}{q}) \leftrightarrow W(h(\left\lceil [e](\overset{m\vee}{p_a} + \overset{n\vee}{q_a}) \right\rceil, \sim \overline{g})$$

$$(7)\ \overset{m}{p} \underset{a}{\perp} \overset{n}{q} \quad \leftrightarrow W(h([e](\overset{m\wedge}{p_a}), \sim \overset{n\wedge}{q_a}))$$

$$(8)\ \overset{m}{p} \underset{a}{\perp\!\!\perp} \overset{n}{q} \quad \leftrightarrow W(h(\left\lceil h([e](\overset{m\wedge}{p_a}), \sim \overset{n\wedge}{q_a}) \right\rceil \sqcup \left\lceil h([e](\overset{n\wedge}{q_a}), \sim \overset{m\wedge}{p_a}), \overline{w}))$$

$$(9)\ \overset{m}{p} \underset{a}{\overset{\vee}{\Rightarrow}} \overset{n}{q} \quad \leftrightarrow W(h([e](\overset{m\wedge}{p_a}), \overset{n\vee}{q_a})){}^{20}$$

[19] TS §164, II 142: "2) [. . .] When we claim that *M*, *N*, *O*, . . . are deducible from *A*, *B*, *C*, . . . with respect to the ideas *i*, *j*, . . . , we at bottom really only say the following (cf. § 155): 'Every collection of ideas which, when put in the places of *i*, *j*, . . . in the propositions *A*, *B*, *C*, . . . , *M*, *N*, *O*, . . . , makes all of *A*, *B*, *C*, . . . true has the attribute of making all of *M*, *N*, *O*, . . . true'."

[20] TS §166, II 146: "5) Finally, when we claim that propositions *M*, *N*, *O*, . . . do not stand in a relation of complementation unconditionally, but only under the assumption of the propositions *A*, *B*, *C*, . . . , and with respect to ideas *i*, *j*, . . . , we express the proposition 'Every collection of ideas which, when put in the places of *i*, *j*, . . . , makes all of *A*, *B*, *C*, . . . true, has the attribute of making the idea of a true proposition among *M*, *N*, *O*, . . . objectual'."

4.10 Generalization to a Finite Set of Ideas

The symbolic treatment of the formal properties and relations among propositions, as developed so far, suffers from an important limitation: it concerns formal properties and relations with respect to *only one idea*. Overcoming this type of limitation is clearly indispensable for providing an adequate image of the authentic power of Bolzano's system and passes through the notion of *relational quality of finite arity* introduced in Chapter 1.

As a special variable for (relational) *s*-ary qualities, we will use 'r^s'.

By means of the usual technique, we begin by postulating that the capability of the components of the ordered *s*-tuple of ideas $< c_1, \ldots, c_s >$ to *make all the propositions* $\overset{n}{p}$ *true* when substituted for certain other ideas $a_1, \ldots, a_s$ *is a quality* of this *s*-tuple, as well as that a *quality* of it is the capability of its components to *make at least one* of those propositions *true*. Second, we postulate that there are ideas that refer to such qualities. Thus

$$\underset{\overset{s}{a}}{\curlywedge}(\overset{n}{p}) \rightarrow \exists r^s \forall \overset{s}{c}(<\overset{s}{c}> \in r^s \leftrightarrow W(\overset{n}{p}[\overset{s}{a}/\overset{s}{c}])) \qquad \text{SCQ8}$$

$$\neg \underset{\overset{s}{a}}{\mathbb{F}}(\overset{n}{p}) \rightarrow \exists r^s \forall \overset{s}{c}(<\overset{s}{c}> \in r^s \leftrightarrow E(\overset{n}{p}[\overset{s}{a}/\overset{s}{c}])) \qquad \text{SCQ9}$$

and then

$$\forall \overset{s}{a} \exists \bar{c} \begin{pmatrix} (\underset{\overset{s}{a}}{\curlywedge}(\overset{n}{p}) \rightarrow \forall r^s(r^s \mathsf{E}\, \bar{c} \leftrightarrow \forall \overset{s}{c}(<\overset{s}{c}> \in r^s \leftrightarrow W(\overset{n}{p}[\overset{s}{a}/\overset{s}{c}])))) \\ \wedge(\neg \underset{\overset{s}{a}}{\mathbb{F}}(\overset{n}{p}) \rightarrow \neg\, G(\bar{c})) \wedge \forall l(l < \bar{c} \leftrightarrow l \leq \mathfrak{p} \vee \bigvee_{k \leq s} l \leq a_k)) \end{pmatrix} \qquad \text{SCB7}$$

$$\forall \overset{s}{a} \exists \bar{c} \begin{pmatrix} (\neg \underset{\overset{s}{a}}{\mathbb{F}}(\overset{n}{p}) \rightarrow \forall r^s(r^s \mathsf{E}\, \overset{n\vee}{p}_{\overset{s}{a}} \leftrightarrow \forall \overset{s}{c}(<\overset{s}{c}> \in r^s \leftrightarrow E(\overset{n}{p}[\overset{s}{a}/\overset{s}{c}])))) \\ \wedge(\underset{\overset{s}{z}}{\mathbb{F}}(\overset{n}{p}) \rightarrow \neg\, G(\bar{c})) \wedge \forall l(l < \bar{c} \leftrightarrow l \leq \overset{n}{p} \vee \bigvee_{k \leq s} l \leq a_k)) \end{pmatrix} \qquad \text{SCB8}$$

$$\overset{n\wedge}{p}_{\overset{s}{a}} = \varepsilon \bar{c} \begin{pmatrix} (\underset{\overset{s}{a}}{\curlywedge}(\overset{n}{p}) \rightarrow \forall r^s(r^s \mathsf{E}\, \bar{c} \leftrightarrow \forall \overset{s}{c}(<\overset{s}{c}> \in r^s \leftrightarrow W(\overset{n}{p}[\overset{s}{a}/\overset{s}{c}])))) \\ \wedge(\neg \underset{\overset{s}{a}}{\mathbb{F}}(\overset{n}{p}) \rightarrow \neg\, G(\bar{c})) \wedge \forall l(l < \bar{c} \leftrightarrow l \leq \overset{n}{p} \vee \bigvee_{k \leq s} l \leq a_k)) \end{pmatrix} \qquad \text{SD4.35}$$

$$\mathfrak{p}^{\vee}_{\overset{s}{a}} = \varepsilon \bar{c} \begin{pmatrix} (\neg \underset{\overset{s}{a}}{\mathbb{F}}(\overset{n}{p}) \rightarrow \forall r^s(r^s \mathsf{E}\, \overset{n\vee}{p}_{\overset{s}{a}} \leftrightarrow \forall \overset{s}{c}(<\overset{s}{c}> \in r^s \leftrightarrow E(\overset{n}{p}[\overset{s}{a}/\overset{s}{c}])))) \\ \wedge(\underset{\overset{s}{z}}{\mathbb{F}}(\overset{n}{p}) \rightarrow \neg\, G(\bar{c})) \wedge \forall l(l < \bar{c} \leftrightarrow l \leq \overset{n}{p} \vee \bigvee_{k \leq s} l \leq a_k)) \end{pmatrix} \qquad \text{SD4.36}$$

so that we have the following:

$$(1.1)\ B(\overset{n\wedge}{p}_{\overset{s}{a}})$$ ST4.43

$$(1.2)\ B(\overset{n\vee}{p}_{\overset{s}{a}})$$

$$(2.1)\ \underset{\overset{s}{a}}{\mathbb{\Lambda}}(\overset{n}{p}) \rightarrow \forall r(r \sqsubseteq \overset{n\wedge}{p}_{\overset{s}{a}} \leftrightarrow \forall \overset{s}{c}(<\overset{s}{c}> \in r^{s} \leftrightarrow W(\overset{n}{p}\left[\overset{s}{a}/\overset{s}{c}\right]))$$

$$(2.2)\ \neg\underset{\overset{s}{a}}{\mathbb{F}}(\overset{n}{p}) \rightarrow \forall r(r \sqsubseteq \overset{n\vee}{p}_{\overset{s}{a}} \leftrightarrow \forall \overset{s}{c}(<\overset{s}{c}> \in r^{s} \leftrightarrow E(\overset{n}{p}\left[\overset{s}{a}/\overset{s}{c}\right]))$$

$$(3.1)\ \neg\underset{\overset{s}{a}}{\mathbb{\Lambda}}(\overset{n}{p}) \rightarrow \neg G(\overset{n\wedge}{p}_{\overset{s}{a}})$$

$$(3.2)\ \underset{\overset{s}{a}}{\mathbb{F}}(\overset{n}{p}) \rightarrow \neg G(\overset{n\vee}{p}_{\overset{s}{a}})$$

$$(4.1)\ \forall l(l < \overset{n\wedge}{p}_{\overset{s}{a}} \leftrightarrow l \leq \overset{n}{p} \vee \bigvee_{k\leq s} l \leq a_k))$$

$$(4.2)\ \forall l(l < \overset{n\vee}{p}_{\overset{s}{a}} \leftrightarrow l \leq \overset{n}{p} \vee \bigvee_{k\leq s} l \leq a_k)$$

It is now easy to see that, writing '$\overset{s}{a}$' and '$\overset{s}{c}$' where above 'a' and respectively 'c' occur, all that was proved remains still provable. We limit ourselves to the statement of the conclusive theorem:

$$(1)\ \overset{m}{p}\underset{\overset{s}{a}}{\mathbb{\Lambda}}\overset{n}{q} \leftrightarrow W(h(\left\lceil [e](\overset{m\vee}{p}_{\overset{s}{a}} + \overset{n\vee}{q}_{\overset{s}{a}})\right\rceil, \overline{g})$$

$$(2)\ \overset{m}{p}\underset{\overset{s}{a}}{\Rightarrow}\overset{n}{q} \leftrightarrow W(h([e](\overset{m\wedge}{p}_{\overset{s}{a}}), \overset{n\wedge}{q}_{\overset{s}{a}}))$$

$$(3)\ \overset{m}{p}\underset{\overset{s}{a}}{\Leftrightarrow}\overset{n}{q} \leftrightarrow W(h(\left\lceil h([e](\overset{m\wedge}{p}_{\overset{s}{a}}), \overset{n\wedge}{q}_{\overset{s}{a}})\right\rceil \sqcup \left\lceil h([e](\overset{n\wedge}{q}_{\overset{s}{a}}), \overset{m\wedge}{p}_{\overset{s}{a}})\right\rceil, \overline{w}))$$

$$(4)\ \overset{m}{p}\underset{\overset{s}{a}}{\dot{\Rightarrow}}\overset{n}{q} \leftrightarrow W(h(\left\lceil h([e](\overset{m\wedge}{p}_{\overset{s}{a}}), \overset{n\wedge}{q}_{\overset{s}{a}})\right\rceil \sqcup \left\lceil h(\left\lceil h([e](\overset{n\wedge}{q}_{\overset{s}{a}}), \overset{m\wedge}{p}_{\overset{s}{a}})\right\rceil, \sim\overline{w})\right\rceil, \overline{w}))$$

$$(5)\ \overset{m}{p}\underset{\overset{s}{a}}{\mathbb{T}}\overset{n}{q} \leftrightarrow W(h(\left\lceil h(\left\lceil [e](\overset{m\vee}{p}_{\overset{s}{a}} + \overset{n\vee}{q}_{\overset{s}{a}})\right\rceil, \overline{g})\right\rceil \sqcup \left\lceil h(\left\lceil [e](\overset{m\vee}{p}_{\overset{s}{a}} + \sim\overset{n\vee}{q}_{\overset{s}{a}})\right\rceil, \overline{g})\right\rceil$$
$$\sqcup \left\lceil h(\left\lceil [e](\sim\overset{m\vee}{p}_{\overset{s}{a}} + \overset{n\vee}{q}_{\overset{s}{a}})\right\rceil, \overline{g})\right\rceil, \overline{w}))$$

$$(6)\ \neg(\overset{m}{p}\underset{\overset{s}{a}}{\mathbb{\Lambda}}\overset{n}{q}) \leftrightarrow W(h(\left\lceil [e](\overset{m\vee}{p}_{\overset{s}{a}} + \overset{n\vee}{q}_{\overset{s}{a}})\right\rceil, \sim\overline{g})$$

$$(7)\ \overset{m}{p}\underset{\overset{s}{a}}{\perp}\overset{n}{q} \leftrightarrow W(h([e](\overset{m\wedge}{p}_{\overset{s}{a}}), \sim\overset{n\wedge}{q}_{\overset{s}{a}}))$$

$$(8)\ \overset{m}{p}\underset{\overset{s}{a}}{\perp\!\!\!\perp}\overset{n}{q} \leftrightarrow W(h(\left\lceil h([e](\overset{m\wedge}{p}_{\overset{s}{a}}), \sim\overset{n\wedge}{q}_{\overset{s}{a}})\right\rceil \sqcup \left\lceil h([e](\overset{n\wedge}{q}_{\overset{s}{a}}), \sim\overset{m\wedge}{p}_{\overset{s}{a}})\right\rceil, \overline{w}))$$

$$(9)\ \overset{m}{p}\underset{\overset{s}{a}}{\overset{\vee}{\Rightarrow}}\overset{n}{q} \leftrightarrow W(h([e](\overset{m\wedge}{p}_{\overset{s}{a}}), \overset{n\vee}{q}_{\overset{s}{a}}))$$

ST4.44

4.11 Formal Connectives Among Finite Sets of Propositions

We conclude this discussion with the natural introduction of a 'formal connective' in correspondence to each one of the properties and relations considered in the preceding section. In this regard, it is useful to emphasize that, appearances notwithstanding, this is not completely superfluous, at least from a practical point of view. Indeed, when one goes over (as will occur in the final sections of the next chapter) to the consideration of propositions that enunciate derivability relations among propositions which, on their part, enunciate the subsistence of some property or relation in question, it turns out very convenient to have simple notations for the latter. We thus define

(1) *Compatibility* $\quad : \boxed{\substack{\bigwedge\!\!\bigwedge \\ s \\ a}}\,(\overset{n}{p}) \quad = h(\left\lceil [e](\overset{n\wedge}{p}{}^{s}_{a}) \right\rceil, \bar{g})$ SD4.37

(2) *Integration* $\quad : \boxed{\substack{\bigvee\!\!\bigvee \\ s \\ a}}\,(\overset{n}{p}) \quad = h(\left\lceil [e](\sim \overset{n\vee}{p}{}^{s}_{a}) \right\rceil, \sim\bar{g})$

(3) *Formal truths* $\quad : \boxed{\substack{\mathbb{W} \\ s \\ a}}\,(\overset{n}{p}) \quad = h(\left\lceil [e](\sim \overset{n\wedge}{p}{}^{s}_{a}) \right\rceil, \sim\bar{g})$

(4) *Formal falsities* $\quad : \boxed{\substack{\mathbb{F} \\ s \\ a}}\,(\overset{n}{p}) \quad = h(\left\lceil [e](\overset{n\vee}{p}{}^{s}_{a}) \right\rceil, \sim\bar{g})$

(5) *Incompatibility* $\quad : \boxminus \boxed{\substack{\bigwedge\!\!\bigwedge \\ s \\ a}}\,(\overset{n}{p}) \quad = h(\left\lceil [e](\overset{n\wedge}{p}{}^{s}_{a}) \right\rceil, \sim\bar{g})$

(6) *Global falsifiability* $\quad : \boxminus \boxed{\substack{\bigvee\!\!\bigvee \\ s \\ a}}\,(\overset{n}{p}) = h(\left\lceil [e](\sim \overset{n\vee}{p}{}^{s}_{a}) \right\rceil, \bar{g})$

(7) *Local falsifiability* $\quad : \boxminus \boxed{\substack{\mathbb{W} \\ s \\ a}}\,(\overset{n}{p}) = h(\left\lceil [e](\sim \overset{n\wedge}{p}{}^{s}_{a}) \right\rceil, \bar{g})$

(8) *Local verifiability* $\quad : \boxminus \boxed{\substack{\mathbb{F} \\ s \\ a}}\,(\overset{n}{p}) = h(\left\lceil [e](\overset{n\vee}{p}{}^{s}_{a}) \right\rceil, \bar{g})$

From these, by simple application of the definitions and of the theorems, we have

(1) $W(\boxed{\substack{\bigwedge\!\!\bigwedge \\ s \\ a}}\,(\overset{n}{p})) \quad \leftrightarrow \exists \overset{s}{c} W(\overset{n}{p}[\overset{s}{a}/\overset{s}{c}]))$ ST4.45

(2) $W(\boxed{\substack{\bigvee\!\!\bigvee \\ s \\ a}}\,(\overset{n}{p})) \quad \leftrightarrow \forall \overset{s}{c} E(\overset{n}{p}[\overset{s}{a}/\overset{s}{c}]))$

(3) $W(\boxed{\substack{\mathbb{W} \\ s \\ a}}\,(\overset{n}{p})) \quad \leftrightarrow \forall \overset{s}{c} W(\overset{n}{p}[\overset{s}{a}/\overset{s}{c}]))$

(4) $W(\boxed{\substack{\mathbb{F} \\ s \\ a}}\,(\overset{n}{p})) \quad \leftrightarrow \forall \overset{s}{c} F(\overset{n}{p}[\overset{s}{a}/\overset{s}{c}]))$

(5) $W(\boxminus \boxed{\substack{\bigwedge\!\!\bigwedge \\ s \\ a}}\,(\overset{n}{p})) \leftrightarrow \forall \overset{s}{c} E(\overset{n}{p}[\overset{s}{a}/\overset{s}{c}]))$

$$(6)\ W(\boxminus \boxed{\substack{\mathbb{W}\\ s\\ a}}\ (\overset{n}{p})) \ \leftrightarrow \exists\overset{s}{c}F(\overset{n}{p}[\overset{s}{a}/\overset{s}{c}]))$$

$$(7)\ W(\boxminus \boxed{\substack{\mathbb{WW}\\ s\\ a}}\ (\overset{n}{p})) \leftrightarrow \exists\overset{s}{c}\neg W(\overset{n}{p}[\overset{s}{a}/\overset{s}{c}]))$$

$$(8)\ W(\boxminus \boxed{\substack{\mathbb{F}\\ s\\ a}}\ (\overset{n}{p})) \ \leftrightarrow \exists\overset{s}{c}E(\overset{n}{p}[\overset{s}{a}/\overset{s}{c}]))$$

Ad (1) $W(\boxed{\substack{\wedge\!\!\wedge\\ s\\ a}}\ (\overset{n}{p})) \leftrightarrow W(h(\left\lceil [e](\overset{n\wedge}{p}_{\substack{s\\a}}) \right\rceil, \overline{g})) \leftrightarrow G([e](\overset{n\wedge}{p}_{\substack{s\\a}})) \leftrightarrow \underset{\substack{s\\a}}{\wedge\!\!\wedge}(\overset{n}{p}) \leftrightarrow \exists\overset{s}{c}W$ $(\overset{n}{p}[\overset{s}{a}/\overset{s}{c}]))$.

Ad (2) $W(\boxed{\substack{\mathbb{W}\\ s\\ a}}\ (\overset{n}{p})) \leftrightarrow W(h(\left\lceil [e](\sim \overset{n\vee}{p}_{\substack{s\\a}}) \right\rceil, \sim\overline{g})) \leftrightarrow \neg G([e](\sim \overset{m\vee}{p}_{\substack{s\\a}}))$ (T4.14).

Analogously, we define

$$(1)\ \overset{m}{p}\boxed{\substack{\wedge\!\!\wedge\\ s\\ a}}\overset{n}{q} \quad = h(\left\lceil [e](\overset{m\vee}{p}_{\substack{s\\a}} + \overset{n\vee}{q}_{\substack{s\\a}}) \right\rceil, \overline{g})$$

$$(2)\ \overset{m}{p}\boxed{\substack{\Rightarrow\\ s\\ a}}\overset{n}{q} \quad = h([e](\overset{m\wedge}{p}_{\substack{s\\a}}), \overset{n\wedge}{q}_{\substack{s\\a}})$$

$$(3)\ \overset{m}{p}\boxed{\substack{\Leftrightarrow\\ s\\ a}}\overset{n}{q} \quad = h(\left\lceil h([e](\overset{m\wedge}{p}_{\substack{s\\a}}), \overset{n\wedge}{q}_{\substack{s\\a}}) \right\rceil \sqcup \left\lceil h([e](\overset{n\wedge}{q}_{\substack{s\\a}}), \overset{m\wedge}{p}_{\substack{s\\a}}) \right\rceil, \overline{w})$$

$$(4)\ \overset{m}{p}\boxed{\substack{\dot{\Rightarrow}\\ s\\ a}}\overset{n}{q} \quad = h(\left\lceil h([e](\overset{m\wedge}{p}_{\substack{s\\a}}), \overset{n\wedge}{q}_{\substack{s\\a}}) \right\rceil \sqcup \left\lceil h(\left\lceil h([e](\overset{n\wedge}{q}_{\substack{s\\a}}), \overset{m\wedge}{p}_{\substack{s\\a}}) \right\rceil, \sim\overline{w}) \right\rceil, \overline{w})$$

$$(5)\ \overset{m}{p}\boxed{\substack{\top\!\!\top\\ s\\ a}}\overset{n}{q} \quad = h(\left\lceil h(\left\lceil [e](\overset{m\vee}{p}_{\substack{s\\a}} + \overset{n\vee}{q}_{\substack{s\\a}}) \right\rceil, \overline{g}) \right\rceil \sqcup \left\lceil h(\left\lceil [e](\overset{m\vee}{p}_{\substack{s\\a}} + \sim \overset{n\vee}{q}_{\substack{s\\a}}) \right\rceil, \overline{g}) \right\rceil$$
$$\sqcup \left\lceil h(\left\lceil [e](\sim \overset{m\vee}{p}_{\substack{s\\a}} + \overset{n\vee}{q}_{\substack{s\\a}}) \right\rceil, \overline{g}) \right\rceil, \overline{w})$$

$$(6)\ \boxminus\ (\overset{m}{p}\boxed{\substack{\wedge\!\!\wedge\\ s\\ a}}\overset{n}{q}) = h(\left\lceil [e](\overset{m\wedge}{p}_{\substack{s\\a}} + \overset{n\wedge}{q}_{\substack{s\\a}}) \right\rceil, \sim \overline{g})$$

$$(7)\ \overset{m}{p}\boxed{\substack{\perp\\ s\\ a}}\overset{n}{q} \quad = h([e](\overset{m\wedge}{p}_{\substack{s\\a}}), \sim \overset{n\wedge}{q}_{\substack{s\\a}})$$

$$(8)\ \overset{m}{p}\boxed{\substack{\perp\!\!\perp\\ s\\ a}}\overset{n}{q} \quad = h(\left\lceil h([e](\overset{m\wedge}{p}_{\substack{s\\a}}), \sim \overset{n\wedge}{q}_{\substack{s\\a}}) \right\rceil \sqcup \left\lceil h([e](\overset{n\wedge}{q}_{\substack{s\\a}}), \sim \overset{m\wedge}{p}_{\substack{s\\a}}) \right\rceil, \overline{w})$$

$$(9)\ \overset{m}{p}\boxed{\substack{\overset{\vee}{\Rightarrow}\\ s\\ a}}\overset{n}{q} \quad = h([e](\overset{m\wedge}{p}_{\substack{s\\a}}), \overset{n\vee}{q}_{\substack{s\\a}})$$

SD4.38

We then obtain

$$(1)\ W(\overset{m}{p}\boxed{\substack{\wedge\!\!\wedge\\ s\\ a}}\ \overset{n}{q}) \leftrightarrow \exists\overset{s}{c}(E(\overset{m}{p}[\overset{s}{a}/\overset{s}{c}]) \wedge E(\overset{n}{q}[\overset{s}{a}/\overset{s}{c}]))$$ ST4.46

$$(2)\ W(\overset{m}{p}\,\boxed{\overset{\Rightarrow}{\overset{s}{a}}}\,\overset{n}{q}) \quad \leftrightarrow \underset{\overset{s}{a}}{\bigwedge}(\overset{m}{p}) \wedge \forall\overset{s}{c}(W(\overset{m}{p}[\overset{s}{a}/\overset{s}{c}])) \rightarrow W(\overset{n}{q}[\overset{s}{a}/\overset{s}{c}]))$$

$$(3)\ W(\overset{m}{p}\,\boxed{\overset{\Leftrightarrow}{\overset{s}{a}}}\,\overset{n}{q}) \quad \leftrightarrow \underset{\overset{s}{a}}{\bigwedge}(\overset{m}{p}) \wedge \forall\overset{s}{c}(W(\overset{m}{p}[\overset{s}{a}/\overset{s}{c}])) \leftrightarrow W(\overset{n}{q}[\overset{s}{a}/\overset{s}{c}])))$$

$$(4)\ W(\overset{m}{p}\,\boxed{\overset{\dot{\Rightarrow}}{\overset{s}{a}}}\,\overset{n}{q}) \quad \leftrightarrow \underset{\overset{s}{a}}{\bigwedge}(\overset{m}{p}) \wedge \forall\overset{s}{c}(W(\overset{m}{p}[\overset{s}{a}/\overset{s}{c}])) \rightarrow W(\overset{n}{q}[\overset{s}{a}/\overset{s}{c}])))\wedge$$
$$\wedge\exists\overset{s}{c}(W(\overset{n}{q}[\overset{s}{a}/\overset{s}{c}])) \wedge \neg W(\overset{m}{p}[\overset{s}{a}/\overset{s}{c}]))$$

$$(5)\ W(\overset{m}{p}\,\boxed{\overset{\top\!\!\top}{\overset{s}{a}}}\,\overset{n}{q}) \quad \leftrightarrow \underset{\overset{s}{a}}{\bigwedge}(\overset{m}{p}) \wedge \exists\overset{s}{c}(W(\overset{m}{p}[\overset{s}{a}/\overset{s}{c}]) \wedge \neg W(q[\overset{s}{a}/\overset{s}{c}]))\wedge$$
$$\wedge\exists\overset{s}{c}(W(\overset{n}{q}[\overset{s}{a}/\overset{s}{c}]) \wedge \neg W(\overset{m}{p}[\overset{s}{a}/\overset{s}{c}]))$$

$$(6)\ W(\boxminus\ (\overset{m}{p}\,\boxed{\overset{\bigwedge}{\overset{s}{a}}}\,\overset{n}{q}) \leftrightarrow \forall\overset{s}{c}(F(\overset{m}{p}[\overset{s}{a}/\overset{s}{c}]) \vee F(\overset{n}{q}[\overset{s}{a}/\overset{s}{c}]))$$

$$(7)\ W(\overset{m}{p}\,\boxed{\overset{\perp}{\overset{s}{a}}}\,\overset{n}{q}) \quad \leftrightarrow \underset{\overset{s}{a}}{\bigwedge}(\overset{m}{p}) \wedge \forall\overset{s}{c}(W(\overset{m}{p}[\overset{s}{a}/\overset{s}{c}])) \rightarrow F(\overset{n}{q}[\overset{s}{a}/\overset{s}{c}])))$$

$$(8)\ W(\overset{m}{p}\,\boxed{\overset{\perp\!\!\perp}{\overset{s}{a}}}\,\overset{n}{q}) \quad \leftrightarrow \underset{\overset{s}{a}}{\bigwedge}(\overset{m}{p}) \wedge \forall\overset{s}{c}(W(\overset{m}{p}[\overset{s}{a}/\overset{s}{c}])) \leftrightarrow F(\overset{n}{q}[\overset{s}{a}/\overset{s}{c}])))$$

$$(9)\ W(\overset{m}{p}\,\boxed{\overset{\overset{\vee}{\Rightarrow}}{\overset{s}{a}}}\,\overset{n}{q}) \quad \leftrightarrow \underset{\overset{s}{a}}{\bigwedge}(\overset{m}{p}) \wedge \forall\overset{s}{c}(W(\overset{m}{p}[\overset{s}{a}/\overset{s}{c}])) \rightarrow E(\overset{n}{q}[\overset{s}{a}/\overset{s}{c}])))$$

4.12 'Arbitrary Sets' of Ideas and Propositions

A remarkable advantage of the symbolic level is the possibility of treating questions involving *arbitrary sets* of ideas and propositions, *without passing the limits of the logical framework we have chosen.* In fact, what we need is not so much the availability of a general notion of a 'set of ideas' or of a 'set of propositions', but rather the availability of ideas that refer to such sets; thus, *in virtue of the fact that the Bolzanian ideas may have a plural reference,* we may indeed be content with the introduction of the notions of '*(objectual) idea of object ideas*' (*VV*) and of '*(objectual) idea of propositions*' (*VS*), which can be very simply be set with the definitions

$$(1)\ \forall x(VV(x) \leftrightarrow G(x) \wedge \forall z(z \mathrel{\mathsf{E}} x \rightarrow V(z))) \qquad \text{D4.39}$$
$$(2)\ \forall x(VS(x) \leftrightarrow G(x) \wedge \forall z(z \mathrel{\mathsf{E}} x \rightarrow S(z)))$$

and the introduction of new special variables for these types of lectological things. With this aim, we take '$\mathfrak{a}$', '$\mathfrak{b}$', ... as special variables for *ideas of object ideas* and '$\mathfrak{p}$','$\mathfrak{q}$',... as special variables for *ideas of propositions.* We remark that from the definitions, there follow

$$(1)\ G(\mathfrak{a}) \qquad \text{T4.47}$$
$$(2)\ G(\mathfrak{p})$$

The tasks we must now accomplish are (1) to generalize to 'arbitrary sets' of ideas and of propositions the properties and the relations, both material and formal, already considered for the finite case; (2) to prove that Bolzano's thesis is valid in this case too. It is, however, opportune to make a preliminary remark that, although it has no significant influence on the case of ideas, has on the contrary a certain weight in the case of propositions. The fact is that the material properties and relations considered up to this point are properties or relations among propositions, or possibly, in the formal case, among propositions and ideas; the derivability of the $q_1, \ldots, q_n$ from the $p_1, \ldots, p_m$ with respect to the $a_1, \ldots, a_s$, for example, is a relation among $m + n$ propositions and s object ideas, which we describe by using the predicate '$\Rightarrow^{m,n,s}$' and (officially) writing

$$`p_1, \ldots, p_m \underset{a_1, \ldots, a_s}{\Rightarrow}{}^{m,n,s} q_1, \ldots, q_n\text{'} \qquad (*)$$

Its generalization, which we will soon analyse and is denoted of course by '$\mathfrak{p} \underset{a_1, \ldots, a_s}{\Rightarrow} \mathfrak{q}$' (or '$\mathfrak{p} \underset{\mathfrak{a}}{\overset{s}{\Rightarrow}} \mathfrak{q}$'), on the contrary, *is not* a relation among propositions and ideas but among two ideas of propositions and s object ideas. Hence the old statement $(*)$ will, indeed, be equivalent, but clearly different from the new statement, which, of course, will be

$$\ulcorner p_1 \urcorner \sqcup \ldots \sqcup \ulcorner p_m \urcorner \underset{\mathfrak{a}}{\overset{s}{\Rightarrow}}{}^{2,s} \ulcorner q_1 \urcorner \sqcup \ldots \sqcup \ulcorner q_n \urcorner$$

The possible ambiguities that may arise from obvious expository simplifications are not, in general, dangerous; nevertheless, it is opportune to be aware of the real right from the outset.

4.13 Generalization of Properties and Material Relations among Ideas

We will begin with the generalization of properties and material relations among ideas.

1. *Compatibility*

   (a) The ideas under $\mathfrak{a}$ are compatible iff they have a common reference:

$$\mathsf{T}(\mathfrak{a}) \leftrightarrow \exists i \forall b (b \mathrel{\varepsilon} \mathfrak{a} \to i \mathrel{\varepsilon} b) \qquad \text{D4.40}$$

(b) $\mathfrak{a}$ and $\mathfrak{c}$ are compatible iff at least one idea under $\mathfrak{a}$ and one idea under $\mathfrak{c}$ are compatible:

$$\mathfrak{a}\mathsf{T}\mathfrak{c} \leftrightarrow \exists i(\exists b(b \,\mathrm{E}\, \mathfrak{a} \wedge i \,\mathrm{E}\, b) \wedge \exists b(b \,\mathrm{E}\, \mathfrak{c} \wedge i \,\mathrm{E}\, b)) \qquad \text{D4.41}$$

2. *Inclusion*

An $\mathfrak{a}$ is included in a $\mathfrak{c}$ iff it is compatible and every object falling under an idea under it also falls under an idea under $\mathfrak{c}$. Thus

$$\mathfrak{a} \sqsubseteq \mathfrak{c} \leftrightarrow \mathsf{T}(\mathfrak{a}) \wedge \forall i(\exists b(i \,\mathrm{E}\, b \wedge b \,\mathrm{E}\, \mathfrak{a}) \rightarrow \exists b(i \,\mathrm{E}\, b \wedge b \,\mathrm{E}\, \mathfrak{c})) \qquad \text{D4.42}$$

i.e., reformulated,

$$\mathfrak{a} \sqsubseteq \mathfrak{c} \leftrightarrow \exists i \forall b(b \,\mathrm{E}\, \mathfrak{a} \rightarrow i \,\mathrm{E}\, b) \wedge \neg\exists i(\exists b(b \,\mathrm{E}\, \mathfrak{a} \wedge i \,\mathrm{E}\, b) \wedge \forall b(b \,\mathrm{E}\, \mathfrak{c} \rightarrow \neg i \,\mathrm{E}\, b)) \qquad \text{T4.48}$$

3. *Equivalence*

$\mathfrak{a}$ and $\mathfrak{c}$ are equivalent iff they include themselves reciprocally:

$$\mathfrak{a} \equiv \mathfrak{c} \leftrightarrow \mathfrak{a} \sqsubseteq \mathfrak{c} \wedge \mathfrak{c} \sqsubseteq \mathfrak{b} \qquad \text{D4.43}$$

4. *Subordination*

$\mathfrak{a}$ is subordinated to $\mathfrak{c}$ iff it is included in $\mathfrak{c}$ but is not equivalent to it:

$$\mathfrak{a} \,\dot{\sqsubseteq}\, \mathfrak{c} \leftrightarrow \mathfrak{a} \sqsubseteq \mathfrak{c} \wedge \neg\mathfrak{c} \sqsubseteq \mathfrak{b} \qquad \text{D4.44}$$

5. *Disparation*

(a) The ideas under $\mathfrak{a}$ are disparate iff they are compatible but none of them is included in another of them:

$$\dot{\mathsf{T}}(\mathfrak{a}) \leftrightarrow \mathsf{T}(\mathfrak{a}) \wedge \forall ab(a \,\mathrm{E}\, \mathfrak{a} \wedge b \,\mathrm{E}\, \mathfrak{a} \wedge a \neq b \rightarrow a\dot{\mathsf{T}}b) \qquad \text{D4.45}$$

(b) $\mathfrak{a}$ and $\mathfrak{c}$ are disparate iff they are compatible but neither of the two includes the other:

$$\mathfrak{a}\dot{\mathsf{T}}\mathfrak{c} \leftrightarrow \mathfrak{a}\mathsf{T}\mathfrak{c} \wedge \neg\mathfrak{a} \sqsubseteq \mathfrak{c} \wedge \neg\mathfrak{c} \sqsubseteq \mathfrak{a} \qquad \text{D4.46}$$

6. *Incompatibility*

(a) The ideas under $\mathfrak{a}$ are incompatible iff they have no common reference:

$$\perp(\mathfrak{a}) \leftrightarrow \forall i \exists b(b \,\mathrm{E}\, \mathfrak{a} \wedge \neg i \,\mathrm{E}\, b) \; [\leftrightarrow \neg\mathsf{T}(\mathfrak{a})] \qquad \text{D4.47}$$

(b) $\mathfrak{a}$ and $\mathfrak{c}$ are incompatible iff no object falling under an idea under $\mathfrak{a}$ falls under an idea under $\mathfrak{c}$:

$$\mathfrak{a} \perp \mathfrak{c} \leftrightarrow \forall i(\exists b(i \,\mathrm{E}\, b \wedge b \,\mathrm{E}\, \mathfrak{a}) \rightarrow \neg\exists b(i \,\mathrm{E}\, b \wedge b \,\mathrm{E}\, \mathfrak{c})) \qquad \text{D4.48}$$

7. *Contradiction*
$\mathfrak{a}$ and $\mathfrak{c}$ contradict each other iff they are incompatible, but every object falls either under an idea under $\mathfrak{a}$ or an idea under $\mathfrak{c}$:

$$\mathfrak{a} \mathrel{\text{⫲}} \mathfrak{c} \leftrightarrow \mathfrak{a} \perp \mathfrak{c} \wedge \forall i \exists b(i \,\mathrm{E}\, b \wedge (b \,\mathrm{E}\, \mathfrak{a} \vee b \,\mathrm{E}\, \mathfrak{c})) \qquad \text{D4.49}$$

8. *Omnilateral incompatibility*
$\mathfrak{a}$ is omnilaterally incompatible iff any two ideas falling under it are incompatible:

$$\perp\!\!\!\perp(\mathfrak{a}) \leftrightarrow \forall ac(a \neq c \wedge a \,\mathrm{E}\, \mathfrak{a} \wedge c \,\mathrm{E}\, \mathfrak{c} \rightarrow a \perp c)) \qquad \text{D4.50}$$

4.14 Two Operations on Ideas of Object Ideas

The validation of Bolzano's thesis for the subsistence of the preceding relations presupposes a further analysis of Bolzano's convictions on the possible *qualities of an object idea*. It seems impossible to seriously doubt that Bolzano, given any aggregate of ideas, accepted the possibility of considering the idea referring to those objects that fall under every idea of the aggregate, as well as the idea altogether, referring to those objects that fall under the various ideas of the aggregate. Once the hindrance constituted by the 'aggregates of ideas' is bypassed by means of ideas of ideas, it becomes natural to recover those Bolzanian convictions through the introduction of two new operations that, for given idea of ideas $\mathfrak{a}$, respectively yield the *object idea* (to be denoted by '$\sqcap\,\mathfrak{a}$'), *which refers to all and only all the objects falling under all the ideas under* $\mathfrak{a}$, and the *object idea* (to be denoted by '$\sqcup\,\mathfrak{a}$'), *which refers to all and only all objects falling under at least one of the ideas under* $\mathfrak{a}$. As proper parts of these new object ideas, we will consider the parts (proper and improper) of the ideas falling under $\mathfrak{a}$.

Let us therefore begin by assuming the two new lectological axioms

$$\forall \mathfrak{a} \exists a(\forall i(i \,\mathrm{E}\, a \leftrightarrow \forall c(c \,\mathrm{E}\, \mathfrak{a}) \rightarrow i \,\mathrm{E}\, c)) \wedge \forall l(l < a \leftrightarrow \exists c(c \,\mathrm{E}\, \mathfrak{a} \wedge l \leq c))) \qquad \text{CV5}$$

$$\forall \mathfrak{a} \exists a(\forall i(i \,\mathrm{E}\, a \leftrightarrow \exists c(c \,\mathrm{E}\, \mathfrak{a}) \wedge i \,\mathrm{E}\, c)) \wedge \forall l(l < a \leftrightarrow \exists c(c \,\mathrm{E}\, \mathfrak{a} \wedge l \leq c))) \qquad \text{CV6}$$

and thus defining

$$\sqcap\,\mathfrak{a} = \varepsilon a(\forall i(i \,\mathrm{E}\, a \leftrightarrow \forall c(c \,\mathrm{E}\, \mathfrak{a}) \rightarrow i \,\mathrm{E}\, c)) \wedge \forall l(l < a \leftrightarrow \exists c(c \,\mathrm{E}\, \mathfrak{a} \wedge l \leq c)))$$

D4.51

$$\bigsqcup \mathfrak{a} = \varepsilon a(\forall i(i \in a \leftrightarrow \exists c(c \in \mathfrak{a}) \wedge i \in c)) \wedge \forall l(l < a \leftrightarrow \exists c(c \in \mathfrak{a} \wedge l \leq c))) \qquad \text{D4.52}$$

With the usual procedure, we obtain

$$(1)\ V(\bigsqcap \mathfrak{a}) \qquad \text{T4.49}$$

$$(2)\ i \in \bigsqcap \mathfrak{a} \leftrightarrow \forall b(b \in \mathfrak{a} \rightarrow i \in b),$$

$$(3)\ l < \bigsqcap \mathfrak{a} \leftrightarrow \exists b(b \in \mathfrak{a} \wedge l \leq b)$$

$$(1)\ V(\bigsqcup \mathfrak{a}) \qquad \text{T4.50}$$

$$(2)\ i \in \bigsqcup \mathfrak{a} \leftrightarrow \exists b(b \in \mathfrak{a} \wedge i \in b),$$

$$(3)\ l < \bigsqcup \mathfrak{a} \leftrightarrow \exists b(b \in \mathfrak{a} \wedge l \leq b)$$

In particular, we have, for $\mathfrak{a} = \lceil a \rceil$,

$$(1)\ \bigsqcap \mathfrak{a} \equiv a \equiv [e](\widehat{a}); \qquad \text{T4.51}$$

$$(2)\ \bigsqcup \mathfrak{a} \equiv a \equiv [e](\widehat{a})$$

Proof *Ad* (1) $i \in \bigsqcap \mathfrak{a} \leftrightarrow \forall b(b \in \lceil a \rceil \rightarrow i \in b) \leftrightarrow \forall b(b = a \rightarrow i \in b) \leftrightarrow i \in a.$

Ad (2) $i \in \bigsqcup \mathfrak{a} \leftrightarrow i \in \bigsqcup \lceil a \rceil \leftrightarrow \exists b(b \in \lceil a \rceil \wedge i \in b) \leftrightarrow \exists b(b = a \wedge i \in b) \leftrightarrow i \in a.$

In general, for $\mathfrak{a} = \lceil a_1 \rceil \sqcup \ldots \sqcup \lceil a_k \rceil$,

$$(1)\ \bigsqcap \mathfrak{a} \equiv a_1 \sqcap \ldots \sqcap a_k \equiv [e](\widehat{a}_1 + \ldots + \widehat{a}_k) \qquad \text{T4.52}$$

$$(2)\ \bigsqcup \mathfrak{a} \equiv a_1 \sqcup \ldots \sqcup a_k \equiv [e](\widehat{a}_1 \cdot \ \ldots \ \cdot \widehat{a}_k)$$

Proof *Ad* (1) $i \in \bigsqcap \mathfrak{a} \leftrightarrow i \in \bigsqcap(\lceil a_1 \rceil \sqcup \ldots \sqcup \lceil a_k \rceil)$
$\leftrightarrow \forall b(b \in \lceil a_1 \rceil \sqcup \ldots \sqcup \lceil a_k \rceil \rightarrow i \in b)$
$\leftrightarrow \forall b(b \in \lceil a_1 \rceil \vee \ldots \vee b \in \lceil a_k \rceil \rightarrow i \in b)$
$\leftrightarrow \forall b(b = a_1 \vee \ldots \vee b = a_k \rightarrow i \in b)$
$\leftrightarrow \forall b((b = a_1 \rightarrow i \in b) \wedge \ldots \wedge (b = a_k \rightarrow i \in b))$
$\leftrightarrow \forall b(b = a_1 \rightarrow i \in b) \wedge \ldots \wedge \forall b(b = a_k \rightarrow i \in b)$
$\leftrightarrow i \in a_1 \wedge \ldots \wedge i \in a_k \leftrightarrow i \in a_1 \sqcap \ldots \sqcap \in a_k$

but also
$i \in a_1 \wedge \ldots \wedge i \in a_k \leftrightarrow i \in [e](\widehat{a}_1) \wedge \ldots \wedge i \in [e](\widehat{a}_k) \leftrightarrow i \in [e](\widehat{a}_1 + \ldots + \widehat{a}_k).$

Ad (2) $i \in \bigsqcup \mathfrak{a} \leftrightarrow \bigsqcup(\ulcorner a_1 \urcorner \sqcup \ldots \sqcup \ulcorner a_k \urcorner)$
$\leftrightarrow \exists b(b \in \ulcorner a_1 \urcorner \sqcup \ldots \sqcup \ulcorner a_k \urcorner \wedge i \in b)$
$\leftrightarrow \exists b((b \in \ulcorner a_1 \urcorner \vee \ldots \vee b \in \ulcorner a_k \urcorner) \wedge i \in b)$
$\leftrightarrow \exists b((b = a_1 \vee \ldots \vee b = a_k) \wedge i \in b)$
$\leftrightarrow \exists b((b = a_1 \wedge i \in b) \vee \ldots \vee (b = a_k \wedge i \in b))$
$\leftrightarrow \exists b(b = a_1 \wedge i \in b) \vee \ldots \vee \exists b(b = a_k \wedge i \in b)$
$\leftrightarrow i \in a_1 \vee \ldots \vee i \in a_k \leftrightarrow i \in a_1 \sqcup \ldots \sqcup i \in a_k$

and also
$i \in a_1 \vee \ldots \vee i \in a_k \leftrightarrow i \in [e](\widehat{a}_1) \vee \ldots \vee i \in [e](\widehat{a}_k) \leftrightarrow i \in [e](\widehat{a}_1 \cdot \ldots \cdot \widehat{a}_k).$

The following theorems illustrate the behaviour of the two new operations with respect to the old ones $\sqcap$ and $\sqcup$. As far as $\bigsqcap$ is concerned, the two principal facts are

$$(1)\ \bigsqcap \mathfrak{a} \sqcap \bigsqcap \mathfrak{c} \equiv \bigsqcap(\mathfrak{a} \sqcup \mathfrak{c}) \qquad \text{T4.53}$$

$$(2)\ \bigsqcap \mathfrak{a} \sqcup \bigsqcap \mathfrak{c} \sqsubseteq \bigsqcap(\mathfrak{a} \sqcap \mathfrak{c})$$

Proof *Ad* (1) $i \in \bigsqcap \mathfrak{a} \sqcap \bigsqcap \mathfrak{c} \leftrightarrow i \in \bigsqcap \mathfrak{a} \wedge i \in \bigsqcap \mathfrak{c}$
$\leftrightarrow \forall b(b \in \mathfrak{a} \to i \in b) \wedge \forall b(b \in \mathfrak{c} \to i \in b)$
$\leftrightarrow \forall b((b \in \mathfrak{a} \to i \in b) \wedge (b \in \mathfrak{c} \to i \in b)) \leftrightarrow \forall b(b \in \mathfrak{a} \vee b \in \mathfrak{c} \to i \in b)$
$\leftrightarrow \forall b(b \in \mathfrak{a} \sqcup b \to i \in b) \leftrightarrow i \in \bigsqcap(\mathfrak{a} \sqcup \mathfrak{c}).$

Ad (2) $i \in \bigsqcap \mathfrak{a} \sqcup \bigsqcap \mathfrak{c} \leftrightarrow i \in \bigsqcap \mathfrak{a} \vee i \in \bigsqcap \mathfrak{c}$
$\leftrightarrow \forall b(b \in \mathfrak{a} \to i \in b) \vee \forall b(b \in \mathfrak{c} \to i \in b)$
$\to \forall b((b \in \mathfrak{a} \to i \in b) \vee (b \in \mathfrak{c} \to i \in b)) \leftrightarrow \forall b(b \in \mathfrak{a} \wedge b \in \mathfrak{c} \to i \in b)$
$\leftrightarrow \forall b(b \in \mathfrak{a} \sqcap \mathfrak{c} \to i \in b) \leftrightarrow i \in \bigsqcap(\mathfrak{a} \sqcap \mathfrak{c}).$

As, trivially,

$$\bigsqcap \mathfrak{a} \sqcap \bigsqcap \mathfrak{c} \sqsubseteq \bigsqcap \mathfrak{a} \sqcup \bigsqcap \mathfrak{c} \qquad \text{T4.54}$$

we also have

$$\bigsqcap \mathfrak{a} \sqcap \bigsqcap \mathfrak{c} \sqsubseteq \bigsqcap(\mathfrak{a} \sqcap \mathfrak{c}) \qquad \text{T4.55}$$

and hence also

$$(1)\ \bigsqcap(\mathfrak{a} \sqcup \mathfrak{c}) \sqsubseteq \bigsqcap(\mathfrak{a} \sqcap \mathfrak{c}) \qquad \text{T4.56}$$

$$(2)\ \bigsqcap \mathfrak{a} \sqcap \bigsqcap \mathfrak{c} \sqsubseteq \bigsqcap(\mathfrak{a} \sqcap \mathfrak{c})$$

As far as $\bigsqcup$ is concerned, the two principal facts are

$$(1)\ \bigsqcup \mathfrak{a} \sqcup \bigsqcup \mathfrak{c} \equiv \bigsqcup (\mathfrak{a} \sqcup \mathfrak{c}) \qquad \text{T4.57}$$
$$(2)\ \bigsqcup (\mathfrak{a} \sqcap \mathfrak{c}) \sqsubseteq \bigsqcup \mathfrak{a} \sqcap \bigsqcup \mathfrak{c}$$

Proof *Ad* (1) $i \varepsilon \bigsqcup \mathfrak{a} \sqcup \bigsqcup \mathfrak{c} \leftrightarrow i \varepsilon \bigsqcup \mathfrak{a} \vee i \varepsilon \bigsqcup \mathfrak{c}$
$\leftrightarrow \exists b(b \varepsilon \mathfrak{a} \wedge i \varepsilon b) \vee \exists b(b \varepsilon \mathfrak{c} \wedge i \varepsilon b)$
$\leftrightarrow \exists b((b \varepsilon \mathfrak{a} \wedge i \varepsilon b) \vee (b \varepsilon \mathfrak{c} \wedge i \varepsilon b)) \leftrightarrow \exists b((b \varepsilon \mathfrak{a} \vee b \varepsilon \mathfrak{c}) \wedge i \varepsilon b))$
$\leftrightarrow \exists b(b \varepsilon \mathfrak{a} \sqcup \mathfrak{c} \wedge i \varepsilon b)) \leftrightarrow i \varepsilon \bigsqcup (\mathfrak{a} \sqcup \mathfrak{c})$.

Ad (2) $i \varepsilon \bigsqcup (\mathfrak{a} \sqcap \mathfrak{c}) \leftrightarrow \exists b(b \varepsilon \mathfrak{a} \sqcap \mathfrak{c} \wedge i \varepsilon b) \leftrightarrow \exists b(b \varepsilon \mathfrak{a} \wedge b \varepsilon \mathfrak{c} \wedge i \varepsilon b)$
$\rightarrow \exists b(b \varepsilon \mathfrak{a} \wedge i \varepsilon b) \wedge \exists b(b \varepsilon \mathfrak{c} \wedge i \varepsilon b) \leftrightarrow i \varepsilon \bigsqcup \mathfrak{a} \sqcap \bigsqcup \mathfrak{c}$.

Also here, because, trivially,

$$\bigsqcup \mathfrak{a} \sqcap \bigsqcup \mathfrak{c} \sqsubseteq \bigsqcup \mathfrak{a} \sqcup \bigsqcup \mathfrak{c} \qquad \text{T4.58}$$

we have

$$(1)\ \bigsqcup \mathfrak{a} \sqcap \bigsqcup \mathfrak{c} \sqsubseteq \bigsqcup (\mathfrak{a} \sqcup \mathfrak{c}) \qquad \text{T4.59}$$
$$(2)\ \bigsqcup (\mathfrak{a} \sqcup \mathfrak{c}) \sqsubseteq \bigsqcup \mathfrak{a} \sqcup \bigsqcup \mathfrak{c}$$
$$(3)\ \bigsqcup (\mathfrak{a} \sqcap \mathfrak{c}) \sqsubseteq \bigsqcup (\mathfrak{a} \sqcup \mathfrak{c})$$

Obviously, all the preceding inclusions are, in general, proper. Hence these relations as a whole can be synoptically presented as follows:

$\bigsqcap \mathfrak{a} \sqcap \bigsqcap \mathfrak{c} \equiv \bigsqcap (\mathfrak{a} \sqcup \mathfrak{c})$	$\bigsqcup (\mathfrak{a} \sqcup \mathfrak{c})$
$\sqcap$	$\sqcap$
$\bigsqcap \mathfrak{a} \sqcup \bigsqcap \mathfrak{c}$	$\bigsqcup \mathfrak{a} \sqcap \bigsqcup \mathfrak{c}$
$\sqcap$	$\sqcap$
$\bigsqcap (\mathfrak{a} \sqcap \mathfrak{c})$	$\bigsqcup \mathfrak{a} \sqcup \bigsqcup \mathfrak{c} \equiv \bigsqcup (\mathfrak{a} \sqcup \mathfrak{c})$

4.15 New Formulation of the Material Relations among Ideas of Ideas

These new instruments allow us to reformulate the various material relations among ideas of ideas considered above in terms of objectuality.

(a) *Compatibility* of a set of ideas:

$\mathfrak{a}$ is compatible iff the idea of an object that falls under every idea under $\mathfrak{a}$ has objectuality:

$$\mathsf{T}(\mathfrak{a}) \leftrightarrow G(\bigsqcap \mathfrak{a}) \qquad \text{T4.60}$$

Proof $\mathsf{T}(\mathfrak{a}) \leftrightarrow \exists i \forall b(b \,\mathrm{E}\, \mathfrak{c} \rightarrow i \,\mathrm{E}\, b) \leftrightarrow \exists i(i \,\mathrm{E} \bigsqcap \mathfrak{a}) \leftrightarrow G(\bigsqcap \mathfrak{a})$.

(b) *Compatibility* between two sets of ideas:

$\mathfrak{a}$ is compatible with $\mathfrak{c}$ iff the idea of an object falling simultaneously under one of the ideas under $\mathfrak{a}$ and one of the ideas under $\mathfrak{c}$ has objectuality:

$$\mathfrak{a}\mathsf{T}\mathfrak{c} \leftrightarrow G(\bigsqcup \mathfrak{a} \sqcap \bigsqcup \mathfrak{c}) \qquad \text{T4.61}$$

Proof $\mathfrak{a}\mathsf{T}\mathfrak{c} \leftrightarrow \exists i(\exists b(b \,\mathrm{E}\, \mathfrak{a} \wedge i \,\mathrm{E}\, b) \wedge \exists b(b \,\mathrm{E}\, \mathfrak{c} \wedge i \,\mathrm{E}\, b)) \leftrightarrow \exists i(i \,\mathrm{E} \bigsqcup \mathfrak{a} \wedge i \,\mathrm{E} \bigsqcup \mathfrak{c})$
$\leftrightarrow \exists i(i \,\mathrm{E} \bigsqcup \mathfrak{a} \sqcap \bigsqcup \mathfrak{c}) \leftrightarrow G(\bigsqcup \mathfrak{a} \sqcap \bigsqcup \mathfrak{c})$.

(c) *Inclusion* between two sets of ideas:

$\mathfrak{a}$ is included in $\mathfrak{c}$ iff $\mathfrak{a}$ is compatible and the idea of an object falling under some idea under $\mathfrak{a}$ but not falling under any idea under $\mathfrak{c}$ is objectless:

$$\mathfrak{a} \sqsubseteq \mathfrak{c} \leftrightarrow G(\bigsqcap \mathfrak{a}) \wedge \neg G(\bigsqcup \mathfrak{a} \sqcap - \bigsqcup \mathfrak{c}) \qquad \text{T4.62}$$

Proof $\mathfrak{a} \sqsubseteq \mathfrak{c} \leftrightarrow \exists i \forall b(b \,\mathrm{E}\, \mathfrak{a} \rightarrow i \,\mathrm{E}\, b) \wedge \forall i(\exists b(b \,\mathrm{E}\, \mathfrak{a} \wedge i \,\mathrm{E}\, b) \rightarrow \exists b(b \,\mathrm{E}\, \mathfrak{c} \wedge i \,\mathrm{E}\, b))$
$\leftrightarrow G(\bigsqcap \mathfrak{a}) \wedge \forall i(i \,\mathrm{E} \bigsqcup \mathfrak{a} \rightarrow i \,\mathrm{E} \bigsqcup \mathfrak{c})$
$\leftrightarrow G(\bigsqcap \mathfrak{a}) \wedge \neg \exists i \neg(i \,\mathrm{E} \bigsqcup \mathfrak{a} \rightarrow i \,\mathrm{E} \bigsqcup \mathfrak{c})$
$\leftrightarrow G(\bigsqcap \mathfrak{a}) \wedge \neg \exists i(i \,\mathrm{E} \bigsqcup \mathfrak{a} \wedge \neg i \,\mathrm{E} \bigsqcup \mathfrak{c})$
$\leftrightarrow G(\bigsqcap \mathfrak{a}) \wedge \neg \exists i(i \,\mathrm{E} \bigsqcup \mathfrak{a} \wedge i \,\mathrm{E} - \bigsqcup \mathfrak{c})$
$\leftrightarrow G(\bigsqcap \mathfrak{a}) \wedge \neg \exists i(i \,\mathrm{E} \bigsqcup \mathfrak{a} \sqcap - \bigsqcup \mathfrak{c}) \leftrightarrow G(\bigsqcap \mathfrak{a}) \wedge \neg \exists i(i \,\mathrm{E} \bigsqcup \mathfrak{a} \sqcap - \bigsqcup \mathfrak{c})$
$\leftrightarrow G(\bigsqcap \mathfrak{a}) \wedge \neg G(\bigsqcup \mathfrak{a} \sqcap - \bigsqcup \mathfrak{c})$.

(d) *Equivalence* between two sets of ideas:

$$\mathfrak{a} \equiv \mathfrak{c} \leftrightarrow G(\bigsqcap \mathfrak{a}) \wedge G(\bigsqcap \mathfrak{c}) \wedge \neg G((\bigsqcup \mathfrak{a} \sqcap - \bigsqcup \mathfrak{c}) \sqcup (\bigsqcup \mathfrak{c} \sqcap - \bigsqcup \mathfrak{a})) \qquad \text{T4.63}$$

(e) *Subordination* between two sets of ideas:

$$\mathfrak{a} \dot{\sqsubseteq} \mathfrak{c} \leftrightarrow G(\bigsqcap \mathfrak{a}) \wedge \neg G(\bigsqcup \mathfrak{a} \sqcap - \bigsqcup \mathfrak{c}) \wedge G(\bigsqcup \mathfrak{c} \sqcap - \bigsqcup \mathfrak{a})) \qquad \text{T4.64}$$

(f) *Disparation* between two sets of ideas:

$$\mathfrak{a} \dot{\top} \mathfrak{c} \leftrightarrow G(\bigsqcup \mathfrak{a} \sqcap \bigsqcup \mathfrak{c}) \wedge G(\bigsqcup \mathfrak{a} \sqcap - \bigsqcup \mathfrak{c}) \wedge G(\bigsqcup \mathfrak{c} \sqcap - \bigsqcup \mathfrak{a}) \qquad \text{T4.65}$$

(g) *Incompatibility* of a set of ideas:

$$\perp (\mathfrak{a}) \leftrightarrow \neg G(\bigsqcap \mathfrak{c}) \qquad \text{T4.66}$$

(h) *Incompatibility* between two sets of ideas:

$$\mathfrak{a} \perp \mathfrak{c} \leftrightarrow \neg G((\bigsqcup \mathfrak{a} \sqcap \bigsqcup \mathfrak{c}). \qquad \text{T4.67}$$

(i) *Contradiction* between two sets of ideas:

$$\mathfrak{a} || \mathfrak{c} \leftrightarrow \neg G((\bigsqcup \mathfrak{a} \sqcap \bigsqcup \mathfrak{c}) \wedge \neg G(e \sqcap -(\bigsqcup \mathfrak{a} \sqcup \bigsqcup \mathfrak{c})) \qquad \text{T4.68}$$

(l) *Contrariety* between two sets of ideas:

$$\mathfrak{a} \dot{\perp} \mathfrak{c} \leftrightarrow \neg G((\bigsqcup \mathfrak{a} \sqcap \bigsqcup \mathfrak{c}) \wedge G(e \sqcap -(\bigsqcup \mathfrak{a} \sqcup \bigsqcup \mathfrak{c})) \qquad \text{T4.69}$$

On the contrary, it does not seem possible to adequately account for the relations of disparation and omnilateral incompatibility in an arbitrary set of ideas; we must content ourselves, as the Bolzanian text itself seems to suggest,[21], with the finite case. That is,

(m) *Disparation* of the set of the $a_1, \ldots, a_n$:

$$\dot{\top}(\overset{n}{a}) \leftrightarrow \bigwedge_{j \neq k \leq n} (G(a_j \sqcap a_k) \wedge G(a_j \sqcap -a_k) \wedge G(-a_j \sqcap a_k)) \qquad \text{T4.70}$$

(n) *Omnilateral incompatibility* of the set of the $a_1, \ldots, a_n$:

$$\perp\!\!\!\perp (\overset{n}{a}) \leftrightarrow \bigwedge_{j \neq k \leq n} \neg G(a_j \sqcap a_k) \qquad \text{T4.71}$$

[21] TS §140, II 42: "7) We may give the name of a *statement of overlapping* to propositions of the form: 'The relation between the ideas *A*, *B*, *C*, ... is one of *overlapping*.' The most common way to express this frequently encountered relation is through a combination of several propositions which are together equivalent to a single one. Thus in order to express the thought that the two ideas *A* and *B* stand in a relation of overlapping, one sets out the following three propositions: Some *A* are *B*, Some *A* are not *B*, and Some *B* are not *A*."

4.16 Bolzano's Thesis for Material Relations Among Sets of Ideas

Once reformulated in term of objectuality or objectlessness of certain object ideas, it is possible to reduce the subsistence or not subsistence of the various relations among ideas to the truth of certain atomic propositions. So,

$$(1)\ \top(\mathfrak{a}) \leftrightarrow W(h(\ulcorner \bigsqcap \mathfrak{a} \urcorner, \bar{g})) \qquad \text{T4.72}$$

$$(2)\ \mathfrak{a} \sqsubseteq \mathfrak{c} \leftrightarrow W(h(\ulcorner \bigsqcap \mathfrak{a} \urcorner, \bar{g})) \wedge W(h(\ulcorner \bigsqcup \mathfrak{a} \sqcap - \bigsqcup \mathfrak{c} \urcorner, \sim \bar{g}))$$

$$(3)\ \mathfrak{a} \equiv \mathfrak{c} \leftrightarrow W(h(\ulcorner \bigsqcap \mathfrak{a} \urcorner, \bar{g})) \wedge W(h(\ulcorner \bigsqcup \mathfrak{a} \sqcap - \bigsqcup \mathfrak{c} \urcorner, \sim \bar{g})) \wedge W(h(\ulcorner \bigsqcup \mathfrak{c} \sqcap - \bigsqcup \mathfrak{a} \urcorner, \sim \bar{g}))$$

$$(4)\ \mathfrak{a} \mathrel{\dot{\sqsubseteq}} \mathfrak{c} \leftrightarrow W(h(\ulcorner \bigsqcap \mathfrak{a} \urcorner, \bar{g})) \wedge W(h(\ulcorner \bigsqcup \mathfrak{a} \sqcap - \bigsqcup \mathfrak{c} \urcorner, \sim \bar{g})) \wedge W(h(\ulcorner \bigsqcup \mathfrak{c} \sqcap - \bigsqcup \mathfrak{a} \urcorner, \bar{g}))$$

$$(5)\ \mathfrak{a} \mathrel{\dot{\top}} \mathfrak{c} \leftrightarrow W(h(\ulcorner \bigsqcup \mathfrak{a} \sqcap \bigsqcup \mathfrak{c} \urcorner, \bar{g})) \wedge W(h(\ulcorner \bigsqcup \mathfrak{a} \sqcap - \bigsqcup \mathfrak{c} \urcorner, \bar{g})) \wedge W(h(\ulcorner - \bigsqcup \mathfrak{a} \sqcap \bigsqcup \mathfrak{c} \urcorner, \bar{g}))$$

$$(6)\ \perp(\mathfrak{a}) \leftrightarrow W(h(\ulcorner \bigsqcap \mathfrak{a} \urcorner, \sim \bar{g}))$$

$$(7)\ \mathfrak{a} \perp \mathfrak{c} \leftrightarrow W(h(\ulcorner \bigsqcup \mathfrak{a} \sqcap \bigsqcup \mathfrak{c} \urcorner, \sim \bar{g}))$$

$$(8)\ \mathfrak{a} || \mathfrak{c} \leftrightarrow W(h(\ulcorner \bigsqcup \mathfrak{a} \sqcap \bigsqcup \mathfrak{c} \urcorner, \sim \bar{g})) \wedge W(h(\ulcorner e \sqcap -(\bigsqcup \mathfrak{a} \sqcup \bigsqcup \mathfrak{c}) \urcorner, \sim \bar{g}))$$

$$(9)\ \mathfrak{a} \mathrel{\dot{\perp}} \mathfrak{c} \leftrightarrow W(h(\ulcorner \bigsqcup \mathfrak{a} \sqcap \bigsqcup \mathfrak{c} \urcorner, \sim \bar{g})) \wedge W(h(\ulcorner e \sqcap -(\bigsqcup \mathfrak{a} \sqcup \bigsqcup \mathfrak{c}) \urcorner, \bar{g}))$$

$$(10)\ \dot{\top}(\overset{n}{a}) \leftrightarrow \bigwedge_{j \neq k \leq n} (W(h(\ulcorner a_j \sqcap a_k \urcorner, \bar{g})) \wedge W(h(\ulcorner a_j \sqcap -a_k \urcorner, \bar{g})) \wedge W(h(\ulcorner -a_j \sqcap a_k \urcorner, \bar{g})))$$

$$(11)\ \perp\!\!\perp(\overset{n}{a}) \leftrightarrow \bigwedge_{j \neq k \leq n} W(h(\ulcorner a_j \sqcap a_k \urcorner, \sim \bar{g}))$$

However, by the theorem $W(\boxed{\triangle}(\overset{n}{p})) \leftrightarrow W(h(\overset{n}{p}, \overline{w}))$, all preceding equivalencies may be reduced, as expected by Bolzano's thesis, to the truth of atomic propositions. That is,

Validity of Bolzano's thesis for the material relations among arbitrary sets of ideas

$$(1)\ \top(\mathfrak{a}) \leftrightarrow W(h(\ulcorner \bigsqcap \mathfrak{a} \urcorner, \bar{g})) \qquad \text{T4.73}$$

$$(2)\ \mathfrak{a} \sqsubseteq \mathfrak{c} \leftrightarrow W(h(\lceil h(\lceil \bigsqcap \mathfrak{a} \rceil, \bar{g}) \rceil \sqcup \lceil h(\lceil \bigsqcup \mathfrak{a} \sqcap - \bigsqcup \mathfrak{c} \rceil, \sim \bar{g}) \rceil, \bar{w}))$$

$$(3)\ \mathfrak{a} \equiv \mathfrak{c} \leftrightarrow W(h(\lceil h(\lceil \bigsqcap \mathfrak{a} \rceil, \bar{g}) \rceil \sqcup \lceil h(\lceil \bigsqcup \mathfrak{a} \sqcap - \bigsqcup \mathfrak{c} \rceil, \sim \bar{g}) \rceil \sqcup \lceil h(\lceil \bigsqcup \mathfrak{c} \sqcap - \bigsqcup \mathfrak{a} \rceil, \sim \bar{g}) \rceil, \bar{w}))$$

$$(4)\ \mathfrak{a} \dot{\sqsubseteq} \mathfrak{c} \leftrightarrow W(h(\lceil h(\lceil \bigsqcap \mathfrak{a} \rceil, \bar{g}) \rceil \sqcup \lceil h(\lceil \bigsqcup \mathfrak{a} \sqcap - \bigsqcup \mathfrak{c} \rceil, \sim \bar{g}) \rceil \sqcup \lceil h(\lceil \bigsqcup \mathfrak{c} \sqcap - \bigsqcup \mathfrak{a} \rceil, \bar{g}) \rceil, \bar{w}))$$

$$(5)\ \mathfrak{a} \dot{\top} \mathfrak{c} \leftrightarrow W(h(\lceil h(\lceil \bigsqcup \mathfrak{a} \sqcap \bigsqcup \mathfrak{c} \rceil, \bar{g}) \rceil \sqcup \lceil h(\lceil \bigsqcup \mathfrak{a} \sqcap - \bigsqcup \mathfrak{c} \rceil, \bar{g}) \rceil \sqcup \lceil h(\lceil - \bigsqcup \mathfrak{a} \sqcap \bigsqcup \mathfrak{c} \rceil, \bar{g}) \rceil, \bar{w}))$$

$$(6)\ \bot(\mathfrak{a}) \leftrightarrow W(h(\lceil \bigsqcap \mathfrak{a} \rceil, \sim \bar{g}))$$

$$(7)\ \mathfrak{a} \bot \mathfrak{c} \leftrightarrow W(h(\lceil \bigsqcup \mathfrak{a} \sqcap \bigsqcup \mathfrak{c} \rceil, \sim \bar{g}))$$

$$(8)\ \mathfrak{a} \| \mathfrak{c} \leftrightarrow W(h(\lceil h(\lceil \bigsqcup \mathfrak{a} \sqcap \bigsqcup \mathfrak{c} \rceil, \sim \bar{g}) \rceil \sqcup \lceil h(\lceil \mathfrak{e} \sqcap -(\bigsqcup \mathfrak{a} \sqcup \bigsqcup \mathfrak{c}) \rceil, \sim \bar{g}) \rceil, \bar{w}))$$

$$(9)\ \mathfrak{a} \dot{\bot} \mathfrak{c} \leftrightarrow W(h(\lceil h(\lceil \bigsqcup \mathfrak{a} \sqcap \bigsqcup \mathfrak{c} \rceil, \sim \bar{g}) \rceil \sqcup \lceil h(\lceil \mathfrak{e} \sqcap -(\bigsqcup \mathfrak{a} \sqcup \bigsqcup \mathfrak{c}) \rceil, \bar{g}) \rceil, \bar{w}))$$

Validity of Bolzano's thesis for the material relations among finite sets of ideas

$$(1)\ \dot{\top}(\overset{n}{a}) \leftrightarrow W(h(\bigsqcup_{j \neq k \leq n}(\lceil h(\lceil a_j \sqcap a_k \rceil, \bar{g}) \rceil \sqcup \lceil h(\lceil a_j \sqcap -a_k \rceil, \bar{g}) \rceil \sqcup \lceil h(\lceil -a_j \sqcap a_k \rceil, \bar{g}) \rceil), \bar{w})) \qquad \text{T4.74}$$

$$(2)\ \perp\!\!\!\perp(\overset{n}{a}) \leftrightarrow W(h(\bigsqcup_{j \neq k \leq n} \lceil h(\lceil a_j \sqcap a_k \rceil, \sim \bar{g}) \rceil, \bar{w}))$$

4.17 Material Conjunctions and Disjunctions

We can now generalize to 'arbitrary sets' of propositions $\mathfrak{p}, \mathfrak{q}, \ldots$ what we have already done for the various $\overset{m}{p}, \overset{n}{q}, \ldots$ This will make it possible to introduce the material conjunction and the material disjunction of any set of propositions $\mathfrak{p}$. $\boxed{\wedge}\,\mathfrak{p}$ will be the proposition that attributes the truth to every proposition under $\mathfrak{p}$, whereas $\boxed{\vee}\,\mathfrak{p}$ will be the proposition that attributes objectuality to the idea of a true proposition among those under $\mathfrak{p}$. Hence

$$(1)\ \boxed{\wedge}\,\mathfrak{p} = h(\mathfrak{p}, \overline{w}) \qquad \text{D4.53}$$
$$(2)\ \boxed{\vee}\,\mathfrak{p} = h(\ulcorner[\mathfrak{p}](\overline{w})\urcorner, \overline{g})$$

Validity of Bolzano's thesis for material conjunctions and disjunctions on arbitrary (but not empty) sets of propositions for which the thesis already holds

$$(1)\ W(\boxed{\wedge}\,\mathfrak{p}) \leftrightarrow \forall p(p \mathrel{\text{E}} \mathfrak{p} \rightarrow W(p)) \qquad \text{T4.75}$$
$$(2)\ W(\boxed{\vee}\,\mathfrak{p}) \leftrightarrow \exists p(p \mathrel{\text{E}} \mathfrak{p} \wedge W(p))$$

Proof *Ad* (1) $W(\boxed{\wedge}\,\mathfrak{p}) \leftrightarrow W(h(\mathfrak{p}, \overline{w}))$
$\leftrightarrow G(\mathfrak{p}) \wedge \forall p\,(p \mathrel{\text{E}} \mathfrak{p} \rightarrow \exists r(p \in r \wedge r \mathrel{\text{E}} \overline{w}))$
$\leftrightarrow \forall p\,(p \mathrel{\text{E}} \mathfrak{p} \rightarrow p \mathrel{\text{E}} [e](\overline{w})) \leftrightarrow \forall p\,(p \mathrel{\text{E}} \mathfrak{p} \rightarrow W(p))$.
Ad (2) $W(\boxed{\vee}\,\mathfrak{p}) \leftrightarrow W(h(\ulcorner[\mathfrak{p}](\overline{w})\urcorner, \overline{g})) \leftrightarrow G([\mathfrak{p}](\overline{w}))$
$\leftrightarrow \exists p(p \mathrel{\text{E}} \mathfrak{p} \wedge \exists r(p \in r \wedge r \mathrel{\text{E}} \overline{w})) \leftrightarrow \exists p(p \mathrel{\text{E}} \mathfrak{p} \wedge W(p))$.

We then accept the presence, for every $\mathfrak{p}$, of the idea of propositions, to be denoted by '$\boxed{\neg}\,\mathfrak{p}$', which refers to all and only all propositions obtained by negating those under $\mathfrak{p}$, as well as the presence for every $\mathfrak{p}$, $\overset{s}{a}$ and $\overset{s}{c}$ of the idea of propositions, to be denoted by '$\mathfrak{p}[\overset{s}{a}/\overset{s}{c}]$', which refers to all and only all the propositions obtained from those under $\mathfrak{p}$ by substituting the $\overset{s}{c}$ for the $\overset{s}{a}$. That is, we have the following:

$$\exists \mathfrak{q} \forall q(q \mathrel{\text{E}} \mathfrak{q} \leftrightarrow \exists p(p \mathrel{\text{E}} \mathfrak{p} \wedge q = \boxed{\neg}\,p) \wedge \forall l(l < \mathfrak{q} \leftrightarrow l \leq \mathfrak{p}) \qquad \text{CS3}$$
$$\exists \mathfrak{q} \forall q(q \mathrel{\text{E}} \mathfrak{q} \leftrightarrow \exists p(p \mathrel{\text{E}} \mathfrak{p} \wedge q = p[\overset{s}{a}/\overset{s}{c}]) \wedge \forall l(l < \mathfrak{q} \leftrightarrow l \leq \mathfrak{p}, \overset{s}{a}, \overset{s}{c})) \qquad \text{CS4}$$

$$\boxed{\neg}\,\mathfrak{p} = \varepsilon \mathfrak{q} \forall q(q \mathrel{\text{E}} \mathfrak{q} \leftrightarrow \exists p(p \mathrel{\text{E}} \mathfrak{p} \wedge q = \boxed{\neg}\,p)) \qquad \text{D4.54}$$

$$\mathfrak{p}[\overset{s}{a}/\overset{s}{c}] = \varepsilon \mathfrak{q} \forall q(q \mathrel{\text{E}} \mathfrak{q} \leftrightarrow \exists p(p \mathrel{\text{E}} \mathfrak{p} \wedge q = p[\overset{s}{a}/\overset{s}{c}])) \qquad \text{SD4.55}$$

$$(1)\ VS(\boxed{\neg}\,\mathfrak{p}) \qquad \text{T4.76}$$
$$(2)\ q \mathrel{\text{E}} \boxed{\neg}\,\mathfrak{p} \leftrightarrow \exists p(p \mathrel{\text{E}} \mathfrak{p} \wedge q = \boxed{\neg}\,p$$
$$(3)\ \forall l(l < \boxed{\neg}\,\mathfrak{p} \leftrightarrow l \leq \mathfrak{p})$$

$$(1)\ VS(\mathfrak{p}[\overset{s}{a}/\overset{s}{c}]) \qquad \text{ST4.77}$$
$$(2)\ q \mathrel{\text{E}} \mathfrak{p}[\overset{s}{a}/\overset{s}{c}] \leftrightarrow \exists p(p \mathrel{\text{E}} \mathfrak{p} \wedge q = p[\overset{s}{a}/\overset{s}{c}])$$
$$(3)\ \forall l(l < \mathfrak{p}[\overset{s}{a}/\overset{s}{c}] \leftrightarrow l \leq \mathfrak{p}, \overset{s}{a}, \overset{s}{c}))$$

Obviously,

$$\boxed{\neg}\, p \mathrel{\varepsilon} \boxed{\neg}\, \mathfrak{p} \leftrightarrow p \mathrel{\varepsilon} \mathfrak{p} \tag{T4.78}$$

$$p[\overset{s}{a}/\overset{s}{c}] \mathrel{\varepsilon} \mathfrak{p}[\overset{s}{a}/\overset{s}{c}] \leftrightarrow p \mathrel{\varepsilon} \mathfrak{p} \tag{ST4.79}$$

Then, in a natural way, we may set

$$\underset{\overset{s}{a}}{\bigwedge\!\!\!\bigwedge}(\mathfrak{p}) \leftrightarrow \exists \overset{s}{c} W(\boxed{\wedge}\, \mathfrak{p}[\overset{s}{a}/\overset{s}{c}]) \tag{SD4.56}$$

$$\underset{\overset{s}{a}}{\mathbb{F}}(\mathfrak{p}) \leftrightarrow \forall \overset{s}{c} F(\boxed{\vee}\, \mathfrak{p}[\overset{s}{a}/\overset{s}{c}]) \tag{SD4.57}$$

$$\underset{\overset{s}{a}}{\bigwedge\!\!\!\bigwedge}(\mathfrak{p}) \leftrightarrow \exists \overset{s}{c} \forall p(p \mathrel{\varepsilon} \mathfrak{p} \rightarrow W(p[\overset{s}{a}/\overset{s}{c}])) \tag{ST4.80}$$

Proof $\underset{\overset{s}{a}}{\bigwedge\!\!\!\bigwedge}(\mathfrak{p}) \leftrightarrow \exists \overset{s}{c} W(\boxed{\wedge}\, \mathfrak{p}[\overset{s}{a}/\overset{s}{c}])) \leftrightarrow \exists \overset{s}{c} \forall q(q \mathrel{\varepsilon} \mathfrak{p}[\overset{s}{a}/\overset{s}{c}] \rightarrow W(q)) \leftrightarrow$
$\exists \overset{s}{c} \forall q(\exists p(p \mathrel{\varepsilon} \mathfrak{p} \wedge q = p[\overset{s}{a}/\overset{s}{c}]) \rightarrow W(q)) \leftrightarrow \exists \overset{s}{c} \forall q \forall p((p \mathrel{\varepsilon} \mathfrak{p} \wedge q = p[\overset{s}{a}/\overset{s}{c}]) \rightarrow W(q)) \leftrightarrow$
$\exists \overset{s}{c} \forall p \forall q((p \mathrel{\varepsilon} \mathfrak{p} \wedge q = p[\overset{s}{a}/\overset{s}{c}]) \rightarrow W(q)) \leftrightarrow \exists \overset{s}{c} \forall p \forall q(q = p[\overset{s}{a}/\overset{s}{c}]) \rightarrow (p \mathrel{\varepsilon} \mathfrak{p} \rightarrow W(q)) \leftrightarrow$
$\exists \overset{s}{c} \forall p(p \mathrel{\varepsilon} \mathfrak{p} \rightarrow W(p[\overset{s}{a}/\overset{s}{c}]))$.

$$\underset{\overset{s}{a}}{\mathbb{F}}(\mathfrak{p}) \leftrightarrow \forall \overset{s}{c} \forall p(p \mathrel{\varepsilon} \mathfrak{p} \rightarrow F(p[\overset{s}{a}/\overset{s}{c}])) \tag{ST4.81}$$

Proof $\underset{\overset{s}{a}}{\mathbb{F}}(\mathfrak{p}) \leftrightarrow \forall \overset{s}{c} F(\boxed{\vee}\, \mathfrak{p}[\overset{s}{a}/\overset{s}{c}])) \leftrightarrow \forall \overset{s}{c} \neg W(\boxed{\vee}\, \mathfrak{p}[\overset{s}{a}/\overset{s}{c}])$
$\leftrightarrow \forall \overset{s}{c} \neg \exists q(q \mathrel{\varepsilon} \mathfrak{p}[\overset{s}{a}/\overset{s}{c}] \wedge W(q))$
$\leftrightarrow \forall \overset{s}{c} \forall q(q \mathrel{\varepsilon} \mathfrak{p}[\overset{s}{a}/\overset{s}{c}] \rightarrow \neg W(q)) \leftrightarrow \forall \overset{s}{c} \forall q(q \mathrel{\varepsilon} \mathfrak{p}[\overset{s}{a}/\overset{s}{c}] \rightarrow F(q))$
$\leftrightarrow \forall \overset{s}{c} \forall q(\exists p(p \mathrel{\varepsilon} \mathfrak{p} \wedge q = p[\overset{s}{a}/\overset{s}{c}]) \rightarrow F(q))$
$\leftrightarrow \forall \overset{s}{c} \forall q \forall p(p \mathrel{\varepsilon} \mathfrak{p} \wedge q = p[\overset{s}{a}/\overset{s}{c}] \rightarrow F(q))$
$\leftrightarrow \forall \overset{s}{c} \forall p \forall q(q = p[\overset{s}{a}/\overset{s}{c}] \rightarrow (p \mathrel{\varepsilon} \mathfrak{p} \rightarrow F(q)) \leftrightarrow \forall \overset{s}{c} \forall p(p \mathrel{\varepsilon} \mathfrak{p} \rightarrow F(p[\overset{s}{a}/\overset{s}{c}]))$.

It will be also possible to ensure, for every $\mathfrak{p}$ and every $\overset{s}{a}$ (with respect to which it is compatible), the presence in the universe of the *s-ary relational quality* that belongs to the *s*-tuple of object ideas $< \overset{s}{c} >$ iff, substituted for the $\overset{s}{a}$, the $\overset{s}{c}$ make true *all* propositions under $\mathfrak{p}$, as well as the presence, for every $\mathfrak{p}$ and every $\overset{s}{a}$ (with respect to which it is compatible), the presence in the universe of the *s-ary relational quality* that belongs to the *s*-tuple of object ideas $< \overset{s}{c} >$ iff, substituted for the $\overset{s}{a}$, the $\overset{s}{c}$ make true *at least one* of the propositions under $\mathfrak{p}$:

$$\underset{\overset{s}{a}}{\bigwedge\!\!\!\bigwedge}(\mathfrak{p}) \rightarrow \exists r^s \forall \overset{s}{c}(< \overset{s}{c} > \in r^s \leftrightarrow W(\mathfrak{p}[\overset{s}{a}/\overset{s}{c}])) \tag{SCQ10}$$

$$\neg\underset{\overset{s}{a}}{\mathbb{F}}(\mathfrak{p}) \rightarrow \exists r^{s}\forall\overset{s}{c}(<\overset{s}{c}>\in r^{s} \leftrightarrow E(\mathfrak{p}[\overset{s}{a}/\overset{s}{c}])) \qquad \text{SCQ11}$$

Moreover, we will accept the presence of the quality ideas, to be denoted respectively by '$\mathfrak{p}^{\wedge}_{\underset{a}{s}}$' and '$\mathfrak{p}^{\vee}_{\underset{a}{s}}$', that, in the case in which the aforesaid relations exist, refer to them, and otherwise are objectless. That is,

$$\forall\mathfrak{p}\forall\overset{s}{a}\exists\bar{c}\begin{pmatrix}(\underset{\overset{s}{a}}{\mathbb{A}}(\mathfrak{p}) \rightarrow \forall r^{s}(r^{s}\mathrm{E}\,\bar{c} \leftrightarrow \forall\overset{s}{c}(<\overset{s}{c}>\in r^{s} \leftrightarrow W(\mathfrak{p}[\overset{s}{a}/\overset{s}{c}]))))\\ \wedge(\neg\underset{\overset{s}{a}}{\mathbb{A}}(\mathfrak{p}) \rightarrow \neg G(\bar{c})) \wedge \forall l(l<\bar{c} \leftrightarrow l \leq \mathfrak{p} \vee \bigvee_{k\leq s} l \leq a_{k}))\end{pmatrix} \qquad \text{SCB9}$$

$$\forall\mathfrak{p}\forall\overset{s}{a}\exists\bar{c}\begin{pmatrix}(\neg\underset{\overset{s}{a}}{\mathbb{F}}(\mathfrak{p}) \rightarrow \forall r^{s}(r^{s}\mathrm{E}\,\mathfrak{p}^{\vee}_{\underset{a}{s}} \leftrightarrow \forall\overset{s}{c}(<\overset{s}{c}>\in r^{s} \leftrightarrow E(\mathfrak{p}[\overset{s}{a}/\overset{s}{c}]))))\\ \wedge(\underset{\overset{s}{z}}{\mathbb{F}}(x) \rightarrow \neg G(\bar{c})) \wedge \forall l(l<\bar{c} \leftrightarrow l \leq \mathfrak{p} \vee \bigvee_{k\leq s} l \leq a_{k}))\end{pmatrix} \qquad \text{SCB10}$$

$$\mathfrak{p}^{\wedge}_{\underset{a}{s}} = \varepsilon\bar{c}\begin{pmatrix}(\underset{\overset{s}{a}}{\mathbb{A}}(\mathfrak{p}) \rightarrow \forall r^{s}(r^{s}\mathrm{E}\,\bar{c} \leftrightarrow \forall\overset{s}{c}(<\overset{s}{c}>\in r^{s} \leftrightarrow W(\mathfrak{p}[\overset{s}{a}/\overset{s}{c}]))))\\ \wedge(\neg\underset{\overset{s}{a}}{\mathbb{F}}(\mathfrak{p}) \rightarrow \neg G(\bar{c})) \wedge \forall l(l<\bar{c} \leftrightarrow l \leq \mathfrak{p} \vee \bigvee_{k\leq s} l \leq a_{k}))\end{pmatrix} \qquad \text{SD4.58}$$

$$\mathfrak{p}^{\vee}_{\underset{a}{s}} = \varepsilon\bar{c}\begin{pmatrix}(\neg\underset{\overset{s}{a}}{\mathbb{F}}(\mathfrak{p}) \rightarrow \forall r^{s}(r^{s}\mathrm{E}\,\mathfrak{p}^{\vee}_{\underset{a}{s}} \leftrightarrow \forall\overset{s}{c}(<\overset{s}{c}>\in r^{s} \leftrightarrow E(\mathfrak{p}[\overset{s}{a}/\overset{s}{c}]))))\\ \wedge(\underset{\overset{s}{z}}{\mathbb{F}}(x) \rightarrow \neg G(\bar{c})) \wedge \forall l(l<\bar{c} \leftrightarrow l \leq \mathfrak{p} \vee \bigvee_{k\leq s} l \leq a_{k}))\end{pmatrix} \qquad \text{SD4.59}$$

so that

$$\begin{aligned}
&(1.1)\ B(\mathfrak{p}^{\wedge}_{\underset{a}{s}}) && \text{ST4.82}\\
&(1.2)\ B(\mathfrak{p}^{\vee}_{\underset{a}{s}})\\
&(2.1)\ \underset{\overset{s}{a}}{\mathbb{A}}(\mathfrak{p}) \rightarrow \forall r(r\,\mathrm{E}\,\mathfrak{p}^{\wedge}_{\underset{a}{s}} \leftrightarrow \forall\overset{s}{c}(<\overset{s}{c}>\in r^{s} \leftrightarrow W(\mathfrak{p}[\overset{s}{a}/\overset{s}{c}])))\\
&(2.2)\ \neg\underset{\overset{s}{a}}{\mathbb{F}}(\mathfrak{p}) \rightarrow \forall r(r\,\mathrm{E}\,\mathfrak{p}^{\vee}_{\underset{a}{s}} \leftrightarrow \forall\overset{s}{c}(<\overset{s}{c}>\in r^{s} \leftrightarrow E(\mathfrak{p}[\overset{s}{a}/\overset{s}{c}])))\\
&(3.1)\ \neg\underset{\overset{s}{a}}{\mathbb{A}}(\mathfrak{p}) \rightarrow \neg G(\mathfrak{p}^{\wedge}_{\underset{a}{s}})\\
&(3.2)\ \underset{\overset{s}{a}}{\mathbb{F}}(\mathfrak{p}) \rightarrow \neg G(\mathfrak{p}^{\vee}_{\underset{a}{s}})\\
&(4.1)\ \forall l(l<\mathfrak{p}^{\wedge}_{\underset{a}{s}} \leftrightarrow l \leq \mathfrak{p} \vee \bigvee_{k\leq s} l \leq a_{k}))\\
&(4.2)\ \forall l(l<\mathfrak{p}^{\vee}_{\underset{a}{s}} \leftrightarrow l \leq \mathfrak{p} \vee \bigvee_{k\leq s} l \leq a_{k})
\end{aligned}$$

The following theorem shows that our operations make just what is expected from them, i.e. that *an s-tuple of object ideas*: (1) *is something that has* $\mathfrak{p}^{\wedge}_{\overset{s}{a}}$ *iff its components substituted for the* $\overset{s}{a}$ *in all propositions under* $\mathfrak{p}$ *make all of them true*; (2) *is something that has* $\mathfrak{p}^{\vee}_{\overset{s}{a}}$ *iff its components substituted for the* $\overset{s}{a}$ *in all propositions under* $\mathfrak{p}$ *make at least one of them true.*

$$(1)\quad <\overset{s}{c}> \mathsf{E}\,[e](\mathfrak{p}^{\wedge}_{\overset{s}{a}}) \leftrightarrow W(\mathfrak{p}[\overset{s}{a}/\overset{s}{c}])\ [\leftrightarrow \forall p(p\,\mathsf{E}\,\mathfrak{p} \rightarrow W(p[\overset{s}{a}/\overset{s}{c}]))] \qquad \text{ST4.83}$$

$$(2)\quad <\overset{s}{c}> \mathsf{E}\,[e](\mathfrak{p}^{\vee}_{\overset{s}{a}}) \leftrightarrow E(\mathfrak{p}[\overset{s}{a}/\overset{s}{c}])\ [\leftrightarrow \exists p(p\,\mathsf{E}\,\mathfrak{p} \wedge W(p[\overset{s}{a}/\overset{s}{c}]))]$$

Proof *Ad* (1) Let $<\overset{s}{c}> \mathsf{E}\,[e](\mathfrak{p}^{\wedge}_{\overset{s}{a}})$. On the one hand, it follows that $[e](\mathfrak{p}^{\wedge}_{\overset{s}{a}})$ is objectual and such is also the case for $\mathfrak{p}^{\wedge}_{\overset{s}{a}}$, and, by ST82.3.1, (i) $\underset{\overset{s}{a}}{\bigwedge\!\!\!\bigwedge}(\mathfrak{p})$; on the other hand, it follows that there is an r such that (ii) $<\overset{s}{c}> \in r$ and (iii) $r\,\mathsf{E}\,\mathfrak{p}^{\wedge}_{\overset{s}{a}}$. It follows from (i) and (iii), by ST82.2.1, that $\forall\overset{s}{c}(<\overset{s}{c}> \in r \leftrightarrow W(\mathfrak{p}[\overset{s}{a}/\overset{s}{c}]))$, and, from this and (iii), as desired, $W(\mathfrak{p}[\overset{s}{a}/\overset{s}{c}]))$. Vice versa, let $W(\mathfrak{p}[\overset{s}{a}/\overset{s}{c}]))$. Then, a fortiori, $\underset{\overset{s}{a}}{\bigwedge\!\!\!\bigwedge}(\mathfrak{p})$ and thus, by axiom SCQ9, there is an r such that (iv) $\forall\overset{s}{c}(<\overset{s}{c}> \in r \leftrightarrow W(\mathfrak{p}[\overset{s}{a}/\overset{s}{c}]))$ and therefore, because $W(\mathfrak{p}[\overset{s}{a}/\overset{s}{c}]))$, (v) $<\overset{s}{c}> \in r$. But, by ST82.2.1, for an r such that $\forall\overset{s}{c}(<\overset{s}{c}> \in r \leftrightarrow W(\mathfrak{p}[\overset{s}{a}/\overset{s}{c}]))$, also (vi) $r\,\mathsf{E}\,\mathfrak{p}^{\wedge}_{\overset{s}{a}}$ holds. From (v) and (vi), it follows precisely that $<\overset{s}{c}> \mathsf{E}\,[e](\mathfrak{p}^{\wedge}_{\overset{s}{a}})$.

Ad (2) Let $<\overset{s}{c}> \mathsf{E}\,[e](\mathfrak{p}^{\vee}_{\overset{s}{a}})$. On the one hand, it follows that $[e](\mathfrak{p}^{\vee}_{\overset{s}{a}})$ is objectual and such is also the case for $\mathfrak{p}^{\vee}_{\overset{s}{a}}$, and, by ST82.3.2, (i) $\neg\underset{\overset{s}{a}}{\mathbb{F}}(\mathfrak{p})$; on the other hand, it follows that there is an r such that (ii) $<\overset{s}{c}> \in r$ and (iii) $r\,\mathsf{E}\,\mathfrak{p}^{\vee}_{\overset{s}{a}}$. It follows from (i) and (iii), by ST82.2.2, that $\forall\overset{s}{c}(<\overset{s}{c}> \in r \leftrightarrow W(\mathfrak{p}[\overset{s}{a}/\overset{s}{c}]))$, and, from this and (iii), as desired $E(\mathfrak{p}[\overset{s}{a}/\overset{s}{c}]))$. Vice versa, let $E(\mathfrak{p}[\overset{s}{a}/\overset{s}{c}]))$. Then, *a fortiori*, $\neg\underset{\overset{s}{a}}{\mathbb{F}}(\mathfrak{p})$ and thus, by axiom SCQ10, there is an r such that (iv) $\forall\overset{s}{c}(<\overset{s}{c}> \in r \leftrightarrow E(\mathfrak{p}[\overset{s}{a}/\overset{s}{c}]))$, and so, because $E(\mathfrak{p}[\overset{s}{a}/\overset{s}{c}]))$, (v) $<\overset{s}{c}> \in r$. But, by ST82.2.2, for an r such that $\forall\overset{s}{c}(<\overset{s}{c}> \in r \leftrightarrow E(\mathfrak{p}[\overset{s}{a}/\overset{s}{c}]))$, also (vi) $r\,\mathsf{E}\,\mathfrak{p}^{\vee}_{\overset{s}{a}}$ holds. From (v) and (vi), it follows precisely that $<\overset{s}{c}> \mathsf{E}\,[e](\mathfrak{p}^{\vee}_{\overset{s}{a}})$.

As we know, it holds that $\neg i\,\mathsf{E}\,[e](\overline{a}) \leftrightarrow i\,\mathsf{E}\,[e](\sim\overline{a})$, and therefore we have the following two corollaries:

$$(1)\ <\overset{s}{c}> \mathsf{E}\,[e](\sim\mathfrak{p}^{\wedge}_{\overset{s}{a}}) \leftrightarrow \neg W(\mathfrak{p}[\overset{s}{a}/\overset{s}{c}]) \leftrightarrow \exists p(p\,\mathsf{E}\,\mathfrak{p} \wedge F(p[\overset{s}{a}/\overset{s}{c}])) \qquad \text{T4.84}$$

$$(2)\ <\overset{s}{c}> \mathsf{E}\,[e](\sim\mathfrak{p}^{\vee}_{\overset{s}{a}}) \leftrightarrow \neg E(\mathfrak{p}[\overset{s}{a}/\overset{s}{c}]) \leftrightarrow \forall p(p\,\mathsf{E}\,\mathfrak{p} \rightarrow F(p[\overset{s}{a}/\overset{s}{c}])$$

4.18 Formal Properties of 'Arbitrary Sets' of Propositions

From the two preceding theorems, by simple application of the definitions, we have the following:

$$(1)\ G([e](\mathfrak{p}^{\wedge}_{\substack{s\\a}})) \leftrightarrow \exists\overset{s}{c}\forall p(p \in \mathfrak{p} \rightarrow W(p[\overset{s}{a}/\overset{s}{c}])) \qquad \text{T4.85}$$

$$(2)\ G([e](\mathfrak{p}^{\vee}_{\substack{s\\a}})) \leftrightarrow \exists\overset{s}{c}\exists p(p \in \mathfrak{p} \wedge W(p[\overset{s}{a}/\overset{s}{c}]))$$

$$(3)\ G([e](\sim\mathfrak{p}^{\wedge}_{\substack{s\\a}})) \leftrightarrow \exists\overset{s}{c}\exists p(p \in \mathfrak{p} \wedge F(p[\overset{s}{a}/\overset{s}{c}]))$$

$$(4)\ G([e](\sim\mathfrak{p}^{\vee}_{\substack{s\\a}})) \leftrightarrow \exists\overset{s}{c}\forall p(p \in \mathfrak{p} \rightarrow F(\mathfrak{p}[\overset{s}{a}/\overset{s}{c}]))$$

Thus, we can reformulate in terms of objectuality or objectlessness of the concretizations of $\mathfrak{p}^{\wedge}_{\substack{s\\a}}$ or of $\mathfrak{p}^{\vee}_{\substack{s\\a}}$ or of their lack, the eight fundamental formal properties of a 'set' of propositions with respect to an idea:

$$\begin{array}{lll}
(1)\ \textit{Compatibility} & : \underset{\substack{s\\a}}{\mathbb{\Lambda}}(\mathfrak{p}) & \leftrightarrow G([e](\mathfrak{p}^{\wedge}_{\substack{s\\a}})) \\
(2)\ \textit{Integration} & : \underset{\substack{s\\a}}{\mathbb{V}}(\mathfrak{p}) & \leftrightarrow \neg G([e](\sim\mathfrak{p}^{\vee}_{\substack{s\\a}})) \\
(3)\ \textit{Formal truths} & : \underset{\substack{s\\a}}{\mathbb{W}}(\mathfrak{p}) & \leftrightarrow \neg G([e](\sim\mathfrak{p}^{\wedge}_{\substack{s\\a}})) \\
(4)\ \textit{Formal falsities} & : \underset{\substack{s\\a}}{\mathbb{F}}(\mathfrak{p}) & \leftrightarrow \neg G([e](\mathfrak{p}^{\vee}_{\substack{s\\a}})) \\
(5)\ \textit{Incompatibility} & : \neg\underset{\substack{s\\a}}{\mathbb{\Lambda}}(\mathfrak{p}) & \leftrightarrow \neg G([e](\mathfrak{p}^{\wedge}_{\substack{s\\a}})) \\
(6)\ \textit{Global falsifiability} & : \neg\underset{\substack{s\\a}}{\mathbb{V}}(\mathfrak{p}) & \leftrightarrow G([e](\sim\mathfrak{p}^{\vee}_{\substack{s\\a}})) \\
(7)\ \textit{Local falsifiability} & : \neg\underset{\substack{s\\a}}{\mathbb{W}}(\mathfrak{p}) & \leftrightarrow G([e](\sim\mathfrak{p}^{\wedge}_{\substack{s\\a}})) \\
(8)\ \textit{Local verifiability} & : \neg\underset{\substack{s\\a}}{\mathbb{F}}(\mathfrak{p}) & \leftrightarrow G([e](\mathfrak{p}^{\vee}_{\substack{s\\a}}))
\end{array} \qquad \text{T4.86}$$

From this, by virtue of the two theorems that allow us to express the belonging or not belonging of objectuality to an idea as the truth of objectuality or objectlessness propositions, we obtain the following:

Validity of Bolzano's thesis for formal properties of an arbitrary set of propositions

$$(1)\ \textit{Compatibility}: \underset{\substack{s\\a}}{\mathbb{\Lambda}}(\mathfrak{p}) \leftrightarrow W(h(\left\lceil [e](\mathfrak{p}^{\wedge}_{\substack{s\\a}}) \right\rceil, \bar{g}))^{22} \qquad \text{T4.87}$$

$$(2)\ \textit{Integration}: \underset{\substack{s\\a}}{\mathbb{V}}(\mathfrak{p}) \leftrightarrow W(h(\left\lceil [e](\sim\mathfrak{p}^{\vee}_{\substack{s\\a}}) \right\rceil, \sim\bar{g}))$$

[22] Cf. TS §164, II 141 (1).

(3) *Formal truths* : $\underset{\substack{s\\a}}{\mathbb{W}}(\mathfrak{p}) \leftrightarrow W(h(\left\lceil [e](\sim \mathfrak{p}^{\wedge}_{\substack{s\\a}}) \right\rceil, \sim\bar{g}))$

(4) *Formal falsities* : $\underset{\substack{s\\a}}{\mathbb{F}}(\mathfrak{p}) \leftrightarrow W(h(\left\lceil [e](\mathfrak{p}^{\vee}_{\substack{s\\a}}) \right\rceil, \sim\bar{g}))$

(5) *Incompatibility* : $\neg\underset{\substack{s\\a}}{\curlywedge}(\mathfrak{p}) \leftrightarrow W(h(\left\lceil [e](\mathfrak{p}^{\wedge}_{\substack{s\\a}}) \right\rceil, \sim\bar{g}))$

(6) *Global falsifiability* : $\neg\underset{\substack{s\\a}}{\mathbb{V}}(\mathfrak{p}) \leftrightarrow W(h(\left\lceil [e](\sim \mathfrak{p}^{\vee}_{\substack{s\\a}}) \right\rceil, \bar{g}))$

(7) *Local falsifiability* : $\neg\underset{\substack{s\\a}}{\mathbb{W}}(\mathfrak{p}) \leftrightarrow W(h(\left\lceil [e](\sim \mathfrak{p}^{\wedge}_{\substack{s\\a}}) \right\rceil, \bar{g}))$

(8) *Local verifiability* : $\neg\underset{\substack{s\\a}}{\mathbb{F}}(\mathfrak{p}) \leftrightarrow W(h(\left\lceil [e](\mathfrak{p}^{\vee}_{\substack{s\\a}}) \right\rceil, \bar{g}))$

4.19 Formal Relations among 'Arbitrary Sets' of Propositions

First of all, we will reformulate the principal formal relations among sets of propositions with respect to an idea in terms of material properties and relations among concretizations of the two new ideas:

(1) *Compatibility* $\mathfrak{p} \underset{\substack{s\\a}}{\curlywedge} \mathfrak{q} \leftrightarrow G([e](\mathfrak{p}^{\vee}_{\substack{s\\a}} + \mathfrak{q}^{\vee}_{\substack{s\\a}}))$ T4.88

(2) *Derivability* $\mathfrak{p} \underset{\substack{s\\a}}{\Rightarrow} \mathfrak{q} \leftrightarrow [e](\mathfrak{p}^{\wedge}_{\substack{s\\a}}) \sqsubseteq [e](\mathfrak{q}^{\wedge}_{\substack{s\\a}})$

(3) *Equivalence* $\mathfrak{p} \underset{\substack{s\\a}}{\Leftrightarrow} \mathfrak{q} \leftrightarrow [e](\mathfrak{p}^{\wedge}_{\substack{s\\a}}) \equiv [e](\mathfrak{q}^{\wedge}_{\substack{s\\a}})$

(4) *Subordination* $\mathfrak{p} \underset{\substack{s\\a}}{\dot{\Rightarrow}} \mathfrak{q} \leftrightarrow [e](\mathfrak{p}^{\wedge}_{\substack{s\\a}}) \sqsubseteq [e](\mathfrak{q}^{\wedge}_{\substack{s\\a}}) \wedge G([e](\mathfrak{q}^{\wedge}_{\substack{s\\a}} + \sim \mathfrak{p}^{\wedge}_{\substack{s\\a}}))$

(5) *Concatenation* $\mathfrak{p} \underset{\substack{s\\a}}{\top\!\!\top} \mathfrak{q} \leftrightarrow \begin{pmatrix} G([e](\mathfrak{p}^{\vee}_{\substack{s\\a}} + \mathfrak{q}^{\vee}_{\substack{s\\a}})) \wedge G([e](\mathfrak{p}^{\wedge}_{\substack{s\\a}} + \sim \mathfrak{q}^{\wedge}_{\substack{s\\a}})) \\ \wedge G([e](\mathfrak{q}^{\wedge}_{\substack{s\\a}} + \sim \mathfrak{p}^{\wedge}_{\substack{s\\a}})) \end{pmatrix}$

(6) *Incompatibility* $\neg(\mathfrak{p} \underset{\substack{s\\a}}{\curlywedge} \mathfrak{q}) \leftrightarrow \neg G([e](\mathfrak{p}^{\vee}_{\substack{s\\a}} + \mathfrak{q}^{\vee}_{\substack{s\\a}}))$

(7) *Exclusion* $\mathfrak{p} \underset{\substack{s\\a}}{\perp} \mathfrak{q} \leftrightarrow [e](\mathfrak{p}^{\wedge}_{\substack{s\\a}}) \sqsubseteq [e](\sim \mathfrak{q}^{\wedge}_{\substack{s\\a}})$

(8) *Reciprocal exclusion* $\mathfrak{p} \underset{\substack{s\\a}}{\perp\!\!\!\perp} \mathfrak{q} \leftrightarrow [e](\sim \mathfrak{p}^{\wedge}_{\substack{s\\a}}) \equiv [e](\mathfrak{q}^{\wedge}_{\substack{s\\a}})$

(9) *Conditioned Integration* $\mathfrak{p} \underset{\substack{s\\a}}{\overset{\vee}{\Rightarrow}} \mathfrak{q} \leftrightarrow [e](\mathfrak{p}^{\wedge}_{\substack{s\\a}}) \sqsubseteq [e](\mathfrak{q}^{\vee}_{\substack{s\\a}})$

Proof *Ad* (1) $\mathfrak{p} \underset{\substack{s\\a}}{\curlywedge} \mathfrak{q} \leftrightarrow \exists \overset{s}{c}(\exists p(p \in \mathfrak{p} \wedge W(p[\overset{s}{a}/\overset{s}{c}])) \wedge \exists q(q \in \mathfrak{q} \wedge W(q[\overset{s}{a}/\overset{s}{c}])))$

$\leftrightarrow \exists \overset{s}{c}(E(\mathfrak{p}[\overset{s}{a}/\overset{s}{c}]) \wedge E(\mathfrak{q}[\overset{s}{a}/\overset{s}{c}])) \leftrightarrow \exists \overset{s}{c}(<\overset{s}{c}> \in [e](\mathfrak{p}^{\vee}_{\substack{s\\a}}) \wedge <\overset{s}{c}> \in [e](\mathfrak{q}^{\vee}_{\substack{s\\a}}))$

$\leftrightarrow \exists \overset{s}{c}(<\overset{s}{c}> \in [e](\mathfrak{p}^{\vee}_{\substack{s\\a}} + \mathfrak{q}^{\vee}_{\substack{s\\a}}))$.

Ad (2)

$\mathfrak{p} \underset{\substack{s\\a}}{\Rightarrow} \mathfrak{q} \leftrightarrow \underset{\substack{s\\a}}{\curlywedge}(\mathfrak{p}) \wedge \forall \overset{s}{c}(\forall p(p \in \mathfrak{p} \to W(p[\overset{s}{a}/\overset{s}{c}])) \to \forall q(q \in \mathfrak{q} \to W(q[\overset{s}{a}/\overset{s}{c}])))$

$\leftrightarrow \underset{\substack{s\\a}}{\curlywedge}(\mathfrak{p}) \wedge \forall \overset{s}{c}(W(\mathfrak{p}[\overset{s}{a}/\overset{s}{c}]) \to W(\mathfrak{q}[\overset{s}{a}/\overset{s}{c}]))$

$\leftrightarrow G([e](\mathfrak{p}^{\wedge}_{\substack{s\\a}})) \wedge \forall \overset{s}{c}(<\overset{s}{c}> \in [e](\mathfrak{p}^{\wedge}_{\substack{s\\a}}) \to <\overset{s}{c}> \in [e](\mathfrak{q}^{\wedge}_{\substack{s\\a}}) \leftrightarrow [e](\mathfrak{p}^{\wedge}_{\substack{s\\a}}) \sqsubseteq [e](\mathfrak{q}^{\wedge}_{\substack{s\\a}})$.

Ad (3) This is trivial.

Ad (4) $\mathfrak{p} \underset{\substack{s\\a}}{\dot{\Rightarrow}} \mathfrak{q} \leftrightarrow \mathfrak{p} \underset{\substack{s\\a}}{\Rightarrow} \mathfrak{q} \wedge \neg(\mathfrak{p} \underset{\substack{s\\a}}{\Rightarrow} \mathfrak{q})$

$\leftrightarrow [e](\mathfrak{p}^{\wedge}_{\substack{s\\a}}) \sqsubseteq [e](\mathfrak{q}^{\wedge}_{\substack{s\\a}}) \wedge \neg([e](\mathfrak{q}^{\wedge}_{\substack{s\\a}}) \sqsubseteq [e](\mathfrak{p}^{\wedge}_{\substack{s\\a}}))$

$\leftrightarrow [e](\mathfrak{p}^{\wedge}_{\substack{s\\a}}) \sqsubseteq [e](\mathfrak{q}^{\wedge}_{\substack{s\\a}}) \wedge \exists \overset{s}{c}(<\overset{s}{c}> \in [e](\mathfrak{q}^{\wedge}_{\substack{s\\a}}) \wedge \neg(<\overset{s}{c}> \in [e](\mathfrak{p}^{\wedge}_{\substack{s\\a}})))$

$\leftrightarrow [e](\mathfrak{p}^{\wedge}_{\substack{s\\a}}) \sqsubseteq [e](\mathfrak{q}^{\wedge}_{\substack{s\\a}}) \wedge \exists \overset{s}{c}(<\overset{s}{c}> \in [e](\mathfrak{q}^{\wedge}_{\substack{s\\a}}) \wedge <\overset{s}{c}> \in [e](\sim \mathfrak{p}^{\wedge}_{\substack{s\\a}})))$

$\leftrightarrow [e](\mathfrak{p}^{\wedge}_{\substack{s\\a}}) \sqsubseteq [e](\mathfrak{q}^{\wedge}_{\substack{s\\a}}) \wedge \exists \overset{s}{c}(<\overset{s}{c}> \in [e](\mathfrak{q}^{\wedge}_{\substack{s\\a}} + \sim \mathfrak{p}^{\wedge}_{\substack{s\\a}})))$

$\leftrightarrow [e](\mathfrak{p}^{\wedge}_{\substack{s\\a}}) \sqsubseteq [e](\mathfrak{q}^{\wedge}_{\substack{s\\a}}) \wedge G([e](\mathfrak{q}^{\wedge}_{\substack{s\\a}} + \sim \mathfrak{p}^{\wedge}_{\substack{s\\a}})))$.

Ad (5) $\mathfrak{p} \underset{\substack{s\\a}}{\top} \mathfrak{q} \leftrightarrow \mathfrak{p} \underset{\substack{s\\a}}{\curlywedge} \mathfrak{q} \wedge \neg(\mathfrak{p} \underset{\substack{s\\a}}{\Rightarrow} \mathfrak{q}) \wedge \neg(\mathfrak{q} \underset{\substack{s\\a}}{\Rightarrow} \mathfrak{p})$

$\leftrightarrow G([e](\mathfrak{p}^{\vee}_{\substack{s\\a}} + \mathfrak{q}^{\vee}_{\substack{s\\a}})) \wedge \neg([e](\mathfrak{p}^{\wedge}_{\substack{s\\a}}) \sqsubseteq [e](\mathfrak{q}^{\wedge}_{\substack{s\\a}})) \wedge \neg([e](\mathfrak{q}^{\wedge}_{\substack{s\\a}}) \sqsubseteq [e](\mathfrak{p}^{\wedge}_{\substack{s\\a}}))$

$\leftrightarrow G([e](\mathfrak{p}^{\vee}_{\substack{s\\a}} + \mathfrak{q}^{\vee}_{\substack{s\\a}})) \wedge G([e](\mathfrak{p}^{\wedge}_{\substack{s\\a}} + \sim \mathfrak{q}^{\wedge}_{\substack{s\\a}})) \wedge G([e](\mathfrak{q}^{\wedge}_{\substack{s\\a}} + \sim \mathfrak{p}^{\wedge}_{\substack{s\\a}})))$.

Ad (6) This is trivial.

Ad (7)

$\mathfrak{p} \underset{\substack{s\\a}}{\perp} \mathfrak{q} \leftrightarrow \underset{\substack{s\\a}}{\curlywedge}(\mathfrak{p}) \wedge \forall \overset{s}{c}(\forall p(p \in \mathfrak{p} \to W(p[\overset{s}{a}/\overset{s}{c}])) \to \forall q(q \in \mathfrak{q} \to \neg W(q[\overset{s}{a}/\overset{s}{c}])))$

$\leftrightarrow G([e](\mathfrak{p}^{\wedge}_{\substack{s\\a}})) \wedge \forall \overset{s}{c}(<\overset{s}{c}> \in [e](\mathfrak{p}^{\wedge}_{\substack{s\\a}}) \to \neg <\overset{s}{c}> \in [e](\mathfrak{q}^{\wedge}_{\substack{s\\a}}))$

$\leftrightarrow G([e](\mathfrak{p}^{\wedge}_{\substack{s\\a}})) \wedge \forall \overset{s}{c}(<\overset{s}{c}> \in [e](\mathfrak{p}^{\wedge}_{\substack{s\\a}}) \to <\overset{s}{c}> \in [e](\sim \mathfrak{q}^{\wedge}_{\substack{s\\a}}))$

$\leftrightarrow [e](\mathfrak{p}^{\wedge}_{\substack{s\\a}}) \sqsubseteq [e](\sim \mathfrak{q}^{\wedge}_{\substack{s\\a}})$.

Ad (8) $\mathfrak{p} \underset{\substack{s\\a}}{\perp\!\!\perp} \mathfrak{q} \leftrightarrow [e](\mathfrak{p}^{\wedge}_{\substack{s\\a}}) \equiv [e](\sim \mathfrak{q}^{\wedge}_{\substack{s\\a}}) \leftrightarrow [e](\sim \mathfrak{p}^{\wedge}_{\substack{s\\a}}) \equiv [e](\mathfrak{q}^{\wedge}_{\substack{s\\a}})$.

Ad (9)

$$\mathfrak{p} \overset{\vee}{\underset{\substack{s\\a}}{\Rightarrow}} \mathfrak{q} \leftrightarrow \underset{\substack{s\\a}}{\wedge\!\!\!\wedge}(\mathfrak{p}) \wedge \forall \overset{s}{c}(\forall p(p \in \mathfrak{p} \rightarrow W(p[\overset{s}{a}/\overset{s}{c}])) \rightarrow \exists q(q \in \mathfrak{q} \wedge W(q[\overset{s}{a}/\overset{s}{c}])))$$

$$\leftrightarrow \underset{\substack{s\\a}}{\wedge\!\!\!\wedge}(\mathfrak{p}) \wedge G([e](\mathfrak{p}^{\wedge}_{\substack{s\\a}})) \wedge \forall \overset{s}{c}(\overset{s}{c} \in [e](\mathfrak{p}^{\wedge}_{\substack{s\\a}}) \rightarrow < \overset{s}{c} > \in [e](\mathfrak{q}^{\vee}_{\substack{s\\a}})$$

$$\leftrightarrow [e](\mathfrak{p}^{\wedge}_{\substack{s\\a}}) \sqsubseteq [e](\mathfrak{q}^{\vee}_{\substack{s\\a}}).$$

Applying the equivalencies $\overset{s}{a} \sqsubseteq [e](\overline{c}) \leftrightarrow W(h(\overset{s}{a}, \overline{c}))$ and $G(\overset{s}{a}) \leftrightarrow W(h(\ulcorner \overset{s}{a} \urcorner, \overline{g}))$, we reach the following reformulation of those relations:

T4.89

$$(1)\ \mathfrak{p} \underset{\substack{s\\a}}{\wedge\!\!\!\wedge} \mathfrak{q} \leftrightarrow W(h(\left\lceil [e](\mathfrak{p}^{\vee}_{\substack{s\\a}} + \mathfrak{q}^{\vee}_{\substack{s\\a}}) \right\rceil, \overline{g})$$

$$(2)\ \mathfrak{p} \underset{\substack{s\\a}}{\Rightarrow} \mathfrak{q} \leftrightarrow W(h([e](\mathfrak{p}^{\wedge}_{\substack{s\\a}}), \mathfrak{q}^{\wedge}_{\substack{s\\a}}))$$

$$(3)\ \mathfrak{p} \underset{\substack{s\\a}}{\Leftrightarrow} \mathfrak{q} \leftrightarrow W(h([e](\mathfrak{p}^{\wedge}_{\substack{s\\a}}), \mathfrak{q}^{\wedge}_{\substack{s\\a}})) \wedge W(h([e](\mathfrak{q}^{\wedge}_{\substack{s\\a}}), \mathfrak{p}^{\wedge}_{\substack{s\\a}}))$$

$$(4)\ \mathfrak{p} \underset{\substack{s\\a}}{\dot{\Rightarrow}} \mathfrak{q} \leftrightarrow W(h([e](\mathfrak{p}^{\wedge}_{\substack{s\\a}}), \mathfrak{q}^{\wedge}_{\substack{s\\a}})) \wedge \neg W(h([e](\mathfrak{q}^{\wedge}_{\substack{s\\a}}), \mathfrak{p}^{\wedge}_{\substack{s\\a}}))$$

$$(5)\ \mathfrak{p} \underset{\substack{s\\a}}{\top\!\!\top} \mathfrak{q} \leftrightarrow \begin{pmatrix} W(h(\left\lceil [e](\mathfrak{p}^{\vee}_{\substack{s\\a}} + \mathfrak{q}^{\vee}_{\substack{s\\a}}) \right\rceil, \overline{g}) \wedge W(h(\left\lceil [e](\mathfrak{p}^{\vee}_{\substack{s\\a}} + \sim \mathfrak{q}^{\vee}_{\substack{s\\a}}) \right\rceil, \overline{g}) \\ \wedge W(h(\left\lceil [e](\sim \mathfrak{p}^{\vee}_{\substack{s\\a}} + \mathfrak{q}^{\vee}_{\substack{s\\a}}) \right\rceil, \overline{g}) \end{pmatrix}$$

$$(6)\ \neg(\mathfrak{p} \underset{\substack{s\\a}}{\wedge\!\!\!\wedge} \mathfrak{q}) \leftrightarrow W(h(\left\lceil [e](\mathfrak{p}^{\vee}_{\substack{s\\a}} + \mathfrak{q}^{\vee}_{\substack{s\\a}}) \right\rceil, \sim \overline{g})$$

$$(7)\ \mathfrak{p} \underset{\substack{s\\a}}{\perp} \mathfrak{q} \leftrightarrow W(h([e](\mathfrak{p}^{\wedge}_{\substack{s\\a}}), \sim \mathfrak{q}^{\wedge}_{\substack{s\\a}}))$$

$$(8)\ \mathfrak{p} \underset{\substack{s\\a}}{\perp\!\!\perp} \mathfrak{q} \leftrightarrow W(h([e](\mathfrak{p}^{\wedge}_{\substack{s\\a}}), \sim \mathfrak{q}^{\wedge}_{\substack{s\\a}})) \wedge W(h([e](\mathfrak{q}^{\wedge}_{\substack{s\\a}}), \sim \mathfrak{p}^{\wedge}_{\substack{s\\a}}))$$

$$(9)\ \mathfrak{p} \overset{\vee}{\underset{\substack{s\\a}}{\Rightarrow}} \mathfrak{q} \leftrightarrow W(h([e](\mathfrak{p}^{\wedge}_{\substack{s\\a}}), \mathfrak{q}^{\vee}_{\substack{s\\a}}))$$

Finally, using the theorem $W(\boxed{\wedge}(\overset{n}{p})) \leftrightarrow W(h(\overset{n}{p}, \overline{w}))$, we confirm the *validity of Bolzano's thesis for formal relations among arbitrary sets of propositions with respect to an s-tuple of ideas*:

T4.90

$$(1)\ \mathfrak{p} \underset{\substack{s\\a}}{\wedge\!\!\!\wedge} \mathfrak{q} \leftrightarrow W(h(\left\lceil [e](\mathfrak{p}^{\vee}_{\substack{s\\a}} + \mathfrak{q}^{\vee}_{\substack{s\\a}}) \right\rceil, \overline{g})$$

$$(2)\ \mathfrak{p} \underset{\substack{s\\a}}{\Rightarrow} \mathfrak{q} \leftrightarrow W(h([e](\mathfrak{p}^{\wedge}_{\substack{s\\a}}), \mathfrak{q}^{\wedge}_{\substack{s\\a}}))$$ [23]

[23] Cf. WL §164, II 142 (2).

$$(3)\ \mathfrak{p} \underset{\substack{s\\a}}{\Leftrightarrow} \mathfrak{q} \quad \leftrightarrow W(h(\lceil h([e](\mathfrak{p}^{\wedge}_{\substack{s\\a}}), \mathfrak{q}^{\wedge}_{\substack{s\\a}})\rceil \sqcup \lceil h([e](\mathfrak{q}^{\wedge}_{\substack{s\\a}}), \mathfrak{p}^{\wedge}_{\substack{s\\a}})\rceil, \overline{w}))$$

$$(4)\ \mathfrak{p} \underset{\substack{s\\a}}{\dot{\Rightarrow}} \mathfrak{q} \quad \leftrightarrow W(h(\lceil h([e](\mathfrak{p}^{\wedge}_{\substack{s\\a}}), \mathfrak{q}^{\wedge}_{\substack{s\\a}})\rceil \sqcup \lceil h(\lceil h([e](\mathfrak{q}^{\wedge}_{\substack{s\\a}}), \mathfrak{p}^{\wedge}_{\substack{s\\a}})\rceil, \sim\overline{w})\rceil, \overline{w}))$$

$$(5)\ \mathfrak{p} \underset{\substack{s\\a}}{\top\!\!\top} \mathfrak{q} \quad \leftrightarrow W\left(h\left(\begin{matrix}\lceil h(\lceil [e](\mathfrak{p}^{\vee}_{\substack{s\\a}} + \mathfrak{q}^{\vee}_{\substack{s\\a}})\rceil, \overline{g})\rceil \sqcup \lceil h(\lceil [e](\mathfrak{p}^{\vee}_{\substack{s\\a}} + \sim\mathfrak{q}^{\vee}_{\substack{s\\a}})\rceil, \overline{g})\rceil \\ \sqcup\ \lceil h(\lceil [e](\sim\mathfrak{p}^{\vee}_{\substack{s\\a}} + \mathfrak{q}^{\vee}_{\substack{s\\a}})\rceil, \overline{g})\rceil\end{matrix}, \overline{w}\right)\right)$$

$$(6)\ \neg(\mathfrak{p} \underset{\substack{s\\a}}{\wedge\!\!\wedge} \mathfrak{q}) \leftrightarrow W(h(\lceil [e](\mathfrak{p}^{\vee}_{\substack{s\\a}} + \mathfrak{q}^{\vee}_{\substack{s\\a}})\rceil, \sim\overline{g})$$

$$(7)\ \mathfrak{p} \underset{\substack{s\\a}}{\perp} \mathfrak{q} \quad \leftrightarrow W(h([e](\mathfrak{p}^{\wedge}_{\substack{s\\a}}), \sim\mathfrak{q}^{\wedge}_{\substack{s\\a}}))$$

$$(8)\ \mathfrak{p} \underset{\substack{s\\a}}{\perp\!\!\perp} \mathfrak{q} \quad \leftrightarrow W(h(\lceil h([e](\mathfrak{p}^{\wedge}_{\substack{s\\a}}), \sim\mathfrak{q}^{\wedge}_{\substack{s\\a}})\rceil \sqcup \lceil h([e](\mathfrak{q}^{\wedge}_{\substack{s\\a}}), \sim\mathfrak{p}^{\wedge}_{\substack{s\\a}}), \overline{w}))$$

$$(9)\ \mathfrak{p} \underset{\substack{s\\a}}{\overset{\vee}{\Rightarrow}} \mathfrak{q} \quad \leftrightarrow W(h([e](\mathfrak{p}^{\wedge}_{\substack{s\\a}}), \mathfrak{q}^{\vee}_{\substack{s\\a}}))$$ [24]

4.20 Formal Connectives among 'Arbitrary Sets' of Propositions

As in the finite case, we can introduce a 'formal connective' in correspondence to every property and relations considered in the preceding section.

(1) *Compatibility* : $\boxed{\substack{\wedge\!\!\wedge\\s\\a}}(\mathfrak{p}) = h(\lceil [e](\mathfrak{p}^{\wedge}_{\substack{s\\a}})\rceil, \overline{g})$ DD4.60

(2) *Integration* : $\boxed{\substack{\vee\!\!\vee\\s\\a}}(\mathfrak{p}) = h(\lceil [e](\sim\mathfrak{p}^{\vee}_{\substack{s\\a}})\rceil, \sim\overline{g})$

(3) *Formal truths* : $\boxed{\substack{\mathbb{W}\\s\\a}}(\mathfrak{p}) = h(\lceil [e](\sim\mathfrak{p}^{\wedge}_{\substack{s\\a}})\rceil, \sim\overline{g})$

(4) *Formal falsities* : $\boxed{\substack{\mathbb{F}\\s\\a}}(\mathfrak{p}) = h(\lceil [e](\mathfrak{p}^{\vee}_{\substack{s\\a}})\rceil, \sim\overline{g})$

(5) *Incompatibility* : $\boxed{\neg}\,\boxed{\substack{\wedge\!\!\wedge\\s\\a}}(\mathfrak{p}) = h(\lceil [e](\mathfrak{p}^{\wedge}_{a})\rceil, \sim\overline{g})$

(6) *Global falsifiability* : $\boxed{\neg}\,\boxed{\substack{\vee\!\!\vee\\s\\a}}(\mathfrak{p}) = h(\lceil [e](\sim\mathfrak{p}^{\vee}_{a})\rceil, \overline{g})$

(7) *Local falsifiability* : $\boxed{\neg}\,\boxed{\substack{\mathbb{W}\\s\\a}}(\mathfrak{p}) = h(\lceil [e](\sim\mathfrak{p}^{\wedge}_{a})\rceil, \overline{g})$

(8) *Local verifiability* : $\boxed{\neg}\,\boxed{\substack{\mathbb{F}\\s\\a}}(\mathfrak{p}) = h(\lceil [e](\mathfrak{p}^{\vee}_{a})\rceil, \overline{g})$

[24] Cf. TS §166, II 146 (5).

Hence, the simple application of the definitions and the theorems yields the following:

$$(1)\ W(\boxed{\substack{⩓\!\!\!⩓\\ s\\ a}}(\mathfrak{p})) \leftrightarrow \exists\overset{s}{c}\forall p(p \mathrel{\text{E}} \mathfrak{p} \rightarrow W(p[\overset{s}{a}/\overset{s}{c}])) \qquad \text{ST4.91}$$

$$(2)\ W(\boxed{\substack{⩔\!\!\!⩔\\ s\\ a}}(\mathfrak{p})) \leftrightarrow \forall\overset{s}{c}\exists p(p \mathrel{\text{E}} \mathfrak{p} \wedge W(p[\overset{s}{a}/\overset{s}{c}]))$$

$$(3)\ W(\boxed{\substack{\mathbb{W}\\ s\\ a}}(\mathfrak{p})) \leftrightarrow \forall\overset{s}{c}\forall p(p \mathrel{\text{E}} \mathfrak{p} \rightarrow W(p[\overset{s}{a}/\overset{s}{c}]))$$

$$4)\ W(\boxed{\substack{\mathbb{F}\\ s\\ a}}(\mathfrak{p})) \leftrightarrow \forall\overset{s}{c}\forall p(p \mathrel{\text{E}} \mathfrak{p} \rightarrow F(p[\overset{s}{a}/\overset{s}{c}]))$$

$$(5)\ W(\boxed{\neg}\boxed{\substack{⩓\!\!\!⩓\\ s\\ a}}(\mathfrak{p})) \leftrightarrow \forall\overset{s}{c}\exists p(p \mathrel{\text{E}} \mathfrak{p} \wedge F(p[\overset{s}{a}/\overset{s}{c}]))$$

$$(6)\ W(\boxed{\neg}\boxed{\substack{⩔\!\!\!⩔\\ s\\ a}}(\mathfrak{p})) \leftrightarrow \exists\overset{s}{c}\forall p(p \mathrel{\text{E}} \mathfrak{p} \rightarrow F(p[\overset{s}{a}/\overset{s}{c}]))$$

$$(7)\ W(\boxed{\neg}\boxed{\substack{\mathbb{W}\\ s\\ a}}(\mathfrak{p})) \leftrightarrow \exists\overset{s}{c}\exists p(p \mathrel{\text{E}} \mathfrak{p} \wedge F(p[\overset{s}{a}/\overset{s}{c}]))$$

$$(8)\ W(\boxed{\neg}\boxed{\substack{\mathbb{F}\\ s\\ a}}(\mathfrak{p})) \leftrightarrow \exists\overset{s}{c}\exists p(p \mathrel{\text{E}} \mathfrak{p} \wedge W(p[\overset{s}{a}/\overset{s}{c}]))$$

Ad (1) $W(\boxed{\substack{⩓\!\!\!⩓\\ s\\ a}}(\mathfrak{p})) \leftrightarrow W(h(\ulcorner[e](\mathfrak{p}^{\wedge}_{\substack{s\\a}})\urcorner, \bar{g})) \leftrightarrow G([e](\mathfrak{p}^{\wedge}_{\substack{s\\a}})) \leftrightarrow \underset{\substack{s\\a}}{⩓\!\!\!⩓}(\mathfrak{p}) \leftrightarrow$ $\exists\overset{s}{c}\forall p(p \mathrel{\text{E}} \mathfrak{p} \rightarrow W(p[\overset{s}{a}/\overset{s}{c}]))$.

Ad (2) $W(\boxed{\substack{⩔\!\!\!⩔\\ s\\ a}}(\mathfrak{p})) \leftrightarrow W(h(\ulcorner[e](\sim\mathfrak{p}^{\vee}_{\substack{s\\a}})\urcorner, \sim\bar{g})) \leftrightarrow \neg G([e](\sim\mathfrak{p}^{\vee}_{\substack{s\\a}}))$ (T4.14).

Analogously, we define

$$(1)\ \mathfrak{p}\boxed{\substack{⩓\!\!\!⩓\\ s\\ a}}\mathfrak{q} = h(\ulcorner[e](\mathfrak{p}^{\vee}_{\substack{s\\a}} + \mathfrak{q}^{\vee}_{\substack{s\\a}})\urcorner, \bar{g}) \qquad \text{SD4.61}$$

$$(2)\ \mathfrak{p}\boxed{\substack{\Rightarrow\\ s\\ a}}\mathfrak{q} = h([e](\mathfrak{p}^{\wedge}_{\substack{s\\a}}), \mathfrak{q}^{\wedge}_{\substack{s\\a}})$$

$$(3)\ \mathfrak{p}\boxed{\substack{\Leftrightarrow\\ s\\ a}}\mathfrak{q} = h(\ulcorner h([e](\mathfrak{p}^{\wedge}_{\substack{s\\a}}), \mathfrak{q}^{\wedge}_{\substack{s\\a}})\urcorner \sqcup \ulcorner h([e](\mathfrak{q}^{\wedge}_{\substack{s\\a}}), \mathfrak{p}^{\wedge}_{\substack{s\\a}})\urcorner, \bar{w})$$

$$(4)\ \mathfrak{p}\boxed{\substack{\dot{\Rightarrow}\\ s\\ a}}\mathfrak{q} = h(\ulcorner h([e](\mathfrak{p}^{\wedge}_{\substack{s\\a}}), \mathfrak{q}^{\wedge}_{\substack{s\\a}})\urcorner \sqcup \ulcorner h(\ulcorner h([e](\mathfrak{q}^{\wedge}_{\substack{s\\a}}), \mathfrak{p}^{\wedge}_{\substack{s\\a}})\urcorner, \sim\bar{w})\urcorner, \bar{w})$$

$$(5)\ \mathfrak{p}\boxed{\substack{\top\!\!\top\\ s\\ a}}\mathfrak{q} = h(\ulcorner h(\ulcorner[e](\mathfrak{p}^{\vee}_{\substack{s\\a}} + \mathfrak{q}^{\vee}_{\substack{s\\a}})\urcorner, \bar{g})\urcorner \sqcup \ulcorner h(\ulcorner[e](\mathfrak{p}^{\vee}_{\substack{s\\a}} + \sim\mathfrak{q}^{\vee}_{\substack{s\\a}})\urcorner, \bar{g})\urcorner \\ \sqcup \ulcorner h(\ulcorner[e](\sim\mathfrak{p}^{\vee}_{\substack{s\\a}} + \mathfrak{q}^{\vee}_{\substack{s\\a}})\urcorner, \bar{g})\urcorner, \bar{w})$$

$$(6)\ \boxminus\,(\mathfrak{p}\boxed{\substack{\wedge\!\!\wedge\\ s\\ a}}\,\mathfrak{q}) = h\left(\left\lceil [e](\mathfrak{p}^{\wedge}_{\substack{s\\a}} + \mathfrak{q}^{\wedge}_{\substack{s\\a}})\right\rceil, \sim \bar{g}\right)$$

$$(7)\ \mathfrak{p}\boxed{\substack{\perp\\ s\\ a}}\,\mathfrak{q} = h([e](\mathfrak{p}^{\wedge}_{\substack{s\\a}}), \sim \mathfrak{q}^{\wedge}_{\substack{s\\a}})$$

$$(8)\ \mathfrak{p}\boxed{\substack{\perp\!\!\perp\\ s\\ a}}\,\mathfrak{q} = h\left(\left\lceil h([e](\mathfrak{p}^{\wedge}_{\substack{s\\a}}), \sim \mathfrak{q}^{\wedge}_{\substack{s\\a}})\right\rceil \sqcup \left\lceil h([e](\mathfrak{q}^{\wedge}_{\substack{s\\a}}), \sim \mathfrak{p}^{\wedge}_{\substack{s\\a}})\right\rceil, \bar{w}\right)$$

$$9)\ \mathfrak{p}\boxed{\substack{\vee\\ \Rightarrow\\ s\\ a}}\,\mathfrak{q} = h([e](\mathfrak{p}^{\wedge}_{\substack{s\\a}}), \mathfrak{q}^{\vee}_{\substack{s\\a}})$$

From these, we obtain

$$(1)\ W(\mathfrak{p}\boxed{\substack{\wedge\!\!\wedge\\ s\\ a}}\,\mathfrak{q}) \leftrightarrow \exists \overset{s}{\mathring{c}}(\exists p(p \mathrel{\mathsf{E}} \mathfrak{p} \wedge W(\overset{m}{p}[\overset{s}{\mathring{a}}/\overset{s}{\mathring{c}}])) \wedge \exists q(q \mathrel{\mathsf{E}} \mathfrak{q} \wedge W(\overset{n}{q}[\overset{s}{\mathring{a}}/\overset{s}{\mathring{c}}]))) \qquad \text{ST4.92}$$

$$(2)\ W(\mathfrak{p}\boxed{\substack{\Rightarrow\\ s\\ a}}\,\mathfrak{q}) \leftrightarrow \underset{\substack{s\\a}}{\wedge\!\!\wedge}(\mathfrak{p}) \wedge \forall \overset{s}{\mathring{c}}(\forall p(p \mathrel{\mathsf{E}} \mathfrak{p} \rightarrow W(p[\overset{s}{\mathring{a}}/\overset{s}{\mathring{c}}])) \rightarrow \forall q(q \mathrel{\mathsf{E}} \mathfrak{q} \rightarrow W(q[\overset{s}{\mathring{a}}/\overset{s}{\mathring{c}}])))$$

$$(3)\ W(\mathfrak{p}\boxed{\substack{\Leftrightarrow\\ s\\ a}}\,\mathfrak{q}) \leftrightarrow \underset{\substack{s\\a}}{\wedge\!\!\wedge}(\mathfrak{p}) \wedge \forall \overset{s}{\mathring{c}}(\forall p(p \mathrel{\mathsf{E}} \mathfrak{p} \rightarrow W(p[\overset{s}{\mathring{a}}/\overset{s}{\mathring{c}}])) \leftrightarrow \forall q(q \mathrel{\mathsf{E}} \mathfrak{q} \rightarrow W(q[\overset{s}{\mathring{a}}/\overset{s}{\mathring{c}}])))$$

$$(4)\ W(\mathfrak{p}\boxed{\substack{\cdot\\ \Rightarrow\\ s\\ a}}\,\mathfrak{q}) \leftrightarrow \underset{\substack{s\\a}}{\wedge\!\!\wedge}(\mathfrak{p}) \wedge \forall \overset{s}{\mathring{c}}(\forall p(p \mathrel{\mathsf{E}} \mathfrak{p} \rightarrow W(p[\overset{s}{\mathring{a}}/\overset{s}{\mathring{c}}])) \rightarrow \forall q(q \mathrel{\mathsf{E}} \mathfrak{q} \rightarrow W(q[\overset{s}{\mathring{a}}/\overset{s}{\mathring{c}}])))\wedge$$
$$\wedge \exists \overset{s}{\mathring{c}}(\forall q(q \mathrel{\mathsf{E}} \mathfrak{q} \rightarrow W(q[\overset{s}{\mathring{a}}/\overset{s}{\mathring{c}}])) \wedge \exists p(p \mathrel{\mathsf{E}} \mathfrak{p} \wedge F(p[\overset{s}{\mathring{a}}/\overset{s}{\mathring{c}}]))$$

$$(5)\ W(\mathfrak{p}\boxed{\substack{\top\!\!\top\\ s\\ a}}\,\mathfrak{q}) \leftrightarrow \underset{\substack{s\\a}}{\wedge\!\!\wedge}(\mathfrak{p}) \wedge \exists \overset{s}{\mathring{c}}(\forall p(p \mathrel{\mathsf{E}} \mathfrak{p} \rightarrow W(p[\overset{s}{\mathring{a}}/\overset{s}{\mathring{c}}])) \wedge \exists q(q \mathrel{\mathsf{E}} \mathfrak{q} \wedge F(q[\overset{s}{\mathring{a}}/\overset{s}{\mathring{c}}])))\wedge$$
$$\wedge \exists \overset{s}{\mathring{c}}(\forall q(q \mathrel{\mathsf{E}} \mathfrak{q} \rightarrow W(q[\overset{s}{\mathring{a}}/\overset{s}{\mathring{c}}])) \wedge \exists p(p \mathrel{\mathsf{E}} \mathfrak{p} \wedge F(p[\overset{s}{\mathring{a}}/\overset{s}{\mathring{c}}]))$$

$$(6)\ W(\boxminus\,(\mathfrak{p}\boxed{\substack{\wedge\!\!\wedge\\ s\\ a}}\,\mathfrak{q}) \leftrightarrow \forall \overset{s}{\mathring{c}}(\forall p(p \mathrel{\mathsf{E}} \mathfrak{p} \rightarrow F(\overset{m}{p}[\overset{s}{\mathring{a}}/\overset{s}{\mathring{c}}])) \vee \forall q(q \mathrel{\mathsf{E}} \mathfrak{q} \rightarrow F(\overset{n}{q}[\overset{s}{\mathring{a}}/\overset{s}{\mathring{c}}])))$$

$$(7)\ W(\mathfrak{p}\boxed{\substack{\perp\\ s\\ a}}\,\mathfrak{q}) \leftrightarrow \underset{\substack{s\\a}}{\wedge\!\!\wedge}(\mathfrak{p}) \wedge \forall \overset{s}{\mathring{c}}(\forall p(p \mathrel{\mathsf{E}} \mathfrak{p} \rightarrow W(p[\overset{s}{\mathring{a}}/\overset{s}{\mathring{c}}])) \rightarrow \forall q(q \mathrel{\mathsf{E}} \mathfrak{q} \rightarrow F(q[\overset{s}{\mathring{a}}/\overset{s}{\mathring{c}}])))$$

$$(8)\ W(\mathfrak{p}\boxed{\substack{\perp\!\!\perp\\ s\\ a}}\,\mathfrak{q}) \leftrightarrow \underset{\substack{s\\a}}{\wedge\!\!\wedge}(\mathfrak{p}) \wedge \forall \overset{s}{\mathring{c}}(\forall p(p \mathrel{\mathsf{E}} \mathfrak{p} \rightarrow W(p[\overset{s}{\mathring{a}}/\overset{s}{\mathring{c}}])) \leftrightarrow \forall q(q \mathrel{\mathsf{E}} \mathfrak{q} \rightarrow F(q[\overset{s}{\mathring{a}}/\overset{s}{\mathring{c}}])))$$

$$(9)\ W(\mathfrak{p}\boxed{\substack{\vee\\ \Rightarrow\\ s\\ a}}\,\mathfrak{q}) \leftrightarrow \underset{\substack{s\\a}}{\wedge\!\!\wedge}(\mathfrak{p}) \wedge \forall \overset{s}{\mathring{c}}(\forall p(p \mathrel{\mathsf{E}} \mathfrak{p} \rightarrow W(p[\overset{s}{\mathring{a}}/\overset{s}{\mathring{c}}])) \rightarrow \exists q(q \mathrel{\mathsf{E}} \mathfrak{q} \wedge W(q[\overset{s}{\mathring{a}}/\overset{s}{\mathring{c}}])))$$

5

Inferences

5.1 Preliminaries

An *inference* [*Schluß*] is a "proposition which enunciates a relation of derivability" with respect to certain ideas, between certain propositions called *premises* [*Vordersätze*] and certain propositions called *conclusions* [*Schlußsätze*]. Alternative denominations for propositions of this type are *enunciation of a derivation* [*Aussage einer Ableitung*] or *derivation proposition* [*Ableitungssatz*], and, sometimes, the more traditional *hypothetic* or *conditional judgement* [*hypothetischer* or *bedingter Urtheil*].[1]

The treatment of the inferences, which covers §§223–68 of the WL, for a total of 178 pages,[2] is characterized by a high level of systematicity and develops according to the explicative-foundational model that had already been examined and theorized in the *Beyträge: from simple to complex; from general to particular*. So, it begins with the examination of the conclusions that may be inferred from a premise of the simplest and most general form, $h(a, \overline{c})$, and continues with the conclusions from a premise with a composite subject (i.e. of the form $[e](\overline{c}_1 + \ldots + \overline{c}_s)$), then from the negation of such propositions (i.e. from propositions of the form $h(\lceil h(a, \overline{c}) \rceil, \sim \overline{w})$) and then, successively, those of the forms $h(\lceil a \rceil, \overline{g})$, $h(\lceil a \rceil, \sim \overline{g})$, etc.

[1] TS §164, II 143: "Given that both kinds of propositions stating that relations of deducibility obtain occur extremely often, we require a name for them. *Statements of deducibility* would be the most suitable name. For the sake of variety, I shall also call them *inferences* or, as is done in other treatises of logic, use the name of *hypothetical* or *conditional* judgements."

Beside this technical meaning, the term '*Schluß*' also has in the Bolzanian exposition, other values more in accordance with common use; in particular, that of *operation by means of which (on the basis of a derivability relation) one passes from the premises to the conclusion* as well as that of *mere conclusion*; for example in WL §159, II 150, we find "[...] und somit läßt sich [...] sicher der Schluß ableiten [...]" (Cf. TS §159, II 107). Furthermore, it is sometimes used to denote the *verbal expression* of a relation of derivability [*Ausdruck eines Verhältnisses der Ableitbarkeit*].

[2] Bolzano's suggestion is that "Who would find too insensitive investigations of this type, can anyway completely skip this chapter" [TS §223, II 283].

Each time, the process takes into consideration the consequences of more premises having one of the forms examined. The justifications are, very often, direct (i.e. they appeal to the intrinsic characteristics of the premise); in other cases, they proceed through the application of some rules to inferences the validity of which has already been ascertained. The exposition, however, is deliberately limited to *simple inferences* (i.e. as seen at due time, inferences that enunciate a relation of *exact derivability that cannot be obtained by cutting two other relations of exact derivability*). This choice finds its justification not so much in—more or less—practical reasons, connected, for example, with its frequency of use (this can indeed be found also in composed inferences), but much more in the conviction that, "apart from every immediate practical advantage, it remains worth knowing which and how many simple inferences there are". Moreover, even without daring to consider his catalogue a complete one, Bolzano emphasizes that "there are *more* simple inferences than one has believed up to now".[3]

Our exposition of the Bolzanian treatment is based on four conventions.

(1) As we have already indicated, above all, inferences are *propositions* in the precise technical sense that this term has in the Bolzanian system; nevertheless, their presentation as such is often difficult, sometime even very difficult, to understand, whereas, on the contrary, the relations they enunciate are often extremely clear. Thus, their theory may be profitably presented (as in the way Bolzano does himself, even if tacitly) in at least *three* distinct forms of increasing interpretative complexity, of which, however only one is to be considered the official form.

In its first form (form A), it works on the elementary level and appears as that chapter of the logic that deals with the subsistence of certain semantic properties or relations (objectuality, objectlessness, singularity, plurality, compatibility, inclusion, equivalence, truth, falsity, derivability, integration, etc.) among ideas or propositions that follow from the fact that properties or relations of the same type subsist among certain other ideas or propositions.

So, a particularly simple, but sufficiently representative, example would be to point out that the truth of a proposition implies the objectuality of its subject, i.e. that it holds that

$$W(h(a,\bar{c})) \to G(a) \qquad \text{A}$$

In form B, it already works at the symbolic level and appears as that chapter of the logic that deals with the subsistence of the relation of formal derivability among

[3] TS §224, II 283: "[...] but rather because it is worthwhile to know how many and which simple inferences there are in and of themselves, irrespective of any indirect benefits this knowledge might bring. Incidentally, it seems to me that it would be foolhardy to give a definitive answer here. I will content myself with showing that there are *more* such inferences than has been believed hitherto."

propositions. Thus, the preceding example becomes

$$h(a,\bar{c}) \underset{a,\bar{c}}{\Rightarrow} h(\ulcorner a \urcorner, \bar{g}) \qquad \text{B}$$

In form C, the 'official' form, it appears as the theory, oriented by Bolzano's thesis, of those conditional propositions that are logically true. Here the preceding example becomes

$$\mathbb{W}(h([e](\ulcorner h(a,\bar{c}) \urcorner_{a\bar{c}}^{\wedge}), \ulcorner h(\ulcorner a \urcorner, \bar{g}) \urcorner_{a\bar{c}}^{\wedge})) \qquad \text{C}$$

The fundamental point is that the three forms are connected as expressed by the following theorem:

$$\begin{aligned} &\mathbb{W}(h([e](\ulcorner h(a,\bar{c}) \urcorner_{a\bar{c}}^{\wedge}), \ulcorner h(\ulcorner a \urcorner, \bar{g}) \urcorner_{a\bar{c}}^{\wedge})) \\ &\leftrightarrow [h(a,\bar{c}) \underset{a,\bar{c}}{\Rightarrow} h(\ulcorner a \urcorner, \bar{g})] \qquad \text{T5.1} \\ &\leftrightarrow \forall a\bar{c}(W(h(a,\bar{c})) \rightarrow G(a)) \end{aligned}$$

Our exposition will *always* present form B; however, when it may be useful, we will also present form A. Only rarely, and simply as an example, will we also adjoin the almost unreadable form C.

(2) Throughout the treatment of inferences, Bolzano agrees to use the expression '*Whatever has a, has b*' for atomic propositions, instead of the expression '*A has b*', where *a* is the *abstractum* corresponding to *A*. He justifies this on the basis of T2.43. In our notation, this would mean that, instead of '$h(a,\bar{c})$', we should always write '$h([e](\widehat{a}),\bar{c})$', or, better, starting directly with quality ideas, '$h([e](\bar{a}),\bar{c})$'. Our notational set-up, in contrast to Bolzano's, privileges *object* ideas before *quality* ideas (we should not forget that '*A*' is only an abbreviation for '*Etwas, das (die Beschaffenheit) a hat*'); however, it recommends the adoption, where this may help the reading, of the simplest notation, and hence, in general, the writing of atomic propositions in the form '$h(a,\widehat{c})$'. This choice has among its consequences that, for example, we will usually write '$h(-a,\widehat{c})$' and, respectively, '$h(a\sqcap b,\widehat{c})$', which, following Bolzano, should be written '$h([e](\sim\bar{a}),\bar{c})$' or '$h([e](\bar{a}+\bar{b}),\bar{c})$' or also $h([e](\sim\widehat{a}),\widehat{c})$' and '$h([e](\widehat{a}+\widehat{b}),\widehat{c})$'; only in particular cases, when this will appear more illuminating, will we write it in the other way.

(3) Our exposition aims to be exhaustive and thus to report all inferences mentioned by Bolzano; however, to avoid an excessive burdening of the text with footnotes, we will indicate the § and the number where it is stated in square brackets next to the theorem.

(4) Writing '$\Rightarrow$' or '$\Leftrightarrow$' between propositions, we will in general understand that they bring underwritten all the variable ideas explicitly mentioned as occurring in the propositions.

5.2 Rules to Obtain Inferences from Inferences

As already remarked, under the title *Some General Rules by Which Conclusions from Given Premises Can Be Recovered,* Bolzano presents three rules he will later use from time to time to prove the validity of inferences:

(1) a rule that we may call the *cut rule* and that can be presented as[4]

$$\frac{\mathfrak{p} \underset{\mathfrak{a}}{\Rightarrow} \mathfrak{q};\quad \mathfrak{r} \subseteq \mathfrak{q};\quad \mathfrak{r} \underset{\mathfrak{a}}{\Rightarrow} \mathfrak{s}}{\mathfrak{p} \underset{\mathfrak{a}}{\Rightarrow} \mathfrak{s}}\,\mathrm{C}$$

(2) a rule that we may call the *deduction rule* and that can be presented as[5]

$$\frac{\mathfrak{p}, \mathfrak{q} \underset{\mathfrak{a}}{\Rightarrow} \mathfrak{r}}{\mathfrak{p} \underset{\mathfrak{a}}{\Rightarrow} (\mathfrak{q} \rightarrow \mathfrak{r})}\,\mathrm{D}$$

(3) a rule that we may call the *contraposition rule* and that can be presented as[6]

$$\frac{\mathfrak{p} \underset{\mathfrak{a}}{\Rightarrow} \mathfrak{q};\ \mathfrak{r} \subseteq \mathfrak{q};\ \mathfrak{s} \underset{\mathfrak{a}}{\Rightarrow} \mathfrak{r}^{\boxed{\neg}}}{\neg(\mathfrak{p} \underset{\mathfrak{a}}{\Rightarrow} \mathfrak{s})}\,\mathrm{K}$$

It is important to emphasize that, as is clear from the way they are presented, these rules are *cognitive tools* (founded, however, on an inference), but not (as opposed to the inferences on which they are based) parts of the universe of things in themselves; they are only proof instruments for ascertaining or even finding inferences; *for Bolzano, indeed, a proof is a cognitive procedure, not a logical entity.*

[4] TS §224, II 284: "1) If we have already deduced certain conclusions *M, N, O,...* from certain propositions *A, B, C, D,...*, and if we know that from one or more of *M, N, O,...* some new conclusions *R, S, T,...* can be deduced, always with respect to the same variable ideas *i, j,...*, then, it will be permissible [...] to consider the latter to be conclusions from the given premises."

[5] TS §224, II 284: "2) Assume that we are to find conclusions to certain propositions *A, B, C, D,...*; assume, furthermore, that we have on another occasion considered *A, B, C, D,...* together with some other propositions *E, F, G,...*, and that we then deduced from the totality of these two collections the conclusions *M, N, O,...* with respect to ideas *i, j,....* In this case we may say that propositions *M, N, O,...* become true whenever the truth of *E, F, G,...* is added to the truth of propositions *A, B, C, D,....* Hence we may form the following hypothetical judgement as a conclusion which follows from propositions *A, B, C, D,...* alone: 'If *E, F, G,...* are true, then *M, N, O,...* are also true'."

[6] TS §224, II 285: "3) If we have found that certain propositions *H, J, K,...* allow the deduction of the negation of one or more given premises *A, B, C, D,...*, or even the negation of a conclusion *M* or *N* that is deducible from them, we may conclude that propositions *H, J, K,...* are never all true when *A, B, C, D,...* are. Hence, we obtain the conclusion 'The collection of propositions *H, J, K,...* is not a collection of nothing but true propositions'."

5.3 Inferences with One or More Universal Premises

Thus, the systematic inquiry of Bolzano begins with the examination of those inferences that have as premise one or more universal propositions.[7] It develops by taking the following into consideration:

(1) inferences with *only one* premise of this type [§225];

(2) inferences from *two* premises, distinguishing three subcases:

- (a) two premises with the *same* variable ideas [§226];
- (b) two premises withe *only one* variable idea in common [§227];
- (c) two premises with *no* variable idea in common [§228];

(3) inferences from an *arbitrary (finite) number* of premises [§229].

Ad (1) As far as the case with *only one* premise is concerned, Bolzano begins by considering those inferences that are founded on the definition of truth. This implies, first of all, that the objectuality of its subject is derivable from the truth of a proposition:

$$\begin{array}{llll}
W(h(a,\widehat{c})) \rightarrow G(a) & \text{A} & [225.4] & \text{T5.2} \\
h(a,\widehat{c}) \Rightarrow h(\ulcorner a \urcorner, \overline{g}) & & & \\
W(h(\ulcorner h(a,\widehat{c}) \urcorner, \overline{w}) \underset{ac}{\boxed{\mapsto}} h(\ulcorner a \urcorner, \overline{g}) & \text{C} & &
\end{array}$$

Analogously, that definition implies that the objectuality both of its predicate and of the concretization of it predicate are derivable from the truth of a proposition:

$$\begin{array}{llll}
W(h(a,\widehat{c})) \rightarrow G(\widehat{c}) & \text{A} & [225.5] & \text{T5.3} \\
h(a,\widehat{c}) \Rightarrow h(\ulcorner \widehat{c} \urcorner, \overline{g}) & & & \\
W(h(\ulcorner h(a,\widehat{c}) \urcorner, \overline{w}) \underset{ac}{\boxed{\mapsto}} h(\ulcorner \widehat{c} \urcorner, \overline{g})) & \text{C} & &
\end{array}$$

$$\begin{array}{llll}
W(h(a,\widehat{c})) \rightarrow G([e](\widehat{c})) & \text{A} & [225.5] & \text{T5.4} \\
h(a,\widehat{c})) \Rightarrow h(\ulcorner [e](\widehat{c}) \urcorner, \overline{g}) & & & \\
W(h(\ulcorner h(a,\widehat{c}) \urcorner, \overline{w}) \underset{ac}{\boxed{\mapsto}} h(\ulcorner [e](\widehat{c}) \urcorner, \overline{g})) & \text{C} & &
\end{array}$$

As[8] each object under *a* has at least one quality falling under $\widehat{c}$, one may derive that the idea $[a](\widehat{c})$ is objectual whereas the idea $[a](\sim \widehat{c})$ is objectless:

[7] To simplify the exposition, we will sometimes call 'universal' propositions of the form '$h(a,\widehat{c})$', 'negations' those of the form '$\boxed{\neg} h(a,\widehat{c})$', 'existential' those of the form '$h(\ulcorner [a](\widehat{c}) \urcorner, \overline{g})$', and 'inexistential' those of the form '$h(\ulcorner a \urcorner, \sim \overline{g})$'.

[8] We will return to the inference contained in [225.6] when we discuss the 'higher level'.

$$W(h(a,\widehat{c})) \rightarrow G([a](\widehat{c})) \qquad \text{A} \qquad [225.7] \qquad \text{T5.5}$$
$$h(a,\widehat{c}) \Rightarrow h(\lceil [a](\widehat{c}) \rceil, \bar{g})$$

$$W(h(a,\widehat{c})) \rightarrow G([a](\sim \widehat{c})) \qquad \text{A} \qquad [225.7] \qquad \text{T5.6}$$
$$h(a,\widehat{c}) \Rightarrow h(\lceil [a](\sim \widehat{c}) \rceil, \sim \bar{g})$$

Recalling the equivalence among the ideas $[a](\widehat{c})$, $[c](\widehat{a})$ and $[e](\widehat{a}+\widehat{c})]$, one draws from the two preceding relations the following:

$$h(a,\widehat{c}) \Rightarrow h(\lceil [c](\widehat{a}) \rceil, \bar{g}) \qquad [225.8] \qquad \text{T5.7}$$
$$h(a,\widehat{c}) \Rightarrow h(\lceil [e](\widehat{a}+\widehat{c})] \rceil, \bar{g}) \qquad [225.8] \qquad \text{T5.8}$$
$$h(a,\widehat{c}) \Rightarrow h(\lceil [c](\sim \widehat{a}) \rceil, \sim \bar{g}) \qquad [225.8] \qquad \text{T5.9}$$
$$h(a,\widehat{c}) \Rightarrow h(\lceil [e](\widehat{a}+ \sim \widehat{c})] \rceil, \sim \bar{g}) \qquad [225.8] \qquad \text{T5.10}$$

From the truth of $h(a,\widehat{c})$, it follows that c is not subordered to a and that a is included in c:[9]

$$h(a,\widehat{c}) \Rightarrow \boxed{\neg}\,(c \boxed{\dot{\sqsubseteq}}\, a) \qquad [225.9] \qquad \text{T5.11}$$
$$h(a,\widehat{c}) \Rightarrow (a \boxed{\sqsubseteq}\, c) \qquad [225.9] \qquad \text{T5.12}$$

but also that a and $[a](\widehat{c})$ (and hence also $[c](\widehat{a})$ and $[e](\widehat{a}+\widehat{c})$) are all equivalent:

$$h(a,\widehat{c}) \Rightarrow (a \boxed{\equiv} [a](\widehat{c})) \qquad [225.10] \qquad \text{T5.13}$$
$$h(a,\widehat{c}) \Rightarrow (a \boxed{\equiv} [c](\widehat{a})) \qquad [225.10] \qquad \text{T5.14}$$
$$h(a,\widehat{c}) \Rightarrow (a \boxed{\equiv} [e](\widehat{a}+\widehat{c})) \qquad [225.10] \qquad \text{T5.15}$$

Moreover those derivabilities are convertible; i.e.

$$h(a,\widehat{c}) \Leftrightarrow (a \boxed{\equiv} [a](\widehat{c})) \qquad [225.10] \qquad \text{T5.16}$$
$$h(a,\widehat{c}) \Leftrightarrow (a \boxed{\equiv} [c](\widehat{a})) \qquad [225.10] \qquad \text{T5.17}$$
$$h(a,\widehat{c}) \Leftrightarrow (a \boxed{\equiv} [e](\widehat{a}+\widehat{c})) \qquad [225.10] \qquad \text{T5.18}$$

Furthermore, if then c is not equivalent to e, i.e. if $-c$ is objectual, then each of its objects lacks $\widehat{a}$, i.e.[10]

$$h(a,\widehat{c}) \Rightarrow h(\lceil -c \rceil, \sim \bar{g}) \boxed{\vee} h(-c, \sim \widehat{a}) \qquad [225.11] \qquad \text{T5.19}$$

[9] TS §225, II 287: "9) [...] 'The totality of the *B* is not a part of the totality of the *A*' or 'The idea *B* is either equivalent to or higher than *A*, or stands to *A* in the relation of inclusion.'."

[10] TS §225 II 287: "11) [...] Either the idea *B* has the widest extension, that of something in general, or else the proposition 'Whatever lacks *b* lacks *a*' is also valid."

and hence also

$$h(a,\widehat{c}) \Rightarrow h(\ulcorner[e](\sim\widehat{c})]\urcorner, \sim\overline{g}) \boxed{\vee} h([e](\sim\widehat{c}), \sim\widehat{a}) \quad \text{etc.} \qquad [225.11] \quad \text{T5.20}$$

Ad (2) Turning to the analysis of inferences with *two universal* premises, Bolzano distinguishes, as has already been pointed out, three subcases.

Ad (2a.) The two premises have the *same ideas.*

After establishing that given two different ideas a and c there are eight and only eight different propositions, precisely

$$\begin{array}{llll} (1)\ h(a,\widehat{c}) & (2)\ h(c,\widehat{a}) & (3)\ h(a,\sim\widehat{c}) & (4)\ h(c,\sim\widehat{a}) \\ (5)\ h(-a,\widehat{c}) & (6)\ h(-c,\widehat{a}) & (7)\ h(-a,\sim\widehat{c}) & (8)\ h(-c,\sim\widehat{a}) \end{array} \qquad [226.3]$$

Bolzano remarks that it is always possible to assume that one of the premises is of the form $h(a,\widehat{c})$. Therefore it will suffice to discuss the combinations of such a premise with each of the remaining seven. The combinations $(1+3)$, $(1+4)$, and $(1+6)$, however, all consist of two incompatible propositions and thus will never constitute the premises of an inference. Hence, what really remain to be examined are only the four combinations

$$\begin{array}{llll} (1+2) & h(a,\widehat{c}),\ h(c,\widehat{a}) & (1+5) & h(a,\widehat{c}), h(-a,\widehat{c}) \\ (1+7) & h(a,\widehat{c}), h(-a,\sim\widehat{c}) & (1+8) & h(a,\widehat{c}), h(-c,\sim\widehat{a}) \end{array} \qquad [226.4]$$

As far as the combination $(1+2)$ is concerned, Bolzano states that it determines that the two ideas are equivalent and one can thus derive that "every object of one of the two ideas a or c is an object of both" (he even states that it is the *reason for this*[11]); i.e.

$$h(a,\widehat{c}), h(c,\widehat{a}) \Rightarrow h(a \sqcup c, \widehat{a}+\widehat{c}) \qquad [226.5] \qquad \text{T5.21}$$

but, he adjoins, the conclusion is also equivalent to the premises.[12] This sets a question of a certain delicacy, though, a question that, by the way, will also arise in other

[11] The value of this term will constitute the subject of Chapter 7.

[12] TS §226, II 293: "5) The first (1, 2) runs:

Whatever has a has b,

Whatever has b has a.

Obviously, the ideas a and b must be different here, if we are really to have two propositions. But then the first proposition states that all the objects that stand under A also stand under B, while the second states that all objects standing under B also stand under A; together, these specify that A and B are a pair of equivalent ideas. Conversely, whenever A and B are equivalent, these two propositions hold. Thus the proposition 'Every object of one of the ideas A, B is an object of the other' may be cited as one of the conclusions that not only follows exactly from these premises, but in fact stands in a relation of *mutual* deducibility with them and indeed, I believe, also in a genuine relation of ground and consequence to them."

cases. The fact is that accepting the validity of both inferences

$$h(a \sqcup c, \widehat{a} + \widehat{c}) \Rightarrow h(a, \widehat{c}) \text{ and } h(a \sqcup c, \widehat{a} + \widehat{c}) \Rightarrow h(c, \widehat{a})$$

contrasts with the principle that the truth of a proposition is subordered to the objectuality of its subject; the truth of $h(a \sqcup c, \widehat{a} + \widehat{c})$, indeed, implies the objectuality of $a \sqcup c$, but not necessarily the objectuality of a and c, taken separately. Without excluding the fact that Bolzano here, as well in other cases that we will point out, had made a mistake, we may nonetheless suppose that in such circumstances he means that the two ideas are equivalent (and hence, implicitly, objectual).

As far as combination (1 + 5) is concerned, Bolzano remarks that from it one may derive that every object has $\widehat{c}$, i.e.

$$h(a, \widehat{c}), h(-a, \widehat{c}) \Rightarrow h(e, \widehat{c}) \qquad [226.6] \qquad \text{T5.22}$$

and this time, he also adjoins that the converse holds; i.e.

$$h(e, \widehat{c}) \Rightarrow h(a, \widehat{c}), h(-a, \widehat{c}) \qquad \text{T5.23}$$

a statement that clearly is true only if a is not equivalent to e. In this case, the inaccuracy is not so easily remediable.

As far as the combination (1 + 8) is concerned, Bolzano points out that one can derive from the premises that a and $-c$ exclude each other (even that they are the *reason for this*), that the idea $a \sqcap -c$ is objectless; i.e.

$$h(a, \widehat{c}), h(-c, \sim \widehat{a}) \Rightarrow h(\ulcorner a \sqcap -c \urcorner, \sim \overline{g}) \qquad [226.7] \qquad \text{T5.24}$$

He also adjoins that this relation is convertible, which clearly holds only if both a and $-c$ are objectual; possibly Bolzano accepts this, as in case (1 + 2), only because the relation of incompatibility is also defined with respect to objectual ideas.

The discussion of the combination (1 + 7) is very interesting because it shows us some non-trivial applications of the cut rule. First of all, Bolzano establishes the two relations

$$h(a, \widehat{c}), h(-a, \sim \widehat{c}) \Rightarrow h(-c, \sim \widehat{a}) \qquad [226.8] \qquad \text{T5.25}$$

$$h(a, \widehat{c}), h(-a, \sim \widehat{c}) \Rightarrow h(c, \widehat{a}) \qquad [226.8] \qquad \text{T5.26}$$

Then, he remarks that the first premise and the first conclusion, as well as the second premise and the second conclusion, are type (1 + 8) situations and hence, by cut, he obtains two new relations; in particular,

$$h(a, \widehat{c}), h(-a, \sim \widehat{c}) \Rightarrow h(\ulcorner a \sqcap -c \urcorner, \sim \overline{g}) \qquad [226.9] \qquad \text{T5.27}$$

Proof

$$\frac{h(a,\widehat{c}), h(-a,\sim\widehat{c}) \Rightarrow h(-c,\sim\widehat{a});\quad h(a,\widehat{c}), h(-c,\sim\widehat{a}) \Rightarrow h(\lceil a \sqcap -c \rceil, \sim\overline{g})}{h(a,\widehat{c}), h(-a,\sim\widehat{c}) \Rightarrow h(\lceil a \sqcap -c \rceil, \sim\overline{g})}\,\mathrm{C}$$

$h(a,\widehat{c}), h(-a,\sim\widehat{c}) \Rightarrow h(\lceil -a \sqcap c \rceil, \sim\overline{g})$ [226.9] T5.28

Proof

$$\frac{h(a,\widehat{c}), h(-a,\sim\widehat{c}) \Rightarrow h(c,\widehat{a});\quad h(-a,\sim\widehat{c}), h(c,\widehat{a}) \Rightarrow h(\lceil -a \sqcap c \rceil, \sim\overline{g})}{h(a,\widehat{c}), h(-a,\sim\widehat{c}) \Rightarrow h(\lceil -a \sqcap c \rceil, \sim\overline{g})}\,\mathrm{C}$$

Bolzano remarks again that the first premise and the second conclusion $[h(a,\widehat{c}), h(c,\widehat{a})]$, as well as the first conclusion and the second premise $[h(-c,\sim\widehat{a}), h(-a,\sim\widehat{c})$, which are equivalent to $h(-c,\widehat{-a}), h(-a,\widehat{-c})]$, are situations of type $(1+2)$. Therefore

$h(a,\widehat{c}), h(-a,\sim\widehat{c}) \Rightarrow h(a \sqcup c, \widehat{a}+\widehat{c})$ [226.9] T5.29

Proof

$$\frac{h(a,\widehat{c}), h(-a,\sim\widehat{c}) \Rightarrow h(c,\widehat{a});\quad h(a,\widehat{c}), h(c,\widehat{a}) \Rightarrow h(a \sqcup c, \widehat{a}+\widehat{c})}{h(a,\widehat{c}), h(-a,\sim\widehat{c}) \Rightarrow h(a \sqcup c, \widehat{a}+\widehat{c})}$$

$h(a,\widehat{c}), h(-a,\sim\widehat{c}) \Rightarrow h(-a \sqcup -c, \sim\widehat{a}+\sim\widehat{c})$ [226.9] T5.30

Proof

$$\frac{h(a,\widehat{c}), h(-a,\sim\widehat{c}) \Rightarrow h(-c,\sim\widehat{a});\quad h(-c,\sim\widehat{a}), h(-a,\sim\widehat{c}) \Rightarrow h(-a \sqcup -c, \sim\widehat{a}+\sim\widehat{c})}{h(a,\widehat{c}), h(-a,\sim\widehat{c}) \Rightarrow h(-a \sqcup -c, \sim\widehat{a}+\sim\widehat{c})}$$

But $h(a \sqcup c, \widehat{a}+\widehat{c}) \Rightarrow h(\lceil a \sqcup c \rceil, \overline{g})$ and $h(-a \sqcup -c, \sim\widehat{a}+\sim\widehat{c}) \Rightarrow h(\lceil -a \sqcup -c \rceil, \overline{g})$, so

$h(a,\widehat{c}), h(-a,\sim\widehat{c}) \Rightarrow h(\lceil a \sqcup c \rceil, \overline{g})$ [226.9] T5.31

$h(a,\widehat{c}), h(-a,\sim\widehat{c}) \Rightarrow h(\lceil -a \sqcup -c \rceil, \overline{g})$ [226.9] T5.32

Note the use of the easily provable lemma $h(a,\sim\widehat{c}) \Leftrightarrow h(a,\widehat{-c})$, i.e. $W(h(a,\sim\widehat{c})) \leftrightarrow W(h(a,\widehat{-c}))$.

It is worth quoting the concluding *remark* of this section, which, among other things, contributes to shedding some light on the doubts that arose above with reference to certain 'equivalencies':

> "Even though the inferences of this section have not been quoted up to now in the treatises of logic, many of them frequently occur not only in scientific expositions but also in common life. Indeed, that certain two ideas are *equivalent* (i.e. that every object falling under one falls also under the other) is a remark that we are very often induced to make. So, in particular, the mathematician uses to diligently annotate which of his concepts are equivalent; indeed, his *equations* are just assertions that two ideas (the members of the equation) are equivalent. When he wants to prove that *A* and *B* are a pair of equivalent ideas, he need no do anything else but show either that every *A* is a *B* and every *B* is an *A* or that every *A* is a *B* and every *non-A* is a *non-B*. Then, when he wants us to infer from this two propositions the relation of equivalence between the two concepts *A* and *B*, he tacitly presupposes the types of inference that we have listed in this section under the numbers 5 and 8."[13]

Ad (2b) Premises have *only one idea in common.*

As this idea may be the subject of both premises or the subject of one and the predicate of the other or, finally, the predicate of both, and moreover the idea may be either positive or negative, there are six possible cases:

$$\begin{array}{llll} (1) & h(a,\widehat{b}), h(a,\widehat{c}) & (2) & h(a,\widehat{b}), h(-a,\widehat{c}) \\ (3) & h(a,\widehat{b}), h(c,\widehat{b}) & (4) & h(a,\widehat{b}), h(c,\sim\widehat{b}) \\ (5) & h(a,\widehat{b}), h(c,\widehat{a}) & (6) & h(a,\widehat{b}), h(c,\sim\widehat{a}) \end{array} \qquad [227.3]$$

Case (1) yields

$$h(a,\widehat{b}), h(a,\widehat{c}) \Leftrightarrow h(a,\widehat{b}+\widehat{c}) \qquad \text{(also } \rhd\text{)} \qquad [227.4] \ \text{T5.33}$$

as well as

$$h(a,\widehat{b}), h(a,\widehat{c}) \Rightarrow h(\lceil [e](\widehat{b}+\widehat{c}) \rceil, \bar{g}) \qquad [227.5] \ \text{T5.34}$$

Case (2) yields

$$h(a,\widehat{b}), h(-a,\widehat{c}) \Rightarrow (b \boxed{\equiv} e) \boxed{\vee} (c \boxed{\equiv} e) \boxed{\vee} (b \sqcup c \boxed{\equiv} e) \qquad [227.6] \ \text{T5.35}$$

Case (3) yields

$$h(a,\widehat{b}), h(c,\widehat{b}) \Leftrightarrow h(a \sqcup c,\widehat{b}) \qquad \text{(also } \rhd\text{; usual 'error' on } \Leftarrow\text{)} \qquad [227.7] \ \text{T5.36}$$

Case (4) yields

$$h(a,\widehat{b}), h(c,\sim\widehat{b}) \Rightarrow h(a,\sim\widehat{c}) \qquad [227.8] \ \text{T5.37}$$

[13] Cf. TS § 226, II 294.

$$h(a,\widehat{b}), h(c,\sim\widehat{b}) \Rightarrow h(c,\sim\widehat{a}) \qquad [227.8] \qquad \text{T5.38}$$

$$h(a,\widehat{b}), h(c,\sim\widehat{b}) \Rightarrow h(a \sqcup c,\sim\overline{g}) \qquad [227.8] \qquad \text{T5.39}$$

Case (5) yields

$$h(a,\widehat{b}), h(c,\widehat{a}) \Rightarrow h(c,\widehat{b}) \qquad [227.9] \qquad \text{T5.40}$$

Finally, case (6) yields

$$h(a,\widehat{b}), h(c,\sim\widehat{a}) \Rightarrow \boxed{\neg}\; h(b,\widehat{c}) \qquad [227.10] \qquad \text{T5.41}$$

Moreover, since $W(h(a,\widehat{b}))$ implies $G(a)$ and $W(h(c,\sim\widehat{a}))$ implies $G(-a)$, and also

$$h(a,\widehat{b}), h(c,\sim\widehat{a}) \Rightarrow a \boxed{\dot{\sqsubseteq}}\; e \qquad [227.11] \qquad \text{T5.42}$$

it also yields

$$h(a,\widehat{b}), h(c,\sim\widehat{a}) \Rightarrow h(a,\sim\widehat{c}) \qquad [227.12] \qquad \text{T5.43}$$

as well as

$$h(a,\widehat{b}), h(a,\sim\widehat{c}) \Rightarrow h(a,(\widehat{b+\sim c})) \qquad [227.13] \qquad \text{T5.44}$$

from which

$$h(a,\widehat{b}), h(a,\sim\widehat{c}) \Rightarrow h(\lceil [e](\widehat{b+\sim c}) \rceil, \overline{g}) \qquad [227.14] \qquad \text{T5.45}$$

Ad (2c) Premises that have *no ideas in common.*
This case refers to the following type of compositions: $h(a,\widehat{b}), h(c,\widehat{d})$. Bolzano emphasizes that, usually, inferences with such premises are not taken into consideration because one thinks that premises must always contain a '*middle term* [*Mittelbegriff*]'; nonetheless, he provides the following as a possible inference: "A whole, in which the subjects of the propositions *A has b* and *C has d*, occur as parts, never lacks parts falling under *B* and *D* (i.e., under the *concreta* pertaining to the predicates of those propositions)." Maintaining that "nothing is more common than this inference", he illustrates it by using an example: "Having noticed that academician *A* is involved in mathematical investigations and academician *B* is involved in naturalistic investigations, one may conclude that in the society of scholars of which they are members neither mathematicians nor naturalists are missing."

The most adequate transcription of this inference seems to be

$$h(a,\widehat{b}), h(c,\widehat{d}) \Rightarrow h(a \sqcup c, \widehat{b} \cdot \widehat{d}) \qquad [228.1] \quad T5.46$$

(i.e. $h(a,\widehat{b}), h(c,\widehat{d}) \Rightarrow h(a \sqcup c, \widehat{b \sqcup d})$).

As a matter of fact, let us note the following chain of equivalencies:

$$W(h(a \sqcup c, \widehat{b} \cdot \widehat{d})) \leftrightarrow G(a \sqcup c) \wedge a \sqcup c \sqsubseteq [e](\widehat{b} \cdot \widehat{d}) \overset{[\text{by } [e](\widehat{a}) \sqcup [e](\widehat{c}) \equiv [e](\widehat{a \cdot c})]}{\leftrightarrow}$$
$$G(a \sqcup c) \wedge a \sqcup c \sqsubseteq [e](\widehat{b}) \sqcup [e](\widehat{d}) \leftrightarrow G(a \sqcup c) \wedge a \sqcup c \sqsubseteq b \sqcup d \leftrightarrow a \sqcup c \sqsubseteq b \sqcup d$$

Based on the fact that even if the subjects of premises are interchanged, the conclusion remains unaltered, Bolzano drew the reasons to maintain that this is an authentic, simple inference and not a mere verbal reformulation of the premises.

Ad (3) Lastly, in the general case, in which *more than two premises* are involved, Bolzano preliminarily remarked that when the ideas occurring in the propositions are the same (a and c), there is, in the case of three propositions, only one interesting combination:

$$h(a,\widehat{c}), h(c,\widehat{a}), h(-c, \sim\widehat{a})$$

The reason is that this is the only one in which the propositions are at the same time compatible but not equivalent. For it, one has

$$h(a,\widehat{c}), h(c,\widehat{a}), h(-c, \sim\widehat{a}) \Rightarrow \boxminus\, h(e, \widehat{a} \cdot \widehat{c}), \boxminus\, h(e, \sim\widehat{a} \cdot \sim\widehat{c}) \qquad [229.1] \quad T5.47$$
$$W(h(a,\widehat{c})) \wedge W(h(c,\widehat{a})) \wedge W(h(-c, \sim\widehat{a})) \rightarrow \neg(a \sqcup c \equiv e) \wedge \neg(-a \sqcup -c \equiv e)$$

In the general case, Bolzano took the following four cases into consideration:

(3a) They *all* have *the same subject*; in this case, he accepted the conclusion that every object falling under the subject also falls under each of the concretizations of the predicates, i.e.

$$h(a,\widehat{c}_1), \ldots, h(a,\widehat{c}_n) \Rightarrow h(a, \widehat{c}_1 + \ldots + \widehat{c}_n) \qquad [229.3] \quad T5.48$$

and that the sum of their predicates is objectual, i.e.

$$h(a,\widehat{c}_1), \ldots, h(a,\widehat{c}_n) \Rightarrow h(\lceil [e](\widehat{c}_1 + \ldots + \widehat{c}_n) \rceil, \overline{g}) \qquad [229.3] \quad T5.49$$

(3b) They *all have the same predicate*; in this case, Bolzano accepted the conclusion that every object falling under one of the subjects also falls under the concretization of the predicate, i.e.

$$h(a_1,\widehat{c}), \ldots, h(a_n,\widehat{c}) \Rightarrow h(a_1 \sqcup \ldots \sqcup a_n, \widehat{c}) \qquad [229.4] \quad T5.50$$

(3c) They can be ordered in such a way that the *concretization of the predicate of the one is the subject of the next*; in this case, he accepted the ('enthymematic–syllogistic') conclusion

$$h(a_1,\widehat{a_2}), h(a_2,\widehat{a_3}),\ldots, h(a_{n-1},\widehat{a_n}) \Rightarrow h(a_1,\widehat{a_n}) \qquad [229.5]\ \text{T5.51}$$

(3d) They have different *subjects and different predicates*. Here Bolzano drew the generalization of what he had already seen in the case of two propositions:

$$h(a_1,\widehat{c_1}),\ldots, h(a_n,\widehat{c_n}) \Rightarrow h(a_1 \sqcup \ldots \sqcup a_n, \widehat{c_1} \cdot \ldots \cdot \widehat{c_n}) \qquad [229.6]\ \text{T5.52}$$

$$W(h(a_1 \sqcup \ldots \sqcup a_n, \widehat{c_1} \cdot \ldots \cdot \widehat{c_n}) \leftrightarrow a_1 \sqcup \ldots \sqcup a_n \sqsubseteq c_1 \sqcup \ldots \sqcup c_n) \qquad \text{NB.}$$

5.4 Inferences with at Least One Negated Universal Premise

Having exhausted the investigation of inferences from universal premises, Bolzano turned to inferences in premises for which at least the negation of a universal proposition occurs. Bolzano addresses

(1) inferences with *only one* premise of this type [§230];
(2) inferences with *two* premises: a universal one and the negation of a universal one, re-proposing the distinctions considered above: (a) both have the *same* ideas; (b) they have *only one idea* in common; (c) they have *no ideas* in common [§231];
(3) inferences from *two negations* of universal propositions [§232].

Naturally, eight possible forms of negation correspond to the eight possible forms of a universal proposition:

$$(1)\ \boxed{\neg}\ h(a,\widehat{c}) \quad (2)\ \boxed{\neg}\ h(c,\widehat{a}) \quad (3)\ \boxed{\neg}\ h(a,\sim\widehat{c}) \quad (4)\ \boxed{\neg}\ h(c,\sim\widehat{a})$$
$$(5)\ \boxed{\neg}\ h(-a,\widehat{c}) \quad (6)\ \boxed{\neg}\ h(-c,\widehat{a}) \quad (7)\ \boxed{\neg}\ h(-a,\sim\widehat{c}) \quad (8)\ \boxed{\neg}\ h(-c,\sim\widehat{a})$$

Ad (1) Only one premise: the *negation of a universal*.

As to the possible conclusions that could be derivable from a premise of this type, Bolzano remarked that, obviously, it does not hold, $\boxed{\neg}\ h(a,\widehat{c}) \Rightarrow h(a,\sim\widehat{c})$, but that neither the conclusion that some a is not c (i.e. that $[a](\sim\widehat{c})$ is objectual) is allowed, because the falsity of $h(a,\widehat{c})$ could also depend on the objectlessness of a. [230.2]

If, however, a is objectual, then such a conclusion is allowed. This can be presented in two forms:

$$\boxed{\neg}\ h(a,\widehat{c}) \Rightarrow h(\ulcorner a \urcorner, \sim\bar{g})\ \boxed{\vee}\ h(\ulcorner [a](\sim\widehat{c}) \urcorner, \bar{g}) \qquad \text{T5.53}$$

$$\boxed{\neg}\ h(a,\widehat{c}), h(\lceil a \rceil, \sim \bar{g}) \Rightarrow h(\lceil [a](\sim \widehat{c}) \rceil, \bar{g}) \qquad [230.3]$$

Using the equivalencies $[a](\sim \widehat{c}) \equiv -[c](\widehat{a}) \equiv [e](\widehat{a}+ \sim \widehat{c})$, we also obtain

$$\boxed{\neg}\ h(a,\widehat{c}) \Rightarrow h(\lceil a \rceil, \sim \bar{g})\ \boxed{\vee}\ h(\lceil -[a](\widehat{c}) \rceil, \bar{g}) \qquad \text{T5.54}$$

$$\boxed{\neg}\ h(a,\widehat{c}), h(\lceil a \rceil, \sim \bar{g}) \Rightarrow h(\lceil -[a](\widehat{c}) \rceil, \bar{g}) \qquad [230.4]$$

$$\boxed{\neg}\ h(a,\widehat{c}) \Rightarrow h(\lceil a \rceil, \sim \bar{g})\ \boxed{\vee}\ h(\lceil [e](\widehat{a}+ \sim \widehat{c}) \rceil, \bar{g}) \qquad \text{T5.55}$$

$$\boxed{\neg}\ h(a,\widehat{c}), h(\lceil a \rceil, \sim \bar{g}) \Rightarrow h(\lceil [e](\widehat{a}+ \sim \widehat{c}) \rceil, \bar{g}) \qquad [230.4]$$

By virtue of the contraposition rule, we know that from the falsity of $h(a,\widehat{c})$ we can also derive the falsity of any proposition from which $h(a,\widehat{c})$ is derivable; hence all inferences of the form $\boxed{\neg}\ h(a,\widehat{c}) \Rightarrow \boxed{\neg}\ p$ are valid for those p for which $p \Rightarrow h(a,\widehat{c})$ is already known. [230.5]

In greater detail, the falsity of $h(a,\widehat{c})$, in the presence of the objectuality of a, implies either the objectlessness of c or its subordination to a or its concatenation with a or its exclusion of a:

$$\boxed{\neg}\ h(a,\widehat{c}), h(\lceil a \rceil, \bar{g}) \Rightarrow h(\lceil c \rceil, \sim \bar{g})\ \boxed{\vee}\ (c \boxed{\dot{\sqsubseteq}}\ a)\ \boxed{\vee}\ (c \boxed{\top}\ a)\ \boxed{\vee}\ (c \boxed{\bot}\ a) \qquad [230.6] \quad \text{T5.56}$$

The falsity of $h(a,\widehat{c})$ also implies the falsity of any other proposition obtained by summing to $\widehat{c}$ any quality idea; i.e.

$$\boxed{\neg}\ h(a,\widehat{c}) \Rightarrow \boxed{\neg}\ h(a,\widehat{c}+\widehat{b}) \qquad [230.7] \quad \text{T5.57}$$

In the concluding note of the section, Bolzano rejects the two pairs of traditional inferences

$$\frac{\text{It is false that all } a \text{ are } c}{\text{It is false that all } -c \text{ are } -a} \qquad \frac{\text{It is false that all } a \text{ are } c}{\text{Some } a \text{ are } -c}$$

$$\frac{\text{It is false that no } a \text{ is } c}{\text{It is false that no } c \text{ is } a} \qquad \frac{\text{It is false that no } a \text{ is } c}{\text{Some } a \text{ are } c}$$

Ad (2) Two premises: a *universal* and a *negation of a universal.*
Three cases are distinguished: (a) they contain the same ideas; (b) they have only one idea in common; (c) they have no common idea.

Ad (2a) When both premises contain *the same ideas,* Bolzano points out that, of the possible correct combinations

(1) $h(a,\widehat{c}), \boxed{\neg}\ h(a,\widehat{c})$ (2) $h(a,\widehat{c}), \boxed{\neg}\ h(c,\widehat{a})$
(3) $h(a,\widehat{c}), \boxed{\neg}\ h(a,\sim\widehat{c})$ (4) $h(a,\widehat{c}), \boxed{\neg}\ h(c,\sim\widehat{a})$
(5) $h(a,\widehat{c}), \boxed{\neg}\ h(-a,\widehat{c})$ (6) $h(a,\widehat{c}), \boxed{\neg}\ h(-c,\widehat{a})$
(7) $h(a,\widehat{c}), \boxed{\neg}\ h(-a,\sim\widehat{c})$ (8) $h(a,\widehat{c}), \boxed{\neg}\ h(-c,\sim\widehat{a})$,

(1), (3), (4), and (6) are not interesting—(1) because it is a contradictory combination and the other three because the second proposition is already derivable from the first. [231.1]

As to the remaining ones (i.e. (2), (5), (7), and (8)), one has, respectively,

$$h(a,\widehat{c}),\ \boxed{\neg}\ h(c,\widehat{a}) \Rightarrow a \boxed{\dot{\sqsubseteq}}\ c \qquad [231.2]\ \ \text{T5.58}$$

$$h(a,\widehat{c}),\ \boxed{\neg}\ h(-a,\widehat{c}) \Rightarrow (a \boxed{\equiv} c) \boxed{\vee} ((a \boxed{\dot{\sqsubseteq}} c) \boxed{\wedge} (c \boxed{\top} - a)) \qquad [231.3]\ \ \text{T5.59}$$

$$h(a,\widehat{c}),\ \boxed{\neg}\ h(-a,\sim\widehat{c}) \Rightarrow (a \boxed{\equiv} e) \boxed{\vee} (a \boxed{\dot{\sqsubseteq}} c) \qquad [231.4]\ \ \text{T5.60}$$

$$h(a,\widehat{c}),\ \boxed{\neg}\ h(-c,\sim\widehat{a}) \Rightarrow h(e,\widehat{c}) \boxed{\vee} (c \boxed{\equiv} e) \qquad [231.5]\ \ \text{T5.61}$$

Ad (2b) When the two premises have only *one idea in common*, of the eight possible combinations

(1) $h(a,\widehat{c}),\ \boxed{\neg}\ h(a,\widehat{b})$	(2) $h(a,\widehat{c}),\ \boxed{\neg}\ h(-a,\widehat{b})$
(3) $h(a,\widehat{c}),\ \boxed{\neg}\ h(c,\widehat{b})$	(4) $h(a,\widehat{c}),\ \boxed{\neg}\ h(-c,\widehat{b})$
(5) $h(a,\widehat{c}),\ \boxed{\neg}\ h(b,\widehat{a})$	(6) $h(a,\widehat{c}),\ \boxed{\neg}\ h(b,\sim\widehat{a})$
(7) $h(a,\widehat{c}),\ \boxed{\neg}\ h(b,\widehat{c})$	(8) $h(a,\widehat{c}),\ \boxed{\neg}\ h(b,\sim\widehat{c})$

(1), (4), (6), and (7) give rise to notable inferences. [231.6]

(1) yields

$$h(a,\widehat{c}),\ \boxed{\neg}\ h(a,\widehat{b}) \Rightarrow h(\lceil [a](\sim\widehat{b})\rceil, \bar{g}); \qquad [231.7]\ \ \text{T5.62}$$

$$h(a,\widehat{c}),\ \boxed{\neg}\ h(a,\widehat{b}) \Rightarrow \boxed{\neg}\ h(c,\widehat{b}); \qquad [231.8]\ \ \text{T5.63}$$

$$h(a,\widehat{c}),\ \boxed{\neg}\ h(a,\widehat{b}) \Rightarrow \boxed{\neg}\ h(a,\widehat{c}+\widehat{b}) \qquad [231.9]\ \ \text{T5.64}$$

(4) yields

$$W(h(a,\widehat{c}) \wedge F(h(-c,\widehat{b})) \rightarrow (G(-c) \rightarrow F(h(-a,\widehat{b}) \wedge F(h(-b,\widehat{a}))) \ \text{A} \qquad [231.10]\ \ \text{T5.65}$$

$$h(a,\widehat{c}),\ \boxed{\neg}\ h(-c,\widehat{b}) \Rightarrow h(\lceil -c\rceil, \bar{g}) \boxed{\rightarrow} (\boxed{\neg}\ h(-a,\widehat{b}) \boxed{\wedge} \boxed{\neg}\ h(-b,\widehat{a}))$$

(6) yields

$$h(a,\widehat{c}),\ \boxed{\neg}\ h(b,\sim\widehat{a}) \Rightarrow \boxed{\neg}\ h(b,\sim\widehat{c}) \qquad [231.11]\ \ \text{T5.66}$$

$$h(a,\widehat{c}),\ \boxed{\neg}\ h(b,\sim\widehat{a}) \Rightarrow (h(\lceil b\rceil, \bar{g}) \boxed{\rightarrow} \boxed{\neg}\ h(c,\sim\widehat{b})) \qquad [231.11]\ \ \text{T5.67}$$

Lastly, (7) yields

$$h(a,\widehat{c}),\ \boxed{\neg}\ h(b,\widehat{c}) \Rightarrow \boxed{\neg}\ h(b,\widehat{a}) \qquad [231.12]\ \ \text{T5.68}$$

Ad (2c) When the two premises have *no variable in common*, there is apparently no interesting inference.

Ad (3) In the case of *two negations of universal propositions*, the only interesting cases are those in which the variables are the same are; in these cases, we obtain

$$\boxed{\neg}\ h(a,\widehat{c}),\ \boxed{\neg}\ h(-a,\widehat{c}) \Rightarrow \neg(c \boxed{\equiv} e) \qquad [232.1]\ \text{T5.69}$$

$$\boxed{\neg}\ h(a,\widehat{c}),\ \boxed{\neg}\ h(-a,\sim\widehat{c}), h(\lceil a \rceil,\overline{g}), h(\lceil -a \rceil,\overline{g}) \Rightarrow (a \boxed{\top} c) \boxed{\vee} (a \boxed{\bot}\ c) \qquad [232.2]\ \text{T5.70}$$

$$\boxed{\neg}\ h(a,\widehat{c}),\ \boxed{\neg}\ h(-a,\sim\widehat{c}) \Rightarrow (-a \boxed{\top} - c) \boxed{\vee} (-a \boxed{\bot} - c) \qquad [232.2]\ \text{T5.71}$$

$$\boxed{\neg}\ h(a,\widehat{c}),\ \boxed{\neg}\ h(c,\widehat{a}), h(\lceil a \rceil,\overline{g}), h(\lceil c \rceil,\overline{g}) \Rightarrow (a \boxed{\top} c) \boxed{\vee} (a \boxed{\bot} c) \qquad [232.3]\ \text{T5.72}$$

5.5 Inferences with One or More Existential or Inexistential Premises

Having exhausted the discussion of the inferences the premises of which are only universal or negations of universal propositions, Bolzano turns to those inferences in the premises of which existential or inexistential propositions occur. Bolzano articulated his investigation as follows: (1) a single existential premise [§233]; (2) a single inexistential premise [§234]; (3) one or more universals and an existential premise [§235]; (4) one or more universals and an inexistential premise [§236]; (5) a negated universal and an existential premise [§237]; (6) a negated universal and an inexistential [§237]; (7) two existential premises [§238]; (8) an existential and an inexistential premise [§238]; (9) two inexistential premises [§238].

Ad (1) *A single existential premise.*

Two case are distinguished: (a) the variable idea is *simple*; (b) the variable idea is *composite*.

Ad (1a) The first Bolzanian remark in this context is that if the idea *a* is objectual, then certainly not all objects fall under the idea $-a$; i.e.

$$h(\lceil a \rceil,\overline{g}) \Rightarrow \boxed{\neg}\ (h(e,\sim\widehat{a}) \qquad [233.3] \qquad \text{T5.73}$$

$$h(\lceil a \rceil,\overline{g}) \Rightarrow \boxed{\neg}\ (e \boxed{\equiv} - a) \qquad [233.3] \qquad \text{T5.74}$$

Taking as given the principle of the excluded middle, which states that, for any object idea, every object either falls under it or under its negation,[14] Bolzano also established that if *a* is objectual, then, given any quality idea, either there is an *a*

[14] TS 233, II 310: "4) [. . .] every thing must stand under *X* or non-*X* [. . .]."

enjoying that quality or there is an a that lacks that quality; i.e.

$$h(\ulcorner a \urcorner, \bar{g}) \Rightarrow \boxminus\, (h(\ulcorner [a](\widehat{c}) \urcorner, \sim \bar{g}) \boxed{\wedge}\, h(\ulcorner [a](\sim \widehat{c}) \urcorner, \sim \bar{g}))^{15} \quad [233.4] \quad T5.75$$

Ad (1b) Concerning the enunciation of objectuality of an idea of the type $[e](\widehat{c_1} + \ldots + \widehat{c_s})$, Bolzano remarks, first of all, that such a proposition is equivalent to all those concerning any equivalent reformulation of that idea, in particular those reformulations that we exemplify here for the case of three ideas c_1, c_2, c_3:

$$h(\ulcorner [e](\widehat{c_1} + \widehat{c_2} + \widehat{c_3}) \urcorner, \bar{g}) \Leftrightarrow \begin{cases} h(\ulcorner [c_1](\widehat{c_2} + \widehat{c_3}) \urcorner, \bar{g}) \\ h(\ulcorner [c_2](\widehat{c_1} + \widehat{c_3}) \urcorner, \bar{g}) \\ h(\ulcorner [c_3](\widehat{c_1} + \widehat{c_2}) \urcorner, \bar{g}) \\ h(\ulcorner [[c_1](\widehat{c_2})](\widehat{c_3}) \urcorner, \bar{g}) \\ h(\ulcorner [[c_1](\widehat{c_3})](\widehat{c_2}) \urcorner, \bar{g}) \\ h(\ulcorner [[c_2](\widehat{c_1})](\widehat{c_3}) \urcorner, \bar{g}) \\ h(\ulcorner [[c_2](\widehat{c_3})](\widehat{c_1}) \urcorner, \bar{g}) \\ h(\ulcorner [[c_3](\widehat{c_1})](\widehat{c_2}) \urcorner, \bar{g}) \\ h(\ulcorner [[c_3](\widehat{c_2})](\widehat{c_1}) \urcorner, \bar{g}) \end{cases} \quad [233.5] \quad T5.76$$

It is worth noticing that if we think of $[e](\widehat{c_1} + \widehat{c_2})$ as $c_1 \sqcap c_2$, then we may say that Bolzano thereby establishes the *commutative* and *associative* properties of $\sqcap$.

Naturally, the objectuality of a composite idea also implies the objectuality of any idea obtained from it by eliminating some components; i.e.

If $\{a_1, \ldots, a_r\} \subset \{c_1, \ldots, c_s\}$, then T5.77

$$h(\ulcorner [e](\widehat{c_1} + \ldots + \widehat{c_s}) \urcorner, \bar{g}) \Rightarrow h(\ulcorner [e](\widehat{a_1} + \ldots + \widehat{a_r}) \urcorner, \bar{g}) \quad [233.6]$$

and this, of course, for all the reformulations of $[e](\widehat{a_1} + \ldots + \widehat{a_r})$ just considered. [233.6]

If $i, j \leq s$, then $h(\ulcorner [e](\widehat{c_1} + \ldots + \widehat{c_s}) \urcorner, \bar{g}) \Rightarrow c_i \boxed{T}\, c_j$ [233.7] T5.78

$$h(\ulcorner [e](\widehat{c_1} + \ldots + \widehat{c_s}) \urcorner, \bar{g})$$ T5.79

[15] TS 233, II 310: "4) [...] We may accordingly set out the conclusion: 'The collection of ideas $[A]x$ and $[A]nonx$ is not a collection of objectless ideas,'; or 'It is false that each of the ideas $[A]x$ and $[A]nx$ is objectless.'."

is equivalent to the proposition "the aggregate of the ideas $c_1, \ldots, c_s$ is an aggregate of reciprocally compatible ideas",[16] i.e.

$$h(\lceil [e](\widehat{c_1} + \ldots + \widehat{c_s}) \rceil, \bar{g}) \Leftrightarrow h(\lceil c_1 \sqcap \ldots \sqcap c_s \rceil, \bar{g}) \qquad [233.8]$$

If $\{a_1, \ldots, a_r\} \subset \{c_1, \ldots, c_s\}$ and c_i $(i \leq s)$ is not in $\{a_1, \ldots, a_r\}$, then T5.80

$$h(\lceil [e](\widehat{c_1} + \ldots + \widehat{c_s}) \rceil, \bar{g}) \Rightarrow \boxed{\neg}\, h([e](\widehat{a_1} + \ldots + \widehat{a_r}), \sim \widehat{c_i}) \qquad [233.9]$$

Furthermore, if $\{a_1, \ldots, a_r\} \subset \{c_1, \ldots, c_s\}$ and the $b_1, \ldots, b_t$ are in $\{c_1, \ldots, c_s\}$ but not in $\{a_1, \ldots, a_r\}$, then [233.10]

$$h(\lceil [e](\widehat{c_1} + \ldots + \widehat{c_s}) \rceil, \bar{g}) \Rightarrow$$
$$\boxed{\neg}\, h([e](\widehat{c_1} + \ldots + \widehat{c_s}), \widehat{a_1} + \ldots + \widehat{a_r} + \sim \widehat{b_1} + \ldots + \sim \widehat{b_t}) \qquad \text{T5.81}$$

as well as

$$h(\lceil [e](\widehat{c_1} + \ldots + \widehat{c_s}) \rceil, \bar{g}) \Rightarrow$$
$$\boxed{\neg}\, h(e, \widehat{a_1} \cdot \ldots \cdot \widehat{a_r} \cdot \sim \widehat{b_1} \cdot \ldots \cdot \sim \widehat{b_t}) \qquad [233.11] \quad \text{T5.82}$$

The last inference mentioned in §233 states that the objectuality of $[e](\widehat{c_1} + \ldots + \widehat{c_s})$ implies the contradictoriness of $c_1 \sqcap \ldots \sqcap c_s$ with $-c_1 \sqcup \ldots \sqcup -c_s$ (i.e. the fact that an object that does not fall under $c_1 \sqcap \ldots \sqcap c_s$ falls under $-c_1 \sqcup \ldots \sqcup -c_s$), whereas one that does not fall under $-c_1 \sqcup \ldots \sqcup -c_s$, falls under $c_1 \sqcap \ldots \sqcap c_s$; i.e.

$$h(\lceil [e](\widehat{c_1} + \ldots + \widehat{c_s}) \rceil, \bar{g}) \Rightarrow$$
$$(c_1 \sqcap \ldots \sqcap c_s) \boxed{|||} (-c_1 \sqcup \ldots \sqcup -c_s) \qquad [233.12] \quad \text{T5.83}$$

Ad (2) *A single inexistential premise.*
Once again, two cases are distinguished: (a) the variable idea is *simple*; (b) the variable idea is *composite*.

Ad (2a) Bolzano begins by establishing that the objectlessness of an idea implies the equivalence of its negation with e; however, he omits to mention that this relationship is convertible:

$$h(\lceil a \rceil, \sim \bar{g}) \Rightarrow e \boxed{\equiv} -a \qquad [234.1] \quad \text{T5.84}$$

[16] TS 233, II 311: "8) If we unite all the ideas that may be formed from the individual parts $a, b, c, d, \ldots$ into a single proposition, we obtain the conclusion: 'The collection of ideas $A, B, C, D, \ldots$ is one of compatible ideas', which is equivalent to the given premise."

TS 246, II 344: "1) A proposition which states a relation of compatibility and thus stands under the form: 'The collection of ideas $A, B, C, D, \ldots$ has the attribute of a collection of compatible ideas', is obviously equivalent to the proposition: 'The idea [something]$(a + b + c + d + \ldots)$ has objectuality'."

$h(\lceil a \rceil, \sim \bar{g}) \Rightarrow h(e, \sim \widehat{a})$ [17] [234.1] T5.85

The objectlessness of an idea also implies the objectlessness of any of its possible 'sums', as well as the falsity of any proposition in which it occurs as a subject or as a predicate:

$h(\lceil a \rceil, \sim \bar{g}) \Rightarrow h(\lceil [a](\widehat{c}) \rceil, \sim \bar{g})$ [234.2] T5.86

$h(\lceil a \rceil, \sim \bar{g}) \Rightarrow \boxminus\, h(a, \widehat{c})$ [234.3] T5.87

$h(\lceil a \rceil, \sim \bar{g}) \Rightarrow \boxminus\, h(c, \widehat{a})$ [234.4] T5.88

Ad (2b) As to the enunciations of objectlessness of composite ideas, Bolzano begins by remarking that, as in the case of objectuality, also

The objectlessness of a composite idea implies the objectlessness of every equivalent reformulation of that idea. [234.5] T5.89

Moreover, the objectlessness of a composite idea implies the incompatibility of its components:

$h(\lceil a_1 \sqcap \ldots \sqcap a_s \rceil, \sim \bar{g}) \Rightarrow \boxed{\perp}\, (\overset{s}{a})$ [234.6] T5.90

Dually to what has already been seen in [233.9],

$h(\lceil a \sqcap c_1 \sqcap \ldots \sqcap c_s \rceil, \sim \bar{g}) \Rightarrow \boxminus\, h([c_1 \sqcap \ldots \sqcap c_s], \widehat{a})$ [234.7] T5.91

If a composite idea is objectless,[18] then, for each of its components, either it is itself objectless or none of its objects falls simultaneously under all the remaining components; i.e.

$h(\lceil a \sqcap c_1 \sqcap \ldots \sqcap c_s \rceil, \sim \bar{g}) \Rightarrow$
$h(\lceil a \rceil, \bar{g}) \boxed{\rightarrow} h(a, \sim (\widehat{c_1} + \ldots + \widehat{c_s}))$ [234.9] T5.92

More generally, if a subset of components is compatible, then none of the objects falling simultaneously under these components falls also simultaneously under all the remaining ones:

If $\{c_1, \ldots, c_s\} = \{a_1, \ldots, a_r, b_1, \ldots, b_{s-r}\}$, then T5.93

$h(\lceil c_1 \sqcap \ldots \sqcap c_s \rceil, \sim \bar{g}) \Rightarrow$
$h(\lceil a_1 \sqcap \ldots \sqcap a_r \rceil, \bar{g}) \boxed{\rightarrow} h(a_1 \sqcap \ldots \sqcap a_r, \sim (\widehat{b_1} + \ldots + \widehat{b_{s-r}})$ [234.9]

[17] TS 234, II 313: "1) The most general form of a proposition that denies the objectuality of an idea is: 'The idea *A* has non-objectuality'. If the idea *A* is supposed not to have any objects, then every object must stand under the idea 'non-*A*'. Thus we obtain the conclusion: 'The idea non-*A* has the extension of the idea of something in general', or 'Every (arbitrary) something has (the attribute) non *a*'."

[18] This seems to us to be the natural reading of this passage, which, by its ambiguity, has given some problems to J. Berg.

If a composite idea is objectless, but one of its components *a* is objectual, then, first, at least one of the objects falling under *a* also falls under the negation of at least one of the remaining:

$$\begin{aligned}&h(\ulcorner a \sqcap c_1 \sqcap \ldots \sqcap c_s \urcorner, \sim \bar{g}) \Rightarrow \\ &\quad (h(\ulcorner a \urcorner, \bar{g}) \boxed{\rightarrow} h(\ulcorner a \sqcap -c_1 \urcorner, \bar{g}) \boxed{\vee} \ldots \boxed{\vee} h(\ulcorner a \sqcap -c_s \urcorner, \bar{g}))\end{aligned} \qquad [234.11] \quad \text{T5.94}$$

Second, the extension of *a* cannot properly include all extensions of the negations of the remaining components (otherwise there would be an object under *a* that, not falling under any of the negations of the remaining components, would fall under all the remaining components):

$$\begin{aligned}&h(\ulcorner a \sqcap c_1 \sqcap \ldots \sqcap c_s \urcorner, \sim \bar{g}), h(\ulcorner a \urcorner, \bar{g}) \Rightarrow \\ &\quad \boxed{\neg}((-c_1 \sqcup \ldots \sqcup -c_s) \boxed{\dot{\sqsubseteq}}\ a)\end{aligned} \qquad [234.12] \quad \text{T5.95}$$

We now turn to the conclusion that Bolzano derived from the premise of objectlessness of a composite idea in §§234.8 and 234.10—its interpretation is not simple.

In 234.8, Bolzano provides, as a conclusion from that premise,

(a) The idea of a true proposition among the propositions arising when in the propositions *This A has not b*, *This A has not c*, *This A has not d*, . . . , any idea is substituted for the *This* has objectuality.[19]

In 234.10, on the other hand, he gives

(b) The aggregate of propositions arising when in the propositions *This A has b*, *This A has c*, *This A has d*, . . . , any idea is substituted for the *This* is never an aggregate of only true propositions,

as well as its generalization

(c) The aggregate of propositions arising when in the propositions: *This* [Etwas]$(a + b)$ *has c*, *This* [Etwas]$(a + b)$ *has d*, . . . , any idea is substituted for the *This* is never an aggregate of only true propositions,

etc.[20]

[19] TS §234, II 314–15: "8) If the idea [something]$(a + b + c + d + \ldots)$ is supposed to have no object [. . .]. Thus the following conclusion holds: 'The idea of a true proposition among those that arise when any idea you like is put in the place of "this" in the propositions "This does not have *a*", "This does not have *b*", "This does not have *c*", etc., has objectuality'."

[20] TS §234, II 315: "10) If it is supposed to be true that no *A* has the sum of attributes *b*, *c*, *d*, . . . , then it must be possible to claim that no collection of propositions that results when any idea at all is put in the place of 'this' in

Taking into account the argument used by Bolzano to justify this inference—if the idea $[e](\widehat{c_1} + \ldots + \widehat{c_s})$ must lack any object, than every object must lack some of the involved qualities (i.e., equivalently, it must have some of the qualities $\sim\widehat{c_1}, \ldots, \sim\widehat{c_s}$)—and the fact that for Bolzano, '*This*' is the name of a singular idea, it seems possible to interpret the (a) as enunciating that, under the hypothesis, the object underlying any singular idea must lack at least one of the involved qualities; i.e.

$$h(\lceil [e](\widehat{c_1} + \ldots + \widehat{c_s}) \rceil, \sim\overline{g}), h(\lceil a \rceil, \overline{1}) \Rightarrow h(a, \sim\widehat{c_1}) \boxed{\vee} \ldots \boxed{\vee}\, h(a, \sim\widehat{c_s}) \qquad [234.8] \quad \text{T5.96}$$

As to (b) it seems that it may be interpreted as enunciating that, under the hypothesis, it is impossible that an object underlying any idea may simultaneously have any of the involved qualities; i.e.

$$h(\lceil [e](\widehat{c_1} + \ldots + \widehat{c_s}) \rceil, \sim\overline{g}) \Rightarrow \boxed{\neg}\, h(a, \widehat{c_1}) \boxed{\vee} \ldots \boxed{\vee} \boxed{\neg}\, h(a, \widehat{c_s}) \qquad [234.10] \quad \text{T5.97}$$

If this interpretation is correct, then (c) will be

$$h(\lceil [e](\widehat{c_1} + \ldots + \widehat{c_s}) \rceil, \sim\overline{g}) \Rightarrow \boxed{\neg}\, h(a \sqcap b, \widehat{c_1}) \boxed{\vee} \ldots \boxed{\vee} \boxed{\neg}\, h(a \sqcap b, \widehat{c_s}) \qquad [234.10] \quad \text{T5.98}$$

It is, however, clear that, especially in cases (b) and (c), different interpretations may also be possible.

Ad (3) *One or more universal and one existential premise.*

Five cases are distinguished: (a) they have no variable in common; (b) the existential premise contains only one variable and this also occurs in the unique universal premise; (c) the existential premise contains two variables and they also occur in the unique universal premise; (d) the existential premise contains two variables but only one of them occurs in the unique universal premise; (e) more than one universal premise and an existential one.

We must immediately remark that, *mutatis mutandis*, the discussions of cases (4)–(9) below develop according to the same schema.

the propositions 'This *A* has *b*', 'This *A* has *c*', 'This *A* has *d*', etc., consists entirely of true propositions. For if there were even one such collection, then there would be at least one *A* which has the sum of the attributes *b*, *c*, *d*, ... Thus we may form the conclusions:

The collection of propositions that arises when when any idea at all is put in the place of 'this' in the propositions 'This *A* has *b*', 'This *A* has *c*', 'This *A* has *d*', etc. never consists entirely of true propositions.

The collection of propositions that arises when when any idea at all is put in the place of 'this' in the propositions 'This [Something]$(a + b)$ has *c*', 'This [Something]$(a + b)$ has *d*', etc., never consists entirely of true propositions."

Ad (3a) The combination of a universal and an existential premise lacking common variables does not lead to interesting inferences.

Ad (3b) When the premises have an idea in common, but that idea is the only one contained by the existential premise, the simplest possible combinations are

$$(1)\quad h(a,\widehat{c}), h(\lceil a \rceil, \overline{g}) \qquad (2)\quad h(a,\widehat{c}), h(\lceil -a \rceil, \overline{g})$$
$$(3)\quad h(a,\widehat{c}), h(\lceil c \rceil, \overline{g}) \qquad (4)\quad h(a,\widehat{c}), h(\lceil -c \rceil, \overline{g})$$

Among these, (1) and (3) are less interesting, since the existential premise already follows from the universal one (it holds that $h(a,\widehat{c}) \Rightarrow h(\lceil a \rceil, \overline{g})$ and $h(a,\widehat{c}) \Rightarrow h(\lceil c \rceil, \overline{g})$, and so the relative inference is reduced to those from $h(a,\widehat{c})$). (2) is not very interesting either. As to (4):

$$h(a,\widehat{c}), h(\lceil -c \rceil, \overline{g}) \Rightarrow h(\lceil -a \rceil, \overline{g}) \qquad [235.3] \qquad \text{T5.99}$$
$$h(a,\widehat{c}), h(\lceil -c \rceil, \overline{g}) \Rightarrow h(-c, \sim\widehat{a}) \qquad [235.4] \qquad \text{T5.100}$$

Ad (3c) When, however, the idea in the existential premise is not simple, but contains two ideas, both occurring in the universal premise, there are four possible combinations:

$$(1)\quad h(a,\widehat{c}), h(\lceil a \sqcap c \rceil, \overline{g}) \qquad (2)\quad h(a,\widehat{c}), h(\lceil a \sqcap -c \rceil, \overline{g})$$
$$(3)\quad h(a,\widehat{c}), h(\lceil -a \sqcap c \rceil, \overline{g}) \qquad (4)\quad h(a,\widehat{c}), h(\lceil -a \sqcap -c \rceil, \overline{g})$$

(1) and (2) are not very interesting, since in (1) the existential premise already follows from the universal one (It holds that $h(a,\widehat{c}) \Rightarrow h(\lceil a \sqcap c) \rceil, \overline{g}))$, whereas in (2) the two premises are incompatible (it holds that $h(a,\widehat{c}) \perp h(\lceil a \sqcap -c) \rceil, \overline{g}))$.

[235.5]

As to (3), we have

$$h(a,\widehat{c}), h(\lceil -a \sqcap c) \rceil, \overline{g}) \Rightarrow a \boxed{\dot{\sqsubseteq}}\, c \qquad [235.6] \qquad \text{T5.101}$$

whereas, as far as (4) is concerned,

$$h(a,\widehat{c}), h(\lceil -a \sqcap -c \rceil, \overline{g}) \Rightarrow h(-c, \sim\widehat{a}) \qquad [235.7] \qquad \text{T5.102}$$

Indeed,

$$\frac{\overset{[233.6]}{h(\lceil -a \sqcap -c \rceil, \overline{g}) \Rightarrow h(\lceil -c \rceil, \overline{g})} \qquad \overset{[235.4]}{h(a,\widehat{c}), h(\lceil -c \rceil, \overline{g}) \Rightarrow h(-c, \sim\widehat{a})}}{h(a,\widehat{c}), h(\lceil -a \sqcap -c \rceil, \overline{g}) \Rightarrow h(-c, \sim\widehat{a})}\ \text{C}$$

Ad (3d) When, however, the second idea of the existential premise does not occur in the universal one, there are four possible combinations:

$$(1)\ \ h(a,\widehat{c}), h(\lceil a \sqcap b \rceil, \overline{g}) \qquad (2)\ \ h(a,\widehat{c}), h(\lceil c \sqcap b \rceil, \overline{g})$$
$$(3)\ \ h(a,\widehat{c}), h(\lceil -a \sqcap b \rceil, \overline{g}) \qquad (4)\ \ h(a,\widehat{c}), h(\lceil -c \sqcap b \rceil, \overline{g})$$

Notwithstanding their compatibility, nothing can be inferred from (2) and (3). [235.8]

As to the other two, one has the inferences

$$h(a,\widehat{c}), h(\lceil a \sqcap b \rceil, \overline{g}) \Rightarrow h(\lceil c \sqcap b) \rceil, \overline{g}) \qquad [235.9] \quad \text{T5.103}$$
$$h(a,\widehat{c}), h(\lceil a \sqcap b \rceil, \overline{g}) \Rightarrow h(a \sqcap b, \widehat{c}) \qquad [235.10] \quad \text{T5.104}$$

and

$$h(a,\widehat{c}), h(\lceil -c \sqcap b \rceil, \overline{g}) \Rightarrow h(-c, \widehat{a}) \qquad [235.11] \quad \text{T5.105}$$
$$h(a,\widehat{c}), h(\lceil -c \sqcap b \rceil, \overline{g}) \Rightarrow h(-a \sqcap b, \widehat{a}) \qquad [235.12] \quad \text{T5.106}$$

respectively.

Ad (3e) Finally, as far inferences from more universal premises are concerned, Bolzano points out the remarkable particularity

$$\begin{pmatrix} h(a_1,\widehat{c_1}), \ldots, h(a_n, \widehat{c_n}), \\ h(\lceil a_1 \sqcap \ldots \sqcap a_n \rceil, \overline{g}) \end{pmatrix} \Rightarrow \qquad [235.13] \quad \text{T5.107}$$
$$h(a_1 \sqcap \ldots \sqcap a_n, \widehat{c_1} + \ldots + \widehat{c_n})$$

Ad (4) *A universal premise and an inexistential one.*

The treatment follows the same schema applied in the positive case.

Ad (4b)

$$(1)\ \ h(a,\widehat{c}), h(\lceil -a \rceil, \sim \overline{g}) \Rightarrow a \boxed{\equiv} c \boxed{\equiv} e \qquad \text{T5.108}$$

Ad (4c) The possible combinations are

$$(1)\ \ h(a,\widehat{c}), h(\lceil a \sqcap c) \rceil, \sim \overline{g}) \qquad (2)\ \ h(a,\widehat{c}), h(\lceil a \sqcap -c \rceil, \sim \overline{g})$$
$$(3)\ \ h(a,\widehat{c}), h(\lceil -a \sqcap c) \rceil, \sim \overline{g}) \qquad (4)\ \ h(a,\widehat{c}), h(\lceil -a \sqcap -c) \rceil, \sim \overline{g}) \qquad [236.2]$$

Neither (1) nor (2) is interesting: (1) because the premises are incompatible [it holds that $h(a,\widehat{c}) \perp h(\lceil a \sqcap c) \rceil, \sim \overline{g})$]; (2) because the second premise follows from the first [$h(a,\widehat{c}) \Rightarrow h(\lceil a \sqcap -c \rceil, \sim \overline{g})$]. [236.2]

$$h(a,\widehat{c}), h(\lceil -a \sqcap c) \rceil, \sim \overline{g}) \Rightarrow h(c, \widehat{a}) \qquad [236.3] \quad \text{T5.109}$$

$$h(a,\widehat{c}), h(\ulcorner -a \sqcap c)\urcorner, \sim \bar{g}) \Rightarrow a \boxed{\equiv} c \qquad [236.3]$$

$$h(a,\widehat{c}), h(\ulcorner -a \sqcap -c)\urcorner, \sim \bar{g}) \Rightarrow c \boxed{\equiv} e \qquad [236.4] \qquad \text{T5.110}$$

Ad (4d) The possible combinations are[21]

$$(1)\ \ h(a,\widehat{c}), h(\ulcorner a \sqcap b\urcorner, \sim \bar{g}) \qquad (2)\ \ h(a,\widehat{c}), h(\ulcorner c \sqcap b\urcorner, \sim \bar{g})$$
$$(3)\ \ h(a,\widehat{c}), h(\ulcorner -a \sqcap b\urcorner, \sim \bar{g}) \qquad (4)\ \ h(a,\widehat{c}), h(\ulcorner -c \sqcap b\urcorner, \sim \bar{g})$$

and we obtain the following inferences:

$$h(a,\widehat{c}), h(\ulcorner a \sqcap b\urcorner, \sim \bar{g}) \Rightarrow h(a, \sim \widehat{b}) \qquad [236.5] \qquad \text{T5.111}$$

$$h(a,\widehat{c}), h(\ulcorner a \sqcap b\urcorner, \sim \bar{g}) \Rightarrow h(\ulcorner c \sqcap -b\urcorner, \bar{g}) \qquad [236.5] \qquad \text{T5.112}$$

$$h(a,\widehat{c}), h(\ulcorner c \sqcap b\urcorner, \sim \bar{g}) \Rightarrow h(c, \sim \widehat{b}) \qquad [236.6] \qquad \text{T5.113}$$

$$h(a,\widehat{c}), h(\ulcorner c \sqcap b\urcorner, \sim \bar{g}) \Rightarrow h(a, \sim \widehat{b}) \qquad [236.6] \qquad \text{T5.114}$$

$$h(a,\widehat{c}), h(\ulcorner -a \sqcap b\urcorner, \sim \bar{g}) \Rightarrow h(\ulcorner b\urcorner, \sim \bar{g}) \boxed{\vee} h(b,\widehat{c}) \qquad [236.7] \qquad \text{T5.115}$$

$$h(a,\widehat{c}), h(\ulcorner -c \sqcap b\urcorner, \sim \bar{g}) \Rightarrow \boxed{\neg}\, h(\ulcorner -c\urcorner, \bar{g}) \boxed{\vee} h(-c, \sim \widehat{a} + \sim \widehat{b}) \qquad [236.8] \qquad \text{T5.116}$$

$$h(a,\widehat{c}), h(\ulcorner -c \sqcap b\urcorner, \sim \bar{g}) \Rightarrow \boxed{\neg}\, (h(\ulcorner -c\urcorner, \bar{g}+) \boxed{\wedge} h(\ulcorner b\urcorner, \bar{g})) \boxed{\vee} h(a \sqcup b, \widehat{c}) \qquad [236.9] \qquad \text{T5.117}$$

Ad (4e) Bolzano uses the term '*complete induction* [*vollständige Induction*]' to introduce the generalization of the inference according to which, if all the objects falling under a certain aggregate of ideas have a certain quality and this aggregate exhausts the universe, then every object has that quality; i.e.

$$h(a_1,\widehat{c}), \ldots, h(a_n,\widehat{c}), h(\ulcorner -a_1 \sqcap \ldots \sqcap -a_n)\urcorner, \sim \bar{g}) \Rightarrow h(e,\widehat{c})$$

$$[236.10] \qquad \text{T5.118}$$

$$h(a \sqcap a_1,\widehat{c}), \ldots, h(a \sqcap a_n,\widehat{c}), h(\ulcorner a \sqcap -a_1 \sqcap \ldots \sqcap -a_n\urcorner, \sim \bar{g}) \Rightarrow h(a,\widehat{c})$$

$$[236.10] \qquad \text{T5.119}$$

Moreover, if, in addition to the hypothesis that all objects falling under a certain aggregate of ideas have a certain quality, one also has a hypothesis stating that there are no objects with that quality not falling under any of the ideas of the aggregate, then every object has that quality iff it falls under at least one of the ideas of the aggregate; i.e.

$$h(a_1,\widehat{c}), \ldots, h(a_n,\widehat{c}), h(\ulcorner c \sqcap -a_1 \sqcap \ldots \sqcap -a_n\urcorner, \sim \bar{g}) \Rightarrow c \boxed{\equiv} a_1 \sqcup \ldots \sqcup a_n \qquad [236.11] \qquad \text{T5.120}$$

[21] The preliminary indications of the alternatives are wrong in the text; the discussion, however, makes the right distinctions.

Furthermore, if, in addition to these hypotheses, one also assumes that the ideas of the aggregate are reciprocally exclusive, then every object with that quality falls at most (and then also exactly) under an idea of the aggregate; i.e.

$$\begin{pmatrix} h(a_1,\widehat{c}),\ldots,h(a_n,\widehat{c}), \quad \boxed{\perp\!\!\!\perp}\,(\overset{n}{a}), \\ h(\ulcorner c \sqcap -a_1 \sqcap \ldots \sqcap -a_n \urcorner, \sim \overline{g}) \end{pmatrix} \Rightarrow \qquad \text{T5.121}$$

$$\left(\begin{pmatrix} h(\ulcorner a \urcorner, \overline{1})\ \boxed{\wedge} \\ h(a, \widehat{c} + \widehat{a}_i) \end{pmatrix} \boxed{\rightarrow} \begin{pmatrix} h(\ulcorner a \sqcap a_1 \urcorner, \sim \overline{g})\ \boxed{\wedge} \ldots \\ \boxed{\wedge}\ h(\ulcorner a \sqcap a_{i-1} \urcorner, \sim \overline{g})\ \boxed{\wedge} \\ h(\ulcorner a \sqcap a_{i+1} \urcorner, \sim \overline{g})\ \boxed{\wedge} \ldots \\ \boxed{\wedge}\ h(\ulcorner a \sqcap a_s \urcorner, \sim \overline{g})) \end{pmatrix} \right) \qquad [236.12]$$

Ad (5) *Negated universal and existential premises.* This follows the usual schema.

Ad (5a) Negated universal and existential premises without common variables hold no interest. [237.1]

Ad (5b) The objectuality enunciation concerns a simple idea occurring in the negation of the universal proposition. Of the four possible combinations

$$(1)\ \boxed{\neg}\ h(a,\widehat{c}), h(\ulcorner a \urcorner, \overline{g}) \qquad (2)\ \boxed{\neg}\ h(a,\widehat{c}), h(\ulcorner -a \urcorner, \overline{g})$$
$$(3)\ \boxed{\neg}\ h(a,\widehat{c}), h(\ulcorner c \urcorner, \overline{g}) \qquad (4)\ \boxed{\neg}\ h(a,\widehat{c}), h(\ulcorner -c \urcorner, \overline{g})$$

only (1) is interesting. It yields

$$\boxed{\neg}\ h(a,\widehat{c}), h(\ulcorner a \urcorner, \overline{g}) \Rightarrow h(\ulcorner a \sqcap -c \urcorner, \overline{g}) \qquad [237.1] \qquad \text{T5.122}$$

$$\boxed{\neg}\ h(a,\widehat{c}), h(\ulcorner a \urcorner, \overline{g}) \Rightarrow \boxed{\neg}\ h(-c, \sim \widehat{a}) \qquad [237.1] \qquad \text{T5.123}$$

Ad (5c) Of the four possible combinations arising when both components of the idea of which the objectuality is stated occur in the negated universal proposition,

$$(1)\ \boxed{\neg}\ h(a,\widehat{c}), h(\ulcorner a \sqcap c \urcorner, \overline{g}) \qquad (2)\ \boxed{\neg}\ h(a,\widehat{c}), h(\ulcorner a \sqcap -c \urcorner, \overline{g})$$
$$(3)\ \boxed{\neg}\ h(a,\widehat{c}), h(\ulcorner -a \sqcap c \urcorner, \overline{g}) \qquad (4)\ \boxed{\neg}\ h(a,\widehat{c}), h(\ulcorner -a \sqcap -c \urcorner, \overline{g})$$

Bolzano remarks that (3) is not interesting because $h(\ulcorner -a \sqcap c \urcorner, \overline{g}) \Rightarrow \boxed{\neg}\ h(a,\widehat{c})$, but he adjoins that neither (2) or (4) seems to give rise to inferences of any interest. As to (1), given that $h(\ulcorner a \sqcap c \urcorner, \overline{g}) \Rightarrow h(\ulcorner a \urcorner, \overline{g})$, we first obtain, as in the preceding case,

$$\boxed{\neg}\ h(a,\widehat{c}), h(\ulcorner a \sqcap c \urcorner, \overline{g}) \Rightarrow h(\ulcorner a \sqcap -c \urcorner, \overline{g}) \qquad [237.2] \qquad \text{T5.124}$$

$$\boxed{\neg}\ h(a,\widehat{c}), h(\ulcorner a \sqcap c \urcorner, \overline{g}) \Rightarrow \boxed{\neg}\ h(-c, \sim \widehat{a}) \qquad [237.2] \qquad \text{T5.125}$$

but then also

$$\begin{array}{l} h(\ulcorner a \sqcap c \urcorner, \overline{g}), h(\ulcorner a \sqcap -c \urcorner, \overline{g}) \Rightarrow \\ \quad (a \sqcap c)\ \boxed{\dot{\sqsubseteq}}\ a\ \boxed{\wedge}\ (a \sqcap -c)\ \boxed{\dot{\sqsubseteq}}\ a \end{array} \qquad [237.2] \qquad \text{T5.126}$$

Ad (5d) When only one component of the idea for which the objectuality is affirmed occurs in the negation of the universal proposition, there again seems to be no interesting inference.

Ad (6) *Negated universal and inexistential premises.* This follows the usual schema.

Ad (6a) Negated universal and inexistential premises without common variable: there is nothing interesting. [237.1]

Ad (6b) Of the four possible combinations

$$(1)\ \boxed{\neg}\, h(a,\widehat{c}), h(\ulcorner a \urcorner, \sim \overline{g}) \qquad (2)\ \boxed{\neg}\, h(a,\widehat{c}), h(\ulcorner -a \urcorner, \sim \overline{g})$$
$$(3)\ \boxed{\neg}\, h(a,\widehat{c}), h(\ulcorner c \urcorner, \sim \overline{g}) \qquad (4)\ \boxed{\neg}\, h(a,\widehat{c}), h(\ulcorner -c \urcorner, \sim \overline{g})$$

satisfying $h(\ulcorner a \urcorner, \sim \overline{g}) \Rightarrow \boxed{\neg}\, h(a,\widehat{c})$ as well as $h(\ulcorner c \urcorner, \sim \overline{g}) \Rightarrow \boxed{\neg}\, h(a,\widehat{c})$, neither (1) nor (3) deserves attention. [237.3]

As to (2), the premise $h(\ulcorner -a \urcorner, \sim \overline{g})$ implies $a \equiv e$ and so, in particular, $h(\ulcorner a \urcorner, \overline{g})$, and therefore it is reduced to the case [237.1]; moreover,

$$\boxed{\neg}\, h(a,\widehat{c}), h(\ulcorner -a \urcorner, \sim \overline{g}) \Rightarrow h(\ulcorner -c \urcorner, \overline{g}) \qquad [237.3] \quad \text{T5.127}$$

$$\boxed{\neg}\, h(a,\widehat{c}), h(\ulcorner -a \urcorner, \sim \overline{g}) \Rightarrow \boxed{\neg}\, h(e,\widehat{c}) \qquad [237.3] \quad \text{T5.128}$$

In (4), however, one has

$$\boxed{\neg}\, h(a,\widehat{c}), h(\ulcorner -c \urcorner, \sim \overline{g}) \Rightarrow h(\ulcorner a \urcorner, \sim \overline{g}) \qquad [237.4] \quad \text{T5.129}$$

Ad (6c) Of the four possible combinations of this case,

$$(1)\ \boxed{\neg}\, h(a,\widehat{c}), h(\ulcorner a \sqcap c \urcorner, \sim \overline{g}) \qquad (2)\ \boxed{\neg}\, h(a,\widehat{c}), h(\ulcorner a \sqcap -c \urcorner, \sim \overline{g})$$
$$(3)\ \boxed{\neg}\, h(a,\widehat{c}), h(\ulcorner -a \sqcap c \urcorner, \sim \overline{g}) \qquad (4)\ \boxed{\neg}\, h(a,\widehat{c}), h(\ulcorner -a \sqcap -c \urcorner, \sim \overline{g})$$

(1) deserves no attention because $h(\ulcorner a \sqcap c \urcorner, \sim \overline{g}) \Rightarrow \boxed{\neg}\, h(a,\widehat{c})$, whereas for (3) one has

$$\boxed{\neg}\, h(a,\widehat{c}), h(\ulcorner -a \sqcap c) \urcorner, \sim \overline{g}), h(\ulcorner c \urcorner, \overline{g}) \Rightarrow c\, \boxed{\dot{\sqsubseteq}}\, a \qquad [237.5] \quad \text{T5.130}$$

As to (2), one has

$$\boxed{\neg}\, h(a,\widehat{c}), h(\ulcorner a \sqcap -c \urcorner, \sim \overline{g}) \Rightarrow h(\ulcorner a \urcorner, \sim \overline{g}) \qquad [237.6] \quad \text{T5.131}$$

whereas for (4) one has

$$\boxed{\neg}\, h(a,\widehat{c}), h(\ulcorner -a \sqcap -c \urcorner, \sim \overline{g}), h(\ulcorner a \urcorner, \overline{g}) \Rightarrow h(-c,\widehat{a}) \qquad [237.7] \quad \text{T5.132}$$

Ad (6d) When only one component of the idea for which the objectlessness is affirmed is present in the negation of the universal proposition, no interesting inferences seem to arise. [237.8]

Ad (7) *Two existential premises.* The usual schema applies.

Ad 7a) There is nothing interesting here.

Ad 7b) If both ideas a and $-a$ are objectual, none of the two exhausts the realm of objects; i.e.

$h(\ulcorner a \urcorner, \bar{g}), h(\ulcorner -a \urcorner, \bar{g}) \Rightarrow \boxminus\, h(e, \widehat{a})$ [238.1] T5.133

$h(\ulcorner a \urcorner, \bar{g}), h(\ulcorner -a \urcorner, \bar{g}) \Rightarrow \boxminus\, h(e, \sim \widehat{a})$ [238.1] T5.134

Ad (7c1) Of the two possible combinations

(1) $h(\ulcorner a \urcorner, \bar{g}), h(\ulcorner a \sqcap c \urcorner, \bar{g})$ (2) $h(\ulcorner a \urcorner, \bar{g}), h(\ulcorner -a \sqcap c \urcorner, \bar{g})$

(1) is not interesting, because $h(\ulcorner a \sqcap c \urcorner, \bar{g}) \Rightarrow h(\ulcorner a \urcorner, \bar{g})$. [238.2]
For (2), however, one has

$h(\ulcorner a \urcorner, \bar{g}), h(\ulcorner -a \sqcap c \urcorner, \bar{g}) \Leftrightarrow a \boxed{\perp} (-a \sqcap c)$ [238.2] T5.135

$h(\ulcorner a \urcorner, \bar{g}), h(\ulcorner -a \sqcap c \urcorner, \bar{g}) \Leftrightarrow \boxminus\, c \boxed{\sqsubseteq} a$ [238.2] T5.136

Ad (7c2) Of the two possible combinations

1) $h(\ulcorner a \sqcap c \urcorner, \bar{g}), h(\ulcorner a \sqcap -c \urcorner, \bar{g})$ 2) $h(\ulcorner a \sqcap c \urcorner, \bar{g}), h(\ulcorner -a \sqcap -c \urcorner, \bar{g})$

(1) has been already considered in [237.2] ($h(\ulcorner a \sqcap -c \urcorner, \bar{g}) \Rightarrow \boxminus\, h(a, \widehat{c})$). [238.3]

As to (2),

$h(\ulcorner a \sqcap c \urcorner, \bar{g}), h(\ulcorner -a \sqcap -c \urcorner, \bar{g}) \Rightarrow$
$a \boxed{\dot{\sqsubseteq}}\, e, c \boxed{\dot{\sqsubseteq}}\, e, -a \boxed{\dot{\sqsubseteq}}\, e, -c \boxed{\dot{\sqsubseteq}}\, e$ [238.3] T5.137

Ad (7d) Neither of the two possible combinations

(1) $h(\ulcorner a \sqcap c \urcorner, \bar{g}), h(\ulcorner a \sqcap b \urcorner, \bar{g})$ (2) $h(\ulcorner a \sqcap c \urcorner, \bar{g}), h(\ulcorner -a \sqcap b \urcorner, \bar{g})$

seems to give rise to any interesting inference. [238.4]

Ad (8) *An existential and an inexistential premise.* The usual schema applies.

Ad (8a) There is nothing interesting here.

Ad (8b) Neither of the two possible combinations

(1) $h(\ulcorner a \urcorner, \bar{g}), h(\ulcorner a \urcorner, \sim \bar{g})$ (2) $h(\ulcorner a \urcorner, \bar{g}), h(\ulcorner -a \urcorner, \sim \bar{g})$

is interesting—(1) because the propositions are contradictory and (2) because the existential premise follows from the inexistential one. Note that $W(h(\ulcorner -a \urcorner, \sim \bar{g})) \leftrightarrow a \equiv e)$. [238.5]

Ad (8c1) Of the possible combinations when one of the premises only contains the common variable,

$$(1)\ h(\lceil a\rceil, \bar{g}), h(\lceil a \sqcap c)\rceil, \sim\bar{g}) \qquad (2)\ h(\lceil a\rceil, \bar{g}), h(\lceil -a \sqcap c\rceil, \sim\bar{g})$$
$$(3)\ h(\lceil a\rceil, \sim\bar{g}), h(\lceil a \sqcap c)\rceil, \bar{g}) \qquad (4)\ h(\lceil -a\rceil, \sim\bar{g}), h(\lceil a \sqcap c)\rceil, \bar{g})$$

(3) fails because of incompatibility, whereas nothing interesting follows from (2). [238.5]

However, (1) yields $h(a, \sim\widehat{c})$, i.e.

$$h(\lceil a\rceil, \bar{g}), h(\lceil a \sqcap c)\rceil, \sim\bar{g}) \Rightarrow h(a, \sim\widehat{c}) \qquad [238.5] \quad \text{T5.138}$$

whereas it follows from (4) that $a \sqcap c \equiv c$, i.e.

$$h(\lceil -a\rceil, \sim\bar{g}), h(\lceil a \sqcap c)\rceil, \bar{g}) \Rightarrow (a \sqcap c) \boxed{\equiv} c \qquad [238.6] \quad \text{T5.139}$$

Ad (8c2) When both premises contain the same variables, of the possible combinations

$$(1)\ h(\lceil a \sqcap c\rceil, \bar{g}), h(\lceil a \sqcap c\rceil, \sim\bar{g}) \qquad (2)\ h(\lceil a \sqcap c\rceil, \bar{g}), h(\lceil a \sqcap -c\rceil, \sim\bar{g})$$
$$(3)\ h(\lceil a \sqcap c\rceil, \bar{g}), h(\lceil -a \sqcap c\rceil, \sim\bar{g}) \qquad (4)\ h(\lceil a \sqcap c\rceil, \bar{g}), h(\lceil -a \sqcap -c\rceil, \sim\bar{g})$$

(1) fails because of incompatibility, whereas (2) is equivalent to (3). [238.7]

As to (3), one has

$$h(\lceil a \sqcap c)\rceil, \bar{g}), h(\lceil -a \sqcap c)\rceil, \sim\bar{g}) \Rightarrow h(c, \widehat{a}) \qquad [238.7] \quad \text{T5.140}$$

whereas from (4), reinforced with the assumptions of objectuality of $-a$ and $-c$, one has

$$h(\lceil a \sqcap c)\rceil, \bar{g}), h(\lceil -a \sqcap -c)\rceil, \sim\bar{g}), h(\lceil -a\rceil, \bar{g}), h(\lceil -c\rceil, \bar{g}) \Rightarrow \qquad \text{T5.141}$$
$$h(-a, \widehat{c}) \boxed{\wedge} h(-c, \widehat{a}) \boxed{\wedge} (-c \boxed{\dot{\sqsubseteq}}\ e) \boxed{\wedge} (-a \boxed{\dot{\sqsubseteq}}\ e) \qquad [238.8]$$

Ad (8d) When both ideas in question are composite, but with only one component in common, of the two possible combinations

$$(1)\ h(\lceil a \sqcap c\rceil, \bar{g}), h(\lceil a \sqcap b\rceil, \sim\bar{g}) \qquad (2)\ h(\lceil a \sqcap c\rceil, \bar{g}), h(\lceil -a \sqcap b\rceil, \sim\bar{g})$$

only (1) is interesting: one has

$$h(\lceil a \sqcap c\rceil, \bar{g}), h(\lceil a \sqcap b\rceil, \sim\bar{g}) \Rightarrow h(a, \sim\widehat{b}) \qquad [238.9] \quad \text{T5.142}$$
$$h(\lceil a \sqcap c\rceil, \bar{g}), h(\lceil a \sqcap b\rceil, \sim\bar{g}) \Rightarrow h(a \sqcap c, \sim\widehat{b}) \qquad [238.9] \quad \text{T5.143}$$

Indeed,

$$\dfrac{\dfrac{\dfrac{h(\lceil a \sqcap c \rceil, \bar{g})}{h(\lceil a \rceil, \bar{g})} \qquad h(\lceil a \sqcap b \rceil, \sim \bar{g})}{h(a, \sim \widehat{b})} \qquad h(\lceil a \sqcap c \rceil, \bar{g})}{h(a \sqcap c, \sim \widehat{b})}$$

Ad (9) *Two inexistential premises.* The usual schema applies.

Ad (9a) There is nothing interesting here. [238.15]

Ad (9b)

$$h(\lceil a \rceil, \sim \bar{g}) \perp h(\lceil -a \rceil, \sim \bar{g}) \qquad [238.10] \quad \text{T5.144}$$

Ad (9c1) of the two possible combinations

$$(1)\ h(\lceil a \rceil, \sim \bar{g}), h(\lceil a \sqcap c \rceil, \sim \bar{g}) \qquad (2)\ h(\lceil a \rceil, \sim \bar{g}), h(\lceil -a \sqcap c \rceil, \sim \bar{g})$$

the first is not interesting, because $h(\lceil a \rceil, \sim \bar{g}) \Rightarrow h(\lceil a \sqcap c \rceil, \sim \bar{g})$. [238.10] As to the second,[22] one has

$$h(\lceil a \rceil, \sim \bar{g}), h(\lceil -a \sqcap c \rceil, \sim \bar{g}) \Rightarrow h(\lceil c \rceil, \sim \bar{g}) \qquad [238.10] \quad \text{T5.145}$$

Ad (9c2)

$$h(\lceil a \sqcap c \rceil, \sim \bar{g}), h(\lceil -a \sqcap c \rceil, \sim \bar{g}) \Rightarrow h(\lceil c \rceil, \sim \bar{g}) \qquad [238.11] \quad \text{T5.146}$$

$$h(\lceil a \sqcap c \rceil, \sim \bar{g}), h(\lceil -a \sqcap -c \rceil, \sim \bar{g}) \Rightarrow (a \boxed{\equiv} -c) \boxed{\vee} (-a \boxed{\equiv} c) \qquad [238.12] \quad \text{T5.147}$$

Ad (9d)

$$h(\lceil a \sqcap c \rceil, \sim \bar{g}), h(\lceil a \sqcap b \rceil, \sim \bar{g}), h(\lceil a \rceil, \bar{g}) \Rightarrow h(a, \sim \widehat{c} + \sim \widehat{b}) \qquad [238.13] \quad \text{T5.148}$$

$$\dfrac{\dfrac{h(\lceil a \sqcap c \rceil, \sim \bar{g}) \qquad h(\lceil a \rceil, \bar{g})}{h(a, \sim \widehat{c})} \qquad \dfrac{h(\lceil a \rceil, \bar{g}) \qquad h(\lceil a \sqcap b \rceil, \sim \bar{g})}{h(a, \sim \widehat{b})}}{h(a, \sim \widehat{c} + \sim \widehat{b})}$$

[22] In the justification of this inference, Bolzano at a certain moment says: "For since the idea *A* has no object at all, and hence non-*A* represents any arbitrary object, the idea [Something]($na + b$) can only be objectless because the idea *b* designates an attribute that belongs to no object" (Cf. TS §238, II, 333). These words induced J. Berg to uphold that Bolzano accepted the existence of the empty quality. This will be discussed in Section 6.2.

$$h(\ulcorner a \sqcap c \urcorner, \sim \bar{g}), h(\ulcorner -a \sqcap b \urcorner, \sim \bar{g}), h(\ulcorner a \urcorner, \bar{g}), h(\ulcorner b \urcorner, \bar{g}) \Rightarrow h(a \sqcup b, \sim \widehat{c}) \qquad [238.14] \qquad \text{T5.149}$$

$$\dfrac{\dfrac{\dfrac{h(\ulcorner a \sqcap c \urcorner, \sim \bar{g}) \qquad h(\ulcorner a \urcorner, \bar{g})}{h(a, \sim \widehat{c})} \qquad \dfrac{h(\ulcorner b \urcorner, \bar{g}) \qquad h(\ulcorner -a \sqcap b \urcorner, \sim \bar{g})}{h(b, \widehat{a})}}{h(b, \sim \widehat{c})}}{h(a \sqcup b, \sim \widehat{c})}$$

Concluding this discussion, Bolzano emphasized that his drawing conclusions from two or more 'particular' propositions did not violate the principle of the syllogistic tradition *ex mere particularibus nil sequetur*, since his conclusions are not of the type intended, in this context, by that tradition.

5.6 Syllogistic Inferences

In two notes, adjoined to §§227 and 235, respectively, Bolzano considers the traditional syllogisms, examining those having two universal premises in the first note and those having a universal and a particular premise in the second. Among the motivations for these treatments there is certainly, besides the all-pervading aspiration to an expository completeness, the intent to point out the superiority of his approach over the traditional one, a superiority that is highlighted in two ways: first by showing that in both cases the various acceptable syllogisms are really only particular cases of a far smaller number of 'inferences' with the same premises and, second, by emphasizing how many other conclusions may follow from the pair of premises considered in the syllogisms. On the whole, Bolzano recalls the following[23] syllogisms that we present both in the traditional notation and 'translated into his language':[24]

Figure I							
Barbara		Celarent		Darii		Ferio	
$M\mathbb{A}P$	$h(b, \widehat{c})$	$M\mathbb{E}P$	$h(b, \sim \widehat{c})$	$M\mathbb{A}P$	$h(b, \widehat{c})$	$M\mathbb{E}P$	$h(b, \sim \widehat{c})$
$S\mathbb{A}M$	$h(a, \widehat{b})$	$S\mathbb{A}M$	$h(a, \widehat{b})$	$S\mathbb{I}M$	$h(\lceil [a](\widehat{b}) \rceil, \bar{g}$	$S\mathbb{I}M$	$h(\ulcorner [a](\widehat{b}) \urcorner, \bar{g})$
$S\mathbb{A}P$	$h(a, \widehat{c})$	$S\mathbb{E}P$	$h(a, \sim \widehat{c})$	$S\mathbb{I}P$	$h(\lceil [a](\widehat{c}) \rceil, \bar{g}$	$S\mathbb{O}P$	$h(\lceil [a](\sim \widehat{c}) \rceil, \bar{g})$

[23] Note that Bolzano sometimes uses an unusual denomination—in particular in the following cases:

Usual name:	Camenes	Bramantip	Baroco	Dimaris	Fresison
Bolzano:	Calentes	Baralip	Barocco	Dimatis	Fresiso

[24] TS §235, II 322: "*Note.* [...] If we render these inferences in our language [...]."

Figure II							
Cesare		Camestres		Festino		Baroco	
$P\mathbb{E}M$	$h(c, \sim\widehat{b})$	$P\mathbb{A}M$	$h(c, \widehat{b})$	$P\mathbb{E}M$	$h(c, \sim\widehat{b})$	$P\mathbb{A}M$	$h(c, \widehat{b})$
$S\mathbb{A}M$	$h(a, \widehat{b})$	$S\mathbb{E}M$	$h(a, \sim\widehat{b})$	$S\mathbb{I}M$	$h(\lceil[a](\widehat{b})\rceil, \overline{g})$	$S\mathbb{O}M$	$h(\lceil[a](\sim\widehat{b})\rceil, \overline{g})$
$S\mathbb{E}P$	$h(a, \sim\widehat{c})$	$S\mathbb{E}P$	$h(a, \sim\widehat{c})$	$S\mathbb{O}P$	$h(\lceil[a](\sim\widehat{c})\rceil, \overline{g})$	$S\mathbb{O}P$	$h(\lceil[a](\sim\widehat{c})\rceil, \overline{g})$

Figure III					
Darapti		Felapton		Disamis	
$M\mathbb{A}P$	$h(b, \widehat{c})$	$M\mathbb{E}P$	$h(b, \sim\widehat{c})$	$M\mathbb{I}P$	$h([b](\widehat{c}), \overline{g})$
$M\mathbb{A}S$	$h(b, \widehat{a})$	$M\mathbb{A}S$	$h(b, \widehat{a})$	$M\mathbb{A}S$	$h(b, \widehat{a})$
$S\mathbb{I}P$	$h([a](\widehat{c}), \overline{g})$	$S\mathbb{O}P$	$h([a](\sim\widehat{c}), \overline{g})$	$S\mathbb{I}P$	$h([a](\widehat{c}), \overline{g})$
Datisi		Bocardo		Ferison	
$M\mathbb{A}P$	$h(b, \widehat{c})$	$M\mathbb{O}P$	$h([b](\sim\widehat{c}), \overline{g})$	$M\mathbb{E}P$	$h(b, \sim\widehat{c})$
$M\mathbb{I}S$	$h([b](\widehat{a}), \overline{g})$	$M\mathbb{A}S$	$h(b, \widehat{a})$	$M\mathbb{I}S$	$h([b](\widehat{a}), \overline{g})$
$S\mathbb{I}P$	$h([a](\widehat{c}), \overline{g})$	$S\mathbb{O}P$	$h([a](\sim\widehat{c}), \overline{g})$	$S\mathbb{O}P$	$h([a](\sim\widehat{c}), \overline{g})$

Figure IV									
Camenes		Bramantip		Fesapo		Dimaris		Fresison	
$P\mathbb{A}M$	$h(c, \widehat{b})$	$P\mathbb{A}M$	$h(c, \widehat{b})$	$P\mathbb{E}M$	$h(c, \sim\widehat{b})$	$P\mathbb{I}M$	$h([c](\widehat{b}), \overline{g})$	$P\mathbb{E}M$	$h(c, \sim\widehat{b})$
$M\mathbb{E}S$	$h(b, \sim\widehat{a})$	$M\mathbb{A}S$	$h(b, \widehat{a})$	$M\mathbb{A}S$	$h(b, \widehat{a})$	$M\mathbb{A}S$	$h(b, \widehat{a})$	$M\mathbb{I}S$	$h([b](\widehat{a}), \overline{g})$
$S\mathbb{E}P$	$h(a, \sim\widehat{c})$	$S\mathbb{I}P$	$h([a](\widehat{c}), \overline{g})$	$S\mathbb{O}P$	$h([a](\sim\widehat{c}), \overline{g})$	$S\mathbb{I}P$	$h([a](\widehat{c}), \overline{g})$	$S\mathbb{O}P$	$h([a](\sim\widehat{c}), \overline{g})$

For their justification, he proceeds as follows. First of all, he unwraps Camenes; indeed, if a is objectless, both premises can be true, but the conclusion is always false. Regarding the others, he provides *inferences* of which they may be considered 'essentially [*wesentlich*]' only particular cases since they differ from them at most either in the order of the premises or in some of the equivalencies $[e](\widehat{a} + \widehat{c}) \equiv [a](\widehat{c}) \equiv [c](\widehat{a}) \equiv a \sqcap c$.[25]

In order to simplify the reading of the justifications we indicate the inferences as they are written in the respective sections and point out each time the substitutions that must be operated to reach the inferences corresponding to the syllogisms as they are written above.

Concerning the syllogisms with two universal premises we have the following:

$$\frac{a\ b\ c\quad h(a, \widehat{b}), h(c, \widehat{a}) \Rightarrow h(c, \widehat{b}) \quad 227.9}{b\ c\ a\quad h(b, \widehat{c}), h(a, \widehat{b}) \Rightarrow h(a, \widehat{c}) \ \text{Barbara}}$$

[25] TS §227, II 299: "*Note.* [...] recall that the order of the premises is arbitrary [...]."

TS §235, II 322: "*Note.* [...] Thus, these seven inferences differ either not at all or else only in that a pair of equivalent ideas, such as $[A]c$ and $[C]a$ or $[C]b$ and $[B]c$ are exchanged in them."

$$b\ {-c}\ a \quad h(b, \sim\widehat{c}), h(a, \widehat{b}) \Rightarrow h(a, \sim\widehat{c}) \qquad \text{Celarent}$$

$$\frac{a\ b\ c \quad h(a, \widehat{b}), h(c, \sim\widehat{b}) \Rightarrow h(a, \sim\widehat{c}) \qquad 227.8}{\text{But the order of premises is inessential; hence}}$$

$$a\ b\ c \quad h(c, \sim\widehat{b}), h(a, \widehat{b}), \Rightarrow h(a, \sim\widehat{c}) \qquad \text{Cesare}$$

$$\frac{a\ b\ c \quad h(a, \widehat{b}), h(c, \sim\widehat{b}) \Rightarrow h(c, \sim\widehat{a}) \qquad 227.8}{c\ b\ a \quad h(c, \widehat{b}), h(a, \sim\widehat{b}) \Rightarrow h(a, \sim\widehat{c}) \qquad \text{Camestres}}$$

$$\frac{a\ b\ c \quad h(a, \widehat{b}), h(a, \widehat{c}) \Rightarrow h(\lceil [e](\widehat{b} + \widehat{c}) \rceil, \overline{g}) \qquad 227.5}{\text{But } [e](\widehat{b} + \widehat{c}) \equiv [c](\widehat{b})\text{; hence}}$$

$$h(a, \widehat{b}), h(a, \widehat{c}) \Rightarrow h(\lceil [c](\widehat{b}) \rceil, \overline{g})$$

$$b\ c\ a \quad h(b, \widehat{c}), h(b, \widehat{a}) \Rightarrow h(\lceil [a](\widehat{c}) \rceil, \overline{g}) \qquad \text{Darapti}$$

$$b\ {-c}\ a \quad h(b, \sim\widehat{c}), h(b, \widehat{a}) \Rightarrow h(\lceil [a](\sim\widehat{c}) \rceil, \overline{g}) \qquad \text{Felapton}$$

$$\frac{a\ b\ c \quad h(a, \widehat{b}), h(a, \sim\widehat{c}) \Rightarrow h(\lceil [e](\widehat{b} + \sim\widehat{c}) \rceil, \overline{g}) \qquad 227.14}{\text{But the order of premises is inessential and moreover } [e](\widehat{b} + \sim\widehat{c}) \equiv [b](\sim\widehat{b})\text{; hence}}$$

$$h(a, \sim\widehat{c}), h(a, \widehat{b}) \Rightarrow h(\lceil [b](\sim\widehat{c}) \rceil, \overline{g})$$

$$b\ a\ c \quad h(b, \sim\widehat{c}), h(b, \widehat{a}) \Rightarrow h(\lceil [a](\sim\widehat{c}) \rceil, \overline{g}) \qquad \text{Fesapo}$$

As to Bramantip, it is obtained by cut from [227.9] and [225.7]; more precisely,

$$\frac{\dfrac{\dfrac{[227.9]}{h(b, \widehat{c}), h(a, \widehat{b}) \Rightarrow h(a, \widehat{c})}}{h(a, \widehat{b}), h(b, \widehat{c}) \Rightarrow h(a, \widehat{c})} \quad \dfrac{[225.7]}{h(a, \widehat{c*}) \Rightarrow h(\lceil [a](\widehat{c}) \rceil, \overline{g})}}{h(c, \widehat{b}), h(b, \widehat{a}) \Rightarrow h([a](\widehat{c}), \overline{g})}$$

As to the syllogisms with a universal and a particular premise, seven of them are 'essentially' 235.9 and the remaining three 'essentially' 235.12:

$a\ b\ c\quad h(a,\widehat{c}), h(\lceil [a](\widehat{b}) \rceil, \bar{g}) \Rightarrow h(\lceil [b](\widehat{c}) \rceil, \bar{g})$ 235.9

$b\ a\ c\quad h(b,\widehat{c}), h(\lceil [b](\widehat{a}) \rceil, \bar{g}) \Rightarrow h(\lceil [a](\widehat{c}) \rceil, \bar{g})$ Datisi

But $[b](\widehat{a}) \equiv [a](\widehat{b})$; thus also

$h(b,\widehat{c}), h(\lceil [a](\widehat{b}) \rceil, \bar{g}) \Rightarrow h(\lceil [a](\widehat{c}) \rceil, \bar{g})$ Darii

---

$b\ a\ {-c}\quad h(b, \sim\widehat{c}), h(\lceil [b](\widehat{a}) \rceil, \bar{g}) \Rightarrow h(\lceil [a](\sim\widehat{c}) \rceil, \bar{g}$ Ferison

But $[b](\widehat{a}) \equiv [a](\widehat{b})$; thus also

$h(b, \sim\widehat{c}), h(\lceil [a](\widehat{b}) \rceil, \bar{g}) \Rightarrow h(\lceil [a](\sim\widehat{c}) \rceil, \bar{g}$ Ferio

---

$b\ c\ a\quad h(b,\widehat{a}), h(\lceil [b](\widehat{c}) \rceil, \bar{g}) \Rightarrow h(\lceil [c](\widehat{a}) \rceil, \bar{g})$

But the order of premises is inessential
and moreover $[c](\widehat{a}) \equiv [a](\widehat{c})$; thus

$h([b](\widehat{c}), \bar{g}), h(b,\widehat{a}) \Rightarrow h([a](\widehat{c}), \bar{g})$ Disamis

---

$b\ {-c}\ a\quad h(b,\widehat{a}), h([b](\sim\widehat{c}), \bar{g}) \Rightarrow h([-c](\widehat{a}), \bar{g})$

But the order of premises is inessential
and moreover $[-c](\widehat{a}) \equiv [a](\sim\widehat{c})$; thus

$h([b](\sim\widehat{c}), \bar{g}), h(b,\widehat{a}) \Rightarrow h([a](\sim\widehat{c}), \bar{g})$ Bocardo

$a\ b\ c\quad h(a,\widehat{c}), h(\lceil [b](\sim\widehat{c}) \rceil, \bar{g}) \Rightarrow h(\lceil [a](\sim\widehat{b}) \rceil, \bar{g})$ 235.12

$c\ a\ {-b}\quad h(c, \sim\widehat{b}), h(\lceil [a](\sim\sim\widehat{b}) \rceil, \bar{g}) \Rightarrow h(\lceil [c](\sim\widehat{a}) \rceil, \bar{g})$

But $[a](\sim\sim\widehat{b}) \equiv [a](\widehat{b}) e [c](\sim\widehat{a}) \equiv [a](\sim\widehat{c})$; thus

$h(c, \sim\widehat{b}), h(\lceil [a](\widehat{b}) \rceil, \bar{g}) \Rightarrow h(\lceil [a](\sim\widehat{c}) \rceil, \bar{g})$ Festino

---

$c\ a\ b\quad h(c, \sim\widehat{b}), h(\lceil [a](\sim\widehat{b}) \rceil, \bar{g}) \Rightarrow h(\lceil [c](\sim\widehat{a}) \rceil, \bar{g})$

But $[c](\sim\widehat{a}) \equiv [a](\sim\widehat{c})$; thus

$h(c, \widehat{b}), h(\lceil [a](\sim\widehat{b}) \rceil, \bar{g}) \Rightarrow h(\lceil [a](\sim\widehat{c}) \rceil, \bar{g})$ Baroco

---

$c\ a\ {-b}\quad h(c, \sim\widehat{b}), h(\lceil [a](\sim\sim\widehat{b}) \rceil, \bar{g}) \Rightarrow h(\lceil [c](\sim\widehat{a}) \rceil, \bar{g})$

But $[a](\sim\sim\widehat{b}) \equiv [b](\widehat{a})$ and $[c](\sim\widehat{a}) \equiv [a](\sim\widehat{c})$; thus

$h(c, \sim\widehat{b}), h([b](\widehat{a}), \bar{g}) \Rightarrow h(\lceil [a](\sim\widehat{c}) \rceil, \bar{g})$ Fresison

5.7 Inferences from Premises Containing Width Determinations

After the discussion of the inferences from universal premises, existential premises, and negations of such propositions, Bolzano develops a remarkably detailed theory of those inferences whose premises contain width determinations. In particular, he addresses

(1) inferences with a unique singular premise [§239];
(2) inferences where a singular premise occurs jointly with
  (2a) a universal premise [§240];
  (2b) the negation of a universal proposition [§241];
  (2c) a premise of objectuality or of objectlessness [§242];
  (2d) another (or more) singular premise [§243];
(3) inferences with plural premises [§244];
(4) inferences with numerical premises [§245].

Ad (1) *A singular premise*

$$h(\lceil a \rceil, \bar{1}) \Rightarrow h(a, \widehat{c}) \boxed{\vee} h(a, \sim\widehat{c}) \qquad [239.1] \qquad \text{T5.150}$$

$$h(\lceil a \rceil, \bar{1}) \Rightarrow h(\lceil -a \rceil, \bar{+}) \qquad [239.1] \qquad \text{T5.151}$$

In this context, Bolzano makes explicit what could be called his '*Axiom of infinity*': "when the idea *A* has a unique object, then also the idea *Non A* has objects, in fact it must have infinitely many of them"[26]

$$h(\lceil a_1 \sqcap \ldots \sqcap a_n \rceil, \bar{1}) \Rightarrow (-a_1 \sqcup \ldots \sqcup -a_n) \boxed{\equiv} (e - (a_1 \sqcap \ldots \sqcap a_n)) \qquad [239.2] \qquad \text{T5.152}$$

(De Morgan)

Ad (2) *A singular premise with a universal premise*

$$h(\lceil a \rceil, \bar{1}), h(a, \widehat{c}) \Rightarrow h(\lceil [a](\widehat{c}) \rceil, \bar{1}) \qquad [240.1] \qquad \text{T5.153}$$

$$h(\lceil a \rceil, \bar{1}), h(a, \widehat{c}) \Rightarrow [a](\widehat{c}) \boxed{\equiv} a \qquad [240.1] \qquad \text{T5.154}$$

[26] TS §239, II 335: "1) [...] If the idea *A* has only one object, then the idea non-*A* is not objectless, indeed, it must have infinitely many objects."

$h(\lceil a \rceil, \bar{1}), h(c, \widehat{a}) \Rightarrow h(\lceil c \rceil, \bar{1})$	[240.2]	T5.155
$h(\lceil a \rceil, \bar{1}), h(c, \widehat{a}) \Rightarrow c \equiv a$	[240.2]	T5.156
$h(\lceil a \rceil, \bar{1}), h(-a, \widehat{c}), h(\lceil -c \rceil, \bar{g}) \Rightarrow h(\lceil -c \rceil, \bar{1})$	[240.3]	T5.157
$h(\lceil a \rceil, \bar{1}), h(-a, \widehat{c}), h(\lceil -c \rceil, \bar{g}) \Rightarrow -c \boxed{\equiv} a$	[240.3]	T5.158
$h(\lceil a \rceil, \bar{1}), h(c, \sim \widehat{a}) \Rightarrow h(\lceil [a](\sim \widehat{c}) \rceil, \bar{1})$	[240.4]	T5.159
$h(\lceil a \rceil, \bar{1}), h(c, \sim \widehat{a}) \Rightarrow [a](\sim \widehat{c}) \boxed{\equiv} a$	[240.4]	T5.160
$h(\lceil a \sqcap c \rceil, \bar{1}), h(a, \widehat{c}) \Rightarrow h(\lceil a \rceil, \bar{1})$	[240.5]	T5.161
$h(\lceil a \sqcap c \rceil, \bar{1}), h(a, \widehat{c}) \Rightarrow a \sqcap c \boxed{\equiv} a$	[240.5]	T5.162
$h(\lceil a \sqcap c \rceil, \bar{1}), h(-a, \widehat{c}) \Rightarrow$ the width of c is the width of $-a$ plus 1	[240.6]	T5.163
$h(\lceil a \sqcap c \rceil, \bar{1}), h(-a, \sim \widehat{c}) \Rightarrow h(\lceil c \rceil, \bar{1})$	[240.7]	T5.164
$h(\lceil a \sqcap c \rceil, \bar{1}), h(-a, \sim \widehat{c}) \Rightarrow (a \sqcap c) \boxed{\equiv} a$	[240.7]	T5.165
$h(\lceil a \sqcap c \rceil, \bar{1}), h(a, \widehat{b}) \Rightarrow h(\lceil a \sqcap c \sqcap b \rceil, \bar{1})$	[240.8]	T5.166
$h(\lceil a \sqcap c \rceil, \bar{1}), h(a, \widehat{b}) \Rightarrow (a \sqcap c \sqcap b) \boxed{\equiv} (a \sqcap c)$	[240.8]	T5.167
$h(\lceil a \sqcap c \rceil, \bar{1}), h(-a, \widehat{b}) \Rightarrow$ $h(\lceil c \sqcap b \rceil, \sim \bar{g}) \boxed{\vee} h(\lceil c \sqcap b \rceil, \bar{1})$	[240.9]	T5.168
$h(\lceil a \sqcap c \rceil, \bar{1}), h(-a, \widehat{b}), h(\lceil c \sqcap b \rceil, \bar{g}) \Rightarrow c \sqcap b \boxed{\equiv} a \sqcap c$	[240.9]	T5.169
$h(\lceil a \sqcap c \rceil, \bar{1}), h(b, \widehat{a}) \Rightarrow h(\lceil c \sqcap b \rceil, \bar{1})$	[240.10]	T5.170
$h(\lceil a \sqcap c \rceil, \bar{1}), h(b, \widehat{a}) \Rightarrow c \sqcap b \boxed{\equiv} a \sqcap c$	[240.10]	T5.171
$h(\lceil a \sqcap c \rceil, \bar{1}), h(b, \sim \widehat{a}) \Rightarrow h(\lceil a \sqcap c \sqcap -b \rceil, \bar{1})$	[240.11]	T5.172
$h(\lceil a \sqcap c \rceil, \bar{1}), h(b, \sim \widehat{a}) \Rightarrow (a \sqcap c \sqcap -b) \boxed{\equiv} (a \sqcap c)$	[240.11]	T5.173

Ad (2b) *A singular premise with a negated universal premise*

$h(\lceil a \rceil, \bar{1}), \boxed{\neg}\, h(a, \widehat{c}) \Rightarrow h(a, \sim \widehat{c})$	[241.1]	T5.174
$h(\lceil a \rceil, \bar{1}), \boxed{\neg}\, h(c, \sim \widehat{a}) \Rightarrow h(\lceil c \rceil, \sim \bar{g}) \boxed{\vee} h(a, \widehat{c})$	[241.2]	T5.175
$h(\lceil a \sqcap c \rceil, \bar{1}), \boxed{\neg}\, h(a, \widehat{c}) \Rightarrow h(\lceil a \rceil, \overline{>1})$	[241.3]	T5.176
$h(\lceil a \sqcap c \rceil, \bar{1}), \boxed{\neg}\, h(-a, \sim \widehat{c}) \Rightarrow (a \equiv e) \boxed{\vee} \boxed{\neg}\, h(\lceil c \rceil, \bar{1})$	[241.4]	T5.177

Ad (2c) *A singular premise with a premise of objectuality or of objectlessness*

$h(\lceil a \sqcap c \rceil, \bar{1}), h(\lceil -a \rceil, \sim \bar{g}) \Rightarrow h(\lceil c \rceil, \bar{1})$ [242.1] T5.178

$h(\lceil a \sqcap c \rceil, \bar{1}), h(\lceil -a \rceil, \sim \bar{g}) \Rightarrow (a \sqcap c) \boxed{\equiv} c$ [242.1] T5.179

$h(\lceil a \rceil, \bar{1}), h(\lceil a \sqcap c \rceil, \bar{g}) \Rightarrow h(\lceil a \sqcap c \rceil, \bar{1})$ [242.2] T5.180

$h(\lceil a \rceil, \bar{1}), h(\lceil a \sqcap c \rceil, \bar{g}) \Rightarrow (a \sqcap c) \boxed{\equiv} a$ [242.2] T5.181

$h(\lceil a \rceil, \bar{1}), h(\lceil a \sqcap c \rceil, \sim \bar{g}) \Rightarrow h(\lceil a \sqcap -c \rceil, \bar{1})$ [242.3] T5.182

$h(\lceil a \rceil, \bar{1}), h(\lceil a \sqcap c \rceil, \sim \bar{g}) \Rightarrow (a \sqcap -c) \boxed{\equiv} a$ [242.3] T5.183

$h(\lceil a \rceil, \bar{1}), h(\lceil -a \sqcap c \rceil, \sim \bar{g}) \Rightarrow$
$h(\lceil c \rceil, \sim \bar{g}) \boxed{\vee} h(\lceil c \rceil, \bar{1})$ [242.4] T5.184

$h(\lceil a \rceil, \bar{1}), h(\lceil -a \sqcap c) \rceil, \sim \bar{g}), h(\lceil c \rceil, \bar{g}) \Rightarrow c \boxed{\equiv} a$ [242.4] T5.185

$h(\lceil a \sqcap c) \rceil, \bar{1}), h(\lceil -a \sqcap c \rceil, \sim \bar{g}) \Rightarrow h(\lceil c \rceil, \bar{1})$ [242.5] T5.186

$h(\lceil a \sqcap c \rceil, \bar{1}), h(\lceil -a \sqcap c \rceil, \sim \bar{g}) \Rightarrow c \boxed{\equiv} a$ [242.5] T5.187

$h(\lceil a \sqcap c \rceil, \bar{1}), h(\lceil -a \sqcap -c \rceil, \sim \bar{g}) \Rightarrow$ T5.188

if $h(\lceil -a \rceil, \bar{g})$, then c contains a unique object more than $-\ a$ [242.6]

if $h(\lceil -c \rceil, \bar{g})$, then a contains a unique object more than $-\ c$ [242.6]

Ad (2d) *Two or more singular premises*

$h(\lceil a \rceil, \bar{1}), h(\lceil c \rceil, \bar{1}), a \perp c \Rightarrow h(\lceil a \sqcup c \rceil, \bar{2})$ [243.1] T5.189

$h(\lceil a_1 \rceil, \bar{1}), \ldots, h(\lceil a_n \rceil, \bar{1}), \perp\!\!\!\perp (\overset{n}{a}) \Rightarrow$
$h(\lceil a_1 \sqcup \ldots \sqcup a_n \rceil, \bar{n})$ [243.2] T5.190

$h(\lceil a_1 \rceil, \bar{1}), \ldots, h(\lceil a_n \rceil, \bar{1}) \Rightarrow$
$(\boxed{\wedge^{i<j\leq n}} h(\lceil a_i \sqcap a_j \rceil, \sim \bar{g})) \boxed{\rightarrow} h(\lceil a_1 \sqcup \ldots \sqcup a_n \rceil, \bar{n})$ [243.2] T5.191

$h(\lceil a \rceil, \bar{1}), h(\lceil -a \sqcap c \rceil, \bar{1}) \Rightarrow h(\lceil a \sqcup (-a \sqcap c) \rceil, \bar{2})$ [243.3] T5.192

$h(a \sqcap c, \bar{1}), h(\lceil -a \sqcap c \rceil, \bar{1}) \Rightarrow h(\lceil c \rceil, \bar{2})$ [243.4] T5.193

$$\begin{bmatrix} h(\lceil a \sqcap c_1 \rceil, \bar{1}), h(\lceil a \sqcap c_2 \sqcap -c_1 \rceil, \bar{1}), \\ h(\lceil a \sqcap c_3 \sqcap -c_1 \sqcap -c_2 \rceil, \bar{1}), \ldots, \\ h(\lceil a \sqcap c_n \sqcap -c_1 \sqcap \cdots \sqcap -c_{n-1} \rceil, \bar{1}) \end{bmatrix} \Rightarrow h\left(\left\lceil \begin{matrix} (a \sqcap c_1) \sqcup (a \sqcap c_2 \sqcap -c_1) \sqcup \\ (a \sqcap c_3 \sqcap -c_1 \sqcap -c_2) \sqcup \ldots \sqcup \\ (a \sqcap c_n \sqcap -c_1 \sqcap \cdots \sqcap -c_{n-1}) \end{matrix} \right\rceil, \bar{n}\right)$$

[243.5] T5.194

Ad (3) *Plural premises*

Let us make a notational convention:

$$h(\lceil a \rceil, \bar{+}) := h(\lceil a \rceil, \bar{g}) \boxed{\wedge} \boxed{\neg}\, h(\lceil a \rceil, \bar{\imath}) \qquad \text{MD1}$$

from which $W(h(\lceil a \rceil, \bar{+})) \leftrightarrow G(a) \wedge F(h(\lceil a \rceil, \bar{\imath}))$

$h(\lceil a \rceil, \bar{+}), h(a, \widehat{c}) \Rightarrow h(\lceil c \rceil, \bar{+})$ [244.1] T5.195

$h(\lceil a \rceil, \bar{+}), h(c, \sim \widehat{a}) \Rightarrow h(\lceil -c \rceil, \bar{+})$ [244.2] T5.196

$h(\lceil a \sqcap c \rceil, \bar{+}), h(a, \widehat{b}) \Rightarrow h(\lceil b \sqcap c \rceil, \bar{+})$ [244.3] T5.197

$h(\lceil a \sqcap c \rceil, \bar{+}), h(b, \sim \widehat{a}) \Rightarrow h(\lceil b \sqcap -c \rceil, \bar{+})$ [244.4] T5.198

$h(\lceil a \rceil, \bar{+}), h(\lceil a \sqcap c \rceil, \sim \bar{g}) \Rightarrow h(\lceil -c \rceil, \bar{+})$ [244.5] T5.199

$h(\lceil a \rceil, \bar{+}), h(\lceil a \sqcap c \rceil, \bar{\imath}) \Rightarrow h(\lceil a \sqcap -c \rceil, \bar{g})$ [244.6] T5.200

$h(\lceil a \sqcap c \rceil, \bar{+}), h(\lceil -a \sqcap c \rceil, \bar{\imath}) \Rightarrow$
c has an object more than $a \sqcap c$ [244.7] T5.201

$h(\lceil a \sqcap c \rceil, \bar{+}), h(\lceil -a \sqcap c \rceil, \bar{+}) \Rightarrow$
c has at least two objects more than $a \sqcap c$ [244.8] T5.202

Ad (4) *Numerical premises*

$h(\lceil a \rceil, \bar{n}), h(a, \widehat{c}) \Rightarrow h(\lceil c \rceil, \overline{\geq n})$ [245.1] T5.203

$h(\lceil a \rceil, \bar{n}), h(c, \widehat{a}) \Rightarrow h(\lceil c \rceil, \overline{\leq n})$ [245.2] T5.204

$h(\lceil a \rceil, \bar{n}), h(-a, \widehat{c}) \Rightarrow h(\lceil -c \rceil, \overline{\leq n})$ [245.3] T5.205

$h(\lceil a \rceil, \bar{n}), h(c, \sim \widehat{a}) \Rightarrow h(\lceil -c \rceil, \overline{\geq n})$ [245.4] T5.206

$h(\lceil a \sqcap c \rceil, \bar{n}), h(a, \widehat{c}) \Rightarrow h(\lceil a \rceil, \bar{n})$ [245.5] T5.207

$h(\lceil a \sqcap c \rceil, \bar{n}), h(-a, \widehat{c}) \Rightarrow$
the extension of *c* is that of $-a$ plus *n* objects [245.6] T5.208

$h(\lceil a \sqcap c \rceil, \bar{n}), h(-a, \sim \widehat{c}) \Rightarrow h(\lceil c \rceil, \bar{n})$ [245.7] T5.209

$h(\lceil a \sqcap c \rceil, \bar{n}), h(a, \widehat{b}) \Rightarrow h(\lceil b \sqcap c \rceil, \overline{\geq n})$ [245.8] T5.210

$h(\lceil a \sqcap c \rceil, \bar{n}), h(b, \sim \widehat{a}) \Rightarrow h(\lceil -b \sqcap c \rceil, \overline{\geq n})$ [245.9] T5.211

$h(\lceil a \sqcap c \rceil, \bar{n}), \boxed{\neg}\, h(a, \widehat{c}) \Rightarrow h(\lceil a \rceil, \overline{> n})$ [245.10] T5.212

$h(\lceil a \sqcap c \rceil, \bar{n}), \boxed{\neg}\, h(-a, \sim \widehat{c}) \Rightarrow$
$h(\lceil -a \rceil, \sim \bar{g}) \boxed{\vee}\, h(\lceil c \rceil, \overline{> n})$ [245.11] T5.213

$h(\lceil a \sqcap c \rceil, \bar{n}), h(\lceil -a \sqcap c \rceil, \bar{g}) \Rightarrow h(\lceil c \rceil, \overline{> n})$ [245.12] T5.214

$h(\lceil a \sqcap c \rceil, \bar{n}), h(\lceil -a \sqcap c \rceil, \sim \bar{g}) \Rightarrow h(\lceil c \rceil, \bar{n})$ [245.13] T5.215

$h(\lceil a \sqcap c \rceil, \bar{n}), h(\lceil -a \sqcap -c \rceil, \sim \bar{g}) \Rightarrow$ if $h(\lceil -a \rceil, \bar{g})$, then c is $-a$ plus n objects and if $h(\lceil -c \rceil, \bar{g})$, then a is $-c$ plus n objects	[245.14]	T5.216
$h(\lceil a \sqcap c \rceil, \bar{n}), h(\lceil a \sqcap b \rceil, \sim \bar{g}) \Rightarrow h(\lceil -b \sqcap c \rceil, \overline{\geq n})$	[245.15]	T5.217
$h(\lceil a \sqcap c \rceil, \bar{n}), h(\lceil -a \sqcap b \rceil, \sim \bar{g}) \Rightarrow h(\lceil b \sqcap c \rceil, \overline{\leq n})$	[245.16]	T5.218
$h(\lceil a \rceil, \overline{n+2}), h(\lceil [a](\hat{c}) \rceil, \bar{1}) \Rightarrow h(\lceil [a](\sim \hat{c}) \rceil, \overline{n+1})$	[245.17]	T5.219
$h(\lceil a \rceil, \bar{2}), h(\lceil [a](\hat{c}) \rceil, \bar{1}) \Rightarrow h(\lceil [a](\sim \hat{c}) \rceil, \bar{1})$	[245.17]	
$h(\lceil a \rceil, \bar{n}), h(\lceil [-a](\hat{c}) \rceil, \bar{1}) \Rightarrow h(\lceil a \sqcup [-a](\hat{c}) \rceil, \overline{n+1})$	[245.18]	T5.220
$h(\lceil a \rceil, \bar{n}), h(\lceil [a](\hat{c}) \rceil, \bar{m}) \Rightarrow$ $h(\lceil [a](\sim \hat{c}) \rceil, \overline{n-m})$ (if only $n - m > 1$)	[245.19]	T5.221
per $n - m = 1 \Rightarrow h(\lceil [a](\sim \hat{c}) \rceil, \bar{1})$	[245.19]	
$h(\lceil a \sqcap c \rceil, \bar{n}), h(\lceil -a \sqcap c \rceil, \bar{m}) \Rightarrow h(\lceil c \rceil, \overline{n+m})$	[245.20]	T5.222
$h(\lceil a \sqcap c \rceil, \bar{n}), h(\lceil -a \sqcap b \rceil, \bar{m}) \Rightarrow$ $h(\lceil (a \sqcap c) \sqcup (-a \sqcap b) \rceil, \overline{n+m})$	[245.21]	T5.223
$h(\lceil a_1 \rceil, \bar{n}_1), \ldots, h(\lceil a_s \rceil, \bar{n}_s) \Rightarrow$ $h(\lceil a_1 \sqcup \ldots \sqcup a_s \rceil, \overline{\geq max(n_1, \ldots, n_s)})$	[245.22]	T5.224
$h(\lceil a_1 \rceil, \bar{n}_1), \ldots, h(\lceil a_s \rceil, \bar{n}_s) \Rightarrow$ $h(\lceil a_1 \sqcup \ldots \sqcup a_s \rceil, \overline{\leq n_1 + \cdots + n_s})$	[245.22]	T5.225
$h(\lceil a_1 \rceil, \bar{n}_1), \ldots, h(\lceil a_s \rceil, \bar{n}_s)$, the a are two by two disjoint $\Rightarrow$		T5.226
$h(\lceil a_1 \sqcup \ldots \sqcup a_s \rceil, \overline{n_1 + \cdots + n_s})$	[245.23]	

5.8 Inferences from Premises Enunciating Relations among Ideas

The next topic concerns inferences from propositions enunciating [*aussagen*] *relations among ideas.* Bolzano's §246, which is devoted to this, draws on the classification examined in §140. Strictly speaking, the theorems we encounter henceforth are general schemata of theorems; in order to avoid a further burdening of an already rather overloaded text, however, we will neglect this detail and omit the S.

The equivalence between *compatibility of the* $\overset{s}{a}$ and *objectuality of* $[e](\widehat{a}_1 + \ldots + \widehat{a}_s)$, as well as that between their opposites, which were established at due time, renders a new treatment of these relations superfluous. [246.1-2]

As to the inferences from a proposition enunciating the *r-compatibility* of *s* ideas ($r < s$), by combinatorics we know how many and which combinations of *r* of them there are, and therefore the conclusions of the inferences having as a premise each of the propositions enunciating these relations of compatibility are also consequences of the initial proposition. Moreover, because if *r* ideas are compatible, then so are also those of any of their subcollections, the conclusions of the inferences having as premises the enunciations of these compatibilities are also conclusions drawn from the initial proposition.[27] [246.3]

Analogous is the discourse relative to propositions enunciating the *r-incompatibility*, when we recall that the incompatibility of *r* ideas implies that of any expansion of them.[28] [246.4]

As to the *inclusion*, after indicating the definition "every object that falls under one of the ideas *A, B, C, . . .* also falls under one of the ideas *M, N, O, . . .* ",

$$[\overset{m}{a} \sqsubseteq \overset{n}{c} \Leftrightarrow \bigwedge_{k\leq m} \forall i(i \mathrel{\text{E}} a_k \rightarrow \bigvee_{j\leq m} i \mathrel{\text{E}} c_j)],$$

Bolzano states that, from the proposition enunciating its subsistence,

> "derive, first of all, the following inferences, which taken together are on their part equivalent to the proposition itself:
>
> "Among the propositions: This *A* is *M*, This *A* is *N*, etc. there are always one or more true ones, whatever the idea substituted for This, if only the idea This *A* remains objectual.
>
> "Among the propositions: This *B* is *M*, This *B* is *N*, etc. there are always one or more true ones, whatever the idea substituted for This, if only the idea This *B* remains objectual.
> etc."

[27] TS §246, II 345: "3) If a proposition states that any *n* of the ideas *A, B, C, D, . . .* are compatible (§. 94.), then the science of combinatorics informs us how many and which combinations of *n* members of the given number of ideas *A, B, C, D, . . .* may be assembled; if we then state the compatibility of these combinations, we obtain just so many consequences of our proposition. Furthermore, since if every combination of *n* ideas is supposed to be compatible, and $n > 2$, then every $(n - 1)$ of them must also be compatible, and the propositions that state this compatibility are in any case correct consequences of our premise. And so on."

[28] TS §246, II 345: "4) Similar consequences may be derived from a proposition that states that any of the ideas *A, B, C, D, . . .* stand in the relation of incompatibility; only we must infer the incompatibility of larger, rather than smaller, combinations of ideas from the incompatibility of *n* of them."

Accepting *for a moment* that, as Bolzano thought,[29] for every object i, there is a singular idea referring to it and also *temporarily* agreeing to denote such an idea by '$\{i\}$' , Bolzano's idea could be formulated as follows:

$$\overset{m}{a} \sqsubseteq \overset{n}{c} \Leftrightarrow \bigwedge_{k \leq m} \forall i (W(h(\{i\}, \overline{a}_k) \rightarrow \bigvee_{j \leq n} W(h(\{i\}, \overline{c}_j))) \qquad [246.5]$$

The examination of further conclusions that can be obtained from these inference is deferred by Bolzano to the subsequent sections devoted to inferences from enunciations of relations among propositions.

Having preliminarily remarked that the *equivalence* is only a reciprocal inclusion, so that from a proposition that enunciates it, the inferences just considered for inclusions are clearly derivable, Bolzano devotes particular attention to equivalences between *single ideas* and maintains their substitutivity:

$$a \equiv c \Rightarrow (W(p) \leftrightarrow W(p[a/c])) \qquad \text{T5.227}$$

$$\overline{a} \equiv \overline{c} \Rightarrow (W(p) \leftrightarrow W(p[\overline{a}/\overline{c}])) \qquad [246.6]$$

From propositions enunciating the subsistence of a *subordination,* that is from propositions reducible to the form "The totality of the objects which fall under the ideas $A, B, C, \dots$ is a part (of course, proper) of the totality of the objects which fall under the ideas $M, N, O, \dots$", we obtain the following inferences:

> "The totality of the A is only a part of the totality of the $M, N, O, \dots$
> The totality of the B is only a part of the totality of the $M, N, O, \dots$
> etc."

We can summarize this as follows:

$$\overset{m}{a} \mathrel{\dot{\sqsubseteq}} \overset{n}{c} \Rightarrow \bigwedge_{i \leq m} (a_i \mathrel{\dot{\sqsubseteq}} c_1 \sqcup \dots \sqcup c_n) \qquad [246.7] \quad \text{T5.228}$$

However, apparently in error, Bolzano affirms that those inferences "have again the peculiarity that taken together are equivalent to it"; i.e. he states that

$$\overset{m}{a} \mathrel{\dot{\sqsubseteq}} \overset{n}{c} \Leftrightarrow \bigwedge_{i \leq m} (a_i \mathrel{\dot{\sqsubseteq}} c_1 \sqcup \dots \sqcup c_n) \qquad [246.7]$$

[29] Cf. Cap 4.

Still erroneously, he also affirms: "If, however, none of the ideas $M, N, O, \ldots$ has the widest extension, or if those which have such an extension are omitted, the following proposition also holds: 'the totality of the objects falling under the ideas **Non**M, **Non**N, **Non**$O, \ldots$ is only a part of the totality of the objects falling under the ideas **Non**A, **Non**B, **Non** $C, \ldots$' "; i.e.

$$\bigwedge_{j\leq n}\neg(c_j \equiv e) \rightarrow (\overset{m}{a} \mathrel{\dot{\sqsubseteq}} \overset{n}{c} \Rightarrow -\overset{n}{c} \sqsubseteq -\overset{m}{a}) \qquad [246.7]$$

With regard to *concatenation*—or better, as rightly pointed out by J. Berg in his edition of the WL, the *omnilateral concatenation*—Bolzano, after recalling the equivalence

$$Omnicon(\overset{m}{a}) \leftrightarrow \bigwedge_{i\leq m}([e](\overset{m}{\bar{a}}) \mathrel{\dot{\sqsubseteq}} [e](\bar{a}_1 + \cdots + \bar{a}_{i-1} + \bar{a}_{i+1} + \cdots + \bar{a}_m))$$

and hence the inference, for every i,

$$Omnicon(\overset{m}{a}) \rightarrow [e](\overset{m}{\bar{a}}) \mathrel{\dot{\sqsubseteq}} [e](\bar{a}_1 + \cdots + \bar{a}_{i-1} + \bar{a}_{i+1} + \cdots + \bar{a}_m)) \qquad [246.8] \quad \text{T5.229}$$

also presents, always for every i, the inference

$$Omnicon(\overset{m}{a}) \rightarrow G([e](-\bar{a}_1 + \cdots + -\bar{a}_{i-1} + ai + -\bar{a}_{i+1} + \cdots + -\bar{a}_m)) \qquad [246.8] \quad \text{T5.230}$$

as well as, always for every i, the inference

$$Omnicon(\overset{m}{a}) \rightarrow \neg[e](\bar{a}_i) \equiv e \qquad [246.8] \quad \text{T5.231}$$

Finally,

$$Omnicon(\overset{m}{a}) \rightarrow Omnicon(a_1, \ldots, a_{i-1}, a_{i+1}, \ldots, a_m) \qquad [246.8] \quad \text{T5.232}$$

Regarding the *contradictoriness* between the ideas $\overset{m}{a}$ and the ideas $\overset{n}{c}$, Bolzano preliminarily remarks that the proposition enunciating it is equivalent to the conjunction of the following three propositions: (a) each one of the ideas $\overset{m}{a}, \overset{n}{c}$ is objectual; (b) no idea obtained by conjoining any one of the $\overset{m}{a}$ with anyone of the $\overset{n}{c}$ is objectual; (c) even the idea obtained by conjoining the negations of all the $\overset{m}{a}$ and all the $\overset{n}{c}$ is objectual; i.e.

$$\overset{m}{a} \mathbin{+\!\!\!+} \overset{n}{c} \leftrightarrow \left(\begin{array}{l} (\bigwedge_{i\leq m} G(a_i) \wedge \bigwedge_{j\leq n} G(c_j)) \;\wedge\; (\bigwedge_{i\leq m}(\bigwedge_{j\leq n} \neg G(a_i \sqcap c_j))) \\ \wedge\; \neg G(-a_1 \sqcap \cdots \sqcap -a_m \sqcap -c_1 \sqcap \cdots \sqcap -c_n) \end{array}\right) \qquad \text{T5.233}$$

[246.9]

After having recalled that every inference that may be obtained from one or more of these three propositions determines, of course, new inferences from the original proposition, Bolzano reminds us, in particular, that the negation of one, or the conjunction of more or even all $\overset{m}{a}$ (or, symmetrically, of the $\overset{n}{c}$) is always objectual; every object that does not fall under any of the $\overset{m}{a}$ falls under at least one of the $\overset{n}{c}$ and, symmetrically, every object not falling under any $\overset{n}{c}$ falls under at least one of the $\overset{n}{c}$; i.e.

$$\overset{m}{a}, \overset{m'}{b} \text{+\!\!\!+} \overset{n}{c} \rightarrow G(-b_1 \sqcap \cdots \sqcap -b_{m'}) \qquad [246.9] \quad T5.234$$

$$\overset{m}{a} \text{+\!\!\!+} \overset{n}{c}, \overset{n'}{d} \rightarrow G(-d_1 \sqcap \cdots \sqcap -d_{n'}) \qquad [246.9] \quad T5.235$$

$$\overset{m}{a} \text{+\!\!\!+} \overset{n}{c} \rightarrow \forall k(\neg(k \sqsubset a_1 \sqcup \cdots \sqcup a_m) \rightarrow (k \sqsubset c_1 \sqcup \cdots \sqcup c_n)) \quad [246.9] \quad T5.236$$

$$\overset{m}{a} \text{+\!\!\!+} \overset{n}{c} \rightarrow \forall k(\neg(k \sqsubset c_1 \sqcup \cdots \sqcup c_n) \rightarrow (k \sqsubset a_1 \sqcup \cdots \sqcup a_m)) \quad [246.9] \quad T5.237$$

At the end, Bolzano turns to inferences from a premise given by a proposition enunciating the subsistence of the relation of *exhaustive coordination* (*partition*) among certain $\overset{m}{a}$ and a c and he provides this new formulation of this proposition:

> "Among the propositions 'This c is an a_1', 'This c is an a_2', . . . , 'This c is an a_m' there always is one and only one true."

He refers to §252 for a discussion of inferences from this type of proposition. This new formulation could be presented in writing, as has already been done in the case of inclusion,

$$Part(\overset{m}{a}, c) \leftrightarrow \forall k W(h(\{k\} \sqcap c, \widehat{a_1}), \ldots, h(\{k\} \sqcap c, \widehat{a_m})) \qquad [246.10]$$

5.9 Inferences from Premises Enunciating Relations among Propositions

5.9.1 Compatibility

The next section [§247] brings us to inferences from propositions that enunciate [*aussprechen*] *relations* among propositions. Concerning the premises that enunciate the subsistence of a relation of compatibility, with respect to certain ideas, among certain propositions, it will be possible to immediately conclude (1) the compatibility, with respect to the same ideas, of every subcollection of those propositions; (2) the non-formal validity of all of them; (3) the compatibility among the same propositions with respect to subcollections of those ideas; i.e.[30]

$$\underset{\substack{s\\a}}{\curlywedge}(\overset{n}{p},\overset{m}{q}) \rightarrow \underset{\substack{s\\a}}{\curlywedge}(\overset{m}{q}) \qquad \text{A} \qquad [247] \qquad \text{T5.238}$$

$$\underset{\substack{s\\a}}{\boxed{\curlywedge}}(\overset{n}{p},\overset{m}{q}) \Rightarrow \underset{\substack{s\\a}}{\boxed{\curlywedge}}(\overset{m}{q})$$

$$W(\boxed{\underset{\substack{s\\a}}{\curlywedge}}(\overset{n}{p},\overset{m}{q})\boxed{\underset{\substack{n\,m\\p,q}}{\Rightarrow}}\boxed{\underset{\substack{s\\a}}{\curlywedge}}(\overset{m}{q})) \qquad \text{C}$$

$$\underset{\substack{n\,m\\p,q}}{\mathbb{W}}(\underset{\substack{s\\a}}{\boxed{\curlywedge}}(\overset{n}{p},\overset{m}{q}) \boxed{\rightarrow} (\underset{\substack{s\\a}}{\boxed{\curlywedge}}(\overset{m}{q}))$$

$$\underset{\substack{s\\a}}{\curlywedge}(\overset{n}{p}) \rightarrow \bigwedge_{i\leq n}\neg\underset{\substack{s\\a}}{\mathbb{F}}(p_i) \qquad [247] \qquad \text{T5.239}$$

$$\underset{\substack{sr\\ac}}{\curlywedge}(\overset{n}{p}) \rightarrow \underset{\substack{r\\c}}{\curlywedge}(\overset{n}{p}) \qquad \text{A} \qquad [247] \qquad \text{T5.240}$$

$$\underset{\substack{sr\\ac}}{\boxed{\curlywedge}}(\overset{n}{p}) \Rightarrow \underset{\substack{r\\c}}{\boxed{\curlywedge}}(\overset{n}{p})$$

$$W(\underset{\substack{sr\\ac}}{\boxed{\curlywedge}}(\overset{n}{p})\underset{\substack{n\\p}}{\boxed{\rightarrow}}\underset{\substack{r\\c}}{\boxed{\curlywedge}}(\overset{n}{p})) \qquad \text{C}$$

Inferences from propositions enunciating the incompatibility among certain propositions are obviously not taken into consideration, since the precondition for the derivability of a proposition is the compatibility of these propositions.

5.9.2 Derivability

[§§248–9] In the far more interesting [*viel merkwürdiger*] case of inferences from enunciations of derivability, Bolzano begins by recalling the distinction (which has already been reviewed) between the two possible types of derivability relations: *relative* to certain ideas and *absolute*. However, taking into account that conclusions deriving from a proposition enunciating a relation of the second type are among those deriving from a proposition enunciating a relation of the first type, since they are themselves derivable from this last proposition, Bolzano states that he will limit himself to the first type (i.e. to those of the form $\overset{m}{p} \underset{\substack{s\\a}}{\Rightarrow} \overset{n}{q}$).

The first recalled inference concerns the obvious fact that derivability implies compatibility among premises and conclusions:

$$(\overset{m}{p} \underset{\substack{s\\a}}{\Rightarrow} \overset{n}{q}) \rightarrow \underset{\substack{s\\a}}{\curlywedge}(\overset{m}{p},\overset{n}{q}) \qquad \text{A} \qquad [248.1] \qquad \text{T5.241}$$

[30] Cf. ST3.11–13.

$$(\overset{m}{p} \boxed{\overset{\Rightarrow}{\overset{s}{a}}} \overset{n}{q}) \Rightarrow \underset{\overset{s}{a}}{\boxed{\wedge\!\!\wedge}}(\overset{m}{p}, \overset{n}{q}) \qquad \text{B}$$

$$W((\overset{m}{p} \boxed{\overset{\Rightarrow}{\overset{s}{a}}} \overset{n}{q} \underset{\overset{n}{p},\overset{m}{q}}{\boxed{\rightarrow}} \underset{\overset{s}{a}}{\boxed{\wedge\!\!\wedge}}(\overset{m}{p}, \overset{n}{q})) \qquad \text{C.}$$

After this, Bolzano remarks that if every substitution that makes all the $\overset{m}{p}$ true also makes all the $\overset{n}{q}$ true, then every substitution that makes some $\overset{n}{q}$ false also makes some $\overset{m}{p}$ false; i.e.

$$\forall\overset{s}{c}(W(\overset{m}{p}[\overset{s}{a}/\overset{s}{c}]) \rightarrow W(\overset{n}{q}[\overset{s}{a}/\overset{s}{c}])) \rightarrow \forall\overset{s}{c}(\neg W(\overset{n}{q}[\overset{s}{a}/\overset{s}{c}] \rightarrow \neg W(\overset{m}{p}[\overset{s}{a}/\overset{s}{c}]))$$

Indeed, obviously

$$\forall\overset{s}{c}(W(\overset{m}{p}[\overset{s}{a}/\overset{s}{c}]) \rightarrow W(\overset{n}{q}[\overset{s}{a}/\overset{s}{c}])) \rightarrow \forall\overset{s}{c}\neg(W(\overset{n}{q}[\overset{s}{a}/\overset{s}{c}] \wedge \neg W(\overset{m}{p}[\overset{s}{a}/\overset{s}{c}]))$$

But

$$\forall\overset{s}{c}\neg(W(\overset{n}{q}[\overset{s}{a}/\overset{s}{c}] \wedge \neg W(\overset{m}{p}[\overset{s}{a}/\overset{s}{c}])) \leftrightarrow \forall\overset{s}{c}\neg(W(\overset{m}{p}[\overset{s}{a}/\overset{s}{c}]) \wedge E(\boxed{-}\overset{n}{q}[\overset{s}{a}/\overset{s}{c}]))$$

and hence

$$\forall\overset{s}{c}(W(\overset{m}{p}[\overset{s}{a}/\overset{s}{c}]) \rightarrow W(\overset{n}{q}[\overset{s}{a}/\overset{s}{c}])) \rightarrow \forall\overset{s}{c}\neg(W(\overset{m}{p}[\overset{s}{a}/\overset{s}{c}]) \wedge E(\boxed{-}\overset{n}{q}[\overset{s}{a}/\overset{s}{c}]))$$

A fortiori,

$$\exists\overset{s}{c}W(\overset{m}{p}[\overset{s}{a}/\overset{s}{c}]) \wedge \forall\overset{s}{c}(W(\overset{m}{p}[\overset{s}{a}/\overset{s}{c}]) \rightarrow W(\overset{n}{q}[\overset{s}{a}/\overset{s}{c}])) \rightarrow \forall\overset{s}{c}\neg(W(\overset{m}{p}[\overset{s}{a}/\overset{s}{c}]) \wedge E(\boxed{-}\overset{n}{q}[\overset{s}{a}/\overset{s}{c}]))$$

i.e.

$$\overset{m}{p} \underset{\overset{s}{a}}{\Rightarrow} \overset{n}{q} \rightarrow \neg\underset{\overset{s}{a}}{\wedge\!\!\wedge}(\overset{m}{p}, \boxed{\vee^{n}}(\boxed{-}\,\overset{n}{q}))$$

Even better, using henceforth a horizontal line—the Bolzanian notational convention—to separate the premises of an inference from its conclusion,

$$\frac{\overset{m}{p} \underset{\overset{s}{a}}{\Rightarrow} \overset{n}{q}}{\neg\underset{\overset{s}{a}}{\wedge\!\!\wedge}(\overset{m}{p}, \boxed{\vee^{n}}(\boxed{-}\,\overset{n}{q}))} \qquad [248.2] \qquad \text{T5.242}$$

In general, the converse does not hold, because it does not follow from the conclusion that the $\overset{m}{p}$ are compatible among them; it does, however, hold under the additional hypothesis that it makes all the $\overset{m}{p}$ true. Indeed,

$$\forall\overset{s}{c}(W(\overset{m}{p}[\overset{s}{a}/\overset{s}{c}]) \wedge (\neg(W(\overset{m}{p}[\overset{s}{a}/\overset{s}{c}]) \wedge E(\boxed{-}\overset{n}{q}[\overset{s}{a}/\overset{s}{c}])) \rightarrow (W(\overset{m}{p}[\overset{s}{a}/\overset{s}{c}]) \rightarrow W(\overset{n}{q}[\overset{s}{a}/\overset{s}{c}])))$$

so

$$\frac{\neg \underset{\overset{s}{a}}{\bigwedge}(\overset{m}{p}, \boxed{\vee^n}(\boxed{\neg}\,\overset{n}{q}), \underset{\overset{s}{a}}{\bigwedge}(\overset{m}{p}))}{\overset{m}{p} \underset{\overset{s}{a}}{\Rightarrow} \overset{n}{q}} \qquad [248.2] \qquad \text{T5.243}$$

Even less valid, naturally, is that if certain propositions derive from certain other propositions, then from the latter also the negation of one (or more) of the former follows; i.e.

$$\frac{\overset{m}{p} \underset{\overset{s}{a}}{\Rightarrow} \overset{n}{q}, r}{\neg(\overset{m}{p} \underset{\overset{s}{a}}{\Rightarrow} \overset{n}{q}, \boxed{\neg}\, r)} \qquad [248.3] \qquad \text{T5.244}$$

In fact, under the same hypothesis, no substitution that makes the negation of one of the consequences true can make all the premises true:

$$\frac{\overset{m}{p} \underset{\overset{s}{a}}{\Rightarrow} \overset{n}{q}, r}{\neg \underset{\overset{s}{a}}{\bigwedge}(\overset{m}{p}, \overset{n}{q}, \boxed{\neg}\, r)} \qquad [248.4] \qquad \text{T5.245}$$

This inference is further strengthened by establishing that every substitution that makes the premises true transforms the set obtained from the set of conclusions by negating k of them in a set that contains exactly k false propositions. The formalization of this becomes uselessly complicated; if one wanted too, one could introduce a formal connective corresponding to the relation that subsists between certain $\overset{m}{p}$ and certain $\overset{s}{a}$ when every substitution for the latter falsifies exactly k of the former in Chapter 3.11. If the result of this work were similar to a connective $\boxed{*k}$, one could write

$$\frac{\overset{m}{p}, \overset{m'}{r} \underset{\overset{s}{a}}{\Rightarrow} \overset{n}{q}, \overset{n'}{s}}{\overset{m}{p}, \overset{m'}{r} \underset{\overset{s}{a}}{\Rightarrow} \boxed{*n'}(\overset{n}{q}, \boxed{\neg}\, \overset{n'}{s})} \qquad [248.5] \qquad \text{T5.246}$$

Furthermore, "either there is no aggregate of ideas that, substituted for the $i, j, \ldots$ makes some of the propositions $M, N, O, \ldots$ false, or that aggregate that doesn't make all of them true, doesn't even make all the $A, B, C, D, \ldots$ true":

$$\frac{\overset{m}{p} \underset{\overset{s}{a}}{\Rightarrow} \overset{n}{q}}{\underset{\overset{s}{a}}{\mathbb{W}}(\overset{n}{q}) \vee (\boxed{\vee^n}(\boxed{\neg}\, \overset{n}{q}) \underset{\overset{s}{a}}{\Rightarrow} (\boxed{\vee^m}(\boxed{\neg}\, \overset{m}{p})} \qquad [248.6] \qquad \text{T5.247}$$

At this point, Bolzano addresses the special cases in which the derivability relation concerns only two propositions, and, in the most general case, when the two propositions have no variable ideas in common, he establishes

$$\frac{h(a,\widehat{b}) \underset{a,c,b,d}{\Rightarrow} h(c,\widehat{d})}{\mathrm{T}(a,b) \rightarrow [e](\widehat{a}+\widehat{c}) \sqsubseteq [e](\widehat{b}+\widehat{d})} \qquad \text{T5.248}$$

$$\frac{h(a,\widehat{b}) \underset{a,c,b,d}{\Rightarrow} h(c,\widehat{d})}{b \perp d \rightarrow \perp (a,b)} \qquad [248.7] \qquad \text{T5.249}$$

whereas in the special cases in which they have the subject, respectively the predicate, in common, he states that

$$\frac{h(a,\widehat{b}) \underset{a,b,d}{\Rightarrow} h(a,\widehat{d})}{h(b,\widehat{d})} \qquad [248.8] \qquad \text{T5.250}$$

and, respectively,

$$\frac{h(a,\widehat{b}) \underset{a,c,b}{\Rightarrow} h(c,\widehat{b})}{h(c,\widehat{a})} \qquad [248.9] \qquad \text{T5.251}$$

The justification he provides is, in the first case, "indeed, as a is arbitrary, it may also become $= b$, where the truth of the inference becomes evident", whereas in the second case "indeed, as b is arbitrary, it may also become $= a$, where it thus appears that the inference must necessarily sound so."

Bolzano's §248 ends with the examination of the possible inferences starting from an assertion of *non*-derivability. As the derivability of the $\overset{n}{q}$ from the $\overset{m}{p}$ presupposes the compatibility of the latter, the non-subsistence between them of the relation of derivability allows us to infer that either the $\overset{m}{p}$ are not compatible or, if they are, there is at least a substitution that makes all $\overset{m}{p}$ true but falsifies at least one of the $\overset{n}{q}$:

$$\neg(\overset{m}{p} \underset{\overset{s}{a}}{\Rightarrow} \overset{n}{q}) \rightarrow \neg \underset{\overset{s}{a}}{\curlywedge} (\overset{m}{p}) \vee (\underset{\overset{s}{a}}{\curlywedge\!\!\curlywedge} (\overset{m}{p}) \wedge \exists \overset{s}{c}(W(\overset{m}{p}) \wedge \neg W(\overset{n}{q}))$$

i.e., equivalently,

$$\neg(\overset{m}{p} \underset{\overset{s}{a}}{\Rightarrow} \overset{n}{q}) \rightarrow \neg \underset{\overset{s}{a}}{\curlywedge} (\overset{m}{p}) \vee \exists \overset{s}{c}(W(\overset{m}{p}) \wedge \neg W(\overset{n}{q}))$$

that is to say,

$$\neg(\overset{m}{p} \underset{\underset{a}{s}}{\Rightarrow} \overset{n}{q}) \rightarrow \neg \underset{\underset{a}{s}}{\bigwedge}(\overset{m}{p}) \vee \exists \overset{s}{c}(W(\overset{m}{p}) \wedge E(\boxed{\neg}\, \overset{n}{q}))$$

again

$$\neg(\overset{m}{p} \underset{\underset{a}{s}}{\Rightarrow} \overset{n}{q}) \rightarrow \neg \underset{\underset{a}{s}}{\bigwedge} (\overset{m}{p}) \vee \exists \overset{s}{c}(W(\overset{m}{p}) \wedge W(\boxed{\vee^n}(\boxed{\neg}\, q)))$$

and finally

$$\neg(\overset{m}{p} \underset{\underset{a}{s}}{\Rightarrow} \overset{n}{q}) \rightarrow \neg \underset{\underset{a}{s}}{\bigwedge} (\overset{m}{p}) \vee \underset{\underset{a}{s}}{\bigwedge} (\overset{m}{p}, \boxed{\vee^n}(\boxed{\neg}\, \overset{n}{q}))$$

Hence

$$\frac{\neg(\overset{m}{p} \underset{\underset{a}{s}}{\Rightarrow} \overset{n}{q})}{\neg \underset{\underset{a}{s}}{\bigwedge} (\overset{m}{p}) \vee \underset{\underset{a}{s}}{\bigwedge} (\overset{m}{p}, \boxed{\vee^n}(\boxed{\neg}\, \overset{n}{q}))} \qquad [248.10] \qquad \text{T5.252}$$

More generally, the negation of a relation of derivability between the $\overset{m}{p}$ and the $\overset{n}{q}$ allows us to infer that either the $\overset{m}{p}$ and certainly, let us say, the first r among the $\overset{n}{q}$ are incompatible, or that they are compatible but one of the substitutions that makes all of them true falsifies at least one of the remaining $q_{r+1}, \ldots, q_n$:

$$\neg(\overset{m}{p} \underset{\underset{a}{s}}{\Rightarrow} \overset{n}{q}) \rightarrow \neg \underset{\underset{a}{s}}{\bigwedge} (\overset{m}{p}, q_1, \ldots, q_r) \vee (\exists \overset{s}{c}(W(\overset{m}{p}, q_1, \ldots, q_r) \wedge \neg W(q_{r+1}, \ldots, q_n))$$

i.e.

$$\frac{\neg(\overset{m}{p} \underset{\underset{a}{s}}{\Rightarrow} \overset{n}{q}, \overset{t}{r})}{\neg \underset{\underset{a}{s}}{\bigwedge} (\overset{m}{p}, \overset{n}{q}) \vee \underset{\underset{a}{s}}{\bigwedge} (\overset{m}{p}, \overset{n}{q}, \boxed{\vee^n}(\boxed{\neg}\, \overset{t}{r})} \qquad [248.10] \qquad \text{T5.253}$$

Remark. The rules (and their applications) are included in this type of inferences (quite generally, apart from the form of the involved propositions).

The four inferences opening §249 yield two possible interpretations: a purely propositional one and a more in-depth interpretation.

According to the first, the four inferences are

$$\frac{W(\overset{m}{p}) \rightarrow W(\overset{n}{q}),\ W(\overset{m}{p})}{W(\overset{n}{q})} \qquad [249.1]$$

$$\frac{W(\overset{m}{p}) \rightarrow W(\overset{n}{q}, q),\ F(q)}{\neg W(\overset{m}{p})} \qquad [249.2]$$

$$\frac{W(\overset{m}{p}, \overset{s}{r}) \rightarrow W(\overset{n}{q}, q), W(\overset{m}{p}, \neg q)}{\neg W(\overset{s}{r})} \qquad [249.3a]$$

$$\frac{W(\overset{m}{p}, p) \rightarrow W(\overset{n}{q}, q), W(\overset{m}{p}, \neg q)}{F(p)} \qquad [249.3b]$$

Possibly, however, what Bolzano meant when he stated: "if *A*, *B*, *C*, . . . are true, then *M*, *N*, *O*, . . . are true" was in reference to a particular substitution, and hence the four inferences could be intended as follows.

The first, obviously, is

$$\frac{\overset{m}{p} \underset{\overset{s}{a}}{\Rightarrow} \overset{n}{q},\ W(\overset{m}{p}[\overset{s}{a}/\overset{s}{c}])}{W(\overset{n}{q}[\overset{s}{a}/\overset{s}{c}])} \qquad \text{T5.254}$$

Second, if one of the consequences of certain premises is false with respect to a certain substitution, then not all premises are true with respect to this substitution:

$$\frac{\overset{m}{p} \underset{\overset{s}{a}}{\Rightarrow} \overset{n}{q},\ F(q_i[\overset{s}{a}/\overset{s}{c}])}{\neg W(\overset{m}{p}[\overset{s}{a}/\overset{s}{c}])} \qquad [249.2] \qquad \text{T5.255}$$

Third, if for a certain substitution some premises are compatible with the negation of one of the consequences, then the remaining premises are not all true in this substitution:

$$\frac{\overset{m}{p}, \overset{t}{r} \underset{\overset{s}{a}}{\Rightarrow} \overset{n}{q},\ W(\overset{t}{r}[\overset{s}{a}/\overset{s}{c}], \boxminus\ q_i[\overset{s}{a}/\overset{s}{c}])}{\neg W(\overset{m}{p}[\overset{s}{a}/\overset{s}{c}])} \qquad [249.3a] \qquad \text{T5.256}$$

Fourth, if for a certain substitution all but one of the premises are compatible with the negation of one of the consequences, then that premise is false in that substitution:

$$\frac{\overset{m}{p}, p \underset{\overset{s}{a}}{\Rightarrow} \overset{n}{q}, q), W(\overset{m}{p}[\overset{s}{a}/\overset{s}{c}]), \boxminus\ q)[\overset{s}{a}/\overset{s}{c}])}{F(p[\overset{s}{a}/\overset{s}{c}]))} \qquad [249.3b] \qquad \text{T5.257}$$

Bolzano continues with inferences in which the 'variation' is explicitly mentioned.

The consequences of premises compatible among them are consequences of the union of the two systems of premises (naturally, with respect to one and the same system of variables):

$$\frac{\overset{m}{p} \underset{\overset{s}{a}}{\Rightarrow} \overset{n}{q},\ \overset{m'}{r} \underset{\overset{s}{a}}{\Rightarrow} \overset{n'}{s},\ \underset{\overset{s}{a}}{\mathbb{A}}(\overset{m}{p}, \overset{m'}{r})}{\overset{m}{p}, \overset{m'}{r} \underset{\overset{s}{a}}{\Rightarrow} \overset{n}{q}, \overset{n'}{s}} \qquad [249.4a] \qquad T5.258$$

If some premises are always true with respect to the substitution of certain ones among the variables with respect to which certain propositions are derivable from those premises, then the same propositions are already derivable with respect to those variable from the remaining premises:

$$\frac{\overset{m}{p}, \overset{t}{r} \underset{\overset{sr}{ac}}{\Rightarrow} \overset{n}{q},\ \underset{\overset{s}{a}}{\mathbb{W}}(\overset{m}{p})}{\overset{t}{r} \underset{\overset{s}{a}}{\Rightarrow} \overset{n}{q}} \qquad [249.4b] \qquad T5.259$$

If the consequences of two systems of premises are incompatible among them, then their premises are also incompatible:

$$\frac{\overset{m}{p} \underset{\overset{s}{a}}{\Rightarrow} \overset{n}{q},\ \overset{m'}{r} \underset{\overset{s}{a}}{\Rightarrow} \overset{n'}{s},\ \neg\underset{\overset{s}{a}}{\mathbb{A}}(\overset{n}{q}, \overset{n'}{s})}{\neg\underset{\overset{s}{a}}{\mathbb{A}}(\overset{m}{p}, \overset{m'}{r})} \qquad [249.4c] \qquad T5.260$$

If from each one of the $\overset{n}{p}$ (always with respect to the same $\overset{s}{a}$) the negation of each one of the remaining is derivable, then in every substitution for the $\overset{s}{a}$, one and only one of the $\overset{n}{p}$ is true:

$$\frac{\bigwedge_{i\leq n}(p_i \underset{\overset{s}{a}}{\Rightarrow} \bigwedge_{j\neq i\leq n} W(\boxed{\neg}\, p_j))}{\underset{\overset{s}{a}}{\mathbb{W}}(\boxed{\sqcup^n}(\overset{n}{p}))} \qquad [249.4d] \qquad T5.261$$

5.9.3 *Equivalence*

Bolzano's §250 examines inferences from propositions that enunciate a relation of equivalence. First of all, there are the obvious inferences

$$\frac{\overset{m}{p} \underset{\overset{s}{a}}{\Leftrightarrow} \overset{n}{q}}{\overset{m}{p} \underset{\overset{s}{a}}{\Rightarrow} \overset{n}{q}} \qquad \text{T5.262} \qquad\qquad \frac{\overset{m}{p} \underset{\overset{s}{a}}{\Leftrightarrow} \overset{n}{q}}{\overset{n}{q} \underset{\overset{s}{a}}{\Rightarrow} \overset{m}{p}}; \qquad \text{T5.263}$$

and, by applying a cut, also all the conclusions we have already seen from premises enunciating relations of derivability. [250.1]

Then, an inference is stated that is difficult to express symbolically in our framework: from the equivalence, with respect to the $\overset{s}{a}$, between the $\overset{m}{p}$ and the $\overset{n}{q}$ "it also follows that the idea of certain $\overset{s}{b}$ which substituted for the $\overset{s}{a}$ make all the $\overset{m}{p}$ true and the idea of certain $\overset{s}{c}$ which substituted for the $\overset{s}{a}$ make all the $\overset{n}{q}$ true are equivalent".[31] [250.2]

From the same equivalence, it also follows that either all the $\overset{n}{p}$ and the $\overset{m}{q}$ are true or at least one of the $\overset{n}{p}$ and one of the $\overset{m}{q}$ are false:

$$\frac{\overset{m}{p} \underset{\overset{s}{a}}{\Leftrightarrow} \overset{n}{q}}{\neg W(\overset{m}{p}, \overset{n}{q}) \underset{\overset{s}{a}}{\Rightarrow} \neg W(\overset{m}{p}) \wedge \neg W(\overset{n}{q})} \qquad [250.3] \qquad \text{T5.264}$$

In the particular case in which the equivalence only concerns two propositions, from their equivalence with respect to certain ideas there follow (1) the equivalence of their negations, if only they are not formally true with respect to the variables, and (2) the contradictoriness between the one and the negation of the other:

$$\frac{p \underset{\overset{s}{a}}{\Leftrightarrow} q}{\underset{\overset{s}{a}}{\mathbb{W}} (p, q) \vee \boxed{\neg}\, p \Leftrightarrow \boxed{\neg}\, q} \qquad [250.4] \qquad \text{T5.265}$$

$$\frac{p \underset{\overset{s}{a}}{\Leftrightarrow} q}{p \underset{\overset{s}{a}}{\mathrel{+\!\!\!+}} \boxed{\neg}\, q}, \frac{p \underset{\overset{s}{a}}{\Leftrightarrow} q}{\boxed{\neg}\, p \underset{\overset{s}{a}}{\mathrel{+\!\!\!+}} q}. \qquad [250.4] \qquad \text{T5.266}$$

If two collections of propositions are respectively equivalent to another collection, then the union of the first two is equivalent to the union of the second two,

[31] TS §250, II 359: "Now every statement of the equivalence of given propositions $A, B, C, D, \ldots$ and $M, N, O, \ldots$ may be brought under the form: 'Every collection of ideas which, when put in the places of $i, j, \ldots$ in the propositions $A, B, C, D, \ldots$ and $M, N, O, \ldots$, transforms one of these collections into a collection of all true propositions, does the same for both'."

under the hypothesis, of course, of compatibility among the first two:

$$\frac{\overset{m}{p} \underset{\overset{s}{a}}{\Leftrightarrow} \overset{n}{q},\ \overset{m'}{r} \underset{\overset{s}{a}}{\Leftrightarrow} \overset{n'}{s},\ \underset{\overset{s}{a}}{\wedge}(\overset{m}{p}, \overset{m'}{r})}{\overset{m}{p}, \overset{m'}{r} \underset{\overset{s}{a}}{\Leftrightarrow} \overset{n}{q}, \overset{n'}{s}} . \qquad [250.5a] \qquad T5.267$$

What is derivable from certain propositions is also derivable from any other collection of propositions equivalent to the first ones; moreover, if a certain collection of propositions $\overset{m}{p}$ is derivable from certain propositions, then any collection equivalent to the $\overset{m}{p}$ is derivable from the same propositions, too:

$$\frac{\overset{m}{p} \underset{\overset{s}{a}}{\Leftrightarrow} \overset{n}{q},\ \overset{m}{p} \underset{\overset{s}{a}}{\Rightarrow} \overset{u}{r}}{\overset{n}{q} \underset{\overset{s}{a}}{\Rightarrow} \overset{u}{r}} \qquad [250.5b] \qquad T5.268$$

$$\frac{\overset{m}{p} \underset{\overset{s}{a}}{\Leftrightarrow} \overset{n}{q}, \overset{u}{r} \underset{\overset{s}{a}}{\Rightarrow} \overset{m}{p}}{\overset{u}{r} \underset{\overset{s}{a}}{\Rightarrow} \overset{n}{q}} \qquad [250.5c] \qquad T5.269$$

5.9.4 Contradictoriness

Turning to inferences from propositions enunciating [*aussagen*] a relation of contradictoriness (§251), Bolzano gives as obvious the following four:

$$\frac{\overset{m}{p} \underset{\overset{s}{a}}{\mathrel{+\!\!\!\!+}} \overset{n}{q}}{\overset{m}{p} \underset{\overset{s}{a}}{\Rightarrow} \boxminus \overset{n}{q}} \qquad \frac{\overset{m}{p} \underset{\overset{s}{a}}{\mathrel{+\!\!\!\!+}} \overset{n}{q}}{\boxminus \overset{n}{q} \underset{\overset{s}{a}}{\Rightarrow} \overset{m}{p}} \qquad \frac{\overset{m}{p} \underset{\overset{s}{a}}{\mathrel{+\!\!\!\!+}} \overset{n}{q}}{\overset{n}{q} \underset{\overset{s}{a}}{\Rightarrow} \boxminus \overset{m}{p}} \qquad \frac{\overset{m}{p} \underset{\overset{s}{a}}{\mathrel{+\!\!\!\!+}} \overset{n}{q}}{\boxminus \overset{m}{p} \underset{\overset{s}{a}}{\Rightarrow} \overset{n}{q}} \qquad [251]$$

T5.270

as well as the fact that, together, the preceding four conclusions imply the premise

$$\frac{\overset{m}{p} \underset{\overset{s}{a}}{\Rightarrow} \boxminus \overset{n}{q},\ \boxminus \overset{n}{q} \underset{\overset{s}{a}}{\Rightarrow} \overset{m}{p}, \overset{n}{q} \underset{\overset{s}{a}}{\Rightarrow} \boxminus \overset{m}{p},\ \boxminus \overset{m}{p} \underset{\overset{s}{a}}{\Rightarrow} q}{\overset{m}{p} \underset{\overset{s}{a}}{\mathrel{+\!\!\!\!+}} \overset{n}{q}} \qquad [251] \qquad T5.271$$

From the contradictoriness between the $\overset{m}{p}$ and the $\overset{n}{q}$, it follows that there is no substitution for the $\overset{s}{a}$ that simultaneously makes all the $\overset{m}{p}$ and at least one of the $\overset{n}{q}$

true, as well as the fact that there is no substitution that simultaneously makes all the $\overset{m}{p}$ and at least one of the $\overset{n}{q}$ false:

$$\frac{\overset{m}{p} \mathrel{\underset{\substack{s\\a}}{+\!\!\!\!+}} \overset{n}{q}}{\neg \underset{\substack{s\\a}}{\boxed{\wedge\!\!\!\wedge}} (\overset{m}{p}, \boxed{\vee^{n}}\, \overset{n}{q})} \qquad [251.1a] \qquad T5.272$$

$$\frac{\overset{m}{p} \mathrel{\underset{\substack{s\\a}}{+\!\!\!\!+}} \overset{n}{q}}{\neg \underset{\substack{s\\a}}{\boxed{\wedge\!\!\!\wedge}} (\boxed{\neg}\, \overset{m}{p}, \boxed{\vee^{n}}\, \boxed{\neg}\, \overset{n}{q})} \qquad [251.1b] \qquad T5.273$$

The contradictoriness between the $\overset{m}{p}$ and the $\overset{n}{q}$ also implies the equivalence of the first with the negations of the second, as well as the contradictoriness of the respective negations:

$$\frac{\overset{m}{p} \mathrel{\underset{\substack{s\\a}}{+\!\!\!\!+}} \overset{n}{q}}{\overset{m}{p} \mathrel{\underset{\substack{s\\a}}{\Leftrightarrow}} \boxed{\neg}\, \overset{n}{q}} \quad [251.1c] \quad T5.274 \qquad \frac{\overset{m}{p} \mathrel{\underset{\substack{s\\a}}{+\!\!\!\!+}} \overset{n}{q}}{\boxed{\neg}\, \overset{m}{p} \mathrel{\underset{\substack{s\\a}}{+\!\!\!\!+}} \boxed{\neg}\, \overset{n}{q}} \quad [251.1d] \quad T5.275$$

The contradictoriness of two propositions implies their disjunction:

$$\frac{p \mathrel{\underset{\substack{s\\a}}{+\!\!\!\!+}} q}{\boxed{\sqcup}(p, q)} \qquad [251.2] \qquad T5.276$$

Contradictory aggregates contradict the respective equivalents and are equivalent to the contradictories of their respective contradictories:

$$\frac{\overset{m}{p} \mathrel{\underset{\substack{s\\a}}{+\!\!\!\!+}} \overset{n}{q}, \overset{m}{p} \mathrel{\underset{\substack{s\\a}}{\Leftrightarrow}} \overset{u}{r}}{\overset{n}{q} \mathrel{\underset{\substack{s\\a}}{+\!\!\!\!+}} \overset{u}{r}} \quad [251.3a] \quad T5.277 \qquad \frac{\overset{m}{p} \mathrel{\underset{\substack{s\\a}}{+\!\!\!\!+}} \overset{n}{q}, \overset{m}{p} \mathrel{\underset{\substack{s\\a}}{+\!\!\!\!+}} \overset{u}{r}}{\overset{n}{q} \mathrel{\underset{\substack{s\\a}}{\Leftrightarrow}} \overset{u}{r}} \quad [251.3b] \quad T5.278$$

Finally, two pairs of contradictory aggregates may be composed preserving the contradictoriness (under the obvious conditions of compatibility):

$$\frac{\overset{m}{p} \mathrel{\underset{\substack{s\\a}}{+\!\!\!\!+}} \overset{n}{q},\ \overset{m'}{r} \mathrel{\underset{\substack{s\\a}}{+\!\!\!\!+}} \overset{n'}{s},\ \underset{\substack{s\\a}}{\wedge\!\!\!\wedge}(\overset{m}{p}, \overset{m'}{r}),\ \underset{\substack{s\\a}}{\wedge\!\!\!\wedge}(\boxed{\neg}\, \overset{m}{p}, \boxed{\neg}\, \overset{m'}{r})}{\overset{m}{p}, \overset{m'}{r} \mathrel{\underset{\substack{s\\a}}{+\!\!\!\!+}} \overset{n}{q}, \overset{n'}{s}} \qquad [251.3c] \qquad T5.279$$

5.9.5 *Material Integration*

Passing to the examination (§252) of inferences from propositions enunciating a relation of integration among other propositions, Bolzano begins with propositions enunciating a relation of material integration. He does this by following the distinction (which we have already seen) in five (plus five) fundamental types: (1) all true (all false); (2) at least one true (at least one false); (3) one and only one true (one and only one false); (4) more than one true (more than one false); (5) exactly k true (exactly k false).

In the first case (all true (false)), he simply remarks that these two types of relations, as well as their consequences, have been repeatedly taken into consideration. [252.1]

In the second case (at least one true (false)),

$$\frac{E(p, q, \overset{n}{p}),\ F(\overset{n}{p})}{E(p, q)}$$ [252.2a] T5.280

$$\frac{\neg W(p, q, \overset{n}{p}),\ W(\overset{n}{p})}{\neg W(p, q)}$$ [252.2a] T5.281

$$\frac{E(p, \overset{n}{p}),\ F(\overset{n}{p})}{W(p)}$$ [252.2b] T5.282

$$\frac{\neg W(p, \overset{n}{p}),\ W(\overset{n}{p})}{F(p)}$$ [252.2b] T5.283

$$\frac{E(\overset{m}{p}),\ E(\overset{n}{q})}{{}_2W(\overset{m}{p}, \overset{n}{q})}$$ [252.2c] T5.284

$$\frac{\neg W(\overset{m}{p}),\ \neg W(\overset{n}{q})}{{}_2F(\overset{m}{p}, \overset{n}{q})}$$ [252.2c] T5.285

$$\frac{\bigwedge_{i \leq n}(W(p_i) \rightarrow W(q_i)), E(\overset{m}{p})}{E(\overset{n}{q})}$$ [252.2d] T5.286

$$\frac{\bigwedge_{i \leq n}(W(p_i) \rightarrow W(q_i)), \neg W(\overset{n}{q})}{\neg W(\overset{m}{p})}$$ [252.2e] T5.287

There is also its generalization to n pairs of implications:

$$\frac{\bigwedge_{k\leq n}(\bigwedge_{i\leq n_k}(W(p_i^k)\rightarrow W(q_i^k)))\wedge\neg W(\overset{k_1}{q},\ldots,\overset{k_n}{q})}{\neg W(\overset{k_1}{q},\ldots,\overset{k_n}{q})}$$

[252.2e] T5.288

In the third case (one and only one true (false)),

$$\frac{D(\overset{n}{p},q),W(q)}{F(\overset{n}{p})}$$

[252.3a] T5.289

$$\frac{D(\overset{n}{p},q),F(q)}{D(\overset{n}{p})}$$

[252.3b] T5.291

$$\frac{{}^1_1F(\overset{n}{p},q),F(q)}{W(\overset{n}{p})}$$

[252.3a] T5.290

$$\frac{{}^1_1F(\overset{n}{p},q),W(q)}{{}^1_1F(\overset{n}{p})}$$

[252.3b] T5.292

$$\frac{D(\overset{n}{p}),D(\overset{m}{q}),\bigwedge_{i\leq n}(\bigwedge_{j\leq m}(p_i\neq q_j))}{{}^2_2W(\overset{n}{p},\overset{m}{q})}$$

[252.3c] T5.293

$$\frac{D(\overset{n}{p},q),W(q)\rightarrow W(r)}{{}^2_1W((\overset{n}{p},r)}$$

[252.3d] T5.294

$$\frac{D(\overset{n}{p},q),W(q)\rightarrow W(r)}{{}^1W((\overset{n}{p},r)}$$

[252.3e] T5.295

$$\frac{{}^1_1F(\overset{n}{p},\overset{m}{q},\overset{s}{r}),\overset{n}{p}\Rightarrow\overset{m}{q}}{W(\overset{m}{q})\wedge{}^1_1F(\overset{n}{p},\overset{s}{r})}$$

[252.3g] T5.296

$$\frac{D(\overset{n}{p}),W(p_1)\rightarrow W(p_2),W(p_3)\rightarrow W(p_4)}{F(p_1,p_3)}$$

[252.3f] T5.297

$$\frac{{}^k_kF(\overset{n}{p},\overset{m}{q},\overset{s}{r}),(\overset{n}{p}\Rightarrow\overset{m}{q})}{{}^{k-1}F(\overset{m}{q})}$$

T5.298

(the other false are possibly all among the $\overset{n}{p},r$) [252.3h]

In the fourth case (more than one true (false)),

$$\frac{{}_2W(\overset{n}{p},\overset{m}{q}),F(\overset{n}{p})}{{}_2W(\overset{m}{q})}$$

[252.4a] T5.299

$$\frac{{}_2F(\overset{n}{p},\overset{m}{q}),\ W(\overset{n}{p})}{{}_2F(\overset{m}{q})}$$

[252.4a] T5.300

$$\frac{{}_2W(\overset{n}{p},\overset{m}{q}),D(\overset{n}{p})}{{}_1W(\overset{m}{q})}\quad [252.4b]\quad T5.301 \qquad \frac{{}_2W(\overset{n}{p},\overset{m}{q}),{}^1_1F(\overset{n}{p})}{{}_1F(\overset{m}{q})}\quad [252.4b]\quad T5.302$$

$$\frac{{}_2W(\overset{n}{p}),\ \bigwedge_{i\leq n}(W(p_i)\rightarrow W(q_i))}{{}_2W(\overset{n}{q})}\quad [252.4c]\quad T5.303$$

Finally, in the fifth case (exactly k true (false)),

$$\frac{{}^k_kW(\overset{n}{p})}{{}^{n-k}_{n-k}F(\overset{n}{p})}\quad [252.5]\quad T5.304 \qquad \frac{{}^k_kF(\overset{n}{p})}{{}^{n-k}_{n-k}W(\overset{n}{p})}\quad [252.5]\quad T5.305$$

$$\frac{{}^h_hW(\overset{n}{p}),\ {}^k_kW(\overset{m}{q}),\ \bigwedge_{i\leq n}(\bigwedge_{j\leq m}(p_i\neq q_j))}{{}^{k+h}_{k+h}W(\overset{n}{p},\overset{m}{q})}\quad [252.5a]\quad T5.306$$

$$\frac{{}^h_hF(\overset{n}{p}),\ {}^k_kF(\overset{m}{q}),\ \bigwedge_{i\leq n}(\bigwedge_{j\leq m}(p_i\neq q_j))}{{}^{k+h}_{k+h}F(\overset{n}{p},\overset{m}{q})}\quad [252.5a]\quad T5.307$$

$$\frac{{}^{k+h}_{k+h}W(\overset{n}{p},\overset{m}{q}),\ {}^h_hW(\overset{m}{q}),\ \bigwedge_{i\leq n}(\bigwedge_{j\leq m}(p_i\neq q_j))}{{}^k_kW(\overset{n}{p})}\quad [252.5b]\quad T5.308$$

$$\frac{{}^k_kW(\overset{n}{p},\overset{m}{q}),\overset{n}{p}\Leftrightarrow\overset{s}{r}}{{}^{k+s-1}_{k-n+1}W(\overset{s}{r},\overset{m}{q})}\quad [252.5c]\quad T5.309 \qquad \frac{{}^k_kW(\overset{n}{p},\overset{m}{q}),\overset{n}{p}\Leftrightarrow\overset{s}{r}}{{}^{k+s-1}_{k-n+1}W(\overset{s}{r},\overset{m}{q})}\quad [252.5c]\quad T5.310$$

$$\frac{{}^{n+k}_{n+k}W(\overset{n}{p},\overset{k-s}{q},\overset{m}{r}),\ W(\overset{n}{p})\rightarrow W(\overset{k-s}{q})}{\neg W(\overset{n}{p})}\quad [252.5d]\quad T5.311$$

5.9.6 *Formal Integration*

Bolzano then takes into consideration relations of *formal integration*. In this context, he preliminarily remarks that if among the n propositions $p_1, \ldots, p_n$ there subsists, with respect to the ideas $a_1, \ldots, a_s$, any relation of absolute formal integration, then, for every correct substitution for the s ideas in the $p_1, \ldots, p_n$, among the resulting propositions there subsists the corresponding relation of material integration; i.e., writing for a moment '$\mathbb{X}(\overset{n}{p})$' for any of the ten relations of integration, it holds that

$$\underset{\overset{s}{a}}{\mathbb{X}}(\overset{n}{p}) \rightarrow \mathbb{X}(\overset{n}{p}[\overset{s}{a}/\overset{s}{c}]) \qquad [252.6] \qquad \text{T5.312}$$

He concludes from this that—under the hypothesis of the correctness of the substitution—from any relation of absolute formal integration, it is possible to derive all those already pointed out for the corresponding material concretizations.

Remarking that by simply taking the negations of the given propositions it is easy to reduce the number of integration to be considered to five, Bolzano proceeds to examine the situations in which one of $n + 1$ propositions involved in a formal integration is such that certain other m propositions are either derivable from it or in contradiction with it.

In the first case, he proves

$$\frac{\underset{\overset{s}{a}}{\mathbb{W}}(\overset{n}{p}, q),\ q \underset{\overset{s}{a}}{\Rightarrow} \overset{m}{q}}{\underset{\overset{s}{a}}{W}(\overset{n}{p}, \overset{m}{q})} \qquad \text{T5.313} \qquad\qquad \frac{\underset{\overset{s}{a}}{E}(\overset{n}{p}, q),\ q \underset{\overset{s}{a}}{\Rightarrow} \overset{m}{q}}{\underset{\overset{s}{a}}{E}(\overset{n}{p}, \overset{m}{q})} \qquad \text{T5.314}$$

$$\frac{\underset{\overset{s}{a}}{D}(\overset{n}{p}, q),\ q \underset{\overset{s}{a}}{\Rightarrow} \overset{m}{q}}{{}^{m}_{1}\underset{\overset{s}{a}}{\mathbb{W}}(\overset{n}{p}, \overset{m}{q})} \qquad \text{T5.315} \qquad\qquad \frac{{}_{2}\underset{\overset{s}{a}}{\mathbb{W}}(\overset{n}{p}, q),\ q \underset{\overset{s}{a}}{\Rightarrow} \overset{m}{q}}{{}_{2}\underset{\overset{s}{a}}{\mathbb{W}}(\overset{n}{p}, \overset{m}{q})} \qquad \text{T5.316}$$

$$\frac{{}^{k}_{k}\underset{\overset{s}{a}}{\mathbb{W}}(\overset{n}{p}, q),\ q \underset{\overset{s}{a}}{\Rightarrow} \overset{m}{q}}{{}^{k+m}_{k}\underset{\overset{s}{a}}{\mathbb{W}}(\overset{n}{p}, \overset{m}{q})} \qquad [252.6a] \qquad \text{T5.317}$$

and, moreover, remarks that this also holds when, instead of '$\underset{\overset{s}{a}}{\Rightarrow}$', there is '$\underset{\overset{s}{a}}{\Leftrightarrow}$'.

In the second case, he proves

$$\frac{\underset{\overset{s}{a}}{\mathbb{W}}(\overset{n}{p}, q),\ q \underset{\overset{s}{a}}{+\!\!\!+} \overset{m}{q}}{{}^{m}_{m}\underset{\overset{s}{a}}{\mathbb{F}}(\overset{n}{p}, \overset{m}{q})} \qquad \text{T5.318} \qquad\qquad \frac{\underset{\overset{s}{a}}{E}(\overset{n}{p}, q),\ q \underset{\overset{s}{a}}{+\!\!\!+} \overset{m}{q}}{\text{may be } \underset{\overset{s}{a}}{E}(\overset{n}{p}, \overset{m}{q})} \qquad \text{T5.319}$$

$$\frac{\underset{\overset{s}{a}}{D}(\overset{n}{p}, q),\ q \underset{\overset{s}{a}}{+\!\!\!+} \overset{m}{q}}{\underset{\overset{s}{a}}{\mathbb{F}}(\overset{n}{p}, \overset{m}{q}) \vee^{m+1}_{m+1} \underset{\overset{s}{a}}{\mathbb{W}}(\overset{n}{p}, \overset{m}{q})} \qquad \text{T5.320} \qquad\qquad \frac{{}_{2}\underset{\overset{s}{a}}{\mathbb{W}}(\overset{n}{p}, q),\ q \underset{\overset{s}{a}}{+\!\!\!+} \overset{m}{q}}{E(\overset{n}{p}, \overset{m}{q})} \qquad \text{T5.321}$$

$$\frac{{}^{k}_{k}\mathbb{W}_{\substack{s\\a}}(\overset{n}{p},q),\ q \nparallel_{\substack{s\\a}} \overset{m}{q}}{{}^{k-1}_{k-1}\mathbb{W}_{\substack{s\\a}}(\overset{n}{p},\overset{m}{q}) \vee {}^{k+m}_{k+m}\mathbb{W}_{\substack{s\\a}}(\overset{n}{p},\overset{m}{q})} \qquad [252.6b] \qquad T5.322$$

As far as situations of integration with respect to more systems of variables are concerned, he remarks that if the same type of formal integration, with respect to two systems of variables, subsists among the $p_1, \ldots, p_n$, then the same type of relation subsists with respect to the union of the two systems; i.e., using the notation above,

$$\frac{\mathbb{X}_{\substack{s\\a}}(\overset{n}{p}),\ \mathbb{X}_{\substack{r\\c}}(\overset{n}{p})}{\mathbb{X}_{\substack{s\ r\\a,c}}(\overset{n}{p})} \qquad [252.6c] \qquad T5.323$$

Finally, concerning situations of combination of homogeneous integrations, Bolzano recalls the following:

$$\frac{W_{\substack{s\\a}}(\overset{n}{p}),\ W_{\substack{s\\a}}(\overset{m}{q})}{W_{\substack{s\\a}}(\overset{n}{p},\overset{m}{q})} \qquad T5.324 \qquad\qquad \frac{E_{\substack{s\\a}}(\overset{n}{p}),\ E_{\substack{s\\a}}(\overset{m}{q})}{E_{\substack{s\\a}}(\overset{n}{p},\overset{m}{q})} \qquad T5.325$$

$$\frac{D_{\substack{s\\a}}(\overset{n}{p}),\ D_{\substack{s\\a}}(\overset{m}{q})}{{}^{2}_{1}W_{\substack{s\\a}}(\overset{n}{p},\overset{m}{q})} \qquad T5.326 \qquad\qquad \frac{{}_{2}W_{\substack{s\\a}}(\overset{n}{p}),\ {}_{2}W_{\substack{s\\a}}(\overset{m}{q})}{{}_{2}W_{\substack{s\\a}}(\overset{n}{p},\overset{m}{q})} \qquad T5.327$$

$$\frac{{}^{k}_{k}W_{\substack{s\\a}}(\overset{n}{p}),\ {}^{h}_{h}W_{\substack{s\\a}}(\overset{m}{q})}{{}^{k+h}_{max(h,k)}W_{\substack{s\\a}}(\overset{n}{p})(\overset{n}{p},\overset{m}{q})} \qquad [252.6d] \qquad T5.328$$

He then moves on to consideration of the inferences

$$\frac{W(\overset{n}{p}[\overset{s}{a}/\overset{s}{c}]),\ \overset{n}{p} \Leftrightarrow_{\substack{s\\a}} \overset{m}{q}}{W(\overset{m}{q}[\overset{s}{a}/\overset{s}{c}])} \quad [252.7a] \quad T5.329 \qquad \frac{\neg W(\overset{n}{p}[\overset{s}{a}/\overset{s}{c}]),\ \overset{n}{p} \Leftrightarrow_{\substack{s\\a}} \overset{m}{q}}{\neg W(\overset{m}{q}[\overset{s}{a}/\overset{s}{c}])} \quad [252.7a] \quad T5.330$$

$$\frac{W(\overset{n}{p}[\overset{s}{a}/\overset{s}{c}]),\ \overset{n}{p} \mathbin{\pmb{+}}_{\substack{s\\a}} \overset{m}{q}}{F(\overset{m}{q}[\overset{s}{a}/\overset{s}{c}])} \quad [252.7b] \quad T5.331 \qquad \frac{\neg W(\overset{n}{p}[\overset{s}{a}/\overset{s}{c}]),\ \overset{n}{p} \mathbin{\pmb{+}}_{\substack{s\\a}} \overset{m}{q}}{\neg F(\overset{m}{q}[\overset{s}{a}/\overset{s}{c}])} \quad [252.7b] \quad T5.332$$

$$\frac{\underset{\substack{s\,r\\a\,c}}{E}(p,\overset{m}{q}),\ \underset{\substack{r\\c}}{F}(p)}{\underset{\substack{r\\c}}{E}(\overset{m}{q})}\qquad [252.7c]\ \text{T5.333}\qquad\qquad \frac{\underset{\substack{s\,r\\a\,c}}{D}(p,\overset{m}{q}),\ \underset{\substack{r\\c}}{F}(p)}{\underset{\substack{r\\c}}{D}(\overset{m}{q})}\qquad [252.7c]\ \text{T5.334}$$

$$\frac{{}_2\underset{\substack{s\,r\\a\,c}}{W}(p,\overset{m}{q}),\ \underset{\substack{r\\c}}{F}(p)}{{}_2\underset{\substack{r\\c}}{W}(\overset{m}{q})}\qquad [252.7c]\ \text{T5.335}\qquad\qquad \frac{{}^k_k\underset{\substack{s\,r\\a\,c}}{W}(p,\overset{m}{q}),\ \underset{\substack{r\\c}}{F}(p)}{{}^k_k\underset{\substack{r\\c}}{W}(\overset{m}{q})}\qquad [252.7c]\ \text{T5.336}$$

$$\frac{\underset{\substack{s\,r\\a\,c}}{E}(p,\overset{m}{q}),\ \underset{\substack{r\\c}}{W}(p)}{?\underset{\substack{r\\c}}{E}(\overset{m}{q})\,?}\qquad [252.7d]\ \text{T5.337}\qquad\qquad \frac{\underset{\substack{s\,r\\a\,c}}{D}(p,\overset{m}{q}),\ \underset{\substack{r\\c}}{F}(p)}{\underset{\substack{r\\c}}{F}(\overset{m}{q})}\qquad [252.7d]\ \text{T5.338}$$

$$\frac{{}_2\underset{\substack{s\,r\\a\,c}}{W}(p,\overset{m}{q}),\ \underset{\substack{r\\c}}{W}(p)}{\underset{\substack{r\\c}}{E}(\overset{m}{q})}\qquad [252.7d]\ \text{T5.339}\qquad\qquad \frac{{}^k_k\underset{\substack{s\,r\\a\,c}}{W}(p,\overset{m}{q}),\ \underset{\substack{r\\c}}{F}(p)}{{}^{k-1}_{k-1}\underset{\substack{r\\c}}{W}(\overset{m}{q})}\qquad [252.7d]\ \text{T5.340}$$

$$\frac{\underset{\substack{s\\a}}{W}(\overset{n}{p}),\overset{n}{p}\underset{\substack{s\\a}}{\Rightarrow}\overset{m}{q})}{\underset{\substack{s\\a}}{W}(\overset{m}{q})}\qquad [252.7e]\ \text{T5.341}\qquad\qquad \frac{\underset{\substack{s\\a}}{D}(p,\overset{m}{q}),\neg\underset{\substack{s\\a}}{\mathbb{F}}(p)}{p\ \#\ \underset{\substack{s\\a}}{\boxed{\vee}}(\overset{m}{q})}\qquad [252.7f]\ \text{T5.342}$$

The inference [252.7g] is, at first sight, less transparent; it presents itself so:

> "Both the assertions that the propositions $A, B, C, \ldots$ are all true and that the $M, N, O, \ldots$ are not all true, are never both true with respect to the variable ideas $i, j, \ldots$
> The propositions $A, B, C, \ldots$ are compatible with respect to the same ideas.
> ———
> If $A, B, C, \ldots$ is, then also $M, N, O, \ldots$ is, with respect to the same ideas"

It seems interpretable as the contraposition of [248.10]

$$\frac{\neg(\overset{m}{p}\underset{\substack{s\\a}}{\Rightarrow}\overset{n}{q})}{\neg\underset{\substack{s\\a}}{\bigwedge}(\overset{m}{p})\vee\underset{\substack{s\\a}}{\bigwedge}(\overset{m}{p},\boxed{\vee^n}(\boxed{\neg}\,\overset{n}{q}))}$$

by contraposition of it; indeed, we have

$$\frac{\neg(\neg\underset{\substack{s\\a}}{\bigwedge}(\overset{m}{p})\vee\underset{\substack{s\\a}}{\bigwedge}(\overset{m}{p},\boxed{\vee^n}(\boxed{\neg}\,\overset{n}{q})))}{\overset{m}{p}\underset{\substack{s\\a}}{\Rightarrow}\overset{n}{q}},\qquad\text{i.e.}\qquad \frac{\neg\underset{\substack{s\\a}}{\bigwedge}(\overset{m}{p},\boxed{\vee^n}(\boxed{\neg}\,\overset{n}{q})),\underset{\substack{s\\a}}{\bigwedge}(\overset{m}{p})}{\overset{m}{p}\underset{\substack{s\\a}}{\Rightarrow}\overset{n}{q}}$$

But $\neg\underset{\substack{s\\a}}{\bigwedge}(\overset{m}{p},\boxed{\vee^n}(\boxed{\neg}\,\overset{n}{q}))$ states precisely that there is no substitution that simultaneously makes all the $\overset{m}{p}$ true and the proposition asserting that at least one of the $\overset{n}{q}$ is false (i.e. that not all the $\overset{n}{q}$ are true). Hence

$$\frac{\neg \underset{\overset{s}{a}}{\bigwedge}(\overset{m}{p}, \boxed{\vee^{n}}(\boxed{\neg}\ \overset{n}{q})), \underset{\overset{s}{a}}{\bigwedge}(\overset{m}{p})}{\overset{m}{p} \underset{\overset{s}{a}}{\Rightarrow} \overset{n}{q}} \qquad [252.7\text{ g}] \qquad \text{T5.343}$$

Finally, as far as the inferences from premises enunciating the subsistence of one of the four relations of formal conditioned integration are concerned, Bolzano presents the two series of inferences that are immediately understandable when recalling the four definitions introduced earlier and that we reproduce here for the sake of simplicity:

$$\overset{n}{p} \overset{\vee}{\underset{\overset{s}{a}}{\Rightarrow}} q \leftrightarrow \forall\overset{s}{c}(W(\overset{n}{p}[\overset{s}{a}/\overset{s}{c}]) \to E(\overset{m}{q}[\overset{s}{a}/\overset{s}{c}])) \quad \overset{n}{p} \overset{\vee 1}{\underset{\overset{s}{a}}{\Rightarrow}} q \leftrightarrow \forall\overset{s}{c}(W(\overset{n}{p}[\overset{s}{a}/\overset{s}{c}]) \to D(\overset{m}{q}[\overset{s}{a}/\overset{s}{c}]))$$

$$\overset{n}{p} \overset{\vee +}{\underset{\overset{s}{a}}{\Rightarrow}} q \leftrightarrow \forall\overset{s}{c}(W(\overset{n}{p}[\overset{s}{a}/\overset{s}{c}]) \to {}_{2}W(\overset{m}{q}[\overset{s}{a}/\overset{s}{c}])) \quad \overset{n}{p} \overset{\vee k}{\underset{\overset{s}{a}}{\Rightarrow}} q \leftrightarrow \forall\overset{s}{c}(W(\overset{n}{p}[\overset{s}{a}/\overset{s}{c}]) \to {}^{k}_{k} W(\overset{m}{q}[\overset{s}{a}/\overset{s}{c}])).$$

The first series is

$$\frac{\overset{n}{p} \overset{\vee}{\underset{\overset{s}{a}}{\Rightarrow}} \overset{m}{q}, W\left(\overset{n}{p}\left[\overset{s}{a}/\overset{s}{c}\right]\right)}{E\left(\overset{m}{q}\left[\overset{s}{a}/\overset{s}{c}\right]\right)} \qquad [252.8] \qquad \text{T5.344}$$

$$\frac{\overset{n}{p} \overset{\vee 1}{\underset{\overset{s}{a}}{\Rightarrow}} q, W\left(\overset{n}{p}\left[\overset{s}{a}/\overset{s}{c}\right]\right)}{D\left(\overset{m}{q}\left[\overset{s}{a}/\overset{s}{c}\right]\right)} \qquad [252.8] \qquad \text{T5.345}$$

$$\frac{\overset{n}{p} \overset{\vee +}{\underset{\overset{s}{a}}{\Rightarrow}} q, W\left(\overset{n}{p}\left[\overset{s}{a}/\overset{s}{c}\right]\right)}{{}_{2}W\left(\overset{m}{q}\left[\overset{s}{a}/\overset{s}{c}\right]\right)} \qquad [252.8] \qquad \text{T5.346}$$

$$\frac{\overset{n}{p} \overset{\vee k}{\underset{\overset{s}{a}}{\Rightarrow}} q, W\left(\overset{n}{p}\left[\overset{s}{a}/\overset{s}{c}\right]\right)}{{}^{k}_{k}W\left(\overset{m}{q}\left[\overset{s}{a}/\overset{s}{c}\right]\right)} \qquad [252.8] \qquad \text{T5.347}$$

whereas the second series is

$$\frac{\overset{n}{p} \overset{\vee}{\underset{\overset{s}{a}}{\Rightarrow}} \overset{m}{q}, \neg E\left(\overset{m}{q}\left[\overset{s}{a}/\overset{s}{c}\right]\right)}{\neg W\left(\overset{n}{p}\left[\overset{s}{a}/\overset{s}{c}\right]\right)} \qquad [252.9] \qquad \text{T5.348}$$

$$\frac{\overset{n}{p} \underset{\overset{s}{a}}{\overset{\vee 1}{\Longrightarrow}} q, \neg D\left(\overset{m}{q}\left[\overset{s}{a}/\overset{s}{c}\right]\right)}{\neg W\left(\overset{n}{p}\left[\overset{s}{a}/\overset{s}{c}\right]\right)} \quad [252.9] \qquad \text{T5.349}$$

$$\frac{\overset{n}{p} \underset{\overset{s}{a}}{\overset{\vee +}{\Longrightarrow}} q, \neg_2 W\left(\overset{m}{q}\left[\overset{s}{a}/\overset{s}{c}\right]\right)}{\neg W\left(\overset{n}{p}\left[\overset{s}{a}/\overset{s}{c}\right]\right)} \quad [252.9] \qquad \text{T5.350}$$

$$\frac{\overset{n}{p} \underset{\overset{s}{a}}{\overset{\vee k}{\Longrightarrow}} q, \neg_k^k W\left(\overset{m}{q}\left[\overset{s}{a}/\overset{s}{c}\right]\right)}{\neg W\left(\overset{n}{p}\left[\overset{s}{a}/\overset{s}{c}\right]\right)} \quad [252.9] \qquad \text{T5.351}$$

6

Higher Level

6.1 Preamble

As already mentioned in Chapter 1, the higher level is characterized by the assumption that qualities are objects, too; i.e.

$$Q(x) \to O(x) \qquad \text{GA22}$$

Two consequences of this assumption deserve special attention. The first concerns the extraordinary parallelism objectuality–truth identified, at this level, by Bolzano, while the second regards the singular remarks made by Bolzano on the relationship between the particular qualities—often referred to today, with a terminological choice of rare inappropriateness, as 'tropes'—and the (universal) qualities.

However, before illustrating these two consequences, as announced in Section 1.3, we will discuss the problem of vacuous qualities.

6.2 The Problem of Vacuous Qualities

In Chapter 1, we accepted the equivalence between *being a quality* and *belonging to something* and hence also the obvious consequence that there are no *vacuous qualities* in the universe. Objections making, in particular, reference to two passages of the WL have been levelled at this consequence as well as at the aforesaid equivalence. Considering the relevance of this issue, we will discuss it in detail. First we will illustrate the reasons that support this equivalence and then present the reasons that suggest that the two passages do not invalidate the equivalence.

We will begin with the passage in which Bolzano explicitly addresses the *definability* problem of the notions of *quality* and *having*:

Bolzano's Logical System. Ettore Casari.

> "I just recall here that I will take the word *having* always in its first and widest sense, so that I refer it, as a rule, only to qualities and so that I can hence say: *anything that is had* (quodcumque habetur) *must be a quality*. If I am asked about the parts of which I think these two so intimately related notions are composed, I must admit to be still somewhat uncertain about their actual components and the relation holding between them under this respect. It seems to me that one of them is absolutely simple and the other is composed only of it and a few other parts. However, whether the absolutely simple notion is the notion of having and the notion of quality arises from it and precisely so that a quality is always just what is had, or whether, on the contrary, quality is the simpler notion and having has it as its component, I do not dare to decide, although the former seems to me more probable."[1]

Setting aside the question whether his doubt concerns the possibility that the absolutely simple notion is *quality* and not *having*, two elements concerning our issue clearly arise from this passage:

(1) Bolzano does not doubt that anything had is a quality; i.e. he accepts the principle

$$\text{P}\ \exists y(y\varepsilon x) \rightarrow Q(x)$$

[*was immer gehabt werde müsse eine Beschaffenheit seyn*]

(2) He questions whether the addition to this principle of its converse—to obtain the principle

$$\text{Q}\ Q(x) \leftrightarrow \exists y(y\varepsilon x)$$

[*eine Beschaffenheit immer nur dasjenige was gehabt wird sey*],

—provides a *definition* of Q.

Taking into account that, for Bolzano, a *definition of an idea* is in particular *a proposition that determines whether it is simple or composed and, in the second case, out of which parts and through which links it is constituted*,[2] we can interpret his uncertainty as follows:

(2.1) a strong interpretation: he doubts the validity of $Q(x) \rightarrow \exists y(y\varepsilon x)$ and hence of the equivalence Q;

[1] Cf. TS §80, I, 273 (1).

[2] TS §554, IV, 230: "[...] *Definitions*. By this I mean [...] nothing other than propositions which specify whether a given idea or proposition is simple or composed of parts and, in the latter case, what these parts are and how they are combined."

(2.2) a weaker interpretation: he does not doubt its validity, but he does question that this equivalence is a definition in his sense (i.e. he doubts that this equivalence provides the exact parts of the notion or the way these parts are composed in it).

However, we may exclude that his doubt concerns the validity of Q with the following passage:

> "Whereas, indeed, it may be asserted of the concept of a quality that it is, if not identical, yet still equivalent to the concept of something which is had, so . . . "[3]

As to the second possibility, the *Grössenlehre* sheds some light on this issue:

> "In other words: everything of which it may be said that a certain object *has it*, this or the *having* of this is a quality of the object."[4]

Thus, his doubt may be interpreted as follows. Which is really the quality: the *redness* or the *having redness*? Forcing the notation, in $x \in y$, is $Q(y)$ or is $Q(\in y)$?

Hence, setting aside the question of the 'definitional' character in the Bolzanian sense of the equivalence Q, from what precedes it follows, as already emphasized, the absence of a vacuous quality.

The two passages that have been presented to confute this conclusion are the following:

(a) (pointed out by Jan Berg[5]):

> "[. . .] so the [Etwas]$(na + b)$ can be objectless only because the idea b denotes a quality which belongs to no object";[6]

(b) (pointed out by Benjamin Schnieder[7]):

> "[. . .] the idea of a self-contradictory quality, e.g., the idea of a round square [. . .]",[8]

[3] TS §127, II, 8.

[4] BGA, IIA7, 98: "Mit andern Worten: Alles wovon gesagt werden kann, daß ein gewisser Gegenstand es habe, das oder dessen Haben ist eine Beschaffenheit dieses Gegenstandes."

[5] Berg, J., 1990, p.136.

[6] TS §238, II, 333: "[. . .] the idea [Something]$(na+b)$ can only be objectless because the idea b designates an attribute that belongs to no object."

[7] Schnieder, B., 2002, pp. 81–2.

[8] TS §197, II, 239: "[. . .] an idea of a self-contradictory attribute, e.g., that of being a round square [. . .]."

to which we could possibly adjoin the following:

> "The same holds of ideas: round square [. . .]. In these examples it is immediately evident that no object can correspond to them, because they attribute to it qualities that contradict each other."[9]

With regard to the passage (a), there is hardly any doubt that, at least in its verbal presentation, this statement is in open contradiction with the previous conclusions; so the question immediately arises: did Bolzano really contradict himself or did he simply use the wrong words?

That an author may contradict himself is certainly not impossible. Even if we accept this possibility, however, we still need to ask ourselves: in a consistent reconstruction of the Bolzanian logical system, which of the two terms of the contradiction (there is–there is not) appears to better agree with the complex of the remaining ideas that characterize the system?

Well, it seems clear to us that the presence of a vacuous quality is incompatible with certain basic insights of Bolzano's construction. The main point is that whereas the presence of an empty quality appears even unavoidable when one accepts—as happens in our predicate logic—that quality ideas are always singular, it appears to be useless, not to say something worse, when one accepts, as Bolzano did, that the quality ideas may also be objectless. To see this, it will be sufficient to ask oneself: is the idea of 'squareness and roundness' (as any other passage suggests) an objectless idea or it is, on the contrary, an objectual idea that refers to a vacuous quality (and in this case to only one or to more)? Alternatively: what are the objectless quality ideas that are often mentioned in the WL? Again: what is the difference between an objectless quality idea and an (evidently objectual) idea that refers to a vacuous quality?

However, apart from this, everything seems to suggest that Bolzano did not really contradict himself, but simply did not express himself very well.

A main reason for this belief may be found in the fact that, at least so far we know, there is no other mention—either in the WL or in other works—of these *qualities that belong to no object.* Indeed, it seems at least strange that the same man who in the WL (and henceforth everywhere) never missed an opportunity to draw attention to the presence of ideas that refer to no object, did not feel the need to clarify or bolster up, *not even to repeat,* his marginal remark on the presence of qualities that belong to no object. Why, for instance, does he not revert to the question in

[9] TS §67, I, 220: "The same holds of the ideas 'a round square' [. . .] it is evident from these ideas themselves that no object can correspond to them, since they would attribute contradictory properties to it."

the (later!) §272, in which, on the other hand, he insisted at length on his using the word '*Beschaffenheit*' in a sense that reaches far beyond the usual one?

In addition to this *argumentum e silentio*, further, even more significant, support for the idea that Bolzano simply expressed himself badly comes from the remark that a well-known and uncontroversial element of Bolzano's terminology is quite incorrectly used in the passage at stake. Indeed, in Bolzano's terminology, as we will see extensively in Chapter 9, '*bezeichnen* [*to denote*]' is used for the relation between words and ideas in themselves or, in a wider sense, for the relation between words and the objects referred to by the ideas denoted, in the strict sense, by those words—*never*, however, for the relation between ideas in themselves and qualities, for which, instead, as we know, '*sich beziehen auf* [*to refer*]' is used.

Well, it can be easily verified that Bolzano, although in general very careful (sometimes almost pedantic) about his formulations, nevertheless sometimes argues in a less controlled form. This is especially the case with more complicated issues, in which he possibly consciously pursues the aim not to overload an argument, which is already difficult in itself, with other, *in that place*, inessential difficulties. In a letter dated 18.XII.1834 to Exner, for instance, Bolzano wrote: "to every objective idea there is a *corresponding* subjective idea which is its *conception*".[10] Is there anyone who—disregarding God's omniscience, which is clearly not at issue here—will draw from this the conclusion that, besides what we already know about Bolzano's views on this subject, he also thought: *every idea in itself will be conceived*? Or, to take a more official case from the *Wissenschaftslehre*: "Since the representation of the sign α stimulates the representation of the sign a, and this however the idea A that it denotes ..."[11] What Bolzano means, of course, is: "Since the representation of the sign α stimulates the representation of the sign a, and this however that one of the idea A that it denotes ..."[12] Is there anyone who will conclude from this that, besides what we already know, Bolzano also maintained that *sometimes* ideas in itself *have existence*, because they are *stimulated* [*erwecket*] by subjective representations?

As to the passage (b), the context in which such a phrase appears is the Bolzanian attempt to prove that there are *synthetic truths* [*synthetische Wahrheiten*] in the sense he has given to such a term. His argument is as follows: if there are qualities that belong to a whole class of objects, but are nevertheless such that their idea is not

[10] BBS IV, 86: "[...] zu jeder objectiven Vorstellung gibt es eine ihr zugehörige subjective, die ihre Auffassung ist [...]."

[11] WL §285, III, 73: "Denn weil die Vorstellung des Zeichens α die Vorstellung des Zeichens a, diese aber die durch sie bezeichnete Vorstellung A erwecket ..." (Cf. TS §285, III, 48).

[12] "Denn weil die Vorstellung des Zeichens α die Vorstellung des Zeichens a, diese aber *jene der* [die] durch $< a >$ [sie] bezeichnete$< n >$ Vorstellung A erwecket [...]"

part of the idea referring to that class, then there are truths of the form $h(a, \bar{c})$, where the idea $\bar{c}$ is not somehow part of the subject idea a and, among such truths, some must be synthetic. Indeed, if $\bar{c}$ is not equivalent to $\widehat{e}$, then, by substituting in it e for a, $h(a, \bar{c})$ becomes false, although, of course, objectual; and, on the other hand, by substituting in it any *imaginary* quality idea (e.g., that of round squareness) for $\bar{c}$, it remains objectual, but becomes false.

What we formulated as 'any *imaginary* quality idea (e.g. that of round squareness) for $\bar{c}$' is what in Bolzano sounds like 'the idea of a self-contradictory quality (e.g. the idea of a round square)'.

Here is our justification for this rewording. As already alluded to in Section 2.2, Bolzano calls an object idea imaginary if among its components there are ideas referring to qualities that are incompatible.[13] Elsewhere, he emphasizes that being imaginary is a stronger condition than simple objectlessness: in fact, it includes the *impossibility*, on the basis of pure conceptual truths concerning qualities, of having an object.[14] A round square is often quoted as an example of an imaginary object idea. As we have seen, the property of being imaginary is first defined for object ideas and, as far as we know, the extension of this notion to quality ideas has never been explicitly formulated; nevertheless, it is used and precisely in the sense of *a quality idea composed by quality ideas that refer to incompatible qualities*. Interesting enough, in particular, this is exposed in the same section in which the critical phrase under discussion occurs, in the generalization of the analysis involving it.

The argument is that the predicate idea of true proposition, which is either the *simple* quality idea $\bar{b}$ or the *sum* of the *simple* quality ideas $\bar{b}, \bar{c}, \ldots$, will not be freely variable by preserving the truth of the proposition, because we can always substitute an imaginary quality idea for some of them.[15]

[13] TS §70, I, 228: "[...] cases can occur where someone includes constituents in an idea that ascribe attributes to an object it might represent that contradict those that follow from the remaining constituents.."

TS §70, I, 38–9: "*Note.* We should not simply identify such denials of the objectuality of an idea with propositions stating that an idea is imaginary. From the definition I gave in §70, it is clear that by calling an idea imaginary I mean to say far more than that the idea simply lacks an object. Rather,such statements indicate that the assumption that there is such an object is impossible, i.e., contradicts certain purely conceptual truths."

Note that Bolzano regards as *imaginär* also ideas such as green virtue that in Husserl's terminology should be better classified as *Unsinn* rather than *Widersinn*. See, in particular, TS §231, II, 317.

[14] TS §138, II, 55: "*Note.* [...] by calling an idea imaginary I mean to say far more than that the idea simply lacks an object. Rather, such statements indicate that the assumption that there is such an object is impossible, i.e., contradicts certain purely conceptual truths."

[15] TS §197, II, 240: "2a) [...] For if a predicate-idea is simple, or if its composition is of the following sort: 'each of the attributes $x + y + z + \ldots$' where the ideas $x, y, z, \ldots$, we can obviously neither consider the entire predicate-idea, nor any of its parts variable. For if the proposition '*A* has *x*', or the proposition '*A* has each of the attributes $(x + y + z + \ldots)$' is to remain true, then neither the idea x in the first case, nor any of the ideas $x, y, z, \ldots$ in the second case, must be arbitrarily altered, since this proposition could certainly become false if any of these ideas is replaced by an imaginary one."

Thus our rewording seems to agree with Bolzano's point of view and the phrase pointed out by Schnieder may be seen again as a hasty shortening, whereby it apparently loses its critical value as far as the existence of an empty quality is concerned.

6.3 Truth as Enunciation of a Situation

In the presence of the general axiom characterizing the higher level, according to which also *qualities are objects*, we have, in particular, that, given any object i and any quality r, there is always the object $< i, r >$. Of such an object, we will say that it is a *fact* or a *circumstance* [*Umstand*] iff the r belongs to i; i.e.

$$Fct(j) \leftrightarrow \exists ir(j =< i, r > \wedge i \in r) \qquad \text{D6.1}$$

from which we get

$$Fct(< i, r >) \leftrightarrow i \in r) \qquad \text{T6.1}$$

Proof If $Fct(< i, r >)$, then, by definition, there is an object j and a quality s such that (i) $< i, r >=< j, s >$ and (ii) $j \in s$. But, by ST1.11, (i) implies $i = j$ and $r = s$; hence, by (ii), $i \in r$. Vice versa, let i be an object and r a quality such that $i \in r$; then the object $< i, r >$ is, by definition, a fact.

A very important point clearly results from the following passage:

> "If *b* is a quality of *A*, then the circumstance that the object *A* and the quality *b* constitute together a whole consisting in an object and a quality belonging to it [...] is a relation subsisting between *A* and *b*."[16]

We see that Bolzano was convinced that *being a fact is a quality of certain objects*; in other words, he accepted the following comprehension principle:

$$\exists r^2 \forall j(j \in r^2 \leftrightarrow Fct(j)) \qquad \text{CQ6}$$

i.e.

$$\exists r^2 \forall j(j \in r^2 \leftrightarrow \exists is(j = < i, s > \wedge i \in s)) \qquad \text{T6.2}$$

In the further development of the section, we find: p. 241 "2c) [...] For if we replace one $a, b, c, \ldots$ with an idea incompatible with the rest of these, the sign '[Something]$(a + b + c + \ldots)$' still expresses an idea, and thus the proposition has an object, yet it becomes false, since such an idea is objectless."

p. 241 "d) Similar considerations hold for propositions of the forms 'The idea [Something]$(a + b)$ is imaginary', where there are only two ideas *a*, *b* For given that every simple attribute-idea is certainly real, the idea '[Something]$(a + b)$' can only be imaginary if the attributes *a* and *b* are incompatible."

[16] Cf. TS §80, I 274.

We are not, however, aware of any point of the Bolzanian thought where the existence of such a universal quality belonging to *all* (and only all) facts plays any role. On the contrary, what we find is the tacit acceptance of the disposability of a whole set of, let us say, 'sub-qualities' of this universal quality, that is to say, of qualities corresponding to certain collections of facts. This happens, in particular, within the frame of Bolzano's reflections and convictions on the *truth of propositions* that he considered as the exact correspondent of the *objectuality of ideas*:

> "Indeed, the same value that has for ideas the circumstance whether through them a certain object is or is not actually represented, has for propositions the circumstance whether truth belongs or does not belong to them."[17]

In order to clarify the meaning of these extraordinary Bolzanian words, we begin by remarking that, up to now, nothing has authorized us to believe that for *each* collection of facts, there is a corresponding 'sub-quality' of the universal quality of being a fact. Let us call the collections for which this happens '*situations*'; even more precisely, let us call '*situation*' a (binary) quality that belongs to facts, that is a quality that belongs to ordered pairs whose first component is an object whereas their second component is a quality belonging to the first. Hence

$$Sit(r) \leftrightarrow \forall i(i \in r \rightarrow Fct(i)) \qquad \text{D6.2}$$

from which

$$Sit(r) \rightarrow Q^2(r) \qquad \text{T6.3}$$

Let us agree on using '*m*' as a special variable for these situations.

Let us now consider a proposition of the form $h(a, \bar{c})$. According to the Bolzanian definition of truth, the truth of $h(a, \bar{c})$ implies, besides the existence of at least an object under a, the existence of a *collection of facts* having two properties: (1) every object under a is the first component of (at least) one of these facts; (2) the second component of each one of these facts is always under $\bar{c}$. Setting temporarily aside the issue of the objectuality of a, we see that, vice versa, the existence of a collection of facts enjoying the two aforementioned properties implies the truth of $h(a, \bar{c})$. Nobody, however, can assure us that the collection of facts implied by the truth of $h(a, \bar{c})$ is a situation and hence something in our universe; on the other hand, it does not seem that there are hindrances to thinking that it is so. Hence we will postulate it. Once the problem of truth is reduced to situations, the obstacle deriving from the

[17] Cf. TS §154, II 71 (4).

need to ensure the objectuality of the subject a disappears, too. A situation having the two characteristics mentioned above is indeed a *quality*, and therefore it belongs to at least one fact, which means that we are sure, by (1), that at least one object falls under a. Thus, in some way, *the truth of a proposition and the existence of a collection of facts of the above-mentioned type are equivalent.*

Unfortunately, Bolzano is not too clear here. He states:

> "Moreover, as in the case in which the proposition [*sc.*, *A has b*] is true, the circumstance that the quality *b* belongs to the object under *A* may be regarded as a relation between *A* and *b*, this leads to the conclusion: the relation between *A* and *b* is the relation of certain objects to a quality which belongs to them. This conclusion, as can be easily judged, is *equivalent* to the given proposition [. . .]"[18]

Before moving on to a rigorous systematization of this important kernel of Bolzanian insights, let us emphasize that the collection of facts involved in the truth of a proposition is in no way *univocally determined.* Indeed, (1) nothing prevents one and the same object from occurring as the first component of more than one fact and (2) nothing prevents the situation occurring in which under $\bar{c}$ also fall qualities that do not occur in the facts. Hence, in general, *the truth of a proposition may be ensured by more than one situation.*

If now by speaking of the '*objects involved in the situation m*' we mean the first components of the facts that constitute m (i.e. all $\pi_1(k)$ such that $k \in m$) and by speaking of the '*qualities involved in m*' we mean the second components of those facts (i.e. all $\pi_2(k)$, for $k \in m$), then we may characterize the relation of *enunciating* [*aussagen*], which subsists between propositions and situations, by saying that *the proposition $h(a, \bar{c})$ enunciates the situation m iff the objects involved in m are exactly those to which a refers and the qualities involved in m are all among those referred to by* $\bar{c}$. Writing '$m \mathrel{\text{A}} p$' for '*the proposition p enunciates the situation m*', we then set

$$m \mathrel{\text{A}} h(a,\bar{c}) \leftrightarrow \forall i(\exists j(j \in m \wedge i = \pi_1(j)) \leftrightarrow i \mathrel{\text{E}} a) \wedge \forall r(\exists j(j \in m \wedge r = \pi_2(j)) \rightarrow r \mathrel{\text{E}} \bar{c})) \qquad \text{D6.3}$$

We may now prove that *if the proposition enunciates a situation, then it is true*; i.e.

$$\exists m(m \mathrel{\text{A}} h(a,\bar{c})) \rightarrow W(h(a,\bar{c})) \qquad \text{T6.4}$$

Proof Let m be a situation enunciated by the proposition $h(a, \bar{c})$. Then, by definition, we have (i) $\forall i(\exists j(j \in m \wedge i = \pi_1(j)) \leftrightarrow i \mathrel{\text{E}} a)$ and (ii) $\forall r(\exists j(j \in m \wedge r = \pi_2(j)) \rightarrow r \mathrel{\text{E}} \bar{c})$. Being a quality, m belongs to something; in other words, there is at least a j such that $\in m$ and, moreover, being a situation, m belongs to facts, and

[18] Cf. TS §225, II 287.

hence j is a fact; therefore, there is an object i and a quality r such that $j = < i, r >$. But, by (i), every such i falls under a; thus (a) $\exists i(i \mathrel{\varepsilon} a)$. On the other hand, let i be any object such that $i \mathrel{\varepsilon} a$. Then, by (i), there is a fact j such that $j \in m$ and has the form $< i, r >$ for some r. By the definition of 'fact', for such i and r, we have $i \in r$ and, by (ii), the r of every fact of m falls under $\bar{c}$; hence, in particular, $r \mathrel{\varepsilon} \bar{c}$. Summarizing, (b) $\forall i(i \mathrel{\varepsilon} a \rightarrow \exists r(i \in r \wedge r \mathrel{\varepsilon} \bar{c}))$. But (a) and (b) say just $W(h(a, \bar{c}))$.

To prove the converse, as remarked in the informal discussion, definitions are not sufficient. Indeed, with every true proposition is naturally associated a collection of facts that has the two mentioned characteristics; nevertheless, nobody can assure us that such a collection is a situation. Now, the truth of the proposition $h(a, \bar{c})$ implies, in particular, the objectuality of the idea $[a](\bar{c})$ and the reference of this idea—the totality of objects under a to which a quality under $\bar{c}$ belongs—obviously involves the collection of those facts whose first components are the objects falling under a, whereas their second components are the relations under $\bar{c}$ that belong to those objects. The assumption that this collection is a situation (i.e. the assumption that *to each objectual idea of the form* $[a](\bar{c})$ *corresponds (at least) a situation made up from all and only all facts the first components of which is an object falling under a whereas their second component is a quality falling under* $\bar{c}$) is sufficient to prove the converse of the preceding theorem.

We thus assume

$$G([a](\bar{c})) \rightarrow \exists m \forall j(j \in m \leftrightarrow \exists ir(i \mathrel{\varepsilon} a \wedge r \mathrel{\varepsilon} \bar{c} \wedge i \in r \wedge j = < i, r >)) \qquad \text{CQ7}$$

and may use this to prove that *if a proposition is true, then it enunciates a situation*; i.e.

$$W(h(a, \bar{c})) \rightarrow \exists m(m \mathrel{\text{A}} h(a, \bar{c})) \qquad \text{T6.5}$$

Proof Let $W(h(a, \bar{c}))$, that is, by definition, $\exists i(i \mathrel{\varepsilon} a)$ and $\forall i(i \mathrel{\varepsilon} a \rightarrow \exists r(i \in r \wedge r \mathrel{\varepsilon} \bar{c}))$. It follows that $\exists i(i \mathrel{\varepsilon} a \wedge \exists r(i \in r \wedge r \mathrel{\varepsilon} \bar{c}))$ (i.e. that $[a](\bar{c})$ is objectual) and hence, by the axiom, (i) there is a situation m that contains all and only all facts j of the form $< i, r >$ with $i \mathrel{\varepsilon} a$ and $r \mathrel{\varepsilon} \bar{c}$. Let i be the first component of some fact j of m; by virtue of (i), this i falls under a. Thus we know that (ii) $\forall i(\exists k(k \in m \wedge i = \pi_1(k)) \rightarrow i \mathrel{\varepsilon} a)$. Let us, vice versa, suppose $i \mathrel{\varepsilon} a$; then, by the hypothesis, there is an r such that $i \in r$ and $r \mathrel{\varepsilon} \bar{c}$ and hence, by (i), the fact $< i, r >$, which obviously satisfies the condition $i = \pi_1(< i, r >)$, is in m. Therefore, taking (ii) into account, we have (iii) $\forall i(\exists j(j \in m \wedge i = \pi_1(j)) \leftrightarrow i \mathrel{\varepsilon} a)$. Now let r be the second component of a fact $j \in m$; by virtue of (i), such an r is under $\bar{c}$. Hence (iv) $\forall r(\exists j(j \in m \wedge r = \pi_2(j)) \rightarrow r \mathrel{\varepsilon} \bar{c})$. From (iii) and (iv), it follows, by the definition, that $m \mathrel{\text{A}} h(a, \bar{c})$.

We have thus established the following:

(1) *A proposition is true iff it enunciates a situation.*
(2) *A proposition is false iff it does not enunciate a situation*; i.e.

$$(1)\ W(h(a,\overline{c})) \leftrightarrow \exists m(m \,\text{A}\, h(a,\overline{c})) \qquad \text{T6.6}$$
$$(2)\ F(h(a,\overline{c})) \leftrightarrow \neg\exists m(m \,\text{A}\, h(a,\overline{c}))$$

The parallelism between the *truth–falsity* pair *for propositions* and the *objectuality–objectlessness* pair *for ideas* that is so clearly stated by Bolzano in the passage quoted above does not, however, exhaust itself in this. It goes even further. As informally remarked above, indeed, the truth of a proposition implies the *existence*, but not the *uniqueness*, of a situation enunciated by it, just as the objectuality of an idea implies the *existence*, but not the *uniqueness*, of an object to which it refers. This consequence of the Bolzanian conviction that quality ideas may also have a plural reference deserves special attention. Indeed, the truth of one and the same propositions may be ensured by a series of *different* situations; the unique things that remain fixed are the objects *handled* by the proposition, whereas the qualities belonging to these objects may differ from each other, as long as they satisfy the conditions of falling all under the predicate-idea and no object under the subject-idea lacks all of them .

In conclusion, it is worth remarking that, with slight alterations, starting from the obvious reformulation of the general axiom

$$x \,\text{E}\, y \rightarrow (V(y) \vee B(y)) \wedge (V(y) \rightarrow O(x)) \wedge (B(y) \rightarrow Q(x))$$

into the axiom

$$x \,\text{E}\, y \rightarrow (V(y) \vee B(y) \vee S(x)) \wedge (V(y) \rightarrow O(x)) \wedge (B(y) \rightarrow Q(x)) \\ \wedge (S(y) \rightarrow Sit(x))$$

one can avoid the introduction of 'A', making the parallelism ideas–propositions, objectuality–truth even more evident.

6.4 Qualities of Qualities

As is clearly from this famous excerpt of a letter to Exner,

> "*This* red (numerically identical) cannot be on a second rose. The red on another rose may result, if you like, *equal, very equal* to that; but it cannot be the *same* because the rose is not the same; for *two* roses, *two* reds are required",[19]

Bolzano was well aware of the problem to which many refer today—with a terminological choice of rare inappropriateness—as the 'tropes' issue; nevertheless, we are not aware of any passages in the WL in which these remarks find any further development. In *Athanasia*, on the contrary, we find two remarks that seem not to have received the attention they deserve.

Let us begin by introducing an ontological operation that associates with a quality r and an object i—to which that quality belongs—the singular quality 'the r of i', which we will denote by '$r \circ i$' and which must satisfy the conditions (a) of belonging to i and (b) of not belonging to anything else. Hence, writing '$r \ni i$' instead of '$i \in r$' (as we will always do in this section), we assume the two axioms

$$r \ni i \leftrightarrow (r \circ i) \ni i \qquad \text{GA23}$$

$$(r \circ i) \ni j \rightarrow i = j. \qquad \text{GA24}$$

It follows that the r *of* i belongs to a j iff r belongs to j and moreover j coincides with i; i.e.,

$$(r \circ i) \ni j \leftrightarrow r \ni j \wedge j = i \qquad \text{T6.7}$$

Proof Let $(r \circ i) \ni j$. From this, by the second axiom, it follows that (i) $j = i$ and hence also $(r \circ i) \ni i$, from which, by the first axiom, $r \ni i$, and so also (ii) $r \ni j$. Vice versa, let (i) $r \ni j$ and (ii) $j = i$. From these, it follows that $r \ni i$ and hence, by the first axiom, $(r \circ i) \ni i$ and, from this and (ii), $(r \circ i) \ni j$.

Let us now remark that on the higher level, where also qualities are objects, our operation may act on a quality of qualities and a quality to which it belongs. In this regard, the first important remark Bolzano makes is the following:

> "Moreover, it cannot be denied that any quality which we can detect on another quality can be conceived by us, from another point of view, as a quality of the object to which the latter belongs. When, for instance, we remark that the memory of a man has an unusual force, then we must not consider 'unusual force in general' as a quality belonging to that man but 'unusual force of memory' must be a quality which we can truly attribute to him."[20].

[19] Bolzano-Exner [1935, pp. 32–3]: "Dies Roth (numero idem) kann sich an keiner zweyten Rose finden. Das Roth, das sich an einer zweyten Rose findet, kann jenem, wenn Sie wollen, gleich, sehr gleich kommen, aber dasselbe kann es nicht seyn, weil es nicht dieselbe Rose ist; zu zwey Rosen werden zwey Röthen erfordert."

[20] Athanasia 22–3: "Unleugbar ist es ferner, daß wir jede Beschaffenheit, die wir an einer andern Beschaffenheit entdecken, unter einem gewissen Gesichtspunct auch als Beschaffenheit des Gegenstandes, welchem die letzten

In the attempt to rigorously assess this, at least *prima facie* completely convincing Bolzanian remark, let us turn to some examples.

Example 1 (3 terms)

i : intensity
m : memory (mnemonic ability)
p : Pico (della Mirandola)

We have

1	$i \ni (m \circ p)$	intensity *belongs* to the (memory of Pico)
2	$(i \circ m) \ni p)$	(intensity of the memory) *belongs* to Pico)

Example 2 (4 terms)

e : exceptionality
i : intensity
m : memory (mnemonic ability)
p : Pico (della Mirandola)

We have

1	$e \ni (i \circ (m \circ p))$	exceptionality *belongs* to the (intensity of the (memory of Pico))
2	$e \ni ((i \circ m) \circ p)$	exceptionality *belongs* to the ((intensity of the memory) of Pico)
3	$(e \circ i) \ni (m \circ p)$	(exceptionality of the intensity) *belongs* to the (memory of Pico)
4	$(e \circ (f \circ m)) \ni p$	((exceptionality of the intensity) of the memory) *belongs* to Pico
5	$((e \circ i) \circ m) \ni p$	(exceptionality of the (intensity of the memory)) *belongs* to Pico

Example 3 (5 terms)

r : reputation
e : exceptionality
i : intensity
m : memory (mnemonic ability)
p : Pico (della Mirandola)

zukommt, auffassen können. Wenn wir z. B. bemerken, daß das Gedächtniß eines Menschen eine ungewöhnliche Stärke besitze; so muß zwar nicht eben "ungewöhnliche Stärke überhaupt" – als eine diesem Menschen zukommende Beschaffenheit zu betrachten, wohl aber muß "ungewöhnliche Stärke des Gedächtnisses" eine Beschaffenheit sein, welche wir ihm in Wahrheit beilegen können."

We have

1	$r \ni (e \circ (i \circ (m \circ p)))$	reputation *belongs* to the (exceptionality of the (intensity of the (memory of (Pico)))
2	$r \ni (e \circ ((i \circ m) \circ p))$	reputation *belongs* to the (exceptionality of the ((intensity of the memory) of Pico))
3	$r \ni ((e \circ i) \circ (m \circ p))$	reputation *belongs* to the ((exceptionality of the intensity) of the (memory of Pico))
4	$r \ni ((e \circ (i \circ m)) \circ p)$	reputation *belongs* to the ((exceptionality of the (intensity of the memory)) of Pico)
5	$r \ni (((e \circ i) \circ m) \circ p)$	reputation *belongs* to the (((exceptionality of the intensity) of the memory) of Pico)
6	$(r \circ e) \ni (i \circ (m \circ p))$	(reputation of the exceptionality) *belongs* to the (intensity of the (memory of Pico))
7	$(r \circ e) \ni ((i \circ m) \circ p)$	(reputation of the exceptionality) *belongs* to the ((intensity of the memory) of Pico)
8	$(r \circ (e \circ i)) \ni (m \circ p)$	(reputation of the (exceptionality of the intensity)) *belongs* to the (memory of Pico)
9	$((r \circ e) \circ i) \ni (m \circ p)$	((reputation of the exceptionality) of the intensity) *belongs* to the (memory of Pico)
10	$(r \circ (e \circ (i \circ m))) \ni p$	(reputation of the (exceptionality of the (intensity of the memory))) *belongs* to Pico
11	$(r \circ ((e \circ i) \circ m)) \ni p$	(reputation of the ((exceptionality of the intensity) of the memory)) *belongs* to Pico
12	$((r \circ e) \circ (i \circ m)) \ni p$	((reputation of the exceptionality) of the (intensity of the memory)) *belongs* to Pico
13	$((r \circ (e \circ i)) \circ m) \ni p$	((reputation of the (exceptionality of the intensity)) of the memory) *belongs* to Pico
14	$(((r \circ e) \circ i) \circ m) \ni p$	(((reputation of the exceptionality) of the intensity) of the memory) *belongs* to Pico

The obvious ascertainment of the equivalence, in each of these examples, among all the possible 'parenthesizations' compels us to find a justification.

Thus, we first remark that a well-known theorem of combinatorics states that for a given 'word' $x_1 \ldots x_k$ of k (> 1) letters there are exactly

$$\mathbb{C}(k) = \frac{1}{k}\binom{2(k-1)}{k-1} = \frac{1}{k}\frac{(2(k-1))!}{(k-1)!\,(2(k-1)-(k-1))!} = \frac{(2k-2)!}{(k-1)!\,(k-1)!\,k}$$

$$= \frac{(2k-2)!}{(k-1)!\,k!}$$

different ways to insert $k-1$ pairs of parentheses (external included) in such a way that within a pair there are always two and only two words. Numbers of this type are called *Catalan numbers* (from Eugène Charles Catalan [Bruges 1814–Liège 1894]) or also, but less frequently, *Segner numbers* (from Johann Andreas Segner [1704–1777]).

The first 24 Catalan numbers (from the *On-Line Encyclopedia of Integer Sequences*) are as follows:

$k :=$	2	3	4	5	6	7	8	9	10	11	12	13
$\mathbb{C}(k) =$	1	2	5	14	42	132	429	1430	4862	16796	58786	208012

$k :=$	14	15	16	17	18	19
$\mathbb{C}(k) =$	742900	2674440	9694845	35357670	129644790	477638700

$k :=$	20	21	22	23	24
$\mathbb{C}(k) =$	6564120420	24466267020	91482563640	343059613650	1289904147324

Turning to the determination of the principles needed to account for the above equivalence, we first of all accept the schematic reformulation of Bolzano's affirmation, i.e.

$$r \ni (s \circ t) \leftrightarrow (r \circ s) \ni t \qquad \text{GA25}$$

as well as the associativity of $\circ$, which results from the preceding examples, i.e.

$$((r \circ s) \circ t) \ni i \leftrightarrow (r \circ (s \circ t)) \ni i \qquad [\text{i.e.} \quad ((r \circ s) \circ t) \equiv (r \circ (s \circ t))] \qquad \text{GA26}$$

and finally the natural substitutivity of $\equiv$ with respect to $\circ$, i.e.

$$\text{E} \quad r \equiv s \rightarrow (r \circ t) \equiv (s \circ t) \wedge (t \circ r) \equiv (t \circ s)$$

By these principles, we may prove the following theorem:

T6.8 If r and r' are two parenthesizations of $r_1 \circ \cdots \circ r_i$ and $r_{i+1} \circ \cdots \circ r_k$, respectively, and s and s' are two parenthesizations of $r_1 \circ \cdots \circ r_j$ and $r_{j+1} \circ \cdots \circ r_k$, respectively, then

$$r \ni r' \leftrightarrow s \ni s'$$

The proof rests on two lemmata. The first lemma guarantees that every *attribution* may be equivalently reduced to an attribution to the last term; i.e.

L1. If r and s are two parenthesizations of $r_1 \circ \cdots \circ r_i$ and $r_{i+1} \circ \cdots \circ r_k$, respectively, then there is a term t such that $r \ni s \leftrightarrow t \ni r_k$.

Proof This is by induction on the construction of s.

(B) $s = r_k$. There is nothing to prove.

(P) $s = (s_1 \circ s_2)$. By A1, $r \ni (s_1 \circ s_2) \leftrightarrow (r \circ s_1) \ni s_2$. As s_2 is less complex than s, we may apply the induction hypothesis.

The second lemma guarantees that every quality term resulting from any parenthesization of i quality terms is equivalent to the term, to be called *in normal form,* that is obtained from the same terms by left parenthesization.

L2. For every k and $r_1, \ldots, r_k$, if t is any parenthesization of $r_1 \circ \cdots \circ r_k$, then there is a term t' in normal form (i.e. $(\ldots(((r_1 \circ r_2) \circ r_3) \circ \ldots) \circ r_k)$ such that $t \equiv t'$.

Proof This is by induction on the 'length' $\ell(t)$ of t.

(B1) $\ell(t) = 1$. Then $t = r_1$ and hence $t' = r_1$.

(B2) $\ell(t) = 2$. Then $t = (r_1 \circ r_2)$ and hence $t' = (r_1 \circ r_2)$.

(P) $\ell(t) > 2$. Then $t = (r \circ s)$ and the induction hypothesis is that the theorem holds for every u with $\ell(u) < \ell(t) = \ell(r) + \ell(s)$. As $\ell(r) < \ell(t)$, there is an r' in normal form such that $r \equiv r'$ and hence, by E, $t \equiv (r' \circ s)$. We now proceed by induction on $\ell(s)$.

(PB) $\ell(s) = 1$. Then $s = r_k$ and $t \equiv (r' \circ r_k)$ and $t' = (r' \circ x)$ is in normal form.

(PP) $\ell(s) > 1$. Then $s = (s_1 \circ s_2)$ and $t \equiv (r' \circ (s_1 \circ s_2))$. By A2, we then have $t \equiv ((r' \circ s_1) \circ s_2)$ with $\ell((r' \circ s_1)) < \ell(t)$ and thus, by the hypothesis of P, there is an r'' in normal form such that $t \equiv (r'' \circ s_2)$. But $\ell(s_2) < \ell(s)$ and so, by the induction hypothesis of PP, there is a t' in normal form such that $t' \equiv (r'' \circ s_2)$ and hence also such that $t \equiv t'$.

We can now prove the theorem. Let r and r' be any two parenthesizations of $r_1 \circ \cdots \circ r_i$ and $r_{i+1} \circ \cdots \circ r_k$, respectively, and let s and s' be any two parenthesizations of $r_1 \circ \cdots \circ r_j$ and $r_{j+1} \circ \cdots \circ r_k$, respectively. By L1, there are r'' and s'' such that $r \ni r' \leftrightarrow r'' \ni r_k$ and $s \ni s' \leftrightarrow s'' \ni r_k$. By L2, r'' and s'' may be equivalently transformed into a term in normal form, which is, obviously, in both cases the same. Hence

$$r \ni r' \leftrightarrow r'' \ni r_k \leftrightarrow t \ni r_k \leftrightarrow s'' \ni r_k \leftrightarrow s \ni s'$$

It is clear that, apart from notation, the preceding theorem leaves the question about the categorical nature of r_k completely open (i.e. whether it is still a quality or, on the contrary, a pure object). The second remark of Bolzano, is however that, at least in the case of the real world, a descending chain of qualities must always end on a pure, existing object (i.e., on a *substance*). Indeed, the passage quoted above from *Athanasia* continues as follows:

> "From this it becomes clear that for every adherence there must be one or more substances such that we can represent it as adherent to them. Indeed, even if one wants to describe this adherence only as a quality on another quality and then to do the same with this and to continue in this way, one should however admit that there is a last one of these qualities, that is a quality which can not still be regarded as adherent to another quality, but only as being on one or more substances. From this admission, however, it would at once follow that all other qualities, including that one being discussed, may be considered from a point of view from which they appear as qualities immediately belonging to the mentioned substances."[21]

[21] *Athanasia*, 23: "Hieraus erhellt, daß es für jede Adhärenz irgend eine oder mehre Substanzen geben müsse, an welchen wir sie uns als haftend vorstellen können. Denn selbst wenn man uns diese Adhärenz nur als Beschaffenheit von einer andern Beschaffenheit schildern, bei dieser dasselbe thun, und so beständig fortfahren wollte, so müßte doch immer zugegeben werden, daß es eine letzte dieser Beschaffenheiten gebe, eine Beschaffenheit nämlich, die sich nicht wieder als haftend an einer andern Beschaffenheit, sondern nur als befindlich an einer oder mehren Substanzen ansehen läßt. Sobald man uns aber dies zugegeben hätte, so würde folgen, daß auch die übrigen Beschaffenheit alle, mithin auch diejenige, die in der Frage stehet, aus einem Gesichtspunct betrachtet werden können, bei welchem sie als Beschaffenheiten erscheinen, die den erwähnten Substanzen unmittelbar zukommen."

7

Etiology

7.1 Consecutivity

The basic notion of the *pure etiology* is the relation of *consecutivity* [*Abfolge*] *among truths* "by virtue of which some of them are *reasons* [*Gründe*] of certain others and these, in turn, *consequences* [*Folgen*] of those".[1]

It is to the study of this relation, which is introduced in §162 and taken up again in §168, that Bolzano devotes, in particular, his §§ 198–222.

Even though the basic insights of the Theory of Consecutivity appear, at least at first sight, rather clear, its systematic reconstruction presents remarkable difficulties. In fact, the Bolzanian treatment contains ambiguities, too. Thus, what follows should not be intended as a definitive assessment of the issue, but rather as an articulate—albeit not always convincing—attempt to account also for the details of this theory. It must, however, be emphasized, as clearly emerges from the introductory section to the Theory of Consecutivity, that Bolzano did not consider his investigation definitive. Indeed, he wrote:

> "As however the investigations into which I shall now enter are extremely difficult and there is almost no preliminary work which I could use in this sector, I beg in advance the reader not to expect in this chapter, even less than in the preceding ones, something nearing any completeness. Almost all I here expose is imbued with uncertainties for me, too; on certain things, I do not even dare to give a settling judgment. My investigations are, in the best case, only fragments and suggestions which will reach their aim if they will induce somebody else to further reflections on these matters."[2]

[1] TS §198, II 243: "In my opinion, the most important of all relations that can hold between truths obtains when some of them form a *ground* and others their *consequence*."

[2] Cf. TS §195, II 327–8.

The first difficulty one encounters in this attempt arises from the fact that Bolzano, even though he worries about distinguishing—also terminologically—between '*partial reasons* [*Theilgründe*]' and '*complete reason* [*vollständiger Grund*]', respectively between '*partial consequences* [*Theilfolgen*]' and '*complete consequence* [*vollständige Folge*]', does not always scrupulously adhere to this distinction.[3] In fact, we are faced with at least seven relations.

As, however, the first characteristic of these relations is to subsist among truths or aggregates of truths,[4] it is useful to introduce special variables both for truths and for ideas of truths.[5] To this end, we first introduce the predicate '*being an idea of truths*' (i.e. '*being an idea of true propositions*'), to be denoted by '*VW*', setting

$$VW(x) \leftrightarrow VS(x) \wedge \forall z(z \mathrel{\textsc{e}} x \rightarrow W(z)) \qquad \text{D7.1}$$

and we then establish the use (possibly with indices) of 'v' and 'w' as special variables for *truths*[6] and '$\mathfrak{v}$' and '$\mathfrak{w}$' as special variables for *ideas of truths.*

The seven relations already alluded to are the following:

(1) The truth v is a *(partial) reason for the truth* w or, equivalently, the truth w is a *(partial) consequence of the truth* v.

(2) The truths (under) $\mathfrak{v}$ are *(partial) reasons of the truth* w or, equivalently, the truth w is a *(partial) consequence of the truths* (under) $\mathfrak{v}$.

(3) The truth v is a *(partial) reason for the truths* (under) $\mathfrak{w}$ or, equivalently, the truths (under) $\mathfrak{w}$ are *(partial) consequences of the truth* v.

[3] TS §198, II 244: "sometimes it is not just one, but a whole collection of truths, which stand to one or more truths in the relation of ground to consequence; it is permissible in such a case to call the individual truths which make up these classes, *partial grounds* and *partial consequences*, and indeed simply to call them grounds and consequences, provided that we take care to ensure that no misunderstanding stems from the double meaning of these words. For instance, when we are not speaking of grounds or consequences which are merely partial, it is more precise to call them *complete*."

[4] TS §203, II 254: "1) [...] I believe [...] that the relation of ground to consequence has this in common with deducibility, that it applies only to propositions (and thus not to anything that is actual); yet it differs from deducibility in that it can only obtain between *true* propositions or truths."

[5] The adequacy of our treatment of the aggregates of truths through 'ideas of truths' and hence, in particular, the renunciation of any ordering of the involved truths finds an explicit confirmation in §211, where, under the title *Whether the parts of a ground or of a consequence are rank ordered?*, we read: "With the relation of *deducibility*, [...] there is no rule according to which one must consider one of these propositions the first, another the second, and so on, in order to discover that they stand in the relation of deducibility. This is also the case with the relation of ground and consequence, when either the ground or the consequence consists of several truths. Here too it is in no way necessary to think of the individual propositions as standing in a rank ordering, with one of them the first, another the second, and so on; rather, these collections are mere *multitudes* (§. 84)."

[6] It seems that no confusion should be possible with the (constant) quality idea '$\overline{w}$'.

(4) The truths (under) $\mathfrak{v}$ are *(partial) reasons for the truths* (under) $\mathfrak{w}$ or, equivalently, the truths (under) $\mathfrak{w}$ are *(partial) consequences of the truths* (under) $\mathfrak{v}$.

(5) The truths (under) $\mathfrak{v}$ are *the complete reason for the truths* (under) $\mathfrak{w}$.

(6) The truths (under) $\mathfrak{w}$ are *the complete consequence of the truths* (under) $\mathfrak{v}$.

(7) The truths (under) $\mathfrak{v}$ are *the complete reason for the truths* (under) $\mathfrak{w}$, which, on their part, are the *complete consequence of the truths (under)* $\mathfrak{v}$.

Whereas it is clearly obvious that not all these relations are reciprocally independent (i.e. some can be expressed in terms of others), on the contrary, it is not at all clear which ones or which one should be taken as primitive. Among the various possibilities that seem to be at our disposal, we prefer to adopt the first one (to be denoted by '$v \rhd w$') as unique primitive relation and employ it to define all the others. Strictly speaking, for each of these other relations, we should introduce a particular sign, for instance the '$\rhd$' enriched by an index. However, the use of the special variables allows us to avoid, with no loss of univocity, such burdening of the notations in cases (2)–(4); in the remaining three cases, we will adopt the three notations '$\}\!\rhd$', '$\rhd\!\{$', and '$\rhd\!\!\!\rhd$', respectively.[7]

The definitions that we accept follow.

The $\mathfrak{v}$ are *reasons for the w* ($\mathfrak{v} \rhd w$) or, also, the w is *consequence of the* $\mathfrak{v}$, iff *each one* of the $\mathfrak{v}$ is reason for that w (i.e. iff the w is a consequence of *each* $\mathfrak{v}$):

$$\mathfrak{v} \rhd w \leftrightarrow \forall v(v \,\mathrm{E}\, \mathfrak{v} \rightarrow v \rhd w) \qquad \text{D7.2}$$

Convention: When $\mathfrak{v}$ is $\lceil v_1 \rceil \sqcup \ldots \sqcup \lceil v_n \rceil$, instead of '$\lceil v_1 \rceil \sqcup \ldots \sqcup \lceil v_n \rceil \rhd v$', we will simply write '$v_1, \ldots, v_n \rhd v$'; i.e.

$$v_1, \ldots, v_n \rhd v := \lceil v_1 \rceil \sqcup \ldots \sqcup \lceil v_n \rceil \rhd v \qquad \text{MD7.1}$$

The v is a *reason for the* $\mathfrak{w}$ ($v \rhd \mathfrak{w}$) or, also, the $\mathfrak{w}$ are *consequences of the* $\mathfrak{v}$ iff v is a reason for *each* $\mathfrak{w}$ (i.e. iff *each* $\mathfrak{w}$ is consequence of v):

$$v \rhd \mathfrak{w} \leftrightarrow \forall w(w \,\mathrm{E}\, \mathfrak{w} \rightarrow v \rhd w) \qquad \text{D7.3}$$

The $\mathfrak{v}$ are *reasons for the* $\mathfrak{w}$ ($\mathfrak{v} \rhd \mathfrak{w}$) or, also, the $\mathfrak{w}$ *are consequences of the* $\mathfrak{v}$ iff *each* $\mathfrak{v}$ is a reason for *each* $\mathfrak{w}$ (i.e. iff *each* $\mathfrak{w}$ is a consequence of *each* $\mathfrak{v}$):

$$\mathfrak{v} \rhd \mathfrak{w} \leftrightarrow \forall vw(v \,\mathrm{E}\, \mathfrak{v} \wedge w \,\mathrm{E}\, \mathfrak{w} \rightarrow v \rhd w) \qquad \text{D7.4}$$

[7] In the only serious attempt we know of to give a formal assessment to the theory of consecutivity (by A. Tatzel), the primitive relation assumed is, in contrast, the relation $\rhd\!\!\!\rhd$

The $\mathfrak{v}$ are the *complete reason* for the $\mathfrak{w}$ ($\mathfrak{v} \}\!\rhd \mathfrak{w}$) iff they are reasons for the $\mathfrak{w}$ and, moreover, any truth that is a reason for all the $\mathfrak{w}$ is already under $\mathfrak{v}$:

$$\mathfrak{v} \}\!\rhd \mathfrak{w} \leftrightarrow \mathfrak{v} \rhd \mathfrak{w} \wedge \forall v(\forall w(w \,\mathrm{E}\, \mathfrak{w} \rightarrow v \rhd w) \rightarrow v \,\mathrm{E}\, \mathfrak{v}) \qquad \text{D7.5}$$

The $\mathfrak{w}$ are the *complete consequence* of the $\mathfrak{v}$ ($\mathfrak{v} \rhd\!\{ \mathfrak{w}$) iff they are consequences of the $\mathfrak{v}$ and, moreover, any truth that is a consequence of all the $\mathfrak{v}$ is already under $\mathfrak{w}$:

$$\mathfrak{v} \rhd\!\{ \mathfrak{w} \leftrightarrow \mathfrak{v} \rhd \mathfrak{w} \wedge \forall w(\forall v(v \,\mathrm{E}\, \mathfrak{v} \rightarrow v \rhd w) \rightarrow w \,\mathrm{E}\, \mathfrak{w}) \qquad \text{D7.6}$$

A *complete consecutivity* subsists between the $\mathfrak{v}$ and the $\mathfrak{w}$ iff the $\mathfrak{v}$ are the complete reason for the $\mathfrak{w}$ and these, in turn, are the complete consequence of the $\mathfrak{v}$:

$$\mathfrak{v}\} \rhd \{\mathfrak{w} \leftrightarrow \mathfrak{v}\} \rhd \mathfrak{w} \wedge \mathfrak{v} \rhd \{\mathfrak{w} \qquad \text{D7.7}$$

To make these definitions more perspicuous, we begin by noting that a trivial logical transformations allow us to equivalently present the relation $\mathfrak{v} \rhd \mathfrak{w}$ in two forms:

$$\begin{aligned} &(1)\ \mathfrak{v} \rhd \mathfrak{w} \leftrightarrow \forall w(w \,\mathrm{E}\, \mathfrak{w} \rightarrow \forall v(v \,\mathrm{E}\, \mathfrak{v} \rightarrow v \rhd w)) \\ &(2)\ \mathfrak{v} \rhd \mathfrak{w} \leftrightarrow \forall v(v \,\mathrm{E}\, \mathfrak{v} \rightarrow \forall w(w \,\mathrm{E}\, \mathfrak{w} \rightarrow v \rhd w)) \end{aligned} \qquad \text{T7.1}$$

It follows that we may reformulate the relation $\mathfrak{v}\} \rhd \mathfrak{w}$ as follows:

$$\mathfrak{v} \}\!\rhd \mathfrak{w} \leftrightarrow \forall v(v \,\mathrm{E}\, \mathfrak{v} \rightarrow \forall w(w \,\mathrm{E}\, \mathfrak{w} \rightarrow v \rhd w)) \wedge \forall v(\forall w(w \,\mathrm{E}\, \mathfrak{w} \rightarrow v \rhd w) \rightarrow v \,\mathrm{E}\, \mathfrak{v})$$

and hence, also, by trivial logical passages,

$$\mathfrak{v} \}\!\rhd \mathfrak{w} \leftrightarrow \forall v((v \,\mathrm{E}\, \mathfrak{v} \rightarrow \forall w(w \,\mathrm{E}\, \mathfrak{w} \rightarrow v \rhd w)) \wedge (\forall w(w \,\mathrm{E}\, \mathfrak{w} \rightarrow v \rhd w) \rightarrow v \,\mathrm{E}\, \mathfrak{v}))$$

as well as

$$\mathfrak{v} \}\!\rhd \mathfrak{w} \leftrightarrow \forall v(v \,\mathrm{E}\, \mathfrak{v} \leftrightarrow \forall w(w \,\mathrm{E}\, \mathfrak{w} \rightarrow v \rhd w))$$

and finally

$$\mathfrak{v}\} \rhd \mathfrak{w} \leftrightarrow \forall v(v \,\mathrm{E}\, \mathfrak{v} \leftrightarrow v \rhd \mathfrak{w}) \qquad \text{T7.2}$$

Hence, *the $\mathfrak{v}$ are the complete reason for the $\mathfrak{w}$ iff the necessary and sufficient condition to be a $\mathfrak{v}$ is to be a reason for the $\mathfrak{w}$.*

Applying analogous arguments to the relation $\mathfrak{v} \rhd\!\{ \mathfrak{w}$, we obtain successively

$$\mathfrak{v} \rhd\!\{ \mathfrak{w} \leftrightarrow \forall w(w \,\mathrm{E}\, \mathfrak{w} \rightarrow \forall v(v \,\mathrm{E}\, \mathfrak{v} \rightarrow v \rhd w)) \wedge \forall w(\forall v(v \,\mathrm{E}\, \mathfrak{v} \rightarrow v \rhd w) \rightarrow w \,\mathrm{E}\, \mathfrak{w})$$

$$\mathfrak{v} \rhd\{ \mathfrak{w} \leftrightarrow \forall w((w \mathrel{\mathsf{E}} \mathfrak{w} \rightarrow \forall v(v \mathrel{\mathsf{E}} \mathfrak{v} \rightarrow v \rhd w)) \wedge (\forall v(v \mathrel{\mathsf{E}} \mathfrak{v} \rightarrow v \rhd w) \rightarrow w \mathrel{\mathsf{E}} \mathfrak{w})))$$
$$\mathfrak{v} \rhd\{ \mathfrak{w} \leftrightarrow \forall w(w \mathrel{\mathsf{E}} \mathfrak{w} \leftrightarrow \forall v(v \mathrel{\mathsf{E}} \mathfrak{v} \rightarrow v \rhd w))$$

and finally

$$\mathfrak{v} \rhd\!\!\rhd \mathfrak{w} \leftrightarrow \forall w(w \mathrel{\mathsf{E}} \mathfrak{w} \leftrightarrow \mathfrak{v} \rhd w) \qquad \text{T7.3}$$

Hence, *the* $\mathfrak{w}$ *are the complete consequence of the* $\mathfrak{v}$ *iff the necessary and sufficient condition to be a* $\mathfrak{w}$ *is to be a consequence of the* $\mathfrak{v}$.

It follows, as far as the complete consecutivity is concerned, that

$$\mathfrak{v} \rhd\!\!\rhd \mathfrak{w} \leftrightarrow \forall v(v \mathrel{\mathsf{E}} \mathfrak{v} \leftrightarrow v \rhd \mathfrak{w}) \wedge \forall w(w \mathrel{\mathsf{E}} \mathfrak{w} \leftrightarrow \mathfrak{v} \rhd w) \qquad \text{T7.4}$$

Turning to the consideration of the properties of these relations, we begin with the question of their possible univocity. In this context, we remark that, as Bolzano thought, the relation of complete reason enjoys *right univocity* (i.e., an aggregate of truths constitutes the complete reason for (at most) one aggregate of truths):[8]

$$\mathfrak{v} \}\rhd \mathfrak{w} \wedge \mathfrak{v}' \}\rhd \mathfrak{w} \rightarrow \mathfrak{v} \equiv \mathfrak{v}' \qquad \text{T7.5}$$

Proof Let $\mathfrak{v} \}\rhd \mathfrak{w} \wedge \mathfrak{v}' \}\rhd \mathfrak{w}$; then $\forall v(v \mathrel{\mathsf{E}} \mathfrak{v} \leftrightarrow v \rhd \mathfrak{w})$ and $\forall v(v \mathrel{\mathsf{E}} \mathfrak{v}' \leftrightarrow v \rhd \mathfrak{w})$; but then, trivially, $\forall v(v \mathrel{\mathsf{E}} \mathfrak{v} \leftrightarrow v \mathrel{\mathsf{E}} \mathfrak{v}')$ and hence $\mathfrak{v} \equiv \mathfrak{v}'$.

Dually, the complete consequence is *univocal on the left* (i.e. an aggregate of truths constitutes the complete consequence of (at most) one aggregate of truths:[9]

$$\mathfrak{v} \rhd\{ \mathfrak{w} \wedge \mathfrak{v} \rhd\{ \mathfrak{w}' \rightarrow \mathfrak{w} \equiv \mathfrak{w}' \qquad \text{T7.6}$$

Proof Let $\mathfrak{v} \rhd\{ \mathfrak{w}$ and $\mathfrak{v} \rhd\{ \mathfrak{w}'$; then $\forall w(w \mathrel{\mathsf{E}} \mathfrak{w} \leftrightarrow \mathfrak{v} \rhd w)$ and $\forall w(w \mathrel{\mathsf{E}} \mathfrak{w}' \leftrightarrow \mathfrak{v} \rhd w)$; but then, trivially, $\forall w(w \mathrel{\mathsf{E}} \mathfrak{w} \leftrightarrow w \mathrel{\mathsf{E}} \mathfrak{w}')$, i.e. $\mathfrak{w} \equiv \mathfrak{w}'$.

As the complete consecutivity results from the conjunction of the two preceding relations, we have obtained a proof of Bolzano's fundamental conviction that *complete consecutivity is a one-to-one relation between aggregates of truths*:

$$\mathfrak{v} \rhd\!\!\rhd \mathfrak{w} \wedge \mathfrak{v}' \rhd\!\!\rhd \mathfrak{w}' \rightarrow \mathfrak{v} \equiv \mathfrak{v}' \wedge \mathfrak{w} \equiv \mathfrak{w}' \qquad \text{T7.7}$$

Another fundamental conviction held by Bolzano was that consecutivity between truths is an *asymmetric* relation; thus we assume

$$v \rhd w \rightarrow \neg w \rhd v \qquad \text{GA27}$$

[8] TS §206, II 258: "*Whether a ground can have several consequences or a consequence several grounds?* [...] For from §205 it is clear that many consequences cannot belong to a given ground, provided that we do not mean partial consequences. For all the truths which stand in the relation of consequence to ground to given truths are, taken together, their complete consequence."

[9] TS §206, II 258: "It is not so obvious that for a given consequence there is only one ground, and that different grounds always have different consequences."

This, of course, implies that also consecutivity between aggregates of truths is asymmetric:[10]

$$\mathfrak{v} \rhd \mathfrak{w} \rightarrow \neg(\mathfrak{w} \rhd \mathfrak{v}) \qquad \text{T7.8}$$

Proof If $\mathfrak{v} \rhd \mathfrak{w}$, then, by definition, for each pair, the first component of which is a v under $\mathfrak{v}$ and the second component is a w under $\mathfrak{w}$, it holds that $v \rhd w$; if now it were $\mathfrak{w} \rhd \mathfrak{v}$, then, for the same converted pairs, we would have $w \rhd v$, against the axiom.

Another immediate consequence of the axiom is the *irreflexivity* of the consecutivity:

$$\neg v \rhd v \qquad \text{T7.9}$$

Proof Obviously, $(v \rhd v \rightarrow \neg v \rhd v) \rightarrow \neg v \rhd v$ and so, because by the axiom $v \rhd v \rightarrow \neg v \rhd v$, it is also $\neg v \rhd v$.

There then follows the reinforcement of the preceding theorem, which asserts that in a relation of consecutivity (partial or complete), no truth is simultaneously reason and consequence:

$$\mathfrak{v} \rhd \mathfrak{w} \rightarrow \neg G(\mathfrak{v} \sqcap \mathfrak{w}) \qquad \text{T7.10}$$

Proof Suppose, indeed, that for some v, $v \in \mathfrak{v}$ and $v \in \mathfrak{w}$. From $\mathfrak{v} \rhd \mathfrak{w}$, it then follows that $v \rhd v$, which is absurd.[11]

Another basic property of complete consecutivity is its *intransitivity*, i.e.[12]

$$\mathfrak{v} \mathrel{\rhd\!\!\rhd} \mathfrak{u} \wedge \mathfrak{u} \mathrel{\rhd\!\!\rhd} \mathfrak{w} \rightarrow \neg(\mathfrak{v} \mathrel{\rhd\!\!\rhd} \mathfrak{w}) \qquad \text{T7.11}$$

Proof Indeed, let $\mathfrak{v} \mathrel{\rhd\!\!\rhd} \mathfrak{u}$ and $\mathfrak{u} \mathrel{\rhd\!\!\rhd} \mathfrak{w}$. If $\mathfrak{v} \mathrel{\rhd\!\!\rhd} \mathfrak{w}$, then $\mathfrak{w}$ would be a complete consequence both of $\mathfrak{v}$ and of $\mathfrak{u}$. Therefore, for the left univocity of the complete consecutivity, it would hold that $\mathfrak{v} \equiv \mathfrak{u}$ and hence also $\mathfrak{v} \rhd \mathfrak{v}$.

[10] TS §209, II 260: "2) It is clear that if the truths $A, B, C, \ldots$ are the complete ground of the complete consequence $M, N, O, \ldots$, then the latter cannot at the same time be the ground of the former. For if $M, N, O, \ldots$ were the complete consequence of $A, B, C, \ldots$, then they will include the proposition that $A, B, C, \ldots$ are all true. Hence that proposition would have to be a partial ground of the truths $A, B, C, \ldots$, which is absurd."

[11] TS §204, II 256: "[...] we can claim neither of an individual truth A, nor of a collection of truths $A, B, C, D, \ldots$ that they stand to themselves in the relation of ground to consequence, i.e., that they are their own ground, their own consequence."

[12] TS §213, II 267: "*Whether a consequence of a consequence can be considered a consequence of the ground?* [...] we usually have no objection to calling one truth a consequence of another when it is the consequence of its *consequence*; [...] Thus we call all the theorems of geometry consequences of the few principles they are derived from, though mediately. [...] It seems to me that the relation of *ground* to *consequence* is such that one cannot, without altering the concept, call the consequence of a consequence a consequence of a given ground (precisely because it is the consequence of the consequence). If one maintains the usual view, one will have to claim, contrary to §206, that the same consequence has many grounds."

The intransitivity of the complete consecutivity does not, however, imply the intransitivity of the partial consecutivity. In fact, Bolzano admits that the following situation may occur: the truths v and u are reasons for a certain truth w, but v is reason for u; i.e. he accepts the possibility of the situation: $v \rhd u \wedge u, v \rhd w$.[13] This possibility clearly contrasts with the intransitivity of the partial consecutivity; from $u, v \rhd w$, indeed, both $u \rhd w$ and $v \rhd w$ follow and therefore, since also $v \rhd u$, we have a triple of truths that simultaneously satisfies the three conditions $v \rhd u$, $u \rhd w$, and $v \rhd w$. Bolzano gives us an example of the type $\lceil v \rceil \rhd u \wedge \lceil u, v \rceil \rhd w$, from which $\lceil v \rceil \rhd u \wedge u \rhd w \wedge v \rhd w$. Here is Bolzano's example:

> "The proposition 'that every truth is something incorporeal' is a truth from which it follows that also this proposition is something incorporeal. Hence, if somebody should ask us about the reason for the following truth: 'The proposition that every truth is something incorporeal is itself something incorporeal' we could not avoid to answer that such a reason lies in the two following truths: 'Every truth is something incorporeal' and 'The proposition that every truth is something incorporeal is itself a truth.' The second of these two truths is clearly a (partial) consequence of the first one."[14]

What has been encoded up to now does not, however, allow us to know if any consecutivity relation among some v or some $\mathfrak{v}$ and some w or some $\mathfrak{w}$ exists. That such is the case results from the fundamental conviction held by Bolzano that "to the complete consequence of certain truths $A, B, C, D, \ldots$ belongs in particular also the truth that 'each one of the propositions A, B, C, D is true'" and, furthermore, that those truths even constitute the complete reason for this last truth.[15] Remembering that $h(\mathfrak{p}, \overline{w}) = \boxed{\wedge}\, \mathfrak{p}$, we thus assume[16]

$$\mathfrak{v} \,\}\!\rhd \boxed{\wedge}\, \mathfrak{v} \qquad \text{GA28}$$

[13] TS §212, II 288: "*Whether the partial grounds of a truth can be grounds or consequences of each other?* [...] It is important to investigate [...] whether several truths which, combined, form the ground of one or more others, may sometimes be related to one another as ground and consequence, or at least as partial grounds to partial consequences. The contrary claim that this cannot happen certainly recommends itself by its simplicity. Nevertheless, I believe that it is incorrect."

[14] Cf. TS §212, II 266.

[15] TS §206, II 258–9: "[...] The complete consequence of certain truths $A, B, C, D, \ldots$ includes, among others, also the truth that each of the propositions $A, B, C, D, \ldots$ is true. But this is a consequence (a partial consequence) only of the class of truths $A, B, C, D, \ldots$, and of no other."

[16] Let us remark that this formulation of the Bolzanian idea, which implies that every aggregate of truths has at least a consequence, should be coupled with some limitations on $\mathfrak{v}$. Indeed, if $\mathfrak{v}$ refers only to a finite set of truths there is no problem; in the infinite case, on the contrary, a certain amount of caution is necessary, because if every set of truths has consequences, then also the totality of truths should have consequences, in obvious contradiction with the irreflexivity. The question of the antinomies is discussed in Appendix E.

Let us remark at once that from this assumption we immediately obtain

$$(1)\ \lceil v \rceil \,\}\!\rhd\, h(\lceil v \rceil, \overline{w}) \qquad \text{T7.12}$$

$$(2)\ \lceil v_1 \rceil \sqcup \lceil v_2 \rceil \,\}\!\rhd\, \boxed{\wedge}(v_1, v_2),\ \text{i.e.}\ \lceil v_1 \rceil \sqcup \lceil v_2 \rceil \,\}\!\rhd\, h(\lceil v_1 \rceil \sqcup \lceil v_2 \rceil, \overline{w}))$$

$$(3)\ \lceil \overset{n}{v} \rceil \,\}\!\rhd\, \boxed{\wedge}(\overset{n}{v}),\ \text{i.e.}\ \lceil v_1 \rceil \sqcup \ldots \sqcup \lceil v_n \rceil \,\}\!\rhd\, h(\lceil v_1 \rceil \sqcup \ldots \sqcup \lceil v_n \rceil, \overline{w})$$

Moreover, we also remark that this theorem also leads to the succession that, together with the irreflexivity of $\rhd$, is used in one of the proofs given by Bolzano in the first part of the WL of the existence of infinitely many truths[17] and which, taken again in §13 of the *Paradoxien des Unendlichen*, will be used by Dedekind to prove the existence of at least one system of natural numbers:[18]

$$\lceil v \rceil \,\}\!\rhd\, h(\lceil v \rceil, \overline{w});$$
$$\lceil h(\lceil v \rceil, \overline{w}) \rceil \,\}\!\rhd\, h(\lceil h(\lceil v \rceil, \overline{w}) \rceil, \overline{w});$$
$$\lceil h(\lceil h(\lceil v \rceil, \overline{w}) \rceil, \overline{w}) \rceil \,\}\!\rhd\, h(\lceil h(\lceil h(\lceil v \rceil, \overline{w}) \rceil, \overline{w}) \rceil, \overline{w}); \ldots$$

The preceding quotation of Bolzano continues as follows: "This is however a consequence (in particular a partial consequence) which, besides it, has no other aggregate of truths. It becomes so clear that every particular reason has a consequence, which at least in some parts has in itself something peculiar."[19] Hence

$$\mathfrak{v} \,\}\!\rhd\, \boxed{\wedge}\, \mathfrak{v} \wedge \mathfrak{v}' \,\}\!\rhd\, \boxed{\wedge}\, \mathfrak{v} \rightarrow \mathfrak{v} \equiv \mathfrak{v}' \qquad \text{T7.13}$$

Proof This is by T5.

Therefore, on the whole, we have

$$\mathfrak{v} \,\}\!\rhd\, \boxed{\wedge}\, \mathfrak{v}' \leftrightarrow \mathfrak{v} \equiv \mathfrak{v}' \qquad \text{T7.14}$$

It should be noted that, as Bolzano himself emphasizes,[20] it seems that the idea that the conjunction of certain truths be also their unique consequence should be ruled out, i.e. the following statements are to be excluded:

$$(1)\ \mathfrak{v} \rhd\!\{\boxed{\wedge}\, \mathfrak{v} \qquad (2)\ \lceil v \rceil \rhd\!\{\, h(\lceil v \rceil, \overline{w})$$
$$(3)\ \lceil v_1 \rceil \sqcup \lceil v_2 \rceil \rhd\!\{\boxed{\wedge}(v_1, v_2) \qquad (4)\ \lceil \overset{n}{v} \rceil \rhd\!\{\boxed{\wedge}(\overset{n}{v})$$

[17] TS § 32, I, 108–9: "*Proof that there are several, even infinitely many, truths.* Note [...] If one wants to avoid the apagogical form, he may proceed in the following way: If the proposition '*A* is *B*' is true, then the claim 'The proposition that *A* is *B* is true' must undoubtedly also be a true proposition, and since the latter has different parts, it is a different proposition from '*A* is *B*' and thus a second and different truth."

[18] Cf. R. Dedekind, *Was sind und was sollen die Zahlen*, §5 n. 66 with relative Note.

[19] Cf. TS §206, II 258.

[20] TS §205, II 257: "2) [...] I have my doubts concerning whether the complete consequence ever consists of a single truth alone. Rather, I would claim even in cases where the ground is only a single truth, and still more in the case where it consists in several truths, that there are always several truths which, taken together, make up the consequence."

Once we have established that every truth always has at least one consequence, the question arises whether its converse is also true (i.e. whether every truth is also always consequence of at least one truth) or whether, on the contrary, there also are *primitive, basic truths* [*Grundwahrheiten*]. About this question, Bolzano expresses his opinion that there are some such truths—for example, the truth '*There is something* $[h(e, \bar{g})]$'—but he confesses to not having succeeded in finding a proof that could at least satisfy himself.[21] Indeed, he believed that there must be even more than one truth of this kind because he could not understand how every other truth could arise as consequence or consequence of consequence and so on, of a single truth. [22]

Turning to the problem of disposing of a proposition that somehow enunciates this consecutivity relation (i.e. disposing of a *consecutive proposition* [*Consecutivsatz*]), we may simply postulate, as has already been done in the case of the other connectives, the presence of a connective, denoted by '$\boxed{\rhd}$', that, when applied to the 'sets' of truths $\mathfrak{v}$ and $\mathfrak{w}$, generates a proposition that is true iff $\mathfrak{v} \rhd \mathfrak{w}$; i.e.

$$\forall \mathfrak{v}\mathfrak{w} \exists p (W(p) \leftrightarrow \mathfrak{v} \rhd \mathfrak{w}) \qquad \text{CS5}$$

$$\mathfrak{v} \boxed{\rhd} \mathfrak{w} = \varepsilon p (W(p) \leftrightarrow \mathfrak{v} \rhd \mathfrak{w}) \qquad \text{D7.8}$$

$$W(\mathfrak{v} \boxed{\rhd} \mathfrak{w}) \leftrightarrow \mathfrak{v} \rhd \mathfrak{w} \qquad \text{T7.15}$$

The 'symbolic treatment' (i.e. the explicit definition) of this kind of propositions seems, however, to be rather problematic.[23]

[21] TS §214, II 269: "*Whether every truth can be envisaged, not only as ground, but also as consequence of others?* [...] The first part of this question, namely, whether for every truth there are others that *follow* from it, has already been decided above. I have repeatedly assumed that the proposition '*A* is a truth' can be considered a genuine consequence of the truth *A*; if this is true, then it is obvious that for every given truth at least one other truth can be found which is a consequence of it. The only contentious part of our question is whether every truth can also be envisaged as a consequence (at least a partial consequence), i.e., whether for every truth we can find another truth, or collection of them, which can be considered its ground. It is my surmise that this must be answered in the negative; I think that there are and must be some truths which have no further grounds for their truth, but I must confess that I have not been able to find a satisfactory proof of this. [...] One of them seems to be the proposition that there is something, or (as I think it should be put) that the idea of something has objectuality. [...] Once it is admitted that there are truths that have no further ground,they will be considered important enough to give them a special name. Since they are only ground, but not consequences, I will call them *basic truths,* all others *derived truths.*"

[22] TS § 215, II 270: "*Whether there are several basic truths.* If it is already a controversial question whether there is even a single truth which has no further ground, it is even more difficult to decide whether there are several. I believe that this should be affirmed, for I cannot understand how from a single truth all others would follow either as consequences, or consequences of consequences, etc."

[23] TS §168, II 147–8: "*Propositions stating a relation of ground and consequence.* 1) If the relation of ground and consequence which I mentioned in §. 162 is of such importance, the propositions stating that this relation obtains

7.2 Dependence

Having decided to depart from the current linguistic use, by attributing to (complete) consecutivity the property of intransitivity, Bolzano recovered the insights underlying the current use of that term through the notions of *dependence* [*Abhängigkeit*] and of its converse *being an auxiliary truth* [*Hülfswahrheit*].

The starting point of his investigation is what he called '*the procedure of going back from the consequence to its reasons* [*das Geschäft des Aufsteigens von der Folge zu ihrem Grunde*]'. It is the 'analytic' procedure that, for a given truth, attempts the gradual identification of its 'reasons', in the common sense of this term. One begins with the identification of its 'immediate reasons' (if they exist), i.e. of those truths of which it is a *partial consequence* (in the technical sense); then one seeks the 'immediate reasons' for these reasons, and so on until the basic truths (in the cases in which this is possible) are reached.

The parenthetic clause 'in the cases in which this is possible' is not superfluous. The question '*Whether the procedure of going back from the consequence to its reason must find an end for every truth*?' even constitutes the title of a section,[24] in which we read

> "that in the procedure of going back from the consequence to its reason by every truth to which it may be applied, one always finds an end does not yet follow from what has been stated up to now. Indeed, although there are also truths which do not rest on further reasons, there could still be other truths which not only have further reasons, but are such that also their reasons in turn have reasons and so on to infinity."[25]

Although hesitatingly [*wenn ich nicht irre*], Bolzano also put forward what he believed to be a typical example of a process regressing to infinity, an example we will discuss in Section 7.5, when we dispose of the conceptual apparatus required for its formulation.

Moving on to a systematization of these ideas, we preliminarily remark that, although without making it explicit, Bolzano believed in this case, even more than in other situations, that reasons are constituted by a *finite* number of truths and that the relation of dependence subsists between two truths, a starting one and another,

merit consideration. If in such a proposition we merely state that certain expressly indicated truths $M, N, O, \ldots$ are proper consequences of certain others $A, B, C, \ldots$, then the immediate constituents of our claim are easy to guess. They will be none others than will be remarked in the following expression: 'The relation of the truths $M, N, O, \ldots$ to the truths $A, B, C, \ldots$ has the character of a relation of a consequence to its ground.' This proposition may be called a statement of *material consequence*, or a *consecutive proposition*. Clearly, there can also be negative propositions of this sort, which may bear the name of $A, B, C, \ldots$ of material consequence."

[24] TS §216, II 271: "*Whether the ascension from consequence to ground must come to an end for every given truth*"

[25] Cf. TS §217, II 271–2.

called an *auxiliary truth* [*Hülfswahrheit*], which is in some point of the '*tree of truths*' generated by the going-back procedure.

By induction on $k > 0$, we define the *dependence relation of order k* that connects a v and a w when the former is reachable from the latter through a chain of length k of consecutivities; i.e., writing '$v \overset{k}{\rhd} w$' for 'w k-depends on v' or also for '*v is an auxiliary truth of w of order k*'),[26] we set

$$\begin{cases} v \overset{1}{\rhd} w = v \rhd w \\ v \overset{k+1}{\rhd} w = \exists v'(v \overset{k}{\rhd} v' \wedge v' \rhd w) \end{cases} \qquad \text{D7.9}$$

By induction on k, we prove

$$\exists v'(v \rhd v' \wedge v' \overset{k}{\rhd} w) \leftrightarrow \exists w'(v \overset{k}{\rhd} w' \wedge w' \rhd w) \qquad \text{ST1}$$

Proof [B] $k = 1$. This is trivial. [P] $k = h + 1$. One has $\exists v'(v \rhd v' \wedge v' \overset{h+1}{\rhd} w)$

$$\overset{\text{def}}{\leftrightarrow} \exists v'(v \rhd v' \wedge \exists w'(v' \overset{h}{\rhd} w' \wedge w' \rhd w)) \overset{\text{CP}}{\leftrightarrow} \exists w'(\exists v'(v \rhd v' \wedge v' \overset{h}{\rhd} w') \wedge w' \rhd w))$$
$$\overset{\text{Ip.ind.}}{\leftrightarrow} \exists w'(\exists v''(v \overset{h}{\rhd} v'' \wedge v'' \rhd w') \wedge w' \rhd w)) \overset{\text{def}}{\leftrightarrow} \exists w'(v \overset{h+1}{\rhd} w' \wedge w' \rhd w).$$[27]

The relation of 'auxiliarity', considered as the 'union' on the $\overset{k}{\rhd}$ (i.e., in current terminology, as the 'transitive closure' of $\rhd$), is clearly transitive; i.e.

$$\text{for all } k, h > 0, v \overset{k}{\rhd} w \wedge w \overset{h}{\rhd} w' \rightarrow v \overset{k+h}{\rhd} w' \qquad \text{ST2}$$

Proof This is by induction on h. [B] $h = 1$. By the definition of $\overset{k+1}{\rhd}$. [P] $h = i + 1$. Let $v \overset{k}{\rhd} w$ and $w \overset{i+1}{\rhd} w'$; then, by definition, there is a v' such that $v \overset{k}{\rhd} w$ and $w \overset{i}{\rhd} v'$

[26] The term 'auxiliary truth' will be used with a quite different meaning in TS §436.

[27] TS §217, II 271–2: "*What the author means by supporting truths?* The truths which we discover when we ascend from a given truth A as consequence to its grounds have such a noteworthy relation to A that it is worth the trouble to give them their own designation. If it were permissible to consider the ground of a ground as itself a ground (§213), one could call all of these truths *grounds*, at least *partial grounds* of A The first ones encountered could be called immediate grounds and the later ones remote or mediate grounds. However, if the view which struck me as most probable in §213 is correct, one could only speak this way on the understanding that the term ground was being used in an improper sense. It would perhaps be more appropriate to use the name *supporting truths*, where it remains undecided whether the truth is a partial or complete ground, and also whether it is the immediate (proper) ground or a so-called remote (and hence improper) ground. Supporting truths can be said to be *dependent* upon one another or *subordinate* to one another with respect to the truth they support, while the latter may be said to be dependent upon the former truths or subordinate to them. Truths which do not stand in such a relation may be called independent."

and $v' \rhd w'$; by the inductive hypothesis, $v \overset{k}{\rhd} w$ and $w \overset{i}{\rhd} v'$ imply $v \overset{k+i}{\rhd} v'$, and this together with $v' \rhd w'$ gives, by definition, $v \overset{k+i+1}{\rhd} w'$.

Note that, using ST8, the proof can be performed by induction on k.

As an aside, let us remark that we can also directly introduce a 'dependence' relation (denoted by '$\overset{*}{\rhd}$') by assuming

$$\overset{m}{v} \rhd \overset{n}{w} \rightarrow \overset{m}{v} \overset{*}{\rhd} \overset{n}{w} \qquad \text{SA1}$$

$$\overset{m}{v} \overset{*}{\rhd} \overset{n}{w} \wedge \overset{n}{w} \overset{*}{\rhd} \overset{r}{u} \rightarrow \overset{m}{v} \overset{*}{\rhd} \overset{r}{u} \qquad \text{SA2}$$

$$\overset{m}{v} \rhd \overset{n}{w} \wedge \overset{n}{w} \overset{*}{\rhd} \overset{r}{u} \rightarrow \overset{m}{v} \overset{*}{\rhd} \overset{r}{u} \qquad \text{SA3}$$

$$\overset{m}{v} \overset{*}{\rhd} \overset{n}{w} \wedge \overset{n}{w} \rhd \overset{r}{u} \rightarrow \overset{m}{v} \overset{*}{\rhd} \overset{r}{u} \qquad \text{SA4}$$

We have, of course,

$$\overset{m}{v} \overset{k}{\rhd} \overset{n}{w} \rightarrow \overset{m}{v} \overset{*}{\rhd} \overset{n}{w}. \qquad \text{ST3}$$

Proof This is by induction on k. [B] $k := 1$. This is by SA. [P] $k := h + 1$. Let $\overset{m}{v} \overset{h+1}{\rhd} \overset{n}{w}$; then, from the definition and the immediately following theorem, there are $\overset{r}{u}$ such that $\overset{m}{v} \rhd \overset{r}{u}$ and $\overset{r}{u} \overset{h}{\rhd} \overset{n}{w}$; by the inductive hypothesis, $\overset{r}{u} \overset{h}{\rhd} \overset{n}{w} \rightarrow \overset{r}{u} \overset{*}{\rhd} \overset{n}{w}$ and therefore, by SA, $\overset{m}{v} \rhd \overset{r}{u}$ and $\overset{r}{u} \overset{*}{\rhd} \overset{n}{w}$ imply $\overset{m}{v} \overset{*}{\rhd} \overset{n}{w}$.

The logical power necessary to prove that if $\overset{m}{v} \overset{*}{\rhd} \overset{n}{w}$, then there is a k such that $\overset{m}{v} \overset{k}{\rhd} \overset{n}{w}$ goes beyond the limits we have chosen.

While not having a satisfying proof, Bolzano is convinced that no truth can be a proper auxiliary truth; i.e.,

$$\text{for all } k > 0, \neg(v \overset{k}{\rhd} v) \qquad \text{SAE1}$$

It should be noted, however that $\neg(v \overset{1}{\rhd} v)$ holds already by T1 and $\neg(v \overset{2}{\rhd} v)$ holds already by T6.[28]

7.3 Consecutivity and Derivability

Among the most interesting issues discussed in this context by Bolzano is undoubtedly that concerning the relationship between the relation of derivability and

[28] TS §218, II 272: "No truth can be a supporting truth of itself. No matter how numerous and varied the truths we encounter in ascending from a given truth *A* as consequence to its grounds, it is certain that *A* itself shall never appear among them or, to put it otherwise, that no truth is a supporting truth of itself. I say that this is certain, even though I have no other proof than the following to offer."

TS §219, II 272: "*Whether one and the same truth can appear as a supporting truth several times?* Although we shall never encounter the truth *A* we started from in ascending from a consequence to its grounds, nothing prevents us from encountering another truth *B* several times. For [. . .]."

the relation of consecutivity. It is clear that not every relation of derivability is also a relation of consecutivity, and this is not only because the former holds among propositions, whereas the latter holds only among truths, but also because it is easy to find examples that contradict that hypothesis.

A particularly simple and enlightening example is that one constituted by the following three truths:

> *When it is warmer, the thermometer reading is higher*
> *In summer, it is usually warmer than in winter*
> *In summer, the thermometer reading is usually higher than in winter.*

It is clear that the third truth is derivable from the first two and that that the second is derivable from the first and the third, but whereas we are all ready to accept that the first two truths give us the reason for the third, no-one would be disposed to say that the first and the third together give us the reason for the second.

The true problem, however, is whether consecutivity is a particular case of derivability—whether, in other words, $\mathfrak{v} \rhd \mathfrak{w} \rightarrow \mathfrak{v} \Rightarrow \mathfrak{w}$ holds. In this case, Bolzano is possibly even more prudent that in the rest of this chapter. On the one hand, indeed, he says that this subordination seems very probable, even though he confesses to having no idea about how to prove such a supposition;[29] on the other hand, he raises a question that seems—on the contrary—to rise doubts about that supposition. If indeed we accept that when the $v_1, \ldots, v_m$ are reasons for the $w_1, \ldots, w_n$, there always are some $a_1, \ldots, a_s$ with respect to which the $w_1, \ldots, w_n$ are derivable from the $v_1, \ldots, v_m$, one cannot see how to reject that also between the $v'_1, \ldots, v'_m$ and the $w'_1, \ldots, w'_n$, obtained each time from the preceding propositions by substituting the $c_1, \ldots, c_s$ for the $a_1, \ldots, a_s$, the consecutivity relation holds. This would, however, mean that, in fact, in every relation of consecutivity, the particularity of the involved ideas has no weight, which seems to be very difficult to accept.[30]

[29] TS §200, II 249: "[...] It could be held that propositions that are related as ground and consequence must also stand in a relation of deducibility. In this case the relation of ground and consequence would have to be envisaged as a special kind of deducibility; the former concept would be subordinate to the latter. Though this seems quite probable to me, I do not know any proof that would justify such a view."

[30] TS §200, II 249–50: "[...] On this view, it is understood that the same relation of ground and consequence which holds between propositions *A, B, C, D, ...* and *M, N, O, ...* also holds between all those truths that result from them by alteration of the variable ideas *i, j, ...*. From this it follows that for every collection of truths *A, B, C, D, ..., M, N, O, ...*, which are related as ground and consequence, there is an infinite number of other collections of propositions which stand in the same relation, provided that infinitely many truths can be generated from the propositions *A, B, C, D, ...* by alteration of the ideas in them which are considered variable. But is it even probable that for every collection of truths from which another collection follows as from their ground, there exists an infinite number of other collections of truths from which other truths follow in the same way, such that the particular nature of the ideas of which these collections of truths are composed has no influence upon their relation of ground and consequence?"

In this connection, Bolzano posits an argument involving both deontic and alethic modalities—unfortunately anything but lucidly—which should strengthen these doubts.

In the second-to-last extraordinary section of this chapter, under the title "*Some marks from which it is possible to conclude whether certain truths are among them in the dependence relation*", Bolzano makes a series of remarks and suggestions on the derivability–consecutivity relationship, which he often considers daring [. . . *ich wage* . . .] and that, even if not always perfectly understandable, belongs to a circle of ideas and problems that we have only learned through the works of Gentzen.

The first of these remarks (R1) is that, although it is clear in many cases—think, for example, about astronomy—that *conceptual* truths themselves are *known* by derivation from *experimental* truths, it appears difficult to understand that they may *depend* on truths of this type. The reasons for a conceptual truth, if present at all, cannot be other than themselves conceptual truths.[31]

The second remark (R2) is that the relation of consecutivity and, more generally, the relation of dependence among conceptual truths, must satisfy a kind of 'principle of sub-idea'; that is to say, such a relation may never go together with a decreasing in '*complexity* [*Zusammengesetztheit*], here understood to be the aggregate of the involved simple ideas (i.e. none of the truths constituting the reason for a conceptual truth may contain more simple ideas than those occurring in the consequence).[32] Hence, the inference

$$\frac{a \sqsubseteq b, b \sqsubseteq c}{a \sqsubseteq c}$$

can never be a consecutivity when b is more complex than a or c; indeed, if b is more complex than a, then $b \sqsubseteq c$ is more complex than $a \sqsubseteq c$, whereas b is more complex than c, then it is $a \sqsubseteq b$ that is more complex than $a \sqsubseteq c$.[33] We could reformulate the

[31] TS §221, II 276: "1) [. . .] truths which contain only pure concepts (conceptual truths) have the peculiarity that they depend only on other conceptual truths, never upon intuitional propositions. However, it is the case that intuitional truths (experiences) are often helpful in the *discovery* of a purely conceptual truth; but the *objective* ground of such a truth cannot lie in them; if they have any ground at all, it must lie in other conceptual truths."

[32] TS §221, II 276: "2) Furthermore, I believe that a purely *conceptual truth* upon which a second one depends must *never be more complex* than the latter, though it does not have to be simpler. Propositions which form the objective ground of a purely conceptual truth must separately contain no more simple parts than the truth which depends upon them."

[33] TS §221, II 276: "2) [. . .] Consider the common argument form: *Whatever has a has b;* *Whatever has b has c;* *Whatever has a has c* In my opinion, such an argument does not represent a relation of ground and consequence if either the concept b is more complex than a or b more complex than c. In the first case, the second, in the second case, the first, of the two premises would be more complex than the conclusion."

whole idea in modern terms by saying that a cut is never a 'consecution' when the cut-formula is not an interpolant.

The third remark (R3) states that the condition of non-decreasing complexity does not, in general, apply to empirical truths.[34] Indeed, for every intuition d, there are infinitely many $\bar{c}$ such that $h(d, \bar{c})$ is true and among these $\bar{c}$ there will be, in general, some very complex ones. Now, the d is always associated with an $\bar{a}$ (*that which has a-ity*) (i.e. there is an $\bar{a}$ such that $h(d, \bar{a})$ is true) and moreover, every other $h(d, \bar{c})$ will find its reason in the fact that (1) d has a-ity and (2) every object having a-ity also has c-ity (i.e. in the two truths $h(d, \bar{a})$ and $h([e](\bar{a}), \bar{c})$), $[e](\bar{a})$ is in any case more complex than d.[35]

The fourth remark (R4) is that if there only are a finite number of simple conceptual ideas, then for every conceptual truth v, the number of conceptual truths that are not more complex than v is always finite.[36] Under suitable conditions, which Bolzano does not list, but seems to accept as obvious, this affirmation can be strictly proven.

From this, Bolzano (R5) draws the effective proof of the existence of basic truths. If there were not such truths, then every truth v would have an infinite ascending chain of truths on which it depends. Each element of this chain, however, would be, by (R2), not more complex than v, but, by (R4), there is only a finite number of things of this kind.[37]

The sixth remark (R6) is that for every fixed finite limit of complexity, for instance 100 simple ideas, the collection of all conceptual truths that do not exceed that limit is finite and hence every truth in it that is not basic finds in it the complete tree of its dependence. In this collection, there are, on the one hand, the basic truths and, on the other hand, the dependent truths; among them are those derivable from the former and possibly also those that do not enjoy this property. Disregarding the latter, we may consider all possible 'axiomatizations' of the remaining ones (i.e. all those groupings of these truths such that there are among them a certain number of truths from which all others are derivable). Now Bolzano goes so far as to maintain

[34] TS §221, II 276: "2) [...] But I do not wish to generalise my claim and extend it to *intuitional* propositions, since they appear to furnish some counter-examples."

[35] TS §221, II 275–6.

[36] TS §221, III 277: "3) [...] If the number of all simple concepts is finite, then the number of all purely conceptual truths which do not exceed a certain degree of complexity, for example, which do not contain more parts than a given proposition M, can also be only finite."

[37] TS §221, II 277: "3) [...] But if somebody wants to claim that every purely conceptual truth must have a ground, then he must also claim that every truth M rests upon an infinite number of supporting truths; for if every ground requires another ground, *ad infinitum*, then the number of propositions upon which truth M depends is infinite. But if these supporting truths are all purely conceptual propositions and are to be of lower complexity than M itself, then this leads to a contradiction with what we have said above. This then shows that, at least for the domain of purely conceptual truths, there are *genuine basic truths*."

that if we reinforce our idea of 'axiomatization' by requiring that in it no conclusion can ever be less complex than any of its premises, then the one of these 'axiomatizations', in this new sense, that has the least number of 'axioms' and the greatest number of 'theorems' is just the one that coincides with the systematization of the dependencies.[38]

The seventh remark (R7) is that, on the one hand, the reasons for a truth v are always, each one for itself, the *simplest* among all its equivalent propositions, and that, on the other hand, a truth v that is a consequence of certain other truths is also, for itself, the *simplest* among all its equivalent propositions, so that each of these equivalents are in fact not consequences of those reasons, but only of v.[39]

The eighth remark (R8) is that the reasons for a truth v that is also derivable from them constitute the *simplest* aggregate of propositions from which v is derivable, in the sense that for every substitution of the same variable, none of the premises becomes more complex than the conclusion.[40]

The ninth remark (R9) is that the reasons for a truth must not only be the simplest propositions from which it is derivable, but also the *most general* ones, in the sense that their subjects and predicates must be as wide as possible.

[38] TS §221, II 277–8: "3) [. . .] I should like to go on to make the following claim: if we consider the collection of all purely conceptual truths whose degree of complexity does not exceed a certain limit (e.g., all truths which do not contain more than 100 simple parts), then it follows from what has been said that every truth which is a member of this collection and which is not itself a basic truth must find its complete ground in this collection of propositions; this holds not only for immediate, but also for remote grounds up to the final ones which have no further grounds. We also already know that this collection will contain truths $A, B, C, \ldots$ which have no ground at all and in addition to them in all likelihood some other truths $D, E, F, \ldots$ that have a ground, but which are not deducible from this ground (cf. §. 200). Finally, all other truths $M, N, \ldots, Z$, which are also contained in this collection, are consequences that are deducible from their ground. But if we can change the order of these truths at will, then it must obviously be easy to arrange them in groups in such a way that, if only a number of them are accepted, the others are deducible from them. Now I wish to claim that if we only always observe the rule never to deduce a simpler truth from premises, any of which is more complex, then the number of propositions which we must accept outright (i.e., without deduction from others) will be the smallest if we arrange the propositions according to their objective connection. Hence, it seems to me that the relation of ground and consequence has the following peculiarity: if we order propositions according to this relation, then the *smallest number* of premises will allow us to deduce the largest number of conclusions, so long as they are more complex than their premises."

[39] TS §221, II 278: "4) Another peculiarity of this relation seems to lie in the fact that each of the truths $A, B, C, D, \ldots$, which together form the ground of truth M, is always the simplest of the propositions that are individually equivalent to it. Another characteristic is that every truth which must be considered a consequence of the several truths, $A, B, C, \ldots$, is always the simplest of all propositions individually equivalent to it. Thus it follows that all truths which are equivalent to a given individual truth M, and which are more complex than it, are always consequences of this truth alone, etc."

[40] TS §221, II 278: "5) In general, it seems to me that the truths $A, B, C, \ldots$ constituting the ground of truth M, which is also deducible from them, must always be the simplest collection of truths from which M is deducible, provided that the same ideas are considered variable in every case, and that noneof the premises are individually more complex than the conclusion."

Subsequently, then, (R10) a cut

$$\frac{a \sqsubseteq b,\ b \sqsubseteq c}{a \sqsubseteq c}$$

is certainly a consecution only when b is equivalent to c, because in that case the two premises are certainly the most general from which the conclusion is derivable.[41]

Last but not least, according to (R11), it is reasonable to suppose that when a certain v is not only derivable, with respect to certain ideas, but also *exactly* derivable from certain premises, and both premises and v are the simplest among their equivalents, then not only do we have to do with a consecution, but this relation also subsists among the propositions obtained from the preceding ones by means of any substitution for the variables that does not introduce redundancies. Therefore, the inference

$$\frac{a \sqsubseteq b,\ a \sqsubseteq c}{a \sqsubseteq b \sqcap c}$$

is *always* a consecution, because the inference is exact (each premise is derivable from the conclusion) and, moreover, it obviously maintains the simplicity conditions.[42]

The conclusion of this discussion is really worth mentioning:

> "This should be, moreover, the most suitable place to confess to the reader that sometimes I am assailed by the doubt that the notion of consecutivity, which I claimed above to be simple, is possibly, on the contrary, composed and after all nothing else

[41] TS §221, II 278: "6) Not only must propositions *A*, *B*, *C*, ... be the simplest, but they must also be the *most general* propositions from which *M* is deducible. [...] It seems to me that the argument cited in no. 2 indicates a relation of ground and consequence only if *B* and *C* are equivalent ideas; for, if this is the case, the two propositions 'All *A* are *B*', and 'All *B* are *C*', are certainly the widest propositions from which the conclusion 'All *A* are *C*' is deducible."

[42] TS §221, II 278–9: "7) If a proposition *M* stands to other propositions *A*, *B*, *C*, ... in the relation of exact deducibility (cf. §.155. n. 26)) with respect to ideas *i*, *j*, ..., if, moreover, propositions *A*, *B*, *C*, ... and *M* are the simplest propositions among those equivalent to them, and if none of *A*, *B*, *C*, ... is more complex than *M*, then we may assume that *M* stands to *A*, *B*, *C*, ... in a true relation of ground and consequence, in such a way that whenever *i*, *j*, ... are replaced by ideas which make propositions *A*, *B*, *C*, ... not only true, but keep them from becoming redundant, then truth *M* is a proper consequence of truths *A*, *B*, *C*, ... Hence I do not hesitate to claim the relation of ground and consequence for the following propositions.

Every A is B;
Every A is C;
Every A is B as well as C

The last proposition is deducible from the first two, but they are also deducible from it and they all fulfil the above stipulation of simplicity."

than the notion of an ordering of the truths by means of which from the smallest number of simple premises one can derive the greatest possible number of other truths as mere conclusions."[43]

7.4 Causality (A): Ontological–Metaphysical Preliminaries

As we have already alluded to, one of the most interesting ideas of Bolzano's Metaphysics is constituted by the reduction of the causality relation to the consecutivity relation, via the identification of the cause–effect relation with the reason–consequence relation subsisting between the propositions asserting the existence of the causes and the propositions asserting the existence of the effects. Bolzano:

> "Closely related with the notions of reason and consequence are, in my opinion, also the notions of *cause* [*Ursache*] and *effect* [*Wirkung*]. Indeed, I believe that these two words refer, in their proper sense, only to objects that have reality; and this precisely so that we say that a real object α is the *cause* of the truth M when the proposition 'α has existence' is one of the partial reasons on which lies the truth of M and so we also say that the real object μ is an effect of the object α when the proposition 'μ has existence' is one of the consequences following from the proposition 'α has existence.'"[44]

Let us recall that the general principles accepted for the basic notion of '*having existence* [D]' are

$$D(x) \to O(x) \vee Q(x) \qquad \text{GA6}$$
$$D(x) \wedge Q(x) \to \forall y(y \in x \to D(y)) \qquad \text{GA7}$$
$$D(x) \wedge y < x \to D(y) \qquad \text{GA8}$$

Starting from D, we also introduce the notions of *substance* and *adherence*, by setting

$$St(x) \leftrightarrow D(x) \wedge \neg Q(x) \qquad \text{D7.10}$$
$$Ad(x) \leftrightarrow D(x) \wedge Q(x) \qquad \text{D7.11}$$

Concerning some further development of metaphysical ontology, the first thing to remark is that for Bolzano *existence* is not only a 'property', but also a *quality* to which a specific idea refers, an idea that we will denote as '$\overline{d}$'.[45] To this end, we thus

[43] TS §221, II 279 (*Note*).

[44] TS §168, II 148 (3).

[45] TS §142, II 45: "I take *being* [*Seyn*] or *actuality* [*Wirklichkeit*] precisely as what language makes it out to be, namely, an attribute; [. . .]."

introduce, following the usual schema, both the ontological axiom

$$\exists r \forall i(i \in r \leftrightarrow D(i)) \qquad \text{CQ8}$$

and the lectological axiom

$$\exists \bar{c} \forall r(r \mathrel{\varepsilon} \bar{c} \leftrightarrow \forall i(i \in r \rightarrow D(r))) \qquad \text{CB7}$$

with the respective resolution

$$\bar{d} = \varepsilon \bar{c} \forall r(r \mathrel{\varepsilon} \bar{c} \leftrightarrow \forall i(i \in r \rightarrow D(r))) \qquad \text{D7.12}$$

obtaining, in particular

$$(1)\ B(\bar{d}) \qquad \text{T7.16}$$
$$(2)\ r \mathrel{\varepsilon} \bar{d} \leftrightarrow \forall i(i \in r \rightarrow D(i))$$

It should be noted that we have not made any hypothesis concerning the mereological structure of $\bar{d}$; Bolzano probably considered this a simple idea.

Using the proof schema that we have already applied time and time again, we obtain the theorem ensuring that *existing is equivalent to falling under the idea of something that has existence,* namely

$$i \mathrel{\varepsilon} [e](\bar{d}) \leftrightarrow D(i) \qquad \text{T7.17}$$

From this, it easily follows that

$$W(h(a, \bar{d})) \leftrightarrow G(a) \wedge \forall i(i \mathrel{\varepsilon} a \rightarrow D(i)) \qquad \text{T7.18}$$

Proof $W(h(a, \bar{d})) \leftrightarrow a \sqsubseteq [e](\bar{d}) \leftrightarrow G(a) \wedge \forall i(i \mathrel{\varepsilon} a \rightarrow i \mathrel{\varepsilon} [e](\bar{d})) \leftrightarrow G(a) \wedge \forall i(i \mathrel{\varepsilon} a \rightarrow D(i))$.

For all further developments, it is convenient to accept, *as far as the existent objects are concerned,* the Bolzanian principle, already mentioned in §33 (but not accepted there), according to which for every object there is at least one idea 'peculiar' to it.[46] From the technical point of view, however, it is convenient to admit, for *every object* i, the idea, the ε-resolution of which will be denoted by '$\{i\}$', that does not contain i as a proper part and that refers (exclusively) to i only in the case i exists and that is otherwise objectless. Hence

$$\forall i \exists a((D(i) \rightarrow \forall j(j \mathrel{\varepsilon} a \leftrightarrow j = i)) \wedge (\neg D(i) \rightarrow \neg G(a)) \wedge \neg i < a)) \qquad \text{CV7}$$

[46] TS §101, I 339: "For it is certain that for every individual object [...] there is and must be an idea applying exclusively to it [...]."

$$\{i\} := \varepsilon a((D(i) \to \forall j(j \mathrel{\text{E}} a \leftrightarrow j = i)) \land (\neg D(i) \to \neg G(a)) \land \neg i < a)) \qquad \text{D7.13}$$

$$(1)\ V(\{i\}) \qquad \text{T7.19}$$

$$(2)\ D(i) \to \forall j(j \mathrel{\text{E}} \{i\} \leftrightarrow j = i)$$

$$(3)\ \neg D(i) \to \neg G(\{i\})$$

$$(4)\ \neg i < \{i\}$$

$$D(i) \to G(\{i\}) \qquad \text{T7.20}$$

Proof By (2) of the preceding theorem, we have, in particular, $D(i) \to (i \mathrel{\text{E}} \{i\} \leftrightarrow i = i)$ and therefore, since anywhere $i = i$, we have $D(i) \to i \mathrel{\text{E}} \{i\}$ and hence $G(\{i\})$.

From this and from (3) of the same theorem,

$$G(\{i\}) \leftrightarrow D(i) \qquad \text{T7.21}$$

We are thus able to prove that *an object exists iff the proposition enunciating its existence is true*, i.e.

$$W(h(\{i\}, \bar{d})) \leftrightarrow D(i) \qquad \text{T7.22}$$

Proof [$\to$] If $W(h(\{i\}, \bar{d}))$, then, by definition, we have, in particular, $G(\{i\})$ and hence, by the preceding theorem, $D(i)$. [$\leftarrow$] If $D(i)$, then, by the preceding theorem, (i) $G(\{i\})$. Now let j be an object falling under $\{i\}$; then, under the hypothesis $D(i)$, we have, by T19, $i = j$ and hence, being $D(i)$, it is also $D(j)$ (i.e. $j \mathrel{\text{E}} [e](\bar{d})$). Thus, $\forall j(j \mathrel{\text{E}} \{i\} \to j \mathrel{\text{E}} [e](\bar{d}))$ (i.e. (ii) $\forall j(j \mathrel{\text{E}} \{i\} \to \exists r(j \in r \land r \mathrel{\text{E}} \bar{d}))$. From (i) and (ii), by definition, $W(h(\{i\}, \bar{d}))$.

The assumed inexistence of lectological things yields, in particular,

$$(1)\ \neg G(\{a\}) \qquad \text{T7.23}$$

$$(2)\ \neg G(\{p\})$$

as well as

$$(1) \neg W(h(\{a\}, \bar{d}))[\leftrightarrow W(\boxminus h(\{a\}, \bar{d}))] \qquad \text{T7.24}$$

$$(2) \neg W(h(\{p\}, \bar{d}))[\leftrightarrow W(\boxminus h(\{p\}, \bar{d}))]$$

Moreover, let us also remark that, for every existent object i there is an idea, the $\{i\}$, that refers exclusively to it and that also for every lectological object l there is an idea, the $\ulcorner l \urcorner$, that refers exclusively to it. This does not, however, imply the Bolzanian conviction that for every object there is an idea peculiar to it. Indeed, we have not postulated that lectological objects exhaust the realm of inexistent objects.

7.5 Causality (B): the Relations of Causality

In order to provide a rigorous assessment of the basic ideas behind Bolzano's Theory of Causality, we begin by considering the simplest relation, namely that occurring between an object i (cause) and an object j (effect). From what has been said up to now, it is clear that such relation will be defined as follows:

$$i \succ j \leftrightarrow h(\{i\}, \bar{d}) \rhd h(\{j\}, \bar{d}) \qquad \text{D7.14}$$

The treatment of the other natural causality relations, which involve aggregates of objects in various ways, requires a wider outlook. Following our usual procedure, we treat these aggregates through ideas of objects, so we will say, for instance, that 'the objects (falling under) a are *causes* of the objects (falling under) c', which we also write as '$a \succ c$'. To account for the Bolzanian ideas, we must now state that this happens when the collection of all and only all propositions enunciating the existence of each object falling under a is the reason for the collection of all and only all the propositions enunciating the existence of each object falling under c; that is to say, according to our strategy, *when the idea under which fall all and only all propositions enunciating the existence of each object falling under a is the reason for the idea under which fall all and only all the propositions enunciating the existence of each object falling under c.*

Naturally, a formal assessment requires a definition of *the idea under which fall all and only all propositions enunciating the existence of each object falling under the idea a*. To this end, we postulate the comprehension

$$\forall a \exists c \forall p (p \sqsubset c \leftrightarrow \exists i (i \sqsubset a \wedge p = h(\{i\}, \bar{d}))) \qquad \text{CV8}$$

and we denote by '$\mathfrak{s}(a)$' its ε-resolution, i.e.

$$\mathfrak{s}(a) = \varepsilon c \forall p (p \sqsubset c \leftrightarrow \exists i (i \sqsubset a \wedge p = h(\{i\}, \bar{d}))) \qquad \text{D7.15}$$

and thus obtain

$$(1)\ V(\mathfrak{s}(a)) \qquad \text{T7.25}$$

$$(2)\ p \sqsubset \mathfrak{s}(a) \leftrightarrow \exists i (i \sqsubset a \wedge p = h(\{i\}, \bar{d}))$$

We may now remark that if a is objectual, then there is at least an i that falls under it and hence, by the preceding theorem, the proposition $h(\{i\}, \bar{d})$ falls under $\mathfrak{s}(a)$, which is then objectual; on the other hand, if $\mathfrak{s}(a)$ is objectual, then each object falling under it is a proposition and hence is an idea of propositions; finally, if $\mathfrak{s}(a)$ is an idea of propositions, then each thing referred to by it is a proposition and precisely a proposition of the form $h(\{i\}, \overline{d})$ with one i falling under a, which is therefore objectual. Summing up,

$$G(a) \leftrightarrow G(\mathfrak{s}(a)) \leftrightarrow VS(\mathfrak{s}(a)) \qquad \text{T7.26}$$

These tools allow us to provide a satisfactory systematization of the various causality relations. It will suffice to set

D7.16

(1) The objects (falling under) a are *causes* of the object j and j is an *effect of the a's*:

$$a \succ j \leftrightarrow \mathfrak{s}(a) \rhd h(\{j\}, \overline{d})$$

(2) The object i is a *cause* of the objects (falling under) c and the c's are *effects* of i:

$$i \succ c \leftrightarrow h(\{i\}, \overline{d}) \rhd \mathfrak{s}(c)$$

(3) The objects (falling under) a are *causes* of the objects (falling under) c and the c's are *effects* of the a's:

$$a \succ c \leftrightarrow \mathfrak{s}(a) \rhd \mathfrak{s}(c)$$

(4) The objects (falling under) a are the *complete cause* of the objects (falling under) c:

$$a]\!\succ c \leftrightarrow \mathfrak{s}(a) \,{}_{\}}\!\rhd \mathfrak{s}(c)$$

(5) The objects (falling under) c are the *complete effect* of the objects (falling under) a:

$$a \succ\![c \leftrightarrow \mathfrak{s}(a) \rhd_{\{} \mathfrak{s}(c)$$

(6) Between the objects (falling under) a and the objects (falling under) c subsists a *complete causation*

$$a \succ\!\!\succ c \leftrightarrow \mathfrak{s}(a) \rhd\!\!\rhd \mathfrak{s}(c)$$

At this point, we are also in a position to present the example that led Bolzano to exclude that the search for reasons must always end after a finite number of steps. Each 'variable state' (i.e. each quality r as a function of time) of a *created* substance (i.e. a substance different from God) has a cause that, at least in part, lies in the preceding state (i.e. there is a sequence of causes regressing into infinity: $\ldots \succ r_k \succ \ldots \succ r_2 \succ r_1$). Therefore, the sequence of partial reasons–consequences associated with it via the preceding theorems regresses itself into infinity; i.e. $\ldots \rhd h(\{r_k\}, \overline{d}) \rhd \ldots \rhd h(\{r_2\}, \overline{d}) \rhd h(\{r_1\}, \overline{d})$.[47]

7.6 Alethic Modalities

The treatment of the problem of existence developed in the initial part of the preceding section allows us to briefly illuminate Bolzano's idea about alethic modalities. He considers four of them—*necessity* [*Nothwendigkeit*], *impossibility* [*Unmöglichkeit*], *possibility* [*Möglichkeit*], and *contingency* [*Zufälligkeit*]—and he examines them, first as *ontological* properties concerning the existence of the objects, second as *lectological* properties concerning the predication relation, and finally, and only with respect to possibility, as *cognitive* properties of certain assertions. A dutiful, preliminarily remark regarding the assessment of these ideas is that Bolzano believed that modalities did not constitute—contrary to what most of his predecessors and his successors have thought—an authentic *logical* problem, but essentially only a fact, so that they are even completely *defined*.

For systematic reasons, it is convenient to begin with the modalities as *lectological properties* of propositions of the form $h(a, \overline{c})$. As has already been said,[48] for Bolzano an *intuition* was a simple and singular idea, a *concept*, an idea that has no intuition as a part, and a *conceptual truth*, a truth that does not contain intuitions. Now, he identified the *necessity* of the proposition $h(a, \overline{c})$ with its being a conceptual truth; its *impossibility* with the proposition $h(a, \sim\overline{c})$ being a conceptual truth; its *possibility*, of course, with its not being impossible; and finally, and no less naturally, its *contingency* with its simultaneously being true and not necessary; i.e.,

$$Nec(h(a,\overline{c})) \leftrightarrow W(h(a,\overline{c})) \wedge \neg\exists b(A(b) \wedge (b \leq a \vee b \leq \overline{c})) \quad \text{D7.17}$$[49]

$$Imp(h(a,\overline{c})) \leftrightarrow W(h(a,\sim\overline{c})) \wedge \neg\exists b(A(b) \wedge (b \leq a \vee b \leq \overline{c})) \quad \text{D7.18}$$[50]

[47] TS §216, II 271: "If I am not mistaken, an example of such a truth is every proposition describing one of the contingent states or attributes of a created substance. Since every such state has a cause at least part of which lies in the preceding state, we find here a sequence of causes which continues ad infinitum. But if there is an infinite sequence of causes, there must also be infinite sequences of grounds, since the actual thing *M* can be a cause of the actual thing *N* only if the truth that *M* exists is a ground or partial ground of the truth that *N* exists."

[48] See Chapter 2.

[49] TS §182, II 165: "4) [...] Whenever a proposition '*A* has *b*' is a purely conceptual truth, it is customary to say that the attribute *b* belongs *necessarily* to object *A*."

[50] TS §182, II 165: "4) [...] If it is not the proposition '*A* has *b*' but rather the proposition '*A* does not have *b*', which is a purely conceptual truth, we say that attribute *b* is *impossible* for object *A*."

$$Pos(h(a,\bar{c})) \leftrightarrow \neg Imp(h(a,\bar{c}))$$ [51] D7.19

$$Con(h(a,\bar{c})) \leftrightarrow W(h(a,\bar{c})) \wedge \neg Nec(h(a,\bar{c}))$$ [52] D7.20

Starting from these definitions, it is possible to obtain, through new definitions, the modalities as *ontological properties* of objects. So the object i is said to be *necessary* iff there is an idea a under which it falls and the proposition asserting that the a have existence is necessary; *impossible* iff there is an idea a under which it falls and the proposition asserting that the a have existence is impossible; *possible* iff it is not impossible; and, finally, *contingent* iff it exists but it is not necessary; i.e.

$$Nec(i) \leftrightarrow \exists a(i \in a \wedge Nec(h(a,\bar{d})))$$ [53] D7.21

$$Imp(i) \leftrightarrow \exists a(i \in a \wedge Imp(h(a,\bar{d})))$$ [54] D7.22

$$Pos(i) \leftrightarrow \neg Imp(i)$$ [55] D7.23

$$Con(i) \leftrightarrow D(i) \wedge \neg Nec(i)$$ [56] D7.24

As to the possibility as a *cognitive property* of certain assertions, we will inspect this within the frame of the Theory of Knowledge.

Besides the *absolute ontological modalities*, which Bolzano also called *internal*, he also shortly considered, albeit with some inaccuracy in the formulation, the corresponding *relative*, *'external'* ones.

An object i is said to be *necessary relative* to the object j iff neither of them is in itself necessary but the proposition enunciating the existence of i is derivable, with respect to certain ideas (not to all ideas involved), from the proposition enunciating the existence of j; i is, on the contrary, said to be *impossible relative to j* iff it is itself possible (this is tacitly assumed by Bolzano) but from the proposition enunciating the existence of j is derivable, with respect to certain ideas, the negation of the proposition enunciating its existence; furthermore, i is said to be *possible relative to j* iff it is possible itself and moreover not impossible relative to j; finally, i is said to be *contingent relative to j* iff it exists but it is not necessary, neither in itself (tacitly

[51] TS §182, II 165: "4) [. . .] If it is not impossible that the attribute *b* belongs to *A* hence if there is no purely conceptual truth of the sort 'No *A* has *b*', then we say that it is *possible* that *A* has *b*."

[52] TS §182, II 165: "4) [. . .] Finally, if the proposition '*A* has *b*' is true, but is not a purely conceptual truth, i.e., if the attribute *b* belongs to objects *A*, without belonging to them necessarily, then we say that it belongs to them *contingently*."

[53] TS §182, II 164: "1) [. . .] we say that the existence of a certain object *A* is necessary, or has necessity, or that *A* must exist, only if there is a purely conceptual truth of the form: 'A′ is (or has existence)', in which A′ is an idea which represents *A*."

[54] TS §182, II 164: "1) [. . .] Conversely, if the proposition 'A′ does not exist', rather than the proposition 'A′ exists' is a purely conceptual truth, we say that the object *A*, which stands under the idea *A′*, is *impossible*."

[55] TS §182, II 164: "1) [. . .] We call the actuality of an object *possible* if it is not impossible."

[56] TS §182, II 164: "1) [. . .] If an object exists without being necessary, we call it *contingent*."

assumed by Bolzano) nor relative to j; i.e. ('$\Rightarrow$' *does not imply* here all involved ideas)

$$Nec(i,j) \leftrightarrow \neg Nec(i) \wedge \neg Nec(j) \wedge h(\{j\},\bar{d}) \Rightarrow h(\{i\},\bar{d})\ ^{57} \qquad \text{D7.25}$$

$$Imp(i,j) \leftrightarrow Pos(i) \wedge h(\{j\},\bar{d}) \Rightarrow \neg h(\{i\},\bar{d})\ ^{58} \qquad \text{D7.26}$$

$$Pos(i,j) \leftrightarrow Pos(i) \wedge \neg Imp(i,j)\ ^{59} \qquad \text{D7.27}$$

$$Con(i,j) \leftrightarrow Con(i) \wedge \neg Nec(i,j)\ ^{60} \qquad \text{D7.28}$$

The *relative lectological* modalities are also mentioned, if just in passing. It is established that proposition $h(a_1,\bar{c}_1)$ is *necessary relative to* the proposition $h(a_2,\bar{c}_2)$ iff neither of them is in itself necessary but there are ideas in a_1 and in a_2 with respect to which the former is derivable from the latter; i.e. (with the same proviso on '$\Rightarrow$')

$$Nec(h(a_1,\bar{c}_1),h(a_2,\bar{c}_2)) \leftrightarrow \neg Nec(h(a_1,\bar{c}_1)) \wedge \neg Nec(h(a_2,\bar{c}_2)) \wedge (h(a_2,\bar{c}_2) \Rightarrow h(a_1,\bar{c}_1))\ ^{61} \qquad \text{D7.29}$$

As to *relative possibility* and *contingency*, Bolzano limits himself to saying that they are defined in an analogous manner,[62] whereas relative impossibility is not even mentioned. We thus set $h(a_1,\bar{c}_1)$ is *impossible relative to* $h(a_2,\bar{c}_2)$ iff it is in itself possible but its negation is derivable from $h(a_2,\bar{c}_2)$; $h(a_1,\bar{c}_1)$ is *possible relative to* $h(a_2,\bar{c}_2)$ iff it is possible in itself and not impossible relative to $h(a_2,\bar{c}_2)$; and, finally, $h(a_1,\bar{c}_1)$ is *contingent relative to* $h(a_2,\bar{c}_2)$ iff it is true but not necessary relative to $h(a_2,\bar{c}_2)$; i.e.

$$Imp(h(a_1,\bar{c}_1),h(a_2,\bar{c}_2)) \leftrightarrow Pos(h(a_1,\bar{c}_1)) \wedge (h(a_2,\bar{c}_2) \Rightarrow \neg h(a_1,\bar{c}_1)) \qquad \text{D7.30}$$

$$Pos(h(a_1,\bar{c}_1),h(a_2,\bar{c}_2)) \leftrightarrow Pos(h(a_1,\bar{c}_1)) \wedge \neg Imp(h(a_1,\bar{c}_1),h(a_2,\bar{c}_2)) \qquad \text{D7.31}$$

$$Con(h(a_1,\bar{c}_1),h(a_2,\bar{c}_2)) \leftrightarrow W(h(a_1,\bar{c}_1)) \wedge \neg Nec(h(a_1,\bar{c}_1),h(a_2,\bar{c}_2)) \qquad \text{D7.32}$$

[57] TS §182, II 164-5: "3) If neither the object *A* nor the object *M* are in themselves necessary, but it appears that the proposition '*M* is actual' is deducible from the proposition '*A* is actual' with respect to some ideas contained in *M* and *A*, then we say that *M* is *necessary relative* to presupposition *A*. We call this necessity of *M* a *relative*, or *external necessity*; by contrast, the necessity discussed under no. 1 is called *internal*."

[58] TS §182, II 165: "3) [...] Conversely, if it is not the proposition '*M* is actual', but rather the proposition '*M* is not actual' which is deducible from the proposition '*A* is actual', then we say that *M* is impossible relative to *A*. We call this impossibility of *M relative* or *external*, and contrast it with the *internal* impossibility of no. 1."

[59] TS §182, II 165: '3) [...] If object *M* is not impossible not only internally, but also in relation to some other object *A*, then we say that *M* is *possible* with respect to object *A*.'

[60] TS §182, II 165: "3) [...] Finally, if object *M*, which is not necessary relative to *A* , is nonetheless actual, we say that its actuality is *externally contingent* or *contingent* relative to *A*."

[61] TS §182, II 165–6: "4) [...] If it is neither the case that attribute *b* belongs to *A* nor that *p* belongs to *M*, each taken in itself, but if the proposition '*M* has *b*' is deducible from the proposition '*A* has *p*' with respect to certain ideas included in *A* and *M*, then we say that attribute *p* belongs to *M* with *relative necessity*, namely, under the assumption that all *A* have the attribute *b*."

[62] TS §182, II 166: "The concepts of relative *possibility* and *contingency* are defined similarly."

8

Theory of Knowledge

8.1 Having as Material: Representations and Judgements

The basic epistemological notion is *Having as material* [*Stoff*] or *content* [*Inhalt*], which, as established in Chapter 1, will be denoted by '$\lessdot$'; thus we will write '$x \lessdot y$' for 'x is the material (or the content) of y'.[1] In the same chapter, we also established that the two basic properties of $\lessdot$ are

(1) $\lessdot$ *is left-unique*:

$$x \lessdot z \wedge y \lessdot z \rightarrow x = y \qquad \text{GA15}$$

(2) $\lessdot$ *subsists between lectological things and singular adherences*:

$$x \lessdot z \rightarrow L(x) \qquad \text{GA16}$$

$$x \lessdot z \rightarrow Ad^{\circ}(z) \qquad \text{GA17}$$

GA15 allows us to conclude that if we know that z has a material, then we may certainly speak of *the material* (or *the content*) *of* z, which we will denote by '$c(z)$'; i.e. we establish

$$c(x) = \varepsilon z(z \lessdot x) \qquad \text{D8.1}$$

[1] Clearly the term '*Inhalt*' here has a totally different meaning from what it has in the lectological mereology.

from which[2]

$$\exists z(z \lessdot x) \rightarrow \forall y(y \lessdot x \leftrightarrow y = c(x)) \qquad \text{T8.3}$$

On the other hand, as will be remembered, singular adherences are existent qualities for which the belonging relation is left-unique, that is to say, for which the following holds:

$$B(x) \wedge D(x) \wedge (y \in x \wedge z \in x \rightarrow y = z)$$

Hence, if x is a singular adherence, then $\varepsilon z(z \in x)$ is, by reasons analogous to the preceding ones, the unique thing to which it belongs, a thing that will be called *the bearer* [*Träger*] *of* x and denoted by '$t(x)$'. Thus

$$t(x) = \varepsilon z(z \in x) \qquad \text{D8.2}$$

$$Ad^{\circ}(x) \rightarrow \forall y(y \in x \leftrightarrow y = t(x))^{3} \qquad \text{T8.4}$$

Clearly not every singular adherence has a content; those that do are called '*cognitive*' and the predicate 'Ad^*' will be used for them. Thus we establish

$$Ad^*(x) \leftrightarrow Ad^{\circ}(x) \wedge \exists y(y \lessdot x) \qquad \text{D8.3}$$

Cognitive adherences constitute a proper subclass of the singular adherences that Bolzano refers to as '*psychic phenomena* [*psychische Erscheinungen*]' and that also include *sensations* [*Empfindungen*], *wishes, volitions,* etc.[4]

[2] If somebody wants a formal proof:

1.	$z \lessdot x \rightarrow \varepsilon z(z \lessdot x) \lessdot x$	E
2.	$c(x) = \varepsilon z(z \lessdot x)$	D8.1
3.	$z \lessdot x \rightarrow c(x) \lessdot x$	R(L)1, 2
4.	$z \lessdot x \rightarrow (c(x) = y \rightarrow y \lessdot x)$	R(L)3
5.	$y \lessdot x \wedge c(x) \lessdot x \rightarrow y = c(x)$	AG15
6.	$c(x) \lessdot x \rightarrow (y \lessdot x \rightarrow y = c(x))$	R(T)5
7.	$z \lessdot x \rightarrow (y \lessdot x \leftrightarrow y = c(x))$	R(T)3, 6
8.	$z \lessdot x \rightarrow \forall y(y \lessdot x \leftrightarrow y = c(x))$	GP7
9.	$\exists z(z \lessdot x) \rightarrow \forall y(y \lessdot x \leftrightarrow y = c(x))$	PA8

[3] The formal proof follows closely the preceding one, taking into account that (1) $Ad^{\circ}(x) \rightarrow B(x)$ and hence, in particular, $Ad^{\circ}(x) \rightarrow \exists z(z \in x)$; (2) $Ad^{\circ}(x) \rightarrow (y \in x \wedge z \in x \rightarrow y = z)$.

[4] By the way, let us remark that an important difference between Bolzano and the Brentanians and in particular the Husserl of the *Logical Investigations*, lies exactly in the fact that for the latter, *all* psychic phenomena have a content.

Thus, we reach the two main notion of the Bolzanian Theory of Knowledge: representation and judgement. A *(subjective) representation* [*subjektive Vorstellung*] is a cognitive adherence whose material is an idea; a *judgement* [*Urtheil*] is a cognitive adherence whose material is a proposition; i.e., writing '*R*' for 'being a representation' and '*U*' for 'being a judgement',[5]

$$R(x) \leftrightarrow Ad^*(x) \wedge V(c(x)) \qquad \text{D8.4}$$

$$U(x) \leftrightarrow Ad^*(x) \wedge S(c(x)) \qquad \text{D8.5}$$

Hence,

$$R(x) \leftrightarrow B(x) \wedge D(x) \wedge \exists a \forall z (z < x \leftrightarrow z = a)\,^{6} \qquad \text{T8.5}$$

$$U(x) \leftrightarrow B(x) \wedge D(x) \wedge \exists p \forall z (z < x \leftrightarrow z = p)\,^{7} \qquad \text{T8.6}$$

As special variables for representations we will use 'α', possibly with indices or primes, whereas as special variables for judgements we will use 'π', also possibly with indices or primes.

With regard to an x that bears a given representation α, we say that it *has that representation* and we write '$R(x, \alpha)$', whereas of an x that bears a given judgement π, we say that it *formulates that judgement* and we write '$U(x, \pi)$'; i.e.

$$R(x, \alpha) \leftrightarrow t(\alpha) = x \qquad \text{D8.6}$$

$$U(x, \pi) \leftrightarrow t(\pi) = x \qquad \text{D8.7}$$

Sometimes, it is useful to have a notation for the representation whose material is the idea a as well as for the judgement whose material is the proposition p; we will write '$\alpha(a)$' and $\pi(p)$, respectively; i.e. we establish

[5] The most natural level for a discussion of Bolzano's Theory of Knowledge is the higher one. If one wants to move on the elementary level, one must adjoin the predicate 'x is a representation of a quality idea [$R'(x)$]' defined by $R'(x) \leftrightarrow \mathrm{Ad}^*(x) \wedge B(c(x))$.

[6] TS §272, III 5–6: "*Every subjective idea is something actual, but only in that it is a kind of property of a being.* 1) A second attribute that belongs to all subjective ideas is that they are something actual, i.e., that they may make claim to existence at a determinate time in the mind of a determinate being. [...] To me at least it seems obvious that every subjective idea merits the name of an *actual thing*—but one which belongs to the species of actual things which occur only as *attributes*—in some other actual thing, we accordingly do not call substances, but rather adherences. For there can be no subjective ideas if there are no beings which have them, and they exist only as *phenomena* in the minds of these beings."

[7] TS §291, III 72: "Just as we suppose that for every subjective idea there is an objective idea which constitutes its matter, so too we must assume that every judgement is the appearance in the mind of some proposition, which is the *matter* of this judgement. 2) Actuality must be ascribed to judgements just as it is to subjective ideas. They have actuality, namely, in the mind of the being that forms them, and during the time that it does. [...] But just as ideas do not exist of themselves, but are only actual in something else as an *adherence*, so too is this the case for judgements. They exist only in beings which form them."

$$\alpha(a) = \varepsilon x(R(x) \wedge c(x) = a) \qquad \text{D8.8}$$

$$\pi(p) = \varepsilon x(U(x) \wedge c(x) = p) \qquad \text{D8.9}$$

Turning to the relationships between cognitive adherences and the entities of the ontological sphere, we may establish that

(1) of the representation α, we will say that it *represents* [*vorstellen*] the object i (and write '$i\,\mathrm{E}^*\alpha$') iff its material refers to i; i.e.

$$i\,\mathrm{E}^*\alpha \leftrightarrow i\,\mathrm{E}\,c(\alpha) \qquad \text{D8.10}$$

(2) of the judgement π, we will say that it *asserts* [*behauptet*] the *relationship* [*Verhältniß*] or the *situation* [*Umstand*] m (and write '$m\,\mathrm{A}^*\pi$') iff its material enunciates m; i.e.

$$m\,\mathrm{A}^*\pi \leftrightarrow m\,\mathrm{A}\,c(\pi) \qquad \text{D8.11}$$

Thus, *representing* is the composition of *having as material* and *referring*, whereas *asserting* is the composition of *having as material* and *enunciating*; i.e.

$$(1)\ i\,\mathrm{E}^*\alpha \leftrightarrow \exists a(i\,\mathrm{E}\,a \wedge a \lessdot \alpha) \qquad \text{T8.7}$$

$$(2)\ m\,\mathrm{A}^*\pi \leftrightarrow \exists p(m\,\mathrm{A}\,p \wedge p \lessdot \pi) \qquad \text{T8.8}$$

Thus, we may also introduce the predicates *x represents to himself the object i* ($R^*(x, i)$) and *x asserts the situation m* ($U^*(x, m)$) by establishing

$$R^*(x, i) \leftrightarrow \exists\alpha(R(x, \alpha) \wedge i\,\mathrm{E}^*\alpha) \qquad \text{D8.12}$$

$$U^*(x, m) \leftrightarrow \exists p(U(x, p) \wedge m\,\mathrm{A}^*p) \qquad \text{D8.13}$$

8.2 Logical Properties of Representations and Judgements

As we have just seen, representations and judgements are singular space–time-determined psychic phenomena and, properly speaking, unrepeatable. Nonetheless, a fundamental equivalence relation among them is induced by their possible identity of materials. Thus, we may say that two cognitive adherences x and y are *equal* [*gleich*], and we write '$x \approx y$', iff their materials are identical;[8] i.e.

$$A^*(x) \wedge A^*(y) \rightarrow (x \approx y \leftrightarrow c(x) = c(y)) \qquad \text{D8.14}$$

[8] TS §273, III 8: "Thus when I speak of *equal* ideas in the future, the reader should think at the very least that these are ideas belonging to the same objective idea."

TS §292, III 75: "When I speak of *equal* judgements in the sequel, the very least one should gather from this is that they are judgements containing the same proposition [. . .]."

from which

$$\alpha_1 \approx \alpha_2 \leftrightarrow c(\alpha_1) = c(\alpha_2) \quad \text{T8.9}$$

$$\pi_1 \approx \pi_2 \leftrightarrow c(\pi_1) = c(\pi_2) \quad \text{T8.10}$$

as well as, of course,

$$\alpha \approx \alpha; \quad \alpha_1 \approx \alpha_2 \rightarrow \alpha_2 \approx \alpha_1; \quad \alpha_1 \approx \alpha_2 \wedge \alpha_2 \approx \alpha_3 \rightarrow \alpha_1 \approx \alpha_3 \quad \text{T8.11}$$

$$\pi \approx \pi; \quad \pi_1 \approx \pi_2 \rightarrow \pi_2 \approx \pi_1; \quad \pi_1 \approx \pi_2 \wedge \pi_2 \approx \pi_3 \rightarrow \pi_1 \approx \pi_3 \quad \text{T8.12}$$

We will now remark that representations and judgements, as *manifestations* [*Erscheinungen*] or *conceptions* [*Auffassungen*] of ideas and propositions, respectively, enjoy properties 'corresponding' to those of their materials. For convenience, we will call these properties 'logical' and agree to use for them the same symbols used in the logical case, enriched by a star.

Hence we have, for example, that a representation is *objectual, objectless, singular, common*, of *width at least* (*at most, exactly*) *n*, etc. iff such is the idea that constitutes its material; i.e.

$$G^*(\alpha) \leftrightarrow G(c(\alpha)) \quad [\textit{α is an objectual representation}] \quad \text{D8.15}$$

$$\neg G^*(\alpha) \leftrightarrow \neg G(c(\alpha)) \quad [\textit{α is an objectless representation}] \quad \text{D8.16}$$

$$nA^*(\alpha) \leftrightarrow {}_nA(c(\alpha) \quad [\textit{α represents at least n objects}] \quad \text{D8.17}$$

$${}^nA^*(\alpha) \leftrightarrow {}^nA(c(\alpha)) \quad [\textit{α represents at most n objects}] \quad \text{D8.18}$$

$${}^n_nA^*(\alpha) \leftrightarrow {}^n_nA(c(\alpha)) \quad [\textit{α represents exactly n objects}] \quad \text{D8.19}$$

$$Cm^*(\alpha) \leftrightarrow Cm(c(\alpha)) \quad [\textit{α is a common representation}] \quad \text{D8.20}$$

....

From which, for example,

$$G^*(\alpha) \leftrightarrow \exists i(i \in^* \alpha) \quad \text{T8.13}$$

$${}^n_nA^*(\alpha) \leftrightarrow \overset{n}{\exists}!!i(i \in^* \alpha) \quad \text{T8.14}$$

Analogously, a judgement is said to be *true* (better: *correct* [*richtig*]) or also a *knowledge* [*Kenntniß* or *Erkenntniß*], *false* (better: *incorrect* [*unrichtig*] or *erroneous* [*irrig*] or an *error* [*Irrthum*]), *analytic, synthetic, empiric, conceptual*, etc. iff such is the proposition that constitutes its material; i.e.

$$W^*(\pi) \leftrightarrow W(c(\pi)) \quad [\textit{π is a true, correct judgement, a knowledge}] \quad \text{D8.21}$$

$F^*(\pi) \leftrightarrow F(c(\pi))$ [*π is a false, incorrect judgement, an error*] D8.22

$AW^*(\pi) \leftrightarrow AW(c(\pi))$ [*π is an analytically true judgement*] D8.23

$SW^*(\pi) \leftrightarrow SW(c(\pi))$ [*π is a synthetically true judgement*] D8.24

....

It is worth noting that at the symbolic level, a representation α *may represent an idea or a proposition*; this occurs, of course, when its material $c(\alpha)$ refers to an idea—or, respectively to a proposition—when, in other words, it is a *symbolic* idea. *Representing a proposition to oneself* is also called *thinking that proposition*. One should not confuse the relations of *thinking* and *judging*, which can be represented schematically as follows:

	in thinking	in judging
the adherence	representation	judgement
the material	idea of proposition	proposition
the represented	proposition	------
the enunciated	------	situation

8.3 Cognitive and Physical Properties of Representations and Judgements

Beside the 'logical' ones, other peculiar conceptualizations and classifications related to the 'materiality' of representations and judgements now become relevant.

As far as properly 'cognitive' properties are at issue, representations are distinguished with respect to *clearness* [*Klarheit*] and with respect to *distinctness* [*Deutlichkeit*].

In the first case, representations may be *clear* or *obscure* [*klar, dunkel*]. Although Bolzano considers that the greater or smaller clearness of a representation is in general determined through a comparison with another representation (on the clearness or obscurity of which the sense of comparison depends), he nonetheless rejects the idea that clearness is to be regarded as an authentic *magnitude* [*Größe*]. The idea that clearness may be regarded as *awareness* [*bewußt seyn*] of the representation, intended as a formulation of the judgement of *having* the representation, is also discussed, but is rejected. At the end of a rather complex discussion, Bolzano arrives at the conclusion that a representation is *clear* when we, in turn, represent it, and do so precisely through an *intuition*, and is *obscure* in the opposite case.[9]

[9] TS §280, III 17: "3) [...] An idea is *clear* if we represent it to ourselves by way of an *intuition*. It will be called *obscure* whenever this is not the case."

With regard to the second distinction, representations may be *distinct, indistinct,* or *confused* [*deutlich, undeutlich, verworren*]. Following an extensive discussion, Bolzano arrives at the conclusion that a representation is distinct when we are able to indicate whether it is simple or, if it is not, from which parts it is composed; it is *indistinct* if we are not able to do this; and it is *confused* if it is indistinct but also composed.[10]

With regard to relations between these two types of properties, Bolzano establishes that a distinct representation must in all cases also be clear; moreover, if it is composed, then at least its next components must be clear, although not necessarily distinct. Developing this idea, Bolzano introduces a scale of *degrees of distinction* [*Grade der Deutlichkeit*] that culminates with *perfect distinction* [*vollendete Deutlichkeit*] that occurs when the representation is mastered in its entire structure.

The cognitive properties, however, do not exhaust the 'material' properties of the representations. Indeed, the fact that two representations are equal in the sense of $\approx$, does not imply their equality as *real* things (i.e. as things capable of producing effects);[11] in fact, it turns out that equal representations may *act* quite differently. What determines the *effectiveness* [*Wirksamkeit*] of a representation are its *intensity* or *power* or *liveliness* [*Stärke* or *Lebhaftigkeit*]. This intensity is an authentic *magnitude.* For example, half- and three-metre representations of a snake have remarkably different powers.

The 'cognitive' properties of representations are transferred in a natural way to the judgements of which they are part; thus, we may have *clear* and *obscure, distinct* and *confused judgements.*[12]

As a judgement is not merely the sum of the ideas occurring in it, in addition to the natural extension to it of the 'physical' properties of power and liveliness, there is a new important property to be considered: *trust* or *approval* [*Beifall, assensio*]. Even equal judgements having the same intensity may differ deeply with regard to the *degree of trust* [*Grad der Zuversicht*] with which they are formulated, the *degree* with which they *are kept as true* [*Grad des Fürwahrhaltens*]. This will be discussed in Section 6.

[10] TS §281, III 27: "2) [. . .] an idea is called *distinct* if we know either the parts of which it is composed, or else that it has none. If we do not know this, I call the idea *indistinct.* If it is also complex, I call it *confused.*"

[11] Note the correspondences

{	*wirken*	*wirklich*	*Wirkung*	*Wirksamkeit*
	to act, to produce	*real*	*effect*	*effectiveness*

[12] This 'transfer' occupies §§295–6.

8.4 The Formation of Judgements

A chapter of special interest in Bolzano's Theory of Knowledge concerns the *genesis* or the *formation* [*Entstehung*] of judgements.[13] From this point of view, judgements turn out to be essentially of two types: *immediate* or *direct* [*unvermittelt* or *unmittelbar*] and, in the opposite case, *mediate* [*vermittelt*] or, as they are also called, *caused* or *generated* [*verursacht, erzeugt*]. It must be emphasized that *all* judgements, as adherences of a thinking being, are *existent* and hence also necessarily, in a proper sense, *caused* (the only existing uncaused thing is God). Hence, this distinction does not concern in general the *causes* of a judgement, but the presence or not among those causes of other judgements. Therefore, when we refer to the *complete cause of a judgement*, we exclusively mean, in this context, the totality of judgements occurring among its causes.[14] In this context, Bolzano once again makes use of a whole series of current denominations, some of which must, however, be clearly distinguished from those equiform ones that are used in other contexts with rather different values. Thus, Bolzano also refers to *mediating* [*vermittelnd*] judgements as *causing* or *generating* [*verursachenden, erzeugenden*] judgements or *premises* [*Vordersätze*] and even *reasons* [*Gründe*], whereas he also refers to the mental process of mediation as *inferring* or *concluding*, even *drawing consequences* [*Schluß, Schließen, Folgern*]; finally, for the capability of our mind to find, in the keeping as true of certain judgements, the cause for the keeping as true of other judgements, he recalls the denominations of *inferential capability* or *faculty* [*Schlußkraft, Schlußvermögen*].[15]

With respect to judgements caused by other judgements, Bolzano makes a further distinction between those that are truly *mediate* and those that are merely *brought about* [*veranlasst*] by other judgements.[16] The distinction, which is illustrated through an example of a judgement of the latter type ('I judge that Caius is

[13] An interesting comparison of it with various contemporary theories is developed in M. Siebel [2004].

[14] TS §300, III 82: "2) [...] In speaking of a *complete* cause I do not mean that nothing other than the presence of *A, B, C, D,...*, not even an activity of the mind, is required, but only that aside from the indicated judgements *A, B, C, D,...* no other judgement is necessary in order to bring about *M*."

[15] TS §300, III 82: "2) An especially noteworthy kind of origin of judgements is that whereby one or more other judgements *cause* or, if you like, *mediate* a given one. I say that a judgement *M* is caused or mediated by certain others *A, B, C, D,...* if the cause of us forming judgement *M* lies in the fact that we have just formed the judgements *A, B, C, D,...*, The judgement *M* in such cases is said to be *caused, produced, mediated* by *A, B, C, D,...*, which are said to *cause, produce* or *mediate M*. Often, the action of the mind when it moves from judgements *A, B, C, D,...* to judgement *M* is called an *inference*, and *M* is called *inferred, concluded* or *derived*, while the judgements *A, B, C, D,...* are called the *premises*, and frequently (though improperly) the *grounds* of *M*. The soul's ability to find the cause of its belief in the proposition *M* in its belief in the propositions *A, B, C, D,...* is called its *power* or *faculty of inference*."

[16] TS §300, III 82–3: "Judgements which are not mediated, whatever their cause may otherwise be, are called unmediated or immediate. To say truly that a judgement *M* was mediated by judgements *A, B, C, D,...* it is not

an erudite man; this occasions in me the judgement that the erudite men are vain'), is not immediately clear. The further development, however, indirectly provides us with a specification of the distinction: in order to speak of mediation through certain judgements that cause another judgement, there must be a 'logical' relation between the materials of these judgements and the material of the resulting judgement.

Before delving into this last affirmation, however, it is important to underline the radically singular nature of a judgement and, in particular, the fact that also the properties and the relations that will be discussed are not substitutive with respect to equality of judgements (i.e. identity of materials). So, of two *equal* judgements, one may be mediate and the other immediate; moreover, two *equal* judgements may be mediated by unequal judgements and not only in the case of different bearers, but also in the case of one and the same bearer.[17]

Concerning the 'logical' relationship between the *mediating* judgements $\pi_1, \ldots, \pi_n$ [we will also write '$\overset{n}{\pi}$'] and the *mediated* judgement π, Bolzano maintains that it may be of three and only three different types:[18]

(1) the $\pi_1, \ldots, \pi_n$ are true and their materials are *reasons* for the material of π;

(2) the material of π is *derivable* from those of the $\pi_1, \ldots, \pi_n$;

(3) the material of π has, with respect to those of the $\pi_1, \ldots, \pi_n$, a certain *degree* of *probability* $> \frac{1}{2}$.[19]

Now, that certain judgements $\pi_1, \ldots, \pi_n$ are among the causes (in the proper sense) of a certain other judgements π naturally means, on the basis of what has been said up to now, that the propositions asserting the existence of the $\pi_1, \ldots, \pi_n$ are partial reasons for the proposition asserting the existence of π:

– it holds indeed, $\pi_1, \ldots, \pi_n \succ \pi \leftrightarrow h(\{\pi_1\}, \overline{d}), \ldots, h(\{\pi_n\}, \overline{d}) \rhd h(\{\pi\}, \overline{d})$ –

enough that *A, B, C, D, ...* provided an occasion for the formation of the judgement *M*; rather, their presence must be the complete cause for the occurrence of *M*."

[17] TS §300, III 83–4: "5) If certain judgements *A, B, C, D, ...* have mediated a judgement *M* on one occasion, it does not follow that *M* will always be mediated by them, nor that the only way for us to form *M* is via the mediation *A, B, C, D, ...*. Still less may it be claimed that the judgements via which a certain other judgement *M* is mediated in us are necessary for the mediation of *M* in every other thinking being."

[18] TS §300, III 84: "7) [...] I believe that a judgement *M* can only be *mediated* by others *A, B, C, D, ...* if one of the following three cases occurs: (a) All of the propositions *A, B, C, D, ...* are true and are the objective grounds of the proposition *M* (cf. §198) or (b) proposition *M*, if it is not a consequence of must still be *deducible* from them in the sense of §155; or, finally, (c) proposition *M*, if not completely certain, nevertheless has a certain degree of *probability* on the basis of *A, B, C, D, ...* in the sense of §161."

[19] In the discussion that follows, Bolzano always speaks only of '*Wahrscheinlichkeit*'; nevertheless, it is clear that, here, he is using this term not with its value he has called 'scientific', but with the value he has called 'common'.

At present, however, we are interested in situations in which the $\pi_1, \ldots, \pi_n$ are the *unique* judgements occurring among the causes of π, and so, to make our work easier, we agree to denote this kind of relation by '$[\pi_1, \ldots, \pi_n] \succ \pi$'; i.e. we define

$$[\pi_1, \ldots, \pi_n] \succ \pi \leftrightarrow \pi_1, \ldots, \pi_n \succ \pi \wedge \forall x(U(x) \wedge x \succ \pi \rightarrow \bigvee_{i \leq n}(x = \pi_i)) \qquad \text{D8.25}$$

We are now able to provide a precise assessment of the Bolzanian ideas, first by introducing the definitions of three relations $\succ_1$, $\succ_2$, and $\succ_3$ by setting

$$\pi_1, \ldots, \pi_n \succ_1 \pi \leftrightarrow [\pi_1, \ldots, \pi_n] \succ \pi \wedge c(\pi_1), \ldots, c(\pi_n) \rhd c(\pi) \qquad \text{D8.26}$$

$$\pi_1, \ldots, \pi_n \succ_2 \pi \leftrightarrow [\pi_1, \ldots, \pi_n] \succ \pi \wedge c(\pi_1), \ldots, c(\pi_n) \Rightarrow c(\pi) \qquad \text{D8.27}$$

$$\pi_1, \ldots, \pi_n \succ_3 \pi \leftrightarrow [\pi_1, \ldots, \pi_n] \succ \pi \wedge c(\pi_1), \ldots, c(\pi_n) \overset{\circ}{\Rightarrow} c(\pi) \qquad \text{D8.28}$$

and successively by assuming the axiom according to which these are also the only ways in which the $\pi_1, \ldots, \pi_n$ may generate π; i.e.

$$[\pi_1, \ldots, \pi_n] \succ \pi \leftrightarrow (\pi_1, \ldots, \pi_n \succ_1 \pi) \vee (\pi_1, \ldots, \pi_n \succ_2 \pi) \vee (\pi_1, \ldots, \pi_n \succ_3 \pi) \qquad \text{SGA9}$$

In brief,

$$\left[\overset{n}{\pi}\right] \succ \pi \leftrightarrow (\overset{n}{\pi} \succ_1 \pi) \vee (\overset{n}{\pi} \succ_2 \pi) \vee (\overset{n}{\pi} \succ_3 \pi)$$

$$\left[\overset{n}{\pi}\right] \succ \pi \rightarrow (c(\overset{n}{\pi}) \rhd c(\pi)) \vee c(\overset{n}{\pi}) \Rightarrow c(\pi)) \vee c(\overset{n}{\pi}) \overset{\circ}{\Rightarrow} c(\pi)) \qquad \text{T8.15}$$

Proof From the definitions, SGA is equivalent to

$$\left[\overset{n}{\pi}\right] \succ \pi \leftrightarrow (\left[\overset{n}{\pi}\right] \succ \pi \wedge c(\overset{n}{\pi}) \rhd c(\pi)) \vee (\left[\overset{n}{\pi}\right] \succ \pi \wedge c(\overset{n}{\pi}) \Rightarrow c(\pi))$$
$$\vee(\left[\overset{n}{\pi}\right] \succ \pi \wedge c(\overset{n}{\pi}) \overset{\circ}{\Rightarrow} c(\pi))$$

i.e., by distributivity,

$$\left[\overset{n}{\pi}\right] \succ \pi \leftrightarrow (\left[\overset{n}{\pi}\right] \succ \pi \wedge (c(\overset{n}{\pi}) \rhd c(\pi)) \vee c(\overset{n}{\pi}) \Rightarrow c(\pi)) \vee c(\overset{n}{\pi}) \overset{\circ}{\Rightarrow} c(\pi))$$

and therefore

$$\left[\overset{n}{\pi}\right] \succ \pi \rightarrow (c(\overset{n}{\pi}) \rhd c(\pi)) \vee c(\overset{n}{\pi}) \Rightarrow c(\pi)) \vee c(\overset{n}{\pi}) \overset{\circ}{\Rightarrow} c(\pi))$$

Bolzano fulfils as follows his obligation to show that each one of these three hypothesized cases occurs in the reality of the cognitive activity and that they are also the only possible ones in it.[20]

[20] TS §300, III 84: "7) [. . .] In order to show the correctness of this claim, I must first establish that each of the three cases occurs, and then that there is no other kind of mediation of one judgement by others."

He has little to say about the first case, since its occurrence seems to be very clear—at least for those, like him, who "accept that the truths may stay among them in a relation of consecutivity and that we men are sometimes able to recognize the occurrence of such a relation".[21]

The second case presents greater difficulties. Experience, indeed, suggests that there are cases in which, once the truth of the premises of an *inference* (in the technical Bolzanian sense) have been accepted, one immediately draws the truth of the conclusion. The same experience, however, shows us that, in certain cases, that acceptance is subordinated to the recognition of the inference at issue, that is to the acceptance of the proposition that enunciates the subsistence of the derivability relation among the materials of the judgements at issue.[22] One could thereby be induced to believe that such an assumption may *always* in fact be necessary. According to Bolzano, however, this is impossible because if it were so, one would regress into infinity and no judgement caused by other judgements would ever be possible.[23] He states:

> "If, indeed, no conclusion from its premises may be recognized without having recognized before the rule according to which that inference is constituted, then the two mentioned judgements, are still not sufficient to produce the knowledge of the proposition *M*; that not only contradicts the premise just accepted, but also shows that in this way it would be totally impossible to ever draw a judgement from other judgements, because one would here in fact need an infinite set of judgements. Indeed, what I have just said about the necessity that to the two given judgements a third one should be adjoined enunciating that the proposition which we should derive from them is really derivable from them as a conclusion, still holds when this proposition has already been adjoined. Now indeed, a fourth proposition is required that shows that the three proposition just found really behave, with respect to the one proposition to be derived, as premises of their conclusion and so on, to infinity."[24]

Let us agree (a) to write '$\pi(\overset{n}{p} \Rightarrow p)$' to denote *the judgement whose material is constituted by the proposition enunciating the subsistence of the derivability relation*

[21] Cf. TS §300, III 84 (3).

[22] TS §300, III 85: "9) [...] It may seem as if the acceptance of *M* can come about only if, in addition to the acceptance of propositions *A, B, C, D,...* the truth that *M* is deducible from *A, B, C, D,...* I admit that this is often the case."

[23] TS §300, III 85: "9) [...] But this is not always the case. We sometimes advance immediately from the acceptance of certain propositions *A, B, C, D,...* to the acceptance of a deducible proposition *M*, without first thinking of the truth that a proposition of form *M* is deducible from propositions *A, B, C, D,...* or without even being aware that there is such a truth; I think that this can be proved, since otherwise not a single inference could be drawn, i.e., not a single judgement could be based upon other judgements as its cause."

[24] Cf. TS §300, III 85 (9).

between the propositions $p_1, \ldots, p_n$ *and the proposition* p;[25] (b) to vertically write the judgements involved in the relation $\pi_1, \ldots, \pi_n \succ_2 \pi$, using a horizontal line to separate those mediating from that mediated; thus, the Bolzanian argument may be presented as follows.

If every judgement based on a certain inference needed the acceptance of the inference itself, then the mediation

$$(1) \qquad \frac{\pi_1}{\pi_2},$$

being justified by the inference $c(\pi_1) \Rightarrow c(\pi_2)$, would in fact be

$$(2) \qquad \frac{\begin{array}{l}\pi_1\\ \overline{\pi}(c(\pi_1) \Rightarrow c(\pi_2))\end{array}}{\pi_2}$$

but this, being justified by the inference $c(\pi_1), (c(\pi_1) \Rightarrow c(\pi_2)) \Rightarrow c(\pi_2)$, would be

$$(3) \qquad \frac{\begin{array}{l}\pi_1\\ \overline{\pi}(c(\pi_1) \Rightarrow c(\pi_2))\\ \overline{\pi}(c(\pi_1), (c(\pi_1) \Rightarrow c(\pi_2)) \Rightarrow c(\pi_2))\end{array}}{\pi_2}$$

Now, however, the justifying inference of this mediation is

$$c(\pi_1), (c(\pi_1) \Rightarrow c(\pi_2)), (c(\pi_1), (c(\pi_1) \Rightarrow c(\pi_2)) \Rightarrow c(\pi_2)) \Rightarrow c(\pi_2)$$

and hence the mediation itself would be

$$(4) \quad \frac{\begin{array}{l}\pi_1\\ \overline{\pi}(c(\pi_1) \Rightarrow c(\pi_2))\\ \overline{\pi}(c(\pi_1), (c(\pi_1) \Rightarrow c(\pi_2)) \Rightarrow c(\pi_2))\\ \overline{\pi}(c(\pi_1), (c(\pi_1) \Rightarrow c(\pi_2)), (c(\pi_1), (c(\pi_1) \Rightarrow c(\pi_2)) \Rightarrow c(\pi_2)) \Rightarrow c(\pi_2))\end{array}}{\pi_2}$$

and so on to infinity.

With regard to the existence of the third type of mediation, the one that is based on a probability relationship, Bolzano believed that experience shows us that very often, as soon as we have formulated the judgements $\pi_1, \ldots, \pi_n$, the materials of

[25] That is, $\overline{\pi}(\overset{n}{p} \Rightarrow p)$ is $\pi(\overset{n}{p} \boxed{\overset{\cdot}{\underset{\overset{s}{a}}{\Rightarrow}}} p)$.

which are connected with a certain proposition p by a certain probability $> \frac{1}{2}$, we not only formulate the obvious judgement that p has, on the basis of the $\pi_1, \ldots, \pi_n$, a certain probability, but also the judgement $\pi(p)$ itself:

> "So we do not simply say that it is probable that in the next spring the surface of the earth will again be covered with grass and flowers, because this has already happened very often, but we just wait just for it (i.e., we formulate – with a greater or smaller degree of trust—just the judgement that it will happen.)"[26]

After the illustration of the fact that each one of the three hypothesized types of mediation really finds instances in experience, Bolzano faces the problem of the exhaustiveness of his classification.[27] Given his tacit conviction that the relations of consecutivity, derivability, and probability are the unique 'logical' relations that may furnish the basis for the transition from certain judgements to other judgements, his problem is to discover whether there are other 'illogical' relationships on which such a transition might be based. In this regard, he considers two possible cases: (a) one in which the transition is accomplished through a more or less tacit assumption of an erroneous, merely *imagined* [*eingebildet*] 'logical' relation and (b) one in which the transition is accomplished 'immediately', even without the tacit relation assumption.

As far as the first is concerned, Bolzano believed that its deviation is only apparent. In fact, the transition takes place through the further assumption of the judgement asserting the subsistence of that relation. Hence, the mediation does not directly develop starting from the judgements $\pi_1, \ldots, \pi_n$, but from them and the judgement

$$\overline{\pi}(W(c(\pi_1) \wedge \ldots \wedge W(c(\pi_n) \rightarrow W(c(\pi)))$$

and such a mediation is irreproachable, as well as being of the second type.[28]

In the second case, Bolzano believed to have the possibility of excluding it on the basis of the following consideration. In order for it to be real, there should

[26] Cf. TS §300, III 86 (10).

[27] TS §300, III 86: "11) We must now prove that these three ways are the only ways of generating a judgement from other judgements."

[28] TS §300, III 86: "11) [. . .] occasionally we take an incorrect rule of inference for correct and use it in order to proceed from certain judgements *A, B, C,* . . . to a new judgement *M*, where *M* is not actually implied by them; this cannot be denied. Suppose that this advance to the new judgement *M* does not flow immediately from a consideration of propositions *A, B, C, D,* . . . but takes place only because we have mistaken an incorrect rule of inference for a correct one, i.e., because we made the mistake of thinking that a proposition like *M* is deducible from propositions like *A, B, C, D,* . . . alone. It follows that judgement *M* was not generated by *A, B, C, D,* . . . alone, but by them together with the tacitly assumed judgement that a proposition like *M* can be deduced from propositions like *A, B, C, D,* . . . "

be a device [*Einrichtung*] in our minds by means of which we shift directly from certain judgements $\pi_1, \ldots, \pi_n$ to a judgement π without the subsistence of any 'logical' relationship among their materials. Now, the proof of the existence of such a predisposition must go through the exhibition of some judgement that, although formulated with all possible prudence, contradicts other judgements formulated with the same prudence; up to now, however, nobody has succeeded in doing so. Hence, if, simply owing to the fact that nobody has to date proven the *impossibility* of such a device, one is worried about its presence, one might as well distrust all of our judgements.[29]

In front of the mediate judgements, there are, as stated above, the *immediate* judgements. There seem to be two highly relevant points for the latter.

The first is that Bolzano believed that they are to be found especially, even if not exclusively, among those that he calls *perceptive judgements* [*Wahrnehmungsurtheile*], namely those *empirical* judgements that have one of the following two forms:

(1) *I – have – the phenomenon A*

where "the subject is the same being (I) that formulates the judgement, whereas the predicate represents the possession of a phenomenon which is just now occurring in this being; for instance, of a just coming to pass representation, of a just formulated judgement, of a just present sensation, of a will, *et similia*";

(2) *This (what I am just now guessing) – is – an A*

where "the subject representation is an insight just now occurring in the judging being, which subordinates it to a certain concept *A*, saying, for instance, This (what I am just guessing) is something red, a perfume, *et similia*"[30]

[29] TS §300, III 87: "11) [...] But to bring this about, there would have to be some kind of mechanism [Einrichtung] in our minds which brings it about that from judgements of the forms *A, B, C, D,*... a judgement of the form *M* is generated although the latter does not stand to the former in a relation of ground and consequence, nor deducibility, nor even probability. [...] Its existence should be claimed only if it could be shown that certain judgements which we form with every care and precaution nonetheless contradict other judgements which we have formed just as cautiously. But nobody has been able to prove this. On the other hand, if we wanted to assume without *proof* the existence of such an error-producing mechanism in our mind merely because its *impossibility* has not been established, then, for the same reason, we should not trust a single one of our judgements, i.e., we should not form any judgements at all. But this would be a contradiction, since it is a consequence of trusting those judgements which determined that very decision."

[30] TS §300, III 87: "12) [...] a) One of these forms is *I – have – the impression A*; the subject of all these judgements is the person who utters them (*I*) while their predicate indicates the possession of an impression which is just now present in that person, for example the presence of an idea, a just formed judgement, a present sensation, an act of will, etc. (b) The other form is this: *This (what I just now observe) – is– an A*; in this case the

It must, however, be emphasized that Bolzano widely argued about the fact that *not every perceptive judgement is immediate, but that some of them must be.*

The second relevant element pointed out by Bolzano is the fact that, besides other empirical judgements that possibly differ from the perceptive ones, immediate judgements must in any case also be found among the *purely conceptual* ones; otherwise, we should accept that each purely conceptual judgement finds, after all, its reason in empirical judgements, which appears to be incomprehensible. Thus, for instance, from 'I have the intuition *A*', I can draw the conclusion 'There must be something real that produces this intuition in me', but I will never draw from it—or from so many other judgements as one will of the types considered above—that 'Every intuition arising in me or in any other finite being presupposes the existence of something real which produces it'.[31]

Having established that the only possible types of mediation are the three examined above, we still need to explain the undeniable fact that in many cases the mediation process develops in an unconscious way; this problem, however, falls outside the logical sphere.

8.5 Knowledge and Error

As already mentioned, a *true* or *right* or *correct* judgement (i.e. a judgement whose material is true) is also called a *knowledge*; a *false* or *wrong* or *incorrect* judgement is also called an *error*. The main problem is: *How is error possible?* How do we formulate an incorrect judgement?

In brief, referring to the immediate–mediate judgement distinction, Bolzano maintains that error can only occur in mediate judgements and more precisely in mediate judgements of the third type. In greater detail, according to Bolzano, we cannot doubt the correctness of immediate judgements. Indeed, they all arise in the same way; hence, if we doubt one of them, then we should doubt all of them and therefore also the mediate ones, since they draw their legitimacy from these.[32] Therefore, errors may be found only among mediate judgements. The validity of

subject idea is an intuition which is just now present in the judging person, and is subsumed by this person under a certain concept *A*; this is the case, for example, when somebody says: This (what I just now observe) is something red, a pleasant fragrance, etc."

[31] TS §300, III 88: "13) It will become clear, upon some reflection, that there must be far more immediate judgements than we have just described. In particular, among *purely conceptual judgements* there must be some which are not mediated. For, if the judgements described in no. 12 were the only immediate judgements, then all the rest of our judgements would have to be generated either immediately or mediately from them. But it is not conceivable how this can be the case."

[32] TS §309, III 143: "1) [...] None of the judgements formed without the mediation of others can reasonably be suspected of being mistaken; for if we wanted to doubt one of them, we should have to doubt them all, since

the type of inference used in the mediation, however, cannot be questioned, unless it is a rule that we have established ourselves (i.e., in fact, a further premise).[33]

Hence, if all immediate judgements are correct and all types of inference used to reach mediate judgements are valid, then how do errors arise? According to Bolzano, the only possible answer is that they arise in mediations of probabilistic types (in the strict sense, i.e. involving only probabilities different from 0 and 1). The conclusion of a probabilistic inference, indeed, does not state anything more than the fact that a certain proposition p has a given smaller or greater degree of probability and our mind, owing to its finiteness, is induced to keep as true a proposition that seems to it probable with a degree of trust suited to this probability. One is, in other words, led to formulate the judgement π. However, a probable, even a very probable, proposition is not always true and so the mind may also accept false propositions as true.[34]

The subject of error is taken up again along two quite different lines of thought, in the context of a new reading, on the cognitive level, of the problems of probability. In the first, which mainly occupies §317, Bolzano returns to ponder, let us say, the 'logicistic' treatment of probability in terms of the *objective credibility of a proposition by a thinking subject*. In the second, on the contrary, he suggests an exquisitely 'subjectivistic' type of approach to probability centred on the notion of *degree of trust in a judgement* (more on this later).

The *probability for a thinking being of a proposition, with respect to certain premises and certain ideas* [*Wahrscheinlichkeit eines Satzes für ein denkendes Wesen hinsichtlich auf gewisse Voraussetzungen und Vorstellungen*], is introduced as *the proposition p is probable for the thinking being x with respect to the premises $p_1, \ldots, p_n$ and the ideas $a_1, \ldots, a_s$* iff

(1) x accepts as true those propositions (i.e. they are materials of judgements belonging to him);

(2) the p has, with respect to those propositions and ideas, a certain probability (in the logicistic sense).

they all come about in the same way. But if we wanted to doubt all immediate judgements, we should also have to doubt all derived ones; hence we should have to doubt all our judgements, which is absurd."

[33] TS §309, III 143: "1) [...] According to §300, we should not assume that the *mode of inference* by which we derive some of our judgements from others is erroneous, unless it is a rule we have formulated, i.e., a judgement."

[34] TS §309, III 143: "2) I can only think of one way in which this might happen. According to what was said in §§300 and 301 there are inferences of probability, i.e., inferences where the conclusion merely states that a certain proposition M has a greater or lesser degree of probability. Moreover, our soul, on account of its finitude, is obliged to hold this proposition to be true, i.e., to form the judgement M, with a degree of confidence proportionate to the probability. But as a proposition which is probable need not always be true, it is clear how the soul, in elevating all the propositions with a high degree of probability to judgements, can sometimes err."

Hence, if we agree on identifying a thinking being as a thing that formulates at least a judgement, i.e. if we set

$$DW(x) \leftrightarrow \exists z U(x, z) \qquad \text{D8.29}$$

then we may transcribe the Bolzanian definition as

$$P(x; \overset{n}{p}; p; \overset{s}{a}) \leftrightarrow DW(x) \wedge \exists \overset{n}{\pi} (\bigwedge_{i \leq n} U(x, \pi_i(p_i)) \wedge \exists \mu (\overset{n}{p} \overset{\mu}{\Rightarrow} p)) \qquad \text{D8.30}$$

It is important to emphasize that in this relation, the subjectivity of the thinking being *only* enters into his accepting the truth of the premises. In fact, the first of the distinctions introduced by Bolzano with respect to this relation concerns the fact whether the thinking being does or does not know the probability connecting the materials of his judgements with the proposition. Indeed, in this new probabilistic relation, Bolzano distinguished the case of *recognized* [*erkannte*] probability and the case of *ignored* [*unerkannte*] probability. The subsequent distinction concerns the truth or falsity of the judgements made by a thinking being. This provides us with a distinction between *real* or *objective* [*wirklich, objectiv*] probability, which occurs when all premises are true and *alleged* or *subjective* [*vermeintlich, subjectiv*] probability, which, on the contrary, occurs when at least one of the premises is false.

In reality, from a general point of view, there seem to be not two but four possible situations:

(a) the judgements $\pi_1, \ldots, \pi_n$ are all true and a relation of μ-probability subsists between their materials and the material of π; i.e.

$$W(\overset{n}{\pi}) \wedge c(\overset{n}{\pi}) \overset{\mu}{\Rightarrow} c(\pi)$$

(b) the judgements $\pi_1, \ldots, \pi_n$ are all true but there subsists no relation of μ-probability between their materials and the material of π; i.e.

$$W(\overset{n}{\pi}) \wedge \neg(c(\overset{n}{\pi}) \overset{\mu}{\Rightarrow} c(\pi))$$

(c) the judgements $\pi_1, \ldots, \pi_n$ are not all true but a relation of μ-probability subsists between their materials and the material of π; i.e.

$$\neg W(\overset{n}{\pi}) \wedge c(\overset{n}{\pi}) \overset{\mu}{\Rightarrow} c(\pi)$$

(d) the judgements $\pi_1, \ldots, \pi_n$ are not all true and there subsists no relation of μ-probability between their materials and the material of π; i.e.

$$\neg W(\overset{n}{\pi}) \wedge \neg(c(\overset{n}{\pi}) \overset{\mu}{\Rightarrow} c(\pi))$$

However, cases (b) and (d), in which the accepted relation of probability between premises and conclusion does not subsist, may be reduced to case (c). In these two cases, indeed, the thinking being simply adjoins, to the premises accepted by him, a new judgement stating the wrong relationship; hence, his inference is in fact of the form 'from $\pi_1, \ldots, \pi_n, \overline{\pi}(c(\overset{n}{\pi}) \overset{\mu}{\Rightarrow} c(\pi))$ to π', where the $c(\pi_1), \ldots, c(\pi_n)$, $c(\overline{\pi}(c(\overset{n}{\pi}) \overset{\mu}{\Rightarrow} c(\pi)))$ are *not* all true but are now in an authentic probabilistic relation with $c(\pi)$; i.e.

$$\text{b)}\ \neg W(\overset{n}{\pi}, \overline{\pi}(c(\overset{n}{\pi}) \overset{\mu}{\Rightarrow} c(\pi))) \land c(\overset{n}{\pi}), c(\overline{\pi}(c(\overset{n}{\pi}) \overset{\mu}{\Rightarrow} c(\pi))) \overset{\mu'}{\Rightarrow} c(\pi)$$

$$\text{d)}\ \neg W(\overset{n}{\pi}), \overline{\pi}(c(\overset{n}{\pi}) \overset{\mu}{\Rightarrow} c(\pi))) \land c(\overset{n}{\pi}), c(\overline{\pi}(c(\overset{n}{\pi}) \overset{\mu}{\Rightarrow} c(\pi))) \overset{\mu'}{\Rightarrow} c(\pi)$$

The conclusion is just that each *error arises only owing to the fact that a thinking being accepts as true propositions that are with other propositions, which he takes to be true, in a 'logicistic' relation of probability.*[35]

Bolzano strongly emphasizes that the *probability of a proposition for a thinking being* is an extremely relative notion. Indeed, not only may it be, as seen above, recognized or ignored, real or alleged, and even more, but its degree also depends on the aggregate of judgements with respect to which it is determined.[36] With respect to this, Bolzano distinguished between a *complete* or *absolute* [*vollständig, absolut*] probability—that resulting from the involvement of the totality of the propositions that are somehow significant to that proposition which the thinking being accepts as true—and a merely *relative* [*beziehungsweise, relativ*] probability, in which only a part of that totality is involved.[37]

[35] TS §317, III 178–9: "3) On the other hand, propositions about which we may be mistaken are, according to no. 1, only those that stand in a real or fancied relation of probability (cf. §161) to other propositions that have already been accepted as true by the being in question. Hence only the following cases can arise: (a) the premises *A, B, C, D, . . .*, which are taken to be true by the being are indeed true and, proposition *M* stands to them in the relation of probability (cf. §161); or (b) the propositions *A, B, C, D, . . .* are true, but the proposition *M* does not really stand in the relation of probability with respect to them, but the being mistakenly thinks that it does; or (c) the premises *A, B, C, D, . . .* are not all true but a real relation of probability obtains between *M* and them; or, finally, (d) the propositions *A, B, C, D, . . .* are not all true nor is there a real relation of probability between *M* and them.—A moment's thought shows that cases (b) and (d) can be reduced to (c). Since [. . .] Hence we can say that error arises only if a being accepts as true propositions standing in the relation of probability to certain other propositions that it has already accepted as true. It is very important to draw attention to these propositions, and to give them a name; let us call them propositions which have *probability* for this being. Hence we must distinguish probability in this sense of the word from probability as it was defined in §161. There we meant by probability a relation which holds between propositions as such, no matter whether they are in themselves true or false or whether somebody takes them to be true or false. Here, on the other hand, probability is an attribute which a proposition can have only in relation to a certain thinking being, and only if that being accepts as true the propositions relative to which that proposition has probability."

[36] TS §320, III 192: "Hence, with respect to one and the same proposition *M* and to one and the same thinking being, we can distinguish several kinds and degrees of probability, but not several kinds and degrees of confidence."

[37] TS §317, III 180: "the probability which accrues to *M* only on the basis of part of this collection is called a relative probability of *M* for that being; while that which accrues to *M* on the basis of all of these propositions is called the *complete* or *absolute* probability of *M* for this being."

8.6 Trust

Naturally, a greater or lesser *liveliness* [*Lebhaftigkeit*] of the representations occurring in a judgement determines the greater or smaller *liveliness* of the judgement itself. As mentioned above, however, this property, unlike what happens in the case of the representation, does not exhaust the *power* [*Kraft*] or the *effectiveness* [*Wirksamkeit*] of the judgement, which depends not only on the liveliness of the involved representations, but also on their being more or less *intimately connected* [*innig*]. "Thus the effectiveness of a judgement, even with the same material and liveliness of each of its representations, still depends on something else, on the more or less intensive *closeness* [*Innigkeit*] with which its single parts are connected in a whole."[38] It is precisely the degree of this effectiveness that Bolzano decides to call the '*degree of trust* [*Grad der Zuversicht*]' with which a judgement is formulated. In reference to the proposition that constitutes its material, Bolzano also speaks of the '*degree of approval* [*Beifall, assensio*]' or the '*degree of our holding as true* [*Fürwahrhalten*]' or even the '*degree of our certainty* [*Gewißheit*]'. Bolzano dwells at length upon this concept. In particular, he ponders the reason we formulate a judgement with a certain degree of trust and another judgement with another degree and, even recognizing that this may sometimes happen because, before formulating it, we evaluate its degree of probability, he excludes that this is always the case; such an evaluation, indeed, can only be given by a new judgement, which, on its part, would require a new evaluation of its reliability, and so on to infinity. That the degree of trust with which we formulate a judgement does not presuppose a preliminary evaluation about its degree of validity is, in particular, well illustrated by the case of immediate judgements, which if they were to presuppose a preliminary evaluation of their probability, would be anything but immediate.

The conclusion of this examination is as follows:

> "What has been said will suffice to illuminate the distinction existing between the concept of probability, even in the meaning of the word explained in § 317, and the concept of trust. The probability is and always remains only a quality which belongs to propositions in themselves, independently of whether they are believed to be true or to be false, even also only represented, and it hence admits to be divided into known and unknown; the trust is, on the contrary, exclusively a quality of formulated judgements. With reference to one and the same proposition *M* and to one and the same thinking being we may distinguish different types and degrees of probability, not however different types and degrees of trust."[39]

[38] TS §293, III 75: "Since a judgement is not a mere sum of ideas, but consists in a certain efficacious combination of them, and since this combination can be more or less strongly felt, the efficacy of the judgement will depend upon this strength, even if the content and vividness of the individual ideas do not change."

[39] Cf. TS §320, III 192 *Note*.

In §320, bearing the title *How Different Degrees of Trust May Be Presented through Calculus*, Bolzano also outlines a possible mathematical development of this strictly subjectivist conception of probability,[40] stating, however, in advance that

> "Although the knowing how to present with great exactness, through calculus, the degree of trust with which we judge may rarely be of a particular utility, nevertheless the problem of the way this degree should be calculated may not be regarded as a useless thinking exercise; and only with this intention, I will pay here to it some attention."[41]

In fact, these outlines do not go much further than the definition of 'trust'. This is, however, worth mentioning; indeed, it turns out that trust is a function that is defined on *every* judgement and has values in the real interval $[-1, +1]$. Bolzano concludes this on the basis of the following argument. If, under the hypothesis of the certainty for us of the judgements $\pi_1, \ldots, \pi_n$, the judgement π has for us the probability μ and therefore the judgement $\neg\pi$ (i.e., more exactly, the judgement $\pi(\boxminus\, c(\pi)))$ the probability $1 - \mu$, our trust degree in π is given by the difference between the probability of π and the probability of $\neg\pi$ (i.e. by $\mu - (1 - \mu) = 2\mu - 1$). If now $\mu = \frac{1}{2}$, then our trust degree in π is $2\frac{1}{2} - 1 = 1 - 1 = 0$; if, on the contrary $\mu > \frac{1}{2}$, then $2\mu > 1$ and so also $2\mu - 1 > 0$, and hence our trust degree in π is positive; if, finally, $\mu < \frac{1}{2}$, then $2\mu < 1$ and so also $2\mu - 1 < 0$, and hence our trust degree in π is negative or, equivalently, that in $\neg\pi$ is positive. As, on the other hand, μ varies between 0 and 1, in these two limiting cases we have $2 \cdot 0 - 1 = -1$ (i.e. total distrust in π), respectively $2 \cdot 1 - 1 = 1$ (i.e. total trust in π). If, in particular, we can express the probability of π as $m/(m + n)$ (i.e. as the ratio between the m favourable cases and the $m + n$ possible cases), then the probability of $\neg\pi$ is clearly $n/(m + n)$, and hence our trust degree in π (i.e. the difference between the two probabilities) is $m/(m + n) - n/(m + n) = (m - n)/(m + n)$. Hence, "we may also say that the degree of trust with which we formulate a judgement can be found by dividing the difference between the quantities of the reciprocally exclusive cases in which the judgement becomes true and the cases in which it becomes false, by their sum".[42]

8.7 Knowing and Believing

Bolzano's treatment of the Theory of Knowledge ends with an analysis of the difference between "two states exceedingly worth of attention in which our mind

[40] It is worth remarking that the exemplifications used to explain this conception are bets.

[41] Cf. TS §320, III 190.

[42] TS §320, III 191: "2) [...] Hence we can say that the degree of confidence with which we form a judgement can be found by calculating the differences between the number of mutually exclusive cases in which the judgement becomes true and the number of cases in which it becomes false, divided by their sum."

comes to be with respect to its judgements": the *knowing* [*Wissen*] and the *believing* [*Glauben*]. These states do not depend on the *degree* of trust with which we adhere to a certain judgement, but on the consideration of the circumstance whether the demolition of that trust is or is not in our power. *To know* a certain proposition means to be aware that our trust in the corresponding judgement is already independent of our will (i.e. that, even if we wanted to, we could not convince ourselves of the contrary). Once having become familiar with the proof of the Pythagorean theorem, for example, we are so sure of the truth of this proposition that it would turn out impossible for us, even if we wanted to, to convince ourselves of the contrary. *To believe* a proposition, on the contrary, means that, even if we are resolute in keeping present the reasons of its truth, "it doesn't appear impossible to us, that by directing our attention to all true or even only seemingly true reasons which oppose themselves to it, we could be lead to the opposite judgement." So, for instance, Bolzano states that he believes that there are inhabited celestial bodies. Even recognizing that, in general, the degree of trust is greater in the case of knowing than in the case of believing, he does not exclude that there may be cases in which such a difference does not exist; so, he says, it usually occurs with our belief in God.

Bolzano concludes his analysis by introducing some other distinctions. A belief with a rather low degree of trust is called a *holding as true* [*Fürwahrhalten*] or a *thinking* [*Meinen*]; a believing may be *firm* [*fest*] or *swaying* [*schwankend*], according to the firmness of our will to keep present the reasons that have led us to such an act; it may even be *reasonable* [*vernünftig*] or moral [*sittlich*] or, on the contrary, *mad* [*töricht*] or *immoral* [*unsittlich*], depending on its object; finally, it may also be *theoretical* [*doctrinell*] or *historical* [*historisch*], according to the conceptual or empirical nature of its material [*Begriffs- or Erfahrungssatz*].[43]

[43] TS §321, III 193–4: "*Knowledge and belief.* There are two important states of our mind with respect to our own judgements, states which (if I understand them correctly) have to do with a consideration of the confidence with which we are attached to a given judgement; they do not depend upon the *degree* of this confidence, but rather on whether or not we are able to destroy this confidence. Suppose that a truth *M*, [. . .] has become so evident to us that we assume we could not persuade ourselves of the opposite even if we wanted to. [. . .] In this case I should like to call truth *M* an item of *knowledge* [*Wissen*]. [. . .] On the other hand, consider a proposition *M* which we take to be true, but of which we do not have knowledge. In this case it does not seem impossible that we could come to form the opposite judgement, Neg.*M*, by concentrating our attention on true or apparent reasons against *M*. In such a case I call the relation between our mind and proposition *M* a *belief in this proposition* [. . .] . Usually, the degree of confidence is higher in the case of knowledge than in the case of belief, but there are also cases where the confidence of a belief does not fall behind that of knowledge. [. . .] A belief which has a lower degree of confidence is usually called an *opinion*, or *taking something to be true.* Moreover, with respect to the firmness of our resolution to keep the grounds for our proposition in mind, we can distinguish a *solid* and a *wavering* belief, with respect to the ground which produced this resolution, we can distinguish a *reasonable, moral,* as well as a *foolish* or *immoral* belief and, finally, with respect to its object (i.e., whether it is a conceptual or an empirical proposition) a *theoretical* or an *historical* belief."

9

Semiotics

9.1 Semiotic Relations

In a single section, §285, under the title *Denotation of Our Representations* [*Bezeichnung unserer Vorstellungen*], Bolzano develops a general Theory of Signs and in particular a theory of linguistic signs that seems to us to be of singular historical and theoretical interest.

The cognitive basis of his theorization can be found in the fact that he draws inspiration mainly from the German tradition[1] of idea associationism, which he judges "one of the most fruitful discoveries done in the field of psychology after the re-establishment of the sciences"[2] and accepts, deciding to call it the *law of awakening* [*Erweckung*], *renewal* [*Erneuerung*], *linkage* [*Verknüpfung*], or *association* [*Association*] *of the representations*, the following principle:

> "If a pair of representations was once produced simultaneously in our mind and successively somehow arises in us a representation that is equal (has the same matter) to one of them, then, if not always, nevertheless very often, also the other one will repeat itself; i.e. another representation will arise having the same matter as the other."

Confirming that this phenomenon happens *often* but *not always*, he decides to call *awakening* [*weckende*] the representation that gives rise to another and *awaken* [*erweckte*] or *renewed* [*erneuerte*] the latter.[3]

[1] Indeed, he speaks of Hume, but never, for example, of David Hartley.

[2] Cf. TS §284, III 40 Note 2.

[3] TS §284, III 38: "4) [. . .] we may set out the following principle: 'If a pair of ideas has ever been produced simultaneously in our soul, and afterwards an idea is produced in us which is equal to one of them (has the same matter), then, if not always, at least quite often, the other will also be renewed, i.e. an idea will be produced which has the same matter as the other.' Permit me to call this proposition the *law of awakening, renewal,* or *association* of our *ideas,* and to say of an idea responsible for the production of another that it is the *stimulating* [*weckende*] idea

It is just this peculiarity of our mind "according to which representations that have been once linked together renew each other" that leads us to use "certain easily producible representations as a mean to generate, through their production, sometimes in us, sometimes in others, other representations which would be more difficult to obtain, but which are associated with them [*verknüpft*]".[4]

The communication process through signs develops, according to Bolzano, as follows.

When a thinking being *A* wants to transmit to thinking being *B* a certain *idea a*, he *produces* a certain *object* (e.g. a voice) *i*; *B* perceives it (i.e. the *object i stimulates* [*anregen*] in *B* a *representation* β that *represents i* (i.e. a *representation* β that has as *matter* the *idea* that *refers* to *i*)); in *B*, the representation β *is associated* with a *representation* α, the *matter* of which is the idea *a*. In this manner, *A* has transmitted to *B*, through the *sign* [*Zeichen*] *i*, the idea *a* that is *denoted* [*bezeichnet*] by that sign and constitutes the *meaning* [*Bedeutung*] of that sign, its *sense* [*Sinn*], or its *intention* [*Verstand*].[5]

A *sign* [*Zeichen*] is thus, by definition, "an object . . . through the representation of which we want to know renewed in a thinking being a representation associated to it", whereas the *meaning in the strict sense* (or the *sense*, or the *intention*) of the sign is the matter of the representation produced by association (in the interpreting being) of the representation of the sign. If the idea meant in a strict sense from the sign is objectual, then the object or the objects referred to by it constitute the *meaning in a broad sense* of the sign.[6]

while saying of the other that it is *stimulated* or *renewed*. 5) So far, I have intentionally refrained from saying that when two ideas have occurred simultaneously in the soul one may reawaken the other *every time*, saying instead that this occurs *frequently*. Experience teaches no more than this."

[4] Cf. TS § 285, III 44 (1).

[5] Bolzano suggests the possibility of making a distinction between the *meaning* of a sign and its *sense* or *intention*. The former would be the 'right' idea for the denoting of which the sign is determined; the latter would be, on the contrary, the idea that from time to time, in various circumstances, one intends to denote by it. The two things do not necessarily coincide. One can imagine that someone, for instance, may by ignorance use a sign in a completely different sense from its authentic meaning. TS § 285, III 44: "1) [. . .] But a certain distinction could be made between them in the following way. We could call the *signification* of the sign that particular idea which it is intended to produce, and succeeds in producing; while we could confine the words *sense* and *meaning* to those ideas which are merely intended to be produced by a certain sign in a given case. Understandably, somebody who does not understand a sign can take it in a sense or meaning which is quite different from its real signification."

[6] TS §285, III 44: "1) [. . .] An object which is used for this purpose, i.e. one through whose idea we intend to stimulate in a thinking being some other, associated idea, is called a *sign*. The objective idea whose corresponding subjective idea is supposed to be stimulated by the idea of the sign is called the *signified* idea, also the *signification* [*Bedeutung*] of the sign. If the signified idea has an object, then it is often called the *signified* object, or again, the *signification* of the sign. Sometimes the words *sense* [*Sinn*] and *meaning* [*Verstand*] are used synonymously with the word 'signification'."

Neglecting for the moment all other aspects and limiting ourselves to the pure structure of the relationship sign—meaning, we see that it can be described as follows ('s.s.' is for 'strict sense' and 'b.s.' for 'broad sense'):

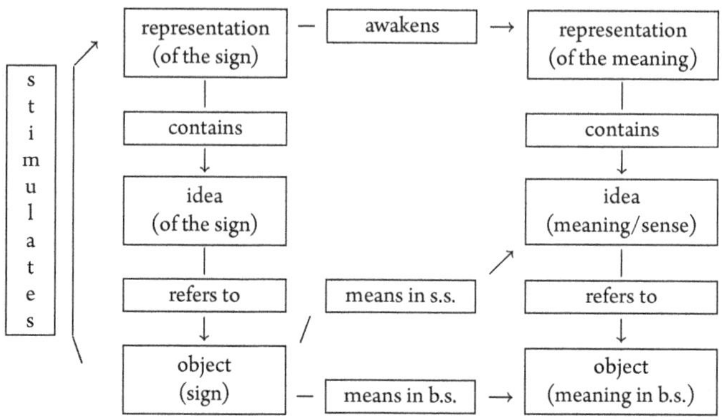

Writing '$s \text{☞} a$' for '*the sign (the object) s means* [or *denotes*] *in a strict sense the idea a*' and '$s\,]\text{☞}\, i$' for '*the sign (the object) s means* [or *denotes*] *in a broad sense the idea a*' allows us to 'define', for the moment,

$$s \text{☞} a \leftrightarrow \forall\alpha(s \,\mathrm{E} * \alpha \rightarrow \exists\beta(\alpha \succ \beta \wedge c(\beta) = a)) \qquad \text{D9.1}$$

$$s\,]\text{☞}\, i \leftrightarrow \exists a(s \text{☞} a \wedge i \,\mathrm{E}\, a) \qquad \text{D9.2}$$

The preceding schema can be then rewritten to highlight all involved relations:

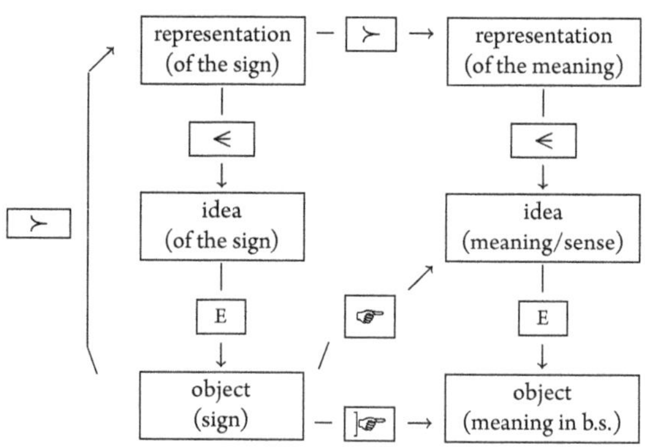

A sign having a meaning in a broad sense may be called '*objectual*'; a sign having no meaning in a broad sense may be referred to as '*objectless*'. An example of an 'objectual sign' is

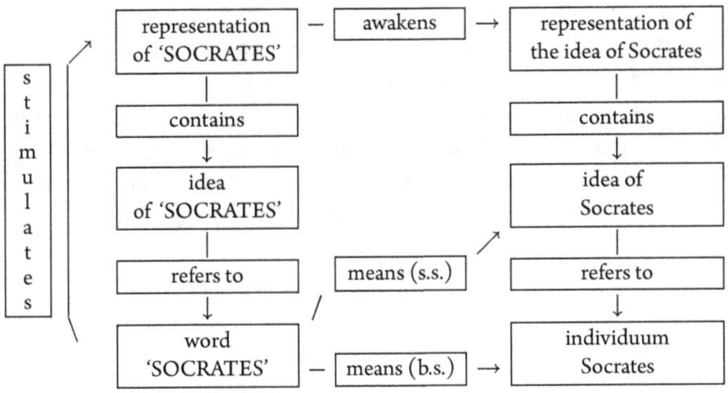

An example of an 'objectless sign' is

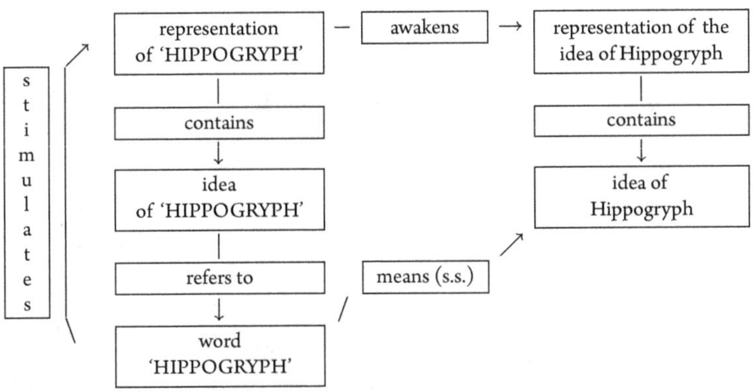

The meaning in a strict sense of a sign is thus always an idea. This idea may of course refer to a vast range of things; in particular, it may refer to another sign (i.e. it may be the *idea of a sign*). This second sign has, as such, a meaning in a strict sense (and possibly also a meaning in a broad sense). In these cases, we say that the first sign means *mediately* in a strict sense (in a broad sense) the meaning in a strict sense (in a broad sense) of the second sign. This is, in general the case of the written signs. A *vox scripta* means *immediately* in s.s. the idea of the 'corresponding' *vox prolata* and the *vox prolata* to which that idea refers means *immediately*, in s.s., a certain idea; this idea constitutes the *mediate* meaning, in s.s., of the first sign and its referents (if occurring) the mediate meaning in b.s. of the first sign.

> "Often, instead of a sign *a* established for a certain idea *A* we use another object *α* which is only a sign of the idea of *a*. Thus, for example, the well-known word 'God' is a sign of the concept of a being whose reality is unconditioned. When, however, the thinking being in whose mind we want to awaken this concept is too far away to hear

our voice, we produce certain visible signs (graphic characters) which designate the single sound of which the word 'God' is composed and which in their linkage denote this word in its entirety. The sight of this signs will at first only awaken the representation of the word: God; subsequently, however, also the representation of the object that this word denotes. Signs of this kind—signs which are signs of other signs—will be called, with respect to the ideas denoted by the latter, *mediate* [*mittelbar*] signs. Signs that are not mediate will be called *immediate* [*unmittelbar*]."[7]

Schematically, we have

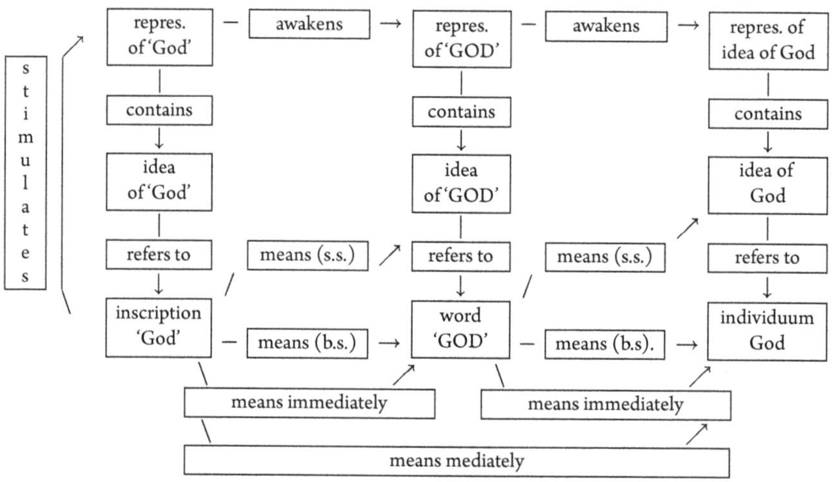

The habit, however, may gradually weaken and reach the complete disappearance of the mediating role of the second sign, thereby setting up an immediate relation between the representation of the first sign and that one of the meaning of the second sign.

"A sign which was originally mediate, can often become, through the continuous usage, an immediate sign. As indeed the representation of the sign α awakens the representation of the sign a and this the idea A denoted by it,[8] through frequent repetitions also α and A become simultaneous and may so immediately associate with each other (i.e., in such a way that the representation of α awakens that of A without the previous interposition of the representation a). Gradually, when it is not relevant for us, this final representation ends by completely coming out from our conscience. So the figure ≏ is, for the beginner astronomer the sign of a balance and hence only

[7] Cf. TS §285, III 47 (10).

[8] This is a somewhat hasty formulation. The 'awaking' relation, as a special case of the relation of cause that may subsist only among existents, cannot really connect representations and ideas. See the discussion on the 'vacuous quality' in Section 6.2.

mediately the sign of a celestial constellation or of the equinox; while an expert astronomer, on the contrary, seeing this figure, will immediately think to one of the two last ideas; the balance does not even enter his mind."[9]

Schematically, we have

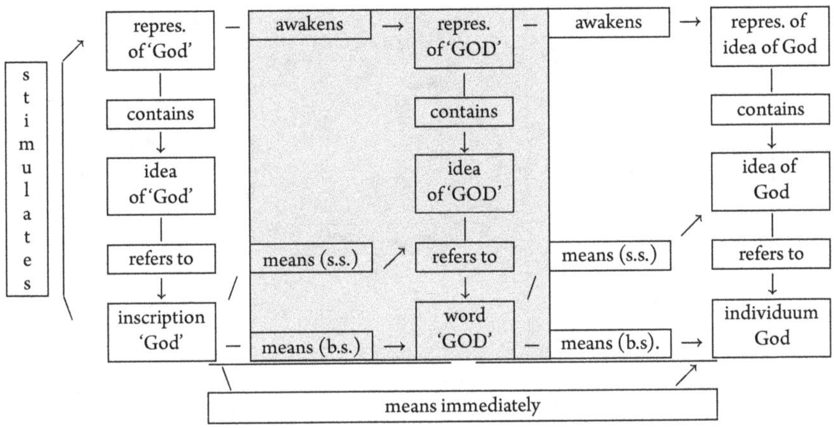

As emphasized already, the meaning in the strict sense of a sign is *always* an idea. Of course, when the idea to be transmitted is mereologically complex, the sign that is employed will somehow reflects this complexity. In this regard, Bolzano makes the following statement:

> "We often use express ideas which are composed from others, for instance ideas of entire propositions, only by using in a certain order those signs of the single ideas from the combination of which those complex ideas result. If we are using audible signs, then we let them follow one another approximately as the simple ideas must follow one another to produce the composite idea. When, for example, we want to produce the idea of an entire proposition, we first produce the sign which serves to generate the subject representation, followed by the sign that means the copula, and finally by the sign which must awaken the predicate representation. If, on the contrary, we are using visible signs, for example certain figures, then we dispose them next to one another in a well determined direction (for instance, from left to right, from upwards to downwards, *et sim.*) and we fix once for all in which temporal sequence they must be seized by the eye."[10]

An important point that arises out of this passage and that markedly differentiates Bolzano's from Frege's theory is that the meaning in the strict sense of a verbal

[9] Cf. TS §285, III 47.
[10] TS §285, III 45 (2).

sentence (and also, by mediation, of a written sentence) *is not a proposition, but an idea of proposition*; it is only the meaning in the broad sense to be constituted by a proposition, the proposition, of course, referred to by that idea. Schematically, we have

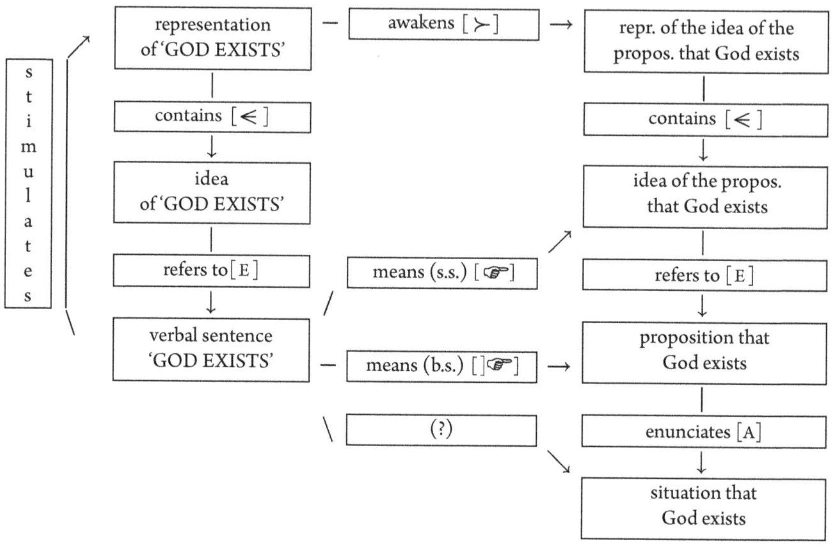

It is, however, worth mentioning that Bolzano, somewhat surprisingly, did not judge this to be particularly relevant:

> "In the formulation of the definition at n.1, I have presupposed that through signs—in the sense we have accepted here—we immediately present only *ideas*, not propositions or judgments. Nevertheless, also this last case seems to occur insofar as we sometimes express a whole proposition not only by means of a combination of more words, but also by means of a single word, for example: *morior*. It is however my opinion that such words or such combinations of words do not present at first the proposition in itself, but only an idea of it, and that they only mediatedly present the proposition in itself, and precisely only as every sign which presents an objectual idea may also be regarded as a sign of the object itself. However, even if things were different here, the question should be of no importance."[11]

9.2 Hermeneutics

After this synthetic examination of the 'logical' moments of the Bolzanian semiotics, let us return to the beginning and review its more properly 'hermeneutic' aspects.

[11] Cf. TS §285, III 49 (*Note* 1).

As has already been stated, a communication process via signs involves two thinking beings *A* and *B* (without excluding the case in which, as we will see, *A* and *B* coincide). The operation carried out by *A* (i.e. *using signs to provoke representations in B*) is called by Bolzano, in a broad sense, '*to speak, to say* [*sprechen, reden*] *to B*', whereas the aggregate of all the signs used by a man, in particular those used by him to communicate his thoughts to others, is called *his language* [*Sprache*]. On the other hand, Bolzano calls '*to read* [*lesen*] *those signs*' the operation carried out by *B* (i.e. *to consider those signs in order to draw from them the representations that the author wanted to provoke in him*). As far as the outcome of his reading—or better of his *interpretation* [*Auslegung*]—is concerned, there is first of all the *understanding* [*verstehen*], which happens when the signs really provoke in *B* the representations that *A* wanted to provoke, and the *misunderstanding* [*missverstehen*], which happens when *B* is erroneously convinced of having drawn such representations. A proposition enunciating that the sense of a certain sign is this or that is called an *interpretation* [*Auslegung*] of that sign;[12] this interpretation may, in turn, either be merely thought or expressed by signs of another type, in which case it is called a *verbal* [*sprachlich*] interpretation.[13] It must, however, be said that a comprehension failure is not necessarily due to misunderstanding; indeed, it may consist in the *indeterminateness* [*Unbestimmtheit*] *of a sign to the reader*, which happens when *B* is not able to decide which idea *A* had wanted to denote. Such indeterminateness may, in turn, have different origins. It may arise from the incapacity of the reader to choose one of the various possible meanings of an ambiguous sign, a case for which Bolzano reserved the term *fluctuation* [*Schwankung*], or it may be a case of a *total incomprehensibility* [*gänzliche Unverständlichkeit*] *for the reader*, which happens when *B* absolutely cannot guess any idea that the author may have denoted by that sign.

> "It is easy to imagine that a sign, which in a certain context or for a certain reader is fluctuating or indeterminate or totally incomprehensible, may in another context or for another reader be comprehensible and determinate. So, various passages of Cicero's *Epistulae* that are indeterminate or totally incomprehensible to us, should have been comprehensible and determinate to the individuals for whom he wrote, even for the whole learned part of his contemporaries."[14]

[12] An *interpretation* must be sharply distinguished from a *definition* [*Erklärung*]. As just said, the former explicates the meaning in s.s. of a *sign*, whereas the latter explicates the mereological structure of an *idea*.

[13] TS §285, III 44: "To use certain signs in order to produce ideas in others is to *speak* or *say something* in the widest sense of these words. To inspect given signs in order to find out which ideas their author wanted to generate is called *reading* in the widest sense. To grasp what ideas their author wanted to produce is to *understand* them. To assume that they have such and such a sense, while they have another, is to *misunderstand* them. Finally, a proposition in which it is stated that certain signs have such and such a sense is called an *interpretation* of these signs. An interpretation may be merely thought, or in its turn expressed with other signs, in which case it is called a *verbal* interpretation."

[14] Cf. TS §285, III 48 (6).

It is, however, important to remark that in a semiotic situation, the *being-a-sign* of an object depends neither on its qualities nor on the interpretative ability of B, but only on its *intentional* use by A.[15]

Thus, the preceding 'definition' of meaning should first of all be relativized to the thinking being B, i.e., taking into account that the representations of B are qualities belonging to him,

$$s \text{ means in s.s } a \text{ (for } B) \leftrightarrow \forall\alpha(B \in \alpha \wedge s \,\mathsf{E}^{*}\, \alpha \rightarrow \exists\beta(B \in \beta \wedge \alpha \succ \beta \wedge c(\beta) = a) \qquad \text{D9.3}$$

$$s \text{ means in b.s. } i \text{ (for } B) \leftrightarrow s \text{ means in s.s } a \text{ (for } B) \wedge i \,\mathsf{E}\, a \qquad \text{D9.4}$$

and then generalized to the various linguistic communities;[16] for example,

$$s \text{ means in s.s } a \text{ (for an anglophone)} \leftrightarrow \qquad \text{D9.5}$$
$$\forall B(B \,\mathsf{E}\, it \rightarrow \forall\alpha(B \in \alpha \wedge s \,\mathsf{E}^{*}\, \alpha \rightarrow \exists\beta(B \in \beta \wedge \alpha \succ \beta \wedge c(\beta) = a))).$$

Of course, the quantifiers '$\forall\alpha$' and '$\forall B$' should be replaced by the less binding ones 'for many α' and 'for many B'.

Actually, as alluded to eariler, the receiver may coincide with the transmitter:

> "Finally, it must still be remarked that, once *accustomed to a language*, we almost always accompany our thoughts with signs, even where we have absolutely no intention of communicating them to others and precisely so that we either really express them, completely or only partially, or, on the contrary, we are already content with their representation. Such a procedure could be called a *speaking with oneself*. Sometimes, however, we do this unintentionally, only because our representations are already so connected with those of their signs that these latter arise for themselves as soon as we form the former; some other times, on the contrary, we do this intentionally, possibly to obtain a more precise connection between our representations and the signs assigned to them, or to give a greater liveliness to our thoughts, and so on."[17]

To these fundamental traits Bolzano adjoins a great number of remarks and conceptual and/or terminological specifications. For instance, with respect to the origins of the single signs, he introduces the distinction between *natural* [*natürlich*] signs and *accidental* [*zufällig*] signs and among the latter *arbitrary* [*willkürlich*] signs. With respect to the nature of the signs, he makes the distinction between

[15] TS §285, III 50: "*Anmerk* 1. [...] Second, it is not necessary that the object used as a sign in fact serve to awaken the signified idea; rather, it is sufficient that it is used with this *intention*. A sign remains just that, even if it is misunderstood or remains unheeded."

[16] TS §285, III 44: "1) [...] If an object is used [...] to designate a given idea [...] is only used [...] by some, e.g., by a certain people, it is said to be merely customary for *them*."

[17] Cf. TS §285, III 48 (11).

signs of hearing or *acoustic signs* [*Gehör-, hörbare Zeichen*] and *visible signs* or *signs of sight* [*sichtbare, Sichtzeichen*], whereas, with respect to the use of signs, he distinguishes between those used with *proper meaning* [*eigentliche Bedeutung*] and those used with *improper* or *borrowed* [*uneigentlich, entlehnt*] *meaning*. The latter should not be confused with the *widened* [*erweitert*] meaning.[18]

Seven notes follow this section. In the first, Bolzano exhaustively discusses the doublenesss of meaning with which the term 'sign' is currently used: the one considered up to here and the one exemplified by the situations in which one says "the smoke is sign of the fire, blushing is a sign (though not completely sure) of a wrong and so on." He believes that the second one, for which he also suggests as a synonym the term '*countermark* [*Kennzeichen*]', is in fact the primary meaning and the second one only borrowed from it, which could be explained by the fact that "sounds and other objects the representations of which are used to awaken in another thinking being certain representations associated with them, are for him, if not always, nevertheless at least in the majority of the cases, a sign (a countermark) that those representations are present in the mind of the being who produces those sounds or other objects".[19] Let us remember that the discussion and study in depth of this distinction will be the starting point of Husserl's *First Logical Investigation*.

In the last of the seven notes, Bolzano discusses at length, but without any particular pretension, the question of the origins of language.

[18] Bolzano's footnote, III 71: "The *borrowed* meaning should not be confused with the *widened* [*erweitert*] meaning which is anyway a proper meaning. When the naturalist counts the bat among the *monkeys*, he takes this last word in a widened meaning. When, however, in certain circumstances we say that a man is a monkey [in German: 'silly'; also: 'fop'], then we are speaking in a borrowed meaning."

[19] Cf. TS §285, III 50 *Anmerk* 1 [. . .].

APPENDIX A

The Epsilon Calculus with Definitory Expansions

LOGICAL POSTULATES

AXIOMS

- Sentences: All tautologies
- Identity:
  - I $t = t$
  - L $t = s \rightarrow (\alpha[z/t] \rightarrow \alpha[z/s])$
- Epsilon:
  - E $\alpha[x] \rightarrow \alpha[x/\varepsilon x\alpha[x]]$
- Quantifiers:
  - D1 $\exists x\alpha[x] \leftrightarrow \alpha[x/\varepsilon x\alpha[x]]$
  - D2 $\forall x\alpha[x] \leftrightarrow \alpha[x/\varepsilon x\neg\alpha[x]]$

RULES

- Separation:
  - RS $\dfrac{\alpha;\ \alpha \rightarrow \beta}{\beta}$
- Substitution:
  - RV $\dfrac{\alpha[x];\quad x \text{ free for } t \text{ in } \alpha}{\alpha[x/t]}$
- Definition:
  - D_f $\dfrac{s \geq 0;\quad fr(t) = \{x_1, \ldots, x_s\};\ f^s \notin L(S);\quad \beta_1, \ldots, \beta_s \in L(S);\ fr(\beta_1) = \{x_1\}, \ldots, fr(\beta_s) = \{x_s\}}{S^f \vdash \forall x_1 \ldots x_s(\bigwedge_{i \leq s} \beta_i(x_i) \rightarrow f(x_1, \ldots, x_s) = t)}$
  - D_P $\dfrac{s \geq 0;\quad fr(\alpha) = \{x_1, \ldots, x_s\};\ P^s \notin L(S);\quad \beta_1, \ldots, \beta_s \in L(S);\ fr(\beta_1) = \{x_1\}, \ldots, fr(\beta_s) = \{x_s\}}{S^f \vdash \forall x_1 \ldots x_s(\bigwedge_{i \leq s} \beta_i(x_i) \rightarrow (P(x_1, \ldots, x_s) \leftrightarrow \alpha))}$

Theorem. *The usual postulate system Q is eliminable.*

Q1 $\forall x\alpha[x] \to \alpha[x/t]$

1	$\neg\alpha[x] \to \neg\alpha[\varepsilon x\neg\alpha[x]]$	E
2	$\neg\alpha[x/t] \to \neg\alpha[\varepsilon x\neg\alpha[x]]$	RV 1
3	$\alpha[\varepsilon x\neg\alpha[x]] \to \alpha[x/t]$	R(C) 2
4	$\forall x\alpha[x] \leftrightarrow \alpha[x/\varepsilon x\neg\alpha[x]$	D2
5	$\forall x\alpha[x] \to \alpha[x/t]$	R(C) 3, 4

Q2 $\alpha[x/t] \to \exists x\alpha[x]$

1	$\alpha[x] \to \alpha[\varepsilon x\alpha[x]]$	E
2	$\alpha[x/t] \to \alpha[\varepsilon x\alpha[x]]$	RV 1
3	$\exists x\alpha[x] \leftrightarrow \alpha[x/\varepsilon x\alpha[x]]$	D1
4	$\alpha[x/t] \to \exists x\alpha[x]$	R(C) 2, 3

GP $\dfrac{\beta \to \alpha[x],\ x \notin \beta}{\beta \to \forall x\alpha[x]}$

1	$\beta \to \alpha[x]$	with $x \notin \beta$
2	$\beta \to \alpha[\varepsilon x\neg\alpha[x]]$	RV 1
3	$\forall x\alpha[x] \leftrightarrow \alpha[x/\varepsilon x\neg\alpha[x]]$	D2
4	$\beta \to \forall x\alpha[x]$	R(C) 2, 3

PA $\dfrac{\alpha[x] \to \beta,\ x \notin \beta}{\exists x\alpha[x] \to \beta}$

1	$\alpha[x] \to \beta$	with $x \notin \beta$
2	$\alpha[\varepsilon x\alpha[x]] \to \beta$	RV 1
3	$\exists x\alpha[x] \leftrightarrow \alpha[x/\varepsilon x\alpha[x]]$	D1
4	$\exists x\alpha[x] \to \beta$	R(C) 2, 3

APPENDIX B
Specific Axioms

I General Axioms

General Axioms (GA)

$x \in y \rightarrow O(x) \wedge Q(y)$	GA1
$O(x) \rightarrow \exists y(x \in y)$	GA2
$Q(x) \rightarrow \exists y(y \in x)$	GA3
$\neg(x < x)$	GA4
$x < y \wedge y < z \rightarrow x < z$	GA5
$D(x) \rightarrow O(x) \vee Q(x)$	GA6
$D(x) \wedge Q(x) \rightarrow \forall y(y \in x \rightarrow D(y))$	GA7
$D(x) \wedge y < x \rightarrow D(y)$	GA8
$V(x) \vee B(x) \rightarrow \neg S(x)$	GA9
$L(x) \rightarrow \neg D(x)$	GA10
$L(x) \rightarrow \neg Q(x)$	GA11
$x \mathsf{E} y \rightarrow (O(x) \wedge V(y)) \vee (Q(x) \wedge B(y))$	GA12
$O(x) \rightarrow \exists y(V(y) \wedge x \mathsf{E} y)$	GA13
$x \mathsf{E} y \rightarrow (O(x) \wedge V(y)) \vee (Q(x) \wedge B(y)) \vee (Sit(y) \wedge S(x))$	GA12′
$Q(x) \rightarrow \exists y(B(y) \wedge x \mathsf{E} y)$	GA14
$Sit(x) \rightarrow \exists y(S(y) \wedge x \mathsf{E} y)$	GA14′
$x \lessdot y \wedge z \lessdot y \rightarrow x = z$	GA15
$x \lessdot y \rightarrow L(x)$	GA16
$x \lessdot y \rightarrow Ad^{\circ}(y)$	GA17
$An(x) \wedge An(y) \rightarrow (z \mathsf{E} x \wedge z \mathsf{E} y \rightarrow x = y)$	GA18
$An(d_n)$	GA19

$$\forall x(V(x) \to O(x)) \qquad \text{GA20}$$

$$\forall x(S(x) \to O(x)) \qquad \text{GA21}$$

$$Q(x) \to O(x) \qquad \text{GA22}$$

$$r \ni i \leftrightarrow (r \circ i) \ni i \qquad \text{GA23}$$

$$(r \circ i) \ni j \to i = j \qquad \text{GA24}$$

$$r \ni (s \circ t) \leftrightarrow (r \circ s) \ni t \qquad \text{GA25}$$

$$((r \circ s) \circ t) \ni i \leftrightarrow (r \circ (s \circ t)) \ni i \qquad \text{GA26}$$

$$v \rhd w \to \neg w \rhd v \qquad \text{GA27}$$

$$\mathfrak{v}_{\}} \rhd \boxed{\wedge}\ \mathfrak{v} \qquad \text{GA28}$$

General Axioms Schemata (SGA)

$$H(x, \substack{x_1, \dots, x_n \\ z_1, \dots, z_n}, z) \to \bigwedge_{i \neq j \leq n} x_i \neq x_j \qquad \text{SGA1}$$

$$H(x, \substack{x_1, \dots, x_n \\ z_1, \dots, z_n}, z) \wedge H(x, \substack{x_1, \dots, x_n \\ z_1, \dots, z_n}, u) \to z = u \qquad \text{SGA2}$$

$$\bigwedge_{i \neq j \leq n} (x_i \leq x \wedge x_j \leq x \to x_i \perp x_j) \to \exists z H(x, \substack{x_1, \dots, x_n \\ z_1, \dots, z_n}, z) \qquad \text{SGA3}$$

$$\neg(x_i \leq x) \to (H(x, \substack{x_1, \dots, x_n \\ z_1, \dots, z_n}, z) \leftrightarrow \qquad \text{SGA4}$$
$$H(x, \substack{x_1, \dots, x_{i-1}, x_i, x_{i+1}, \dots, x_n \\ z_1, \dots, z_{i-1}, x_i, z_{i+1}, \dots, z_n}, z) \leftrightarrow H(x, \substack{x_1, \dots, x_{i-1}, x_{i+1}, \dots, x_n \\ z_1, \dots, z_{i-1}, z_{i+1}, \dots, z_n}, z))$$

$$\bigwedge_{k \leq n} O(x_k) \leftrightarrow O(\pi^n(x_1, \dots, x_n)) \qquad \text{SGA5}$$

$$O^n(z) \leftrightarrow \bigwedge_{k \leq n} O(\pi_{nk}(z)) \qquad \text{SGA6}$$

$$O^n(x) \to \pi^n(\pi_{n1}(x), \dots, \pi_{nn}(x)) = x \qquad \text{SGA7}$$

$$\pi^n(\pi_{n1}(i), \dots, \pi_{nn}(i)) = i \qquad \text{SGA7}'$$

$$\bigwedge_{k \leq n} O(x_k) \to \pi^n_k(\pi(x_1, \dots, x_k, \dots, x_n)) = x_k \qquad \text{SGA8}$$

$$\pi^n_k(\pi(i_1, \dots, i_k, \dots, i_n)) = i_k \qquad \text{SGA8}'$$

$$[\pi_1, \dots, \pi_n] \succ \pi \leftrightarrow (\pi_1, \dots, \pi_n \succ {}_1\pi) \vee (\pi_1, \dots, \pi_n \succ {}_2\pi) \vee (\pi_1, \dots, \pi_n \succ {}_3\pi)$$
SGA9

II. Comprehension Axioms

Axioms of Quality Existence (CQ)

$$G(a) \rightarrow \exists r \forall i (i \in r \leftrightarrow i \mathrel{\mathsf{E}} a) \qquad \text{CQ1}$$

$$\exists i (i \in r \wedge i \in s) \rightarrow \exists t \forall i (i \in t \leftrightarrow i \in r \wedge i \in s) \qquad \text{CQ2}$$

$$\forall i \exists r \forall j (j \in r \leftrightarrow j = i) \qquad \text{CQ3}$$

$$\exists r \forall a (a \in r \leftrightarrow G(a)) \qquad \text{CQ4}$$

$$\exists r (\forall p (p \in r \leftrightarrow W(p)) \qquad \text{CQ5}$$

$$\exists r^2 \forall j (j \in r^2 \leftrightarrow Fct(j)) \qquad \text{CQ6}$$

$$G([a](\overline{c})) \rightarrow \exists m \forall j (j \in m \leftrightarrow \exists ir (i \mathrel{\mathsf{E}} a \wedge r \mathrel{\mathsf{E}} \overline{c} \wedge i \in r \wedge j = < i, r >)) \qquad \text{CQ7}$$

$$\exists r \forall i (i \in r \leftrightarrow D(i)) \qquad \text{CQ8}$$

Axiom Schemata of Quality Existence (SCQ)

$$\forall r^n \forall i \exists s \forall i_1 \ldots i_{k-1} i_{k+1} \ldots i_n (i \in s \leftrightarrow < i_1, \ldots, i_{k-1}, i, i_{k+1}, \ldots, i_n > \ \in r^n) \qquad \text{SCQ1}$$

$$\exists i (\bigwedge_{k \leq n} i \in r_k) \rightarrow \exists r \forall i (i \in r \leftrightarrow \bigwedge_{k \leq n} i \in r_k) \qquad \text{SCQ2}$$

$$\exists r \forall a (a \in r \leftrightarrow {}_{n}A(a)) \qquad \text{SCQ3}$$

$$\exists r \forall a (a \in r \leftrightarrow {}^{n}A(a)) \qquad \text{SCQ4}$$

$$\exists r \forall a (a \in r \leftrightarrow {}^{n}_{n}A(a)) \qquad \text{SCQ5}$$

$$\mathop{\bigwedge\!\!\!\bigwedge}_{a} (\overset{n}{p}) \rightarrow \exists r \forall c (c \in r \leftrightarrow W(\overset{n}{p}\, [a/c])) \qquad \text{SCQ6}$$

$$\neg \mathop{\mathbb{F}}_{a} (\overset{n}{p}) \rightarrow \exists r \forall c (c \in r \leftrightarrow E(\overset{n}{p}\, [a/c])) \qquad \text{SCQ7}$$

$$\mathop{\bigwedge\!\!\!\bigwedge}_{\overset{s}{a}} (\overset{n}{p}) \rightarrow \exists r^s \forall \overset{s}{c} (< \overset{s}{c} > \in r^s \leftrightarrow W(\overset{n}{p}[\overset{s}{a}/\overset{s}{c}])) \qquad \text{SCQ8}$$

$$\neg \mathop{\mathbb{F}}_{\overset{s}{a}} (\overset{n}{p}) \rightarrow \exists r^s \forall \overset{s}{c} (< \overset{s}{c} > \in r^s \leftrightarrow E(\overset{n}{p}[\overset{s}{a}/\overset{s}{c}])) \qquad \text{SCQ9}$$

$$\mathop{\bigwedge\!\!\!\bigwedge}_{\overset{s}{a}} (\mathfrak{p}) \rightarrow \exists r^s \forall \overset{s}{c} (< \overset{s}{c} > \in r^s \leftrightarrow W(\mathfrak{p}[\overset{s}{a}/\overset{s}{c}])) \qquad \text{SCQ10}$$

$$\neg \mathop{\mathbb{F}}_{\overset{s}{a}} (\mathfrak{p}) \rightarrow \exists r^s \forall \overset{s}{c} (< \overset{s}{c} > \in r^s \leftrightarrow E(\mathfrak{p}[\overset{s}{a}/\overset{s}{c}])) \qquad \text{SCQ11}$$

Axioms of Existence of Object Ideas (CV)

$$\exists a (\forall x (x \mathrel{\mathsf{E}} a \leftrightarrow O(x)) \wedge \neg \exists l (l < a)) \qquad \text{CV1}$$

$$\forall a \forall \overline{c} \exists b \forall i (i \mathrel{\mathsf{E}} b \leftrightarrow i \mathrel{\mathsf{E}} a \wedge \exists r (i \in r \wedge r \mathrel{\mathsf{E}} \overline{c})) \wedge \forall l (l < b \leftrightarrow l \leq a \vee l \leq \overline{c}) \qquad \text{CV2}$$

$$\forall a\exists c(\forall i(i \mathrel{\mathsf{E}} c \leftrightarrow i = a) \wedge \forall l(l < c \leftrightarrow l \leq a)) \qquad \text{CV3}$$

$$\forall l\exists a(\forall i(i \mathrel{\mathsf{E}} a \leftrightarrow i = l) \wedge \forall l'(l' < a \leftrightarrow l' \leq l)) \qquad \text{CV4}$$

$$\forall \mathfrak{a}\exists a(\forall i(i \mathrel{\mathsf{E}} a \leftrightarrow \forall c(c \mathrel{\mathsf{E}} \mathfrak{a}) \rightarrow i \mathrel{\mathsf{E}} c)) \wedge \forall l(l < a \leftrightarrow \exists c(c \mathrel{\mathsf{E}} \mathfrak{a} \wedge l \leq c))) \qquad \text{CV5}$$

$$\forall \mathfrak{a}\exists a(\forall i(i \mathrel{\mathsf{E}} a \leftrightarrow \exists c(c \mathrel{\mathsf{E}} \mathfrak{a}) \wedge i \mathrel{\mathsf{E}} c)) \wedge \forall l(l < a \leftrightarrow \exists c(c \mathrel{\mathsf{E}} \mathfrak{a} \wedge l \leq c))) \qquad \text{CV6}$$

$$\forall i\exists a((D(i) \rightarrow \forall j(j \mathrel{\mathsf{E}} a \leftrightarrow j = i)) \wedge (\neg D(i) \rightarrow \neg G(a)) \wedge \neg i < a)) \qquad \text{CV7}$$

$$\forall a\exists c\forall p(p \mathrel{\mathsf{E}} c \leftrightarrow \exists i(i \mathrel{\mathsf{E}} a \wedge p = h(\{i\}, \overline{d}))) \qquad \text{CV8}$$

Axioms of Existence of Quality Ideas (CB)

$$\forall a\exists \overline{c}\begin{pmatrix} (G(a) \rightarrow \forall r(r \mathrel{\mathsf{E}} \overline{c} \leftrightarrow \forall i(i \in r \leftrightarrow i \mathrel{\mathsf{E}} a))) \\ \wedge (\neg G(a) \rightarrow \neg G(\overline{c})) \wedge \forall l(l < \overline{c} \leftrightarrow l \leq a) \end{pmatrix} \qquad \text{CB1}$$

$$\forall \overline{a}\,\overline{c}\exists \overline{b}\begin{pmatrix} (\overline{a} \mathrel{\mathrm{T\!T}} \overline{c} \rightarrow \forall t(t \mathrel{\mathsf{E}} \overline{b} \leftrightarrow \exists r\exists s(r \mathrel{\mathsf{E}} \overline{a} \wedge s \mathrel{\mathsf{E}} \overline{c} \wedge \forall i(i \in t \leftrightarrow i \in r \wedge i \in s))) \\ \wedge \quad (\neg(\overline{a} \mathrel{\mathrm{T\!T}} \overline{c}) \rightarrow \neg G(\overline{b})) \quad \wedge \quad \forall l(l < \overline{b} \leftrightarrow l \leq \overline{a} \vee l \leq \overline{c}) \end{pmatrix}$$

CB2

$$\forall \overline{a}\,\overline{c}\exists \overline{b}\begin{pmatrix} (G(\overline{a}) \vee G(\overline{c}) \rightarrow \forall r(r \mathrel{\mathsf{E}} \overline{b} \leftrightarrow r \mathrel{\mathsf{E}} \overline{a} \vee r \mathrel{\mathsf{E}} \overline{c}) \wedge \\ (\neg(G(\overline{a}) \vee G(\overline{c})) \rightarrow \neg G(\overline{b})) \wedge \forall l(l < \overline{b} \leftrightarrow l \leq \overline{a} \vee l \leq \overline{c}) \end{pmatrix}$$

CB3

$$\forall \overline{a}\exists \overline{c}\left(\begin{matrix} (G(\overline{a}) \rightarrow \begin{pmatrix} \exists i(\neg i \mathrel{\mathsf{E}} [e](\overline{a})) \rightarrow (\forall r(r \mathrel{\mathsf{E}} \overline{c} \leftrightarrow \forall s(s \mathrel{\mathsf{E}} \overline{a} \rightarrow s \perp r))) \\ \wedge \forall i(i \mathrel{\mathsf{E}} [e](\overline{a})) \rightarrow \neg G(\overline{c})) \end{pmatrix}) \\ \wedge (\neg G(\overline{a}) \rightarrow \forall i(i \mathrel{\mathsf{E}} [e](\overline{c}))) \quad \wedge \quad \forall l(l < \overline{c} \leftrightarrow l \leq \overline{a}) \end{matrix} \right)$$

CB4

$$\exists \overline{c}\,\forall r(r \mathrel{\mathsf{E}} \overline{c} \leftrightarrow \forall a(a \in r \rightarrow G(a))) \qquad \text{CB5}$$

$$\exists \overline{c}\,\forall r(r \mathrel{\mathsf{E}} \overline{c} \leftrightarrow \forall p(p \in r \rightarrow W(p))) \qquad \text{CB6}$$

$$\exists \overline{c}\forall r(r \mathrel{\mathsf{E}} \overline{c} \leftrightarrow \forall i(i \in r \rightarrow D(r))) \qquad \text{CB7}$$

Axiom Schemata of Existence of Quality Ideas (SCB)

$$\forall \overset{n}{\overline{a}}\exists \overline{c}\left(\begin{matrix} (\mathrm{T\!T}^n(\overset{n}{\overline{a}}) \rightarrow \forall r(r \mathrel{\mathsf{E}} \overline{c} \leftrightarrow \exists \overset{n}{s}(\bigwedge_{k\leq n}(s_k \mathrel{\mathsf{E}} \overline{a}_k) \wedge \forall i(i \in r \leftrightarrow \bigwedge_{k\leq n}(i \mathrel{\mathsf{E}} s_k))))) \\ \wedge(\neg \mathrm{T\!T}^n(\overset{n}{\overline{a}}) \rightarrow \neg G(\overline{c})) \wedge \forall l(l < \overline{c} \leftrightarrow \bigvee_{k\leq n}(l \leq \overline{a}_k))) \end{matrix} \right)$$

SCB1

$$\exists \overline{c}\,\forall r(r \mathrel{\mathsf{E}} \overline{c} \leftrightarrow \forall a(a \in r \rightarrow {}_nA(u))) \qquad \text{SCB2}$$

$$\exists \overline{c}\,\forall r(r \mathrel{\mathsf{E}} \overline{c} \leftrightarrow \forall a(a \in r \rightarrow {}^nA(u))) \qquad \text{SCB3}$$

$$\exists\bar{c}\forall r(r \mathrel{\mathsf{E}} \bar{c} \leftrightarrow \forall a(a \in r \rightarrow {}^{n}_{n}A(u))) \qquad \text{SCB4}$$

$$\forall a\exists\bar{c}\begin{pmatrix} (\underset{a}{\bigwedge\!\!\bigwedge}(\overset{n}{p}) \rightarrow \forall r(r \mathrel{\mathsf{E}} \bar{c} \leftrightarrow \forall b(b \in r \leftrightarrow W(\overset{n}{p}[a/b]))))\wedge \\ (\neg\underset{a}{\bigwedge\!\!\bigwedge}(\overset{n}{p}) \rightarrow \neg G(\bar{c})) \wedge \forall l(l < \bar{c} \leftrightarrow l \leq \overset{n}{p} \vee l \leq a)) \end{pmatrix} \qquad \text{SCB5}$$

$$\forall a\exists\bar{c}\begin{pmatrix} (\neg\underset{a}{\mathbb{F}}(\overset{n}{p}) \rightarrow \forall r(r \mathrel{\mathsf{E}} \bar{c} \leftrightarrow \forall b(b \in r \leftrightarrow E(\overset{n}{p}[a/b]))))\wedge \\ ((\underset{a}{\mathbb{F}}(\overset{n}{p}) \rightarrow \neg G(\bar{c})) \wedge \forall l(l < \bar{c} \leftrightarrow l \leq \overset{n}{p} \vee l \leq a)) \end{pmatrix} \qquad \text{SCB6}$$

$$\forall\overset{s}{a}\exists\bar{c}\begin{pmatrix} (\underset{\overset{s}{a}}{\bigwedge\!\!\bigwedge}(\overset{n}{p}) \rightarrow \forall r^{s}(r^{s} \mathrel{\mathsf{E}} \bar{c} \leftrightarrow \forall\overset{s}{c}(<\overset{s}{c}> \in r^{s} \leftrightarrow W(\overset{n}{p}\left[\overset{s}{a}/\overset{s}{c}\right])))) \\ \wedge(\neg\underset{\overset{s}{a}}{\mathbb{F}}(\overset{n}{p}) \rightarrow \neg G(\bar{c})) \wedge \forall l(l < \bar{c} \leftrightarrow l \leq \mathfrak{p} \vee \bigvee_{k\leq s} l \leq a_{k})) \end{pmatrix} \qquad \text{SCB7}$$

$$\forall\overset{s}{a}\exists\bar{c}\begin{pmatrix} (\neg\underset{\overset{s}{a}}{\mathbb{F}}(\overset{n}{p}) \rightarrow \forall r^{s}(r^{s} \mathrel{\mathsf{E}} \overset{n\vee}{p}_{\overset{s}{a}} \leftrightarrow \forall\overset{s}{c}(<\overset{s}{c}> \in r^{s} \leftrightarrow E(\overset{n}{p}\left[\overset{s}{a}/\overset{s}{c}\right])))) \\ \wedge(\underset{\overset{s}{z}}{\mathbb{F}}(\overset{n}{p}) \rightarrow \neg G(\bar{c})) \wedge \forall l(l < \bar{c} \leftrightarrow l \leq \overset{n}{p} \vee \bigvee_{k\leq s} l \leq a_{k})) \end{pmatrix} \qquad \text{SCB8}$$

$$\forall\mathfrak{p}\forall\overset{s}{a}\exists\bar{c}\begin{pmatrix} (\underset{\overset{s}{a}}{\bigwedge\!\!\bigwedge}(\mathfrak{p}) \rightarrow \forall r^{s}(r^{s} \mathrel{\mathsf{E}} \bar{c} \leftrightarrow \forall\overset{s}{c}(<\overset{s}{c}> \in r^{s} \leftrightarrow W(\mathfrak{p}\left[\overset{s}{a}/\overset{s}{c}\right])))) \\ \wedge(\neg\underset{\overset{s}{a}}{\bigwedge\!\!\bigwedge}(\mathfrak{p}) \rightarrow \neg G(\bar{c})) \wedge \forall l(l < \bar{c} \leftrightarrow l \leq \mathfrak{p} \vee \bigvee_{k\leq s} l \leq a_{k})) \end{pmatrix} \qquad \text{SCB9}$$

$$\forall\mathfrak{p}\forall\overset{s}{a}\exists\bar{c}\begin{pmatrix} (\neg\underset{\overset{s}{a}}{\mathbb{F}}(\mathfrak{p}) \rightarrow \forall r^{s}(r^{s} \mathrel{\mathsf{E}} \mathfrak{p}^{\vee}_{\overset{s}{a}} \leftrightarrow \forall\overset{s}{c}(<\overset{s}{c}> \in r^{s} \leftrightarrow E(\mathfrak{p}\left[\overset{s}{a}/\overset{s}{c}\right])))) \\ \wedge(\underset{\overset{s}{z}}{\mathbb{F}}(x) \rightarrow \neg G(\bar{c})) \wedge \forall l(l < \bar{c} \leftrightarrow l \leq \mathfrak{p} \vee \bigvee_{k\leq s} l \leq a_{k})) \end{pmatrix} \qquad \text{SCB10}$$

Axioms of Existence of Propositions (CS)

$$\forall a\forall\bar{c}\exists p\forall l(l \leq p \leftrightarrow l \leq a \vee l \leq \bar{c}) \qquad \text{CS1}$$

$$\forall p\exists q((W(q) \leftrightarrow F(p)) \wedge \forall l(l < q \leftrightarrow l \leq p)) \qquad \text{CS2}$$

$$\exists\mathfrak{q}\forall q(q \mathrel{\mathsf{E}} \mathfrak{q} \leftrightarrow \exists p(p \mathrel{\mathsf{E}} \mathfrak{p} \wedge q = \boxminus p) \wedge \forall l(l < \mathfrak{q} \leftrightarrow l \leq \mathfrak{p}) \qquad \text{CS3}$$

$$\exists\mathfrak{q}\forall q(q \mathrel{\mathsf{E}} \mathfrak{q} \leftrightarrow \exists p(p \mathrel{\mathsf{E}} \mathfrak{p} \wedge q = p[\overset{s}{a}/\overset{s}{c}]) \wedge \forall l(l < \mathfrak{q} \leftrightarrow l \leq \mathfrak{p}, \overset{s}{a}, \overset{s}{c})) \qquad \text{CS4}$$

$$\forall\mathfrak{v}\mathfrak{w}\exists p(W(p) \leftrightarrow \mathfrak{v} \rhd \mathfrak{w}) \qquad \text{CS5}$$

Axiom Schemata of Existence of Propositions (SCS)

$$\forall\overset{n}{p}\,\exists q((W(q) \leftrightarrow W(\overset{n}{p})) \wedge \forall l(l < q \leftrightarrow l \leq \overset{n}{p})) \qquad \text{SCS1}$$

$$\forall\overset{n}{p}\,\exists q((W(q) \leftrightarrow E(\overset{n}{p})) \wedge \forall l(l < q \leftrightarrow l \leq \overset{n}{p})) \qquad \text{SCS2}$$

$$\forall \overset{n}{p}\, \exists q((W(q) \leftrightarrow D(\overset{n}{p})) \wedge \forall l(l < q \leftrightarrow l \leq \overset{n}{p})) \qquad \text{SCS3}$$

$$\exists r((W(r) \leftrightarrow (W(\overset{m}{p}) \rightarrow W(\overset{n}{q})) \wedge \forall l(l < r \leftrightarrow l \leq \overset{m}{p}, \overset{n}{q})) \qquad \text{SCS4}$$

$$\forall \overset{m}{p}\, \forall \overset{s}{a}\, \exists q(W(q) \leftrightarrow \forall \overset{s}{c} W(\overset{m}{p}[\overset{s}{a}/\overset{s}{c}])) \qquad \text{SCS5}$$

$$\forall \overset{m}{p}\, \forall \overset{s}{a}\, \exists q(W(q) \leftrightarrow \forall \overset{s}{c} E(\overset{m}{p}[\overset{s}{a}/\overset{s}{c}])) \qquad \text{SCS6}$$

$$\forall \overset{m}{p}\, \forall \overset{n}{q}\, \forall \overset{s}{a}\, \exists q'(W(q') \leftrightarrow \overset{m}{p} \underset{\overset{s}{a}}{\Rightarrow} \overset{n}{q}) \qquad \text{SCS7}$$

$$\forall \overset{m}{p}\, \forall \overset{n}{q}\, \forall \overset{s}{a}\, \exists q'(W(q') \leftrightarrow \overset{m}{p} \underset{\overset{s}{v}}{\overset{\vee}{\Rightarrow}} \overset{n}{q}) \qquad \text{SCS8}$$

APPENDIX C
Aggregates

1 Preamble

As already said, the Bolzanian 'mereology' is a rather complex and not easily interpretable subject. A very great difficulty, even if not the only one, that one must face with regard to it lies in the fact that insights, which we are used to grouping around two quite different notions—*inclusion* and *membership*—still occur inextricably, and possibly not even coherently, interwoven together.

On the other hand, we have succeeded in reconstructing in a sufficiently persuading manner the main features of the Bolzanian logical theory—moving, for instance, on the symbolic level and taking advantage of the fact that the Bolzanian 'ideas' admit a plural reference—without giving a precise account of the two notions of *aggregate* [*Inbegriff*] and *part* [*Theil*], which, on the contrary, are omnipresent in the exposition of the *Wissenschaftslehre*. There are, however, certain aspects of these conceptualizations that, independently from their greater or lesser relevance to the proper logical system, seem to be at least worth mentioning. Let us, however, remark that in the works after the *Wissenschaftslehre*, and in particular in the *Einleitung in die Größenlehre* and the *Paradoxien des Unendlichen*, Bolzano sometimes modifies his terminology, if not even the conceptualizations, of this 'mereology'. Just one example: a 'sum', which, as we will see, denotes in the *Wissenschaftslehre* something near to what we call today a *partially ordered set*, becomes something very close to what we call today a *union of a set of disjoint sets*. Here, however, we will not follow Bolzano in his evolution, but simply limit ourselves to mentioning some of the traits of this issue as they occur in the *Wissenschaftslehre*.

2 The Notion of Aggregate

Now, as we have already seen, Bolzano knew the relation of *being part of* [<] as an *irreflexive* and *transitive* relation that may subsist between arbitrary objects, i.e.

$$\neg(i < i) \tag{AG57}$$

$$i < j \wedge j < k \rightarrow i < k \tag{AG58}$$

and he thought of an *aggregate* as an object having at least two parts; i.e. writing '$In(i)$' for '*the object i is an aggregate*',

$$In(i) \leftrightarrow \exists jk(j, k < i \wedge j \neq k) \tag{CD1}$$

In the following, it also turns out convenient to define as usual

$$i \leq j \leftrightarrow i < j \vee i = j \tag{CD2}$$

As an aside, let us remark that Bolzano also faced up to the question of a possible 'definition', in his pregnant sense, of the notion of aggregate (i.e. he asked himself which are the authentic components and the structure of the notion of aggregate). Not without hesitation, he concluded that an aggregate must be thought of as 'something having composition' and he hypothesized that the truly primitive notion is that of '*having composition* [*Zusammengesetztheit*]'.[1]

One of the most interesting facts that are worth mentioning, however, is that besides these notions of *part* and *aggregate* that we may call *absolute*, Bolzano more or less explicitly used, without really lingering over this, their versions *relative to a given idea*. Among the many examples of this, a particularly significant one may be found in his discussion of a particular instance of an aggregate. He says: "if we want to affirm that each one of the three persons Titius, Caius and Sempronius have the capacity to hold a certain office, then the idea 'each one of the persons T., C. e S.' is the idea of whichever part of the aggregate which Titius, Caius and Sempronius

[1] TS §82, I 282–3: "1) [. . .] I must begin by determining somewhat more precisely the concept that I connect with the word 'collection'. I take this word in the meaning that common usage accords it, that is, by a collection of certain things I mean a combination or union of these things, a gathering of them, a whole of which these things are parts, and the like; furthermore, in the mere idea of a collection it is not presupposed that the order or sequence in which the united objects appear is fixed, nor even that there is or even can be such an order. I will not venture to decide with certainty whether this concept is simple, or, if complex, of which parts it might be composed. I will say, however, that if it isn't completely simple, it must have very few parts. It seems to me that a collection is simply something that has compositeness. But this abstractum, the concept expressed by means of the word *compositeness*, cannot in my opinion be further analysed."

constitute together", and he adjoins: " Under 'part', however, I understand each object from which an aggregate is composed." Worried, however, about a possible misunderstanding of his definition, he immediately specifies that although "one can call 'part of the aggregate constituted by the three person T., C. and S.' also every member which lies in their bodies, for example, the arms of Caius, as well as the aggregate constituted by some of these persons, for example, C. and T.", such parts are certainly not the parts to which one refers with the above statement, because "only each one of the single objects from which we just now think a certain aggregate composed may be regarded as parts of it, whereas the other things which are only parts of these things as well as those which are aggregates of these parts have not properly the right to this denomination".[2] The crucial passage is "we just now think a certain aggregate composed"; disregarding the 'pure mereological structure' of that aggregate, it is clearly *regarded from the point of view of a certain idea,* in this case the idea of a person.

We may try to account for this perspective by beginning with the introduction, for a given idea a, of the ternary relation, which will be called '*being a part relative to a*', that subsists among two objects and the idea a when the first object is (in the absolute sense) part of the second object and moreover it falls under the idea a; i.e., writing '$i <_a j$' for '(*the object*) *i is a part of the* (*object*) *j relative to the idea a*' (in brief, '*i is an a-part of j*'),

$$i <_a j \leftrightarrow i < j \wedge i \,\text{E}\, a \qquad \text{CD3}$$

Informally speaking, '*an aggregate relative to a*' is now an object constituted by a collection of objects falling under a; the specification of this notion, obviously without a more or less explicit use of the notion of set, is anything but simple. The first main difficulty is that, as is clearly evidenced by the Bolzanian remark on people's arms or couples, it is necessary to sharply distinguish the *aggregate object* (which has *also* these as proper 'absolute' parts) and its *structure relative to the idea* under which this object is now being considered. When the idea varies, the object does not change, while, on the contrary, relative to this idea, its structure does. The

[2] TS §83, I 284–5: "1) [...] If, for example, we wish to claim that each of the three persons Caius, Sempronius and Titus is able to fill a certain post, the idea 'each of the persons C., S. and T.' is the idea of one part (whichever you like) of the collection made up of Caius, Sempronius and Titus. In my view, ideas of this sort are composed of the idea of the collection of enumerated objects, and the idea designated by the word 'part'. By a part I mean any object out of which a collection is composed. 2) It might well appear that the kind of ideas to which the above example belongs has been incorrectly grasped by the given definition. For we could call any limb on any of their bodies, or any collection composed of two of them, e.g., Titus and Sempronius, a part of the collection. On the other hand, when we say that each of these persons is an individual capable of filling a certain post, we do not want to say that the arms of Caius, or the two persons Caius and Titus together, are suitable for this office. It seems therefore that the two expressions 'Each of the objects *A, B, C, D,* . . . ', and 'Each part of the collection *A, B, C, D,* . . . ' do not have the same sense."

same object regarded under the idea 'person' or under the idea 'atom of carbon' is a quite different thing. The second difficulty is that in fact we do not know precisely how the 'pure' mereological structure of an object must be understood. We may infer from Bolzano's discussions that he considered it obvious that the union of two parts of an object is still a part of the object, as well as that a part of the object is (if not empty) the common part of two of its parts and finally that a part of it is also what remains when another of its parts is eliminated; in other words, it seems that he regarded that the pure mereological structure of an object is obviously what we would call today a 'Boolean algebra (complete?) without zero' or an 'algebra of Leśniewski'. Fortunately, however, something significant may be said without committing oneself to too hazardous an attribution.

Indeed, it is in any case clear that an object i that is an a-aggregate must contain as its parts, not only no less than two a's, but also *all parts of each of these a's, all the combinations among these parts, and furthermore all the combinations of the a's among themselves and with parts of some a.* In the Bolzanian example, the types of parts with which one has to do can be exemplified as follows. A part is one of the following: T.; or a foot of T.; or a foot of T. and an arm of C.; or T. and C.; or T. and an arm of C. In order to be an a-aggregate, an object i must in any case satisfy the condition $A(i, a)$ that is defined as follows:

$$A(i,a) \leftrightarrow \exists jk(j \neq k <_{a} i) \wedge \forall j(j < i \rightarrow \exists k(k <_{a} i \wedge (k \leq j \vee \exists h(h < j, k)))) \qquad \text{CD4}$$

This condition does not, however, suffice to characterize an object i as an a-aggregate. Indeed, the idea is that an a-aggregate, besides containing the a's, contains only parts that 'have to do' with the a's, that is to say, parts and combinations of parts of the a's, or combinations of the a's among themselves or with parts of them. However, the preceding condition is compatible with the fact that, for instance, i contains as proper parts an object made up from an a and any other object, for example, an object made up from T. and a trout. Hence the following requirement appears reasonable: an a-aggregate must be a minimal object among those that satisfy the condition $A(i, a)$ (i.e. an object that is included in any other object satisfying that condition). Thus, writing '$In(i, a)$' for '*i is an a-aggregate* [*a-Inbegriff*]', we define

$$In(i,a) \leftrightarrow A(i,a) \wedge \forall i' A(i',a) \rightarrow i \leq i') \qquad \text{CD5}$$

We have, obviously,

$$\exists a In(i,a) \rightarrow In(i) \qquad \text{CT1}$$

but its converse is, as will be seen later, more problematic.

One can immediately see, however, that this definition, as far as the *a*'s are concerned that an *i* must contain in order to be an *a*-aggregate, simply requires that *i* contains *at least two a*'s, but it does not say anything more (i.e. an aggregate of persons must contain at least two persons). Bolzano also explicitly considers the case in which an *a*-aggregate has as its *a*-parts *each one* of the *a*'s and he speaks in this case of the *concrete aggregate of all a*'s [*der concrete Inbegriff aller a*] or of the *(concrete) all of the a*'s [*das All der a*] or of the *(concrete) whole of the a*'s [*das Ganze der a*];[3] i.e., writing '$In^*(i, a)$' for '*i* is *the aggregate of the a*'s',

$$In^*(i,a) \leftrightarrow In(i,a) \wedge \forall j(j \sqsubseteq a \rightarrow j < i) \qquad \text{CD6}$$

Obviously,

$$In(i,a) \wedge In^*(j,a) \rightarrow i \leq j \qquad \text{CT2}$$

With a definite theory of the pure mereological structure of objects, one could also prove, as Bolzano believed,

$$In^*(i,a) \wedge In^*(j,a) \rightarrow i = j^{4} \qquad \text{CT3}$$

As can be seen from the last quoted passage, Bolzano somehow opposed the pure '*being a part* [*Theil*]' and the '*being a component* [*Bestandtheil*]'. It turns out useful to fix this distinction. With regard to this, we begin by remarking that the preceding definitions do not in any way exclude that one or more parts of an *a*-part be themselves *a*-parts. To give a Bolzanian example, to which we will later revert: "if in a line we mind only its length, then we regard the line as composed of shorter lines [...] and we may regard the parts of which they are made up (in so far as they are still lines)[5] again as parts of the whole line." On the other hand, it is clear that, in the aggregate of T., C., and S., none of its *a*-parts has other *a*-parts among its parts. In another circumstance, Bolzano had distinguished between *proximate* [*nächste*] parts and *more distant* [*entferntere*] parts and, even if only implicitly, he had relativized to an idea also this relation. He stated:

[3] TS §86, I 292: "3) Finally, the collection which contains all and only objects that are subsumed under the idea *A*, or, more briefly, in which every *A* occurs as a part and in which every part is an *A*, is called the *collection of all A*, or the *totality* or *entirety* of the *A* in the *concrete* sense."

[4] TS §68, I 223: "1) [...] there can be only one collection of *all* objects that have a certain property *b*."

[5] Note that for Bolzano points are not lines ('of length 0') and hence, although being parts of the line, they are not *a*-parts of it.

> "In itself, it is indeed possible to distinguish, in any whole made up by more than two parts, between nearer and more distant parts; indeed if we decompose it in a number of parts which is smaller than the number of all parts, then some of them, possibly all of them, will be still composed of parts and so we may call *proximate*, the parts constituted at first and *more distant* the parts of these parts. Only that, if it were completely arbitrary into how many and into which parts one decomposes, at first, the whole and if the next parts obtained in this way are not distinguished from the more distant ones by some remarkable circumstance, it would be pointless to distinguish them by this denominations."[6]

That 'remarkable circumstance' is indeed what we may somehow connect with a criterion, an *idea*.

Hence we firstly establish the relation of *next part* $[\ll]$ by setting

$$i \ll j \leftrightarrow i < j \wedge \forall k(i \leq k \leq j \rightarrow (i = k \vee k = j)) \qquad \text{CD7}$$

and we then relativize it to an idea a by setting

$$i \ll {}_{a}j \leftrightarrow i < {}_{a}j \wedge \forall k(i < k < j \rightarrow \neg k \sqsubset a) \qquad \text{CD8}$$

$$i \ll {}_{a}j \leftrightarrow i < {}_{a}j \wedge \forall k(k \sqsubset a \wedge i \leq k \leq j \rightarrow i = k \vee k = j) \qquad \text{CT4}$$

Clearly T., C., and S. are next a-parts of their aggregate, whereas, since there does not exist an immediately shorter subline, the line has no next a-parts.

Of course, given an a-aggregate, an a-part of it, and a part of the latter, this last is, in any case, by virtue of the transitivity of $<$, a part of the aggregate; nevertheless, nothing is said, in general, about its falling or not falling under a (i.e. about its being or not being an a-part of the aggregate). Bolzano considers, in fact, the limiting case in which no part of an a-part is in turn an a-part (i.e. the case in which every a-part is next). For such a type of aggregate, he does not explicitly use a particular name, but in contexts that are significant for this notion, he sometimes uses, instead of the term '*part* [*Theil*]', the term '*component* [*Bestandtheil*]'. We are thus led to speak, for this type of aggregate, of '*composition* [*Zusammensetzung*] *of* a's' or of '*a-compound*'; i.e., writing '$Zs(i, a)$' for 'i is an a-compound', we set

$$Zs(i, a) \leftrightarrow In(i, a) \wedge \forall j(j < {}_{a}i \rightarrow j \ll {}_{a}i) \qquad \text{CD9}$$

Obviously,

$$Zs(i, a) \leftrightarrow In(i, a) \wedge \forall j(j < {}_{a}i \leftrightarrow j \ll {}_{a}i) \qquad \text{CT5}$$

[6] Cf. TS §58, I 183.

An *a*-compound is thus an *a*-aggregate the next *a*-parts of which are all and only all its *a*-parts. The *a*-aggregate of T., C., and S. is then properly an *a*-compound, whereas the *a*-aggregate of the sublines is not. The *same line* is, however, an *a*-compound when it is regarded under the idea of point—when, in other words, it is considered as the aggregate of its points.

Of course, as has already been done in the case of the *a*-aggregate, the notion of *a*-compound may be strengthened in the notion of *complete a-compound,* by the assumption that it contains *all the a*'s, and it must be said that this is just this the notion to which Bolzano seems to refer when he speaks, as already mentioned, of the *complete aggregate of all the a*'s.

We thus set

$$Zs^*(i, a) \leftrightarrow Zs(i, a) \wedge \forall j(j \sqsubseteq i \rightarrow j < i) \qquad \text{CD10}$$

$$Zs^*(i, a) \leftrightarrow {}_2A(a) \wedge \forall j(j \sqsubseteq a \leftrightarrow j \ll {}_a i) \qquad \text{CT6}$$

Hence, if only *a* is a plural idea, a *complete a-compound* is an object having as its next *a*-parts all and only all the *a*′s.

Examining the four definitions of *a*-aggregate considered up to now, one can see that they always and only involve *proper parts* of *i* and do not say anything about its falling or not falling under *a* (i.e. they are compatible both with $i \sqsubseteq a$ and with $\neg i \sqsubseteq a$). Now, Bolzano has also touched on this issue. In his discussion of the Kantian antinomies, and in particular of the second one, we read:

> "By composed things, indeed, we can often distinguish parts of two kinds: *homogeneous* [*gleichartig*], parts falling under the same notion under which we are just considering the whole, and *inhomogeneous parts* [*ungleichartig*], those by which this is not the case."

The examples offered by Bolzano are interesting because they furnish, among other things, a clear confirmation of what is affirmed above about the fact that the *same* object may constitute different aggregates according to the ideas under which it is considered. He says:

> "Thus in a piece of saltpetre we show homogeneous parts when we break up it still in saltpetre pieces; and inhomogeneous parts, on the contrary, when we present the nitric acid and the potassium in the combination of which it consists. Therefore, if one must speak only of homogeneous parts, then it is certainly right that an extended space consists, in turn, only of extended spaces and, precisely, the line only of lines and so on. If however one speaks of parts in general, so that also inhomogeneous ones are intended, then one can in no way refuse to admit that every extended space, be it line

or surface or solid is also made up of simple parts, and precisely of points; because it is in fact nothing but an aggregate of points."

In the discussion that follows, Bolzano also introduces a new term: '*integrant parts* [*integrante Theile*]'. From the context, one can conclude that these are those parts from which the whole aggregate is made up (i.e., in fact, what we have above called *a*-parts). The problem at issue is precisely whether an extended space may be regarded as an aggregate of points—whether, in other words, the points may be regarded as the integrant parts of that space.[7]

Let us now try to show that the notion of *a*-aggregate may be regarded as a sufficiently satisfying specification of that notion for which Bolzano uses the name of *sum* [*Summe*], whereas the notion of *a*-compound seems to be in an analogous relationship with what Bolzano calls *set* [*Menge*](*which is not a sum*).

The starting point of the Bolzanian conceptualizations is the idea that we have already encountered in the discussion on the distinction *simple ideas–composed ideas*: in a composed object, one must distinguish both the *component parts* and the *type of composition* [*Zusammensetzung*- or *Verbindungsart*]. Bolzano states:

"In the ideas of aggregate considered up to now, it remained open, in which connection the single parts of the aggregate which is thought of occur in it. [...] The most simple case is when one doesn't establish anything but: '*the type of composition among the parts must be considered something uninteresting*'. [...] The current usage calls these aggregates *sets* [*Mengen*]. Hence, I shall allow myself to call *set* any aggregate in which the type of composition is to be considered uninteresting, also if it contains few, even only two parts."[8]

Continuing along this line of thought, Bolzano introduced the notion of *sum* [*Summe*] by distinguishing two kinds of set: those in which the parts of the parts of the whole may not be considered as parts of the whole and those in which, on the contrary, such a consideration is possible; it is for the latter that he introduced the name of *sums*:

"[...] even for an aggregate in which the type of composition is uninteresting, there may be reasons which forbid to consider the parts of the parts as parts of the whole [...]. However, there also are aggregates for which both things—considering uninteresting the type of composition and as parts of the whole the parts of the parts—are simultaneously allowed. [...] Aggregates of this kind [...], I shall allow myself to call, with a term borrowed from the mathematicians, *sums*."[9]

[7] Cf. TS §315, III 169 (7).
[8] Cf. TS §84, I 286–7 (1).
[9] Cf. TS §84, I 287 (2).

As already remarked, our *a*-aggregates are characterized by the fact that the parts of their *a*-parts *may* in turn be *a*-parts; the example by means of which we have illustrated the questions is just the example proposed in this circumstance by Bolzano. Trying to be precise, we may thus say that the Bolzanian sums are our *a*-aggregates in which not all *a*-parts are next, even, possibly also those in which the *a*-parts are homogeneous to their aggregate; i.e.

$$Sm(i,a) \leftrightarrow In(i,a) \wedge \exists jk(j <_a k <_a i) \wedge i \in a \qquad \text{CD11}$$

Our *a*-compounds are instead clearly the Bolzanian sets that are not sums or, at least, those sets for which "there may be reasons which forbid us to consider as parts of the whole the parts of the parts."

3 Ideas of Aggregates

We still need to cover the *ideas of aggregates* [*Inbegriffsvorstellungen*] that are the true issue that Bolzano intended to discuss. Indeed, the first section of the series devoted to the issue of aggregates is entitled: "*Various types of ideas in which the idea of aggregate occurs* [...]". In it, Bolzano introduces the issue we have discussed up to here with the words: "First of all, however, I must more accurately determine the notion which I connect with the word '*aggregate*'."[10]

Bolzano distinguishes between two types of aggregate ideas: *collective* ideas [*Collectivvorstellungen*] and *distributive* ideas [*distributive oder Theilvorstellungen*].

The first thing to be pointed out is that ideas of aggregates are, no matter in which form, singular ideas.[11] In the case of *idea of an aggregate in the absolute sense* [$InV(c)$], we will of course have

$$InV(c) \leftrightarrow {}^1_1A(c) \wedge \exists ia(In(i,a) \wedge i \in c) \qquad \text{CD12}$$

As we have distinguished four types of relativized aggregates, we will have four corresponding relations: *c is an idea of an a-aggregate* [$InV(c,a)$]; *c is an idea of the complete a-aggregate* [$InV^*(c,a)$]; *c is an idea of an a-compound* [$ZsV(c,a)$]; and *c is an idea of the complete a-compound* [$ZsV^*(c,a)$]. These will be respectively defined as

$$InV(c,a) \leftrightarrow {}^1_1A(c) \wedge \exists i(i \in c \wedge In(i,a)) \qquad \text{CD13}$$

[10] Cf. WL §82, I 282.

[11] TS §68, I 222–3: "1) [...] There is another whole class of ideas that obviously have and can have only one object, namely, ideas of the form '*the totality of all things that have* (the attribute) *b*.' It is the very form of these ideas that makes it impossible that they should represent more than one object (which is always composed of parts, and sometimes of infinitely many of them). The only exception arises if there are no things at all that have the attribute *b*. This is due to the fact that there can be only one collection of all objects that have a certain property *b*."

$$InV^*(c,a) \leftrightarrow {}^1_1A(c) \wedge \exists i(In^*(i,a) \wedge i \in c) \qquad \text{CD14}$$
$$ZsV(c,a) \leftrightarrow {}^1_1A(c) \wedge \exists i(Zs(i,a) \wedge i \in c) \qquad \text{CD15}$$
$$ZsV^*(c,a) \leftrightarrow {}^1_1A(c) \wedge \exists i(Zs^*(i,a) \wedge i \in c) \qquad \text{CD16}$$

Corresponding to these relations, we may then define four ideas, of *an a-aggregate* $[\iota(a)]$, of *the complete a-aggregate* $[\iota^*(a)]$, of *an a-compound* $[\zeta(a)]$, and of *the complete a-compound* $[\zeta^*(a)]$, by setting

$$\iota(a) = \varepsilon c\, InV(c,a) \qquad \text{CD17}$$
$$\iota^*(a) = \varepsilon c\, In^*V(c,a) \qquad \text{CD18}$$
$$\zeta(a) = \varepsilon c\, ZsV(c,a) \qquad \text{CD19}$$
$$\zeta^*(a) = \varepsilon c\, ZsV^*(c,a) \qquad \text{CD20}$$

It should be noted that we have not made the following assumptions:

$$\exists c\, InV(c,a), \quad \exists c\, In^*V(c,a), \quad \exists c\, ZsV(c,a), \; \exists c ZsV^*(c,a)$$

Bolzano dwells at length on what he calls *ideas of an aggregate of named* or *nominally given objects* [*Vorstellung von einem Inbegriffe gennanter* oder *namentlich angegebene Gegenstände*] (i.e. those ideas of complete compounds relative to an idea that is constituted by the union of a collection of singular ideas referring to certain objects). If *t*, *c* and *s* are ideas referring to the individuals Titius, Caius, and Sempronius, the idea of their complete compound is the idea relative to the idea $t \sqcup c \sqcup s$.[12]

In general, *being an idea of the complete compound relative to a finite collection of nominally given objects* [$NaV(c)$] may thus be defined by setting

$$NaV(c) \leftrightarrow \exists a(ZsV^*(c,a) \wedge \exists a_1 \ldots a_n(\bigwedge_{j\leq n} {}^1_1A(a_j) \wedge \bigwedge_{j\neq k\leq n} a_j \not\equiv a_k \wedge a = a_1 \sqcup \ldots \sqcup a_n)) \qquad \text{CD21}$$

As, however, Bolzano explicitly recognized that ideas of this type, "even if not *subjectively* achievable are nevertheless *objectively* present also in the case in which the set of the objects $\alpha, \beta, \gamma, \delta, \ldots$ should be infinite",[13] it is also worth systematizing

[12] TS §82, I 283-4: "2) [...] In the first place, we quite often have occasion to form an idea of a collection of individually listed objects: *A, B, C, D, ...* [...] The question now arises concerning the parts of these ideas, which we may call ideas of a collection of individually listed objects in order to distinguish them from others. [...] And so it seems to me that the idea 'collection of *A, B, C, D, ...*,' has no other parts than the ideas of the individual objects *A, B, C, D, ...* and the concept that is indicated by the word 'collection'."

[13] Cf. TS §101, I 339.

the general formulation, using, of course, our ideas of ideas. Thus,

$$NaV(c) \leftrightarrow \exists \mathfrak{a}(ZsV^{*}(c, \bigsqcup \mathfrak{a}) \wedge \forall a(a \mathrel{\scriptstyle\mathsf{E}} \mathfrak{a} \rightarrow {}^{1}_{1}A(a)) \wedge \forall ac(a \neq c \mathrel{\scriptstyle\mathsf{E}} \mathfrak{a} \rightarrow a \not\equiv c)) \tag{CD22}$$

Bolzano, taking up the medieval distinction between *distributive* and *collective*, saw in the kind of ideas just considered a particularly significant example of distributive ideas, but it must be said that his taking up this genuinely 'suppositional' theme was not a very good decision.

4 Existence of Aggregates

We now refer to what was alluded to after the theorem $\exists aIn(i, a) \rightarrow In(i)$ and touch on the question that for us, coming after Russell, appears crucial and to which Bolzano devotes an entire, interesting section: "§101. *Whether for any set of objects there is a common concept which comprehends all of them?*"[14]

The entire discussion suggests that the objects at issue are *disjoint* (i.e. without common parts), and hence it seems reasonable to believe, within the framework of our reconstruction, that the Bolzanian question can quite certainly be faithfully reformulated as the question whether for their aggregate (i.e. for the least object that contains as parts all of them and, moreover, all that is needed to be an object: the parts of these objects, etc.), there is or is not a concept *a* that allows said aggregate to be regarded as an *a*-aggregate or, better, an *a*-compound. Because, however, as already indicated, it does not seem appropriate to adopt any well-defined theory about the pure mereological structure of an object, we may somehow generalize the issue and consider it as the question whether for every aggregate *i*, there is a concept (i.e. an idea without intuitions) *a* such that *i* may be regarded as an *a*-aggregate or, even better, an *a*-compound.

After some introductory remarks, Bolzano specified that the problem must be intended in the strong sense, namely of the existence of a concept that comprehends all and *only* all the elements of that set, because it seemed that, given the presence of the concept *e*, without such restriction, the answer is trivially positive; an aggregate is in any case an *e*-aggregate.[15]

[14] Cf. TS §101, I 338–9.

[15] TS §101, I 339: "If we merely demand of the concept we seek that it include all the objects before us $\alpha, \beta, \gamma, \delta, \ldots$, regardless of whether it fits others as well, then there is no doubt that there is always one such concept. For be the objects $\alpha, \beta, \gamma, \delta, \ldots$ as heterogeneous as you like, they are certainly all contained at least under the concept of an *object in general*."

Quite different is the issue when the requirement is not simply that the aggregate be some a-aggregate but that it be some *complete a-compound*.[16].

First, Bolzano answers that even in this case there is always an *idea*. This answer relies on the already-mentioned Bolzanian conviction that for every object there is in any case a singular idea referring to it; the union of the singular ideas of each object will therefore always lead to the idea, discussed above, of the nominally given objects of the set.[17]

This answer, however, did not satisfy him. Indeed, even disregarding the subjective difficulties, in particular in the case where the objects at issue are infinitely many, there is the basic fact that this procedure not only nominally involves the objects, but, in general, will produce an *idea*, not a *concept*. However, the fact that there may always be such a concept, i.e. a concept a such that i results in being a complete a-compound, seemed rather doubtful. How can there be, for instance, "a concept which collects under it all and only all objects falling under the three concepts: proposition, triangle and virtue; and a concept which does this without containing as its parts the aforesaid concepts or other concepts equivalent to them or even subordered to them?"[18]

Having thus established that there is not always a concept a that allows one to regard the aggregate as a complete a-compound, Bolzano closed his discussion, on the one hand, by remarking that even in the case when there is no such concept, there still is a 'symbolic' concept of such concept and, on the other hand, by asking himself about the possibility of receiving a positive answer to a weakened form of its question.

With regard to the first remark, he maintained that even if there is no concept that comprehends all and only all objects $\alpha, \beta, \gamma, \delta, \ldots$, nevertheless "this idea 'a concept which comprehends all and only all objects $\alpha, \beta, \gamma, \delta, \ldots$,' is, by §90, a symbolical idea." Therefore we may say that he accepted the axiom

$$\forall ai\exists c(Bg(c) \wedge (< i, a > \text{ E } c \leftrightarrow Zs^*(i, a)))\ ^{19} \qquad \text{CV9}$$

[16] TS §101, I 339: "It is something else altogether to demand that the sought concept represent the objects $\alpha, \beta, \gamma, \delta, \ldots$ exclusively or *exactly* (i.e., that it represent these objects and no others)."

[17] TS §101, I 339: "2) [...] In this case too there is always an idea that answers to this demand, provided it may be composed of any parts we like, e.g., from the ideas that represent the objects $\alpha, \beta, \gamma, \delta, \ldots$ individually. For it is certain that for every individual object, and hence for each of $\alpha, \beta, \gamma, \delta, \ldots$ there is and must be an idea applying exclusively to it, because this object must be the subject of certain truths which concern it alone—for instance, the truth that it is an individual object. In every such truth, the subject-concept must be an idea that fits this object alone. Supposing this to be so, it will certainly be possible to cover all of $\alpha, \beta, \gamma, \delta, \ldots$ with a single idea applying to these objects alone. If in no other way, this can at least be done by constructing ideas $A, B, C, D, \ldots$ that apply exclusively to each of these objects and then connecting them to produce a distributive idea of the form 'Any part of the collection $A, B, C, D, \ldots$ [...]' It is obvious that this idea solves the problem, though it has the inconvenience that it contains ideas of the individual objects $\alpha, \beta, \gamma, \delta, \ldots$ as constituents."

[18] TS §101, I 339–40 (3).

[19] TS §101, I 340 (4).

As to the weakening of the question, he began with the remark that, in general, besides e there will be other concepts less extended than e which nevertheless include all the objects for which a characterization is sought. The question is now whether, at least, there is among them a 'minimal' one (i.e., up to equivalences, a concept with least extension). The answer is, also in this case, "I don't dare to affirm its existence"[20]

[20] TS §101, I 340: "5) Should we fail to find a concept that covers the objects $\alpha, \beta, \gamma, \delta, \ldots$ exclusively, or should this be impossible, then in addition to the concept 'object in general', which certainly covers all of them, there will be a great many others which do so as well, and which are considerably lower than that of an object in general. It is advantageous in many cases to become acquainted at least with these lower concepts, the advantage being greater the lower the concept is. This leads to the thought of the *lowest* of all concepts that cover the objects $\alpha, \beta, \gamma, \delta, \ldots$ along with some others. If by this we mean such a concept that is lower than all the others, we already know that there is no such thing, since for every concept with an extension there are others with the same extension. But even if we mean by the lowest concept covering $\alpha, \beta, \gamma, \delta, \ldots$ simply one for which there is no lower such concept, I still cannot dare to claim that such a thing exists in all cases."

APPENDIX D
Unity and Multiplicity

Some aspects of the issue discussed in Appendix C are taken up again by Bolzano in discussing the notions of *unity* [*Einheit*] and *multiplicity* [*Vielheit*]. In particular, he considers, in connection to any objectual idea *a*, the *property of an object to fall under a* and that of the *quality enjoyed by an object to fall under a*. In the first case, he speaks of *concrete unity of type a*, in the second, of *abstract unity of type a*.

More exactly:

(1) An object *i* is a *concrete unity of type a* [*konkrete Einheit von der Art a*] (or also simply *an a*) iff it falls under *a*;[1] i.e.

$$Uniconcr(i, a) \leftrightarrow i \mathrel{\text{E}} a \qquad \text{DD1}$$

(2) A quality *r* is an *abstract unity of type a* [*abstracte Einheit von der Art a*] iff it belongs to an object provided that it can be subsumed under *a*;[2] i.e.

$$Uniastr(r, a) \leftrightarrow \forall i(i \in r \leftrightarrow i \mathrel{\text{E}} a) \qquad \text{DD2}$$

Remembering T2.38, one has, of course

$$Uniastr(r, a) \leftrightarrow r \mathrel{\text{E}} \widehat{a} \qquad \text{DT1}$$

Finally Bolzano introduced the *abstract notion of unity* [*Einheit überhaupt*], stating that such is the quality of a thing in virtue of which if falls under some idea;[3] i.e.

$$Unit(r) \leftrightarrow \forall i(i \in r \leftrightarrow \exists a Uniconcr(i, a)) \qquad \text{DD3}$$

[1] TS §86, I 292: "1) [...] Each object that has a certain attribute *a*, or is subsumed under the idea 'something that has *a*' (or *A*), I want to call a *unit of kind A* in the *concrete sense* of the word 'unit', or a *concrete unit* of kind *A*, or, more briefly, *one A*."

[2] TS §86, I 292: "1) [...] The property of a thing by virtue of which it can be envisaged as a concrete unit of kind *A*, or can be subsumed under the idea of *A* as an object, I want to call the *abstract unit of kind A*."

[3] TS §86, I 292: "1) [...] By *unit in general* in the *abstract* sense I accordingly mean nothing but the attribute which every thing has by virtue of which there exists an idea under which it can be subsumed as an object."

An *a*-compound is called a *concrete multiplicity of type a* [*konkrete Vielheit von der Art a*] (or a *plurality* [*Mehrheit*] of type *a* or *a set* [*Menge*] *of type a*);[4] i.e.

$$Multconcr(i, a) \leftrightarrow Zs(i, a) \qquad \text{DD4}$$

An *abstract multiplicity* (*plurality*) *of type a* is, on the other hand, the quality of a thing in virtue of which it is a concrete multiplicity of type *a*;[5] i.e.

$$Multastr(r) \leftrightarrow \forall i(i \in r \leftrightarrow \exists a Multconcr(i, a)) \qquad \text{DD5}$$

A *concrete Two of type A* is a concrete multiplicity of type *A* having exactly two *a*-parts; a *concrete Three of type A* is a concrete multiplicity of type *A* having exactly three *a*-parts; and so on;[6] i.e.

$$2concr(i, a) \leftrightarrow Multconcr(i, a) \wedge \overset{2}{\exists}!!j(j <_a i) \qquad \text{DD6}$$

$$3concr(i, a) \leftrightarrow Multconcr(i, a) \wedge \overset{3}{\exists}!!z(j <_a i) \qquad \text{DD7}$$

The corresponding abstracts, which Bolzano did not introduce, are, of course,

$$2astr(r) \leftrightarrow \forall i(i \in r \leftrightarrow \exists a 2concr(i, a)) \qquad \text{DD8}$$

$$3astr(i) \leftrightarrow \forall i(i \in r \leftrightarrow \exists a 3concr(i, a)) \qquad \text{DD9}$$

The *concrete aggregate of all the a's* [*der concrete Inbegriff aller a*] or the *(concrete) totality of the a's* [*das All der a*] or the *(concrete) whole of the a's* [*das Ganze der a*] is the *a*-compound;[7] i.e.

$$Totconcr(i, a) \leftrightarrow Zs^*(i, a) \qquad \text{DD10}$$

[4] TS §86, I 292: "2) A collection whose parts are concrete units of a kind *A* I call a *plurality*, indeed a *concrete plurality of kind A*. Where there is no risk of confusion, one sometimes also uses the word *multitude* [*Menge*] (§84) for the sake of variety (§84)."

[5] TS §86, I 292: "2) [...] The property of a concrete plurality by virtue of which it is a plurality is called its *plurality* in the *abstract* sense or *abstract plurality*."

[6] TS §86, I 292: "2) [...] A plurality of *A* which contains nothing but one *A* and one *A* is called a *Two* of kind *A*. A plurality which consists of Two *A*, and, in addition, one *A*, and which does not contain anything else, is called a *Three* of kind *A*, etc."

[7] TS §86, I 292: "3) Finally, the collection which contains all and only objects that are subsumed under the idea *A*, or, more briefly, in which every *A* occurs as a part and in which every part is an *A*, is called the collection of all *A*, or the totality or entirety of the *A* in the concrete sense."

whereas the *abstract totality* [*abstracte Allheit*] is the quality by virtue of which an aggregate is the concrete totality of some *a*;[8] i.e.

$$Totastr(r) \leftrightarrow \forall i(i \in r \leftrightarrow \exists a Totconcr(i, a)) \qquad \text{DD11}$$

Bolzano underlined that the expression '*All the a*'s' is often used, besides 'distributively' (*All the men are mortal*) and also 'collectively' to indicate the totality of the *a*'s in the sense just specified, and then further illustrated and specified the ambiguities occurring in the current usage of terms of the type considered here.

[8] TS §86, I 292–3: "3) [...] The property, which makes a concrete totality what it is, is called *totality* in the *abstract* sense or *abstract totality*."

APPENDIX E
Antinomies

A discussion of the antinomies in the Bolzanian system may aim at two different type of questions. The first, of an exquisite philological nature, concerns what we could call 'the Bolzanian sensibility about antinomies'; the second, purely of theoretical interest, concerns, on the opposite, the possible presence of antinomies in the Bolzanian system, in particular as it has been reconstructed in this book.

As far as the first type of issue is concerned, is must be said that the WL is somewhat disappointing. Indeed, the only antinomy that is taken into consideration is the *Liar*, and, moreover, all that is devoted to this is a rather short note at the end of the fundamental section: "*What the Author Means by a Proposition in Itself*".[1]. As point of reference for the discussion of this paradox, Bolzano took the section *Insolubile propositum nec est concedendum nec negandum* of the *Compendium logicae* of Gerolamo Savonarola, who, having specified that an *insolubile propositum* is a *propositio se ipsam destruens* (i.e. a proposition for which if it is said to be true then it follows that it is false and if it is said to be false then it follows that it is true) had thought to escape the difficulty maintaining that constructs of this kind are in fact not proper propositions, but only have the form [*figura*] of propositions.[2]

Bolzano first of all rejected the idea of the possibility of doing away with the liar by maintaining that it is not a proposition. In particular, he rejected that one could arrive at such a conclusion by accepting that "the subject of a proposition can't be the proposition itself, just as a part can't exhaust the whole." He based this rejection

[1] TS §19, I 58

[2] TS §19, I 60: "*Note*. After what has been said, it may be quite clear what should and what should not be understood by a proposition in itself. Still, the following question may cause some embarrassment. In *SAVONAROLA's Compendium aureum totius logicae (Leipzig, Bk. 10, no. 18)*, the following example occurs under the heading: ***Insolubile propositum** (h.e. propositio se ipsam destruens)* ***nec est concendum nec negandum: Hoc*** *est falsum, posito quod per subjectum demonstretur ipsamet proposition*, i.e., *This* (what I just now say) *is false*. The question is whether this form of words deserves the name of a proposition, and moreover, whether this proposition is true or false. SAVONAROLA says about such collections of words that one should neither affirm nor deny them: *Et si dicatur, omnis propositio est vera vel falsa: dicendum est, quod* ***non sunt propositiones****. Nam definitio propositionis, quod est oratio vera vel falsa, non competit eis in veritate. Habent tamen* ***figuram*** *propositionum. Sicut* ***homo mortuus*** *habet figuram et similitudinem hominis, non tamen est homo: ita et hae dicuntur* ***propositiones destruentes se ipsas, vel insolubiles,*** *non tamen propositiones simpliciter.*"

on his 'symbolic' reading of the proposition: the subject of the proposition is not the proposition itself, but the idea of this proposition. Once it has been accepted, in this way, that the liar is a proposition, and hence something that is either true or false, he faced the problem of knowing which of the two is here the case, and his conclusion is that the liar is false, i.e., more precisely,

> " 'What I am now affirming is false' is a false proposition, because it is equivalent to the following 'What I am now affirming, *I declare false* and *I do not affirm it.*' And this is certainly not true!"

As, however, the contradictory opposite of a false proposition is always true, he was immediately compelled to specify:

> "From this it does in no way follow that I must state the following: 'What I am now affirming is true'."

To justify this, in fact, he shifted the issue to an essentially linguistic level, maintaining that the contradictory proposition of 'What I am now affirming is false' is not obtained, as usually happens with singular propositions, by prefixing the 'not' to the predicate (Socrates is mortal–Socrates is not mortal); this because, in this case, 'a variation of its predicate involves a variation of its subject'. To confirm this, he presented examples in which the simple application of a 'not' to a proposition does not generate its contradictory opposite.

In particular, he mentions that the two propositions

> 'The second-last word of the phrase I am uttering is *an* article'
> 'The second-last word of the phrase I am uttering is *not an* article'

are both true, whereas, on the contrary, the two propositions

> 'The number of words making up the proposition I am uttering is 14'
> 'The number of words making up the proposition I am uttering is not 14'

are both false.

His conclusion, really not properly proven but only claimed, is that the contradictory opposite 'What I am now affirming, is—or I declare it—false' is in fact 'What I am now affirming, I affirm it'.

The surprising end of this discussion seems to confirm the impression that Bolzano did not really grasp the deep logical question involved in the paradox: "Enough, however, with this captiousness!"[3]

[3] TS §19, I 60–1: "What does the reader think? One could think that SAVONAROLA is right, especially as the subject of a proposition can never be that proposition itself, just as a part cannot constitute the whole. Still, I

On the contrary, it is far more interesting that Bolzano, after completing the WL, but before publishing it,[4] came back to the assumption—which is given as obvious in that work—of the existence of an idea, the idea e, which refers to all objects. He did not seem upset, however, about the fact that if ideas are objects, then also e is an object and, as e refers to all objects, one has $e \sqsubset e$.[5] What upset him, on the contrary, was the related concept of the *totality of objects* (i.e., in our notation, the concept $\zeta^*(e)$).

Indeed, in a preparatory manuscript for the *Größenlehre*—which has been retranslated into German by J. Sebestik from the Czech translation of K. Vecerca—we find that

> "This concept [of the absolute all] is absurd, because in the aggregate of all objects also this aggregate should be counted (if also it should be *something*), and hence the whole would appear as a part of itself."[6]

dare profess the contrary opinion, and I believe that common sense, too, will decide in my favour. Which teacher of languages will hesitate to call the words 'What I now say is false' a proposition with a complete meaning? However, there still is the objection that this proposition is its own subject, which seems as absurd as to claim that a part of a whole constitutes the whole itself. This objection, it seems to me, disappears if a distinction is made between the proposition as such and the mere idea of it. Not the proposition itself, as a proposition, but rather the idea of it is the subject-idea of that proposition. That this distinction is well founded can be seen from the fact that not only here, but in all cases a thing must be distinguished from a concept of it, if one does not want to become entangled in gross absurdities. However, if I call the above utterance a complete proposition, I must also be prepared to call it either true or false. As one might suppose, I do the latter and say that the proposition 'What I now claim is false' is itself a false proposition, for it is equivalent to the following 'What I now claim I declare for false and do not claim it.' And this is, of course, not true. However, from this it does not follow at all that I have to assert the following proposition, 'What I now claim is true.' SAVONAROLA seems to have believed this, and since the latter proposition seemed almost as absurd to him as the former, he preferred to seek refuge in the claim that none of the two combinations of words are proper propositions. I say, on the other hand, that the proposition 'What I now claim is false' merely has the peculiarity that its contradictory cannot be constructed in the same way it is done with many other propositions (whose subject-idea has only one object). That is to say that we cannot find the contradictory of this proposition by merely prefacing its predicate 'false' with the negation 'not.' This cannot be done in our case since a change in the predicate causes at the same time a change in the subject. For this subject, i.e., the concept that is expressed by the words 'what I now claim' is different when I say 'What I now claim is false' than when I say 'What I now claim is not false.' A similar phenomenon is encountered with all propositions whose subjects and predicates contain a reference to the propositions themselves or to any one of their parts. Thus, for example, the following two propositions, which to judge from their wording seem to be contradictory, are both true: 'In the present utterance, the third word from the end is the verb' and 'In the present utterance, the third word from the end is not the verb.' On the other hand, the following proposition are both false: 'The number of words in the present proposition is eleven' and: 'The number of words in the present proposition is not eleven', for the last proposition does in fact consist of eleven words, since it has one more (the word 'not') than the first, etc. Therefore, the contradictory of the proposition 'What I now claim is—or I declare it for—false' is not the proposition 'What I now claim is true', but 'What I now claim, I claim."But enough of such hairsplitting!"

For a deeper analysis of Bolzano's treatment of the Liar, see Künne [2011].

[4] See Sebestik [1992], pp. 314–15.

[5] TS §99, I 333: "[...] the extension [...] of the idea something [...] to which [...] every idea and proposition in itself [...] belongs."

[6] *Manuscript ÖNB s.n. 3463, 74 recto*; retranslated into German by J. Sebestik from the Czech translation of K. Vecerka in *Filozoficky Casopis* 38 (1990), p. 525: "Dieser Begriff [des absoluten Alls] ist ungereimt, weil zum

According to our definitions, indeed, $\zeta^*(e) = \varepsilon c({}_1^1A(c) \wedge \exists i(Zs^*(i,e) \wedge i \sqsubset c))$ (i.e. $\zeta^*(e) = \varepsilon c({}_1^1A(c) \wedge \exists i({}_2A(e) \wedge \forall j(j \sqsubset e \leftrightarrow j \ll_a i) \wedge i \sqsubset c)))$, which, taking into account that $\forall j(j \sqsubset e)$, reduces to $\zeta^*(e) = \varepsilon c({}_1^1A(c) \wedge \exists i(\forall j(j \ll_a i) \wedge i \sqsubset c))$ and therefore, in particular, to $\varepsilon c({}_1^1A(c) \wedge \exists i(\forall j(j < i) \wedge i \sqsubset c))$. Hence, the idea 'the whole of the *e*' would refer to an object of which every object, and so in particular itself, is a proper part. Therefore, such an idea must be objectless.

Moreover, in another manuscript, under the title *Difficulties in the First Concepts of the Größenlehre*, we may read the following:

> "The all of *A*'s, where *A* denotes an idea of limited extension, e.g. 'man', can very well be thought. As soon as, however, one takes as *A* the widest of all ideas, i.e. the idea of something in general, the difficulty arises that the idea '*the all of objects*' or '*the all of all*' or '*the absolute all*' should in fact comprehend also itself, because also this all is again *something*; which is indeed absurd. Hence I think that this idea should be counted among the contradictory (imaginary) ones; just as the idea of the speediest motion already also 'the all of the concepts' should be such a self-contradictory idea."[7]

Bolzano himself, however, returning to this remarks, later adjoined, between the lines, the following:

> "But no! These concepts are not contradictory. Really not! In the concept of any aggregate lies that one and the same object cannot be taken twice; and hence not the finger of the right hand and then the thumb, the forefinger, the middle finger, the ring finger and the little finger."[8]

Finally, in his *Philosophische Tagebücher* of the period 1827–44, Bolzano notes the following:

> "*On the logic.* A remarkable example of self-contradictory concept is the concept of the 'all of things' or of the 'sum of all things'; to this sum, indeed, this sum itself should belong as a part, if it were something. The same holds for the concept of the 'all of

Inbegriffe aller Gegenstände auch dieser Inbegriff selbst gezählt werden sollte (wenn es auch *Etwas* sein sollte), wobei dann das Ganze als ein Teil von sich selbst erscheinen würdet."

[7] "*Schwierigkeiten in den ersten Begriffen der Grössenlehre.* Das *All der A*, wo *A* eine Vorstellung von beschränktem Umfange bezeichnet, z.B. Mensch—lässt sich recht wohl denken. Sobald man aber für *A* die weiteste aller Vorstellungen, nähmlich die eines Etwas Überhaupt setzt, so entsteht die Schwierigkeit, dass die Vorstellung: '*Das All der Gegenstände*' oder das *All von allen* oder das *absolute All*, eigentlich auch sich selbst, weil ja diess All auch wieder *Etwas* ist, umfassen sollte; welches doch ungereimt ist.—Ich glaube deshalb, dass man diese Vorstellung in der That zu den widersprechenden (imaginären) zählen müsse; gerade wie die Vorstellung von der geschwindesten Bewegung. Auch schon das *All der Inbegriffe* wäre eine solche sich selbst widersprechende Vorstellung."

[8] "Nicht doch! Diese Begriffe sind nicht widersprechend. Nicht doch! In den Begriffe eines jeden Inbegriffes liegt es, dass man denselben Gegenstand nicht 2 mal aufnehmen darf, also nicht z.B. die Finger der rechten Hand und dann den Daumen, und auch den Zeigefinger, und den Mittelfinger und den Ringfinger und jenen äussersten."

> real things', because this all should be in its turn something real. Not so is it with the concept: 'all of substances'."[9]

Hence, leaving aside that singular and really not easily interpretable insertion in the manuscript, the Bolzanian opinion seems to be that the idea 'all of things' is an objectless idea. That an analogous destiny that also affects the idea 'all of real things' does not on the contrary affect the 'all of substances' is of course a consequence that this all is not itself a substance.

Also the idea of 'truth' (in the sense of true proposition) raised some problems for him, but, singularly, as seen above, not the *Liar*. Indeed, the preceding preparatory manuscript to the *Größenlehre* continues as follows:

> "Just so is it also with the concept 'the all of truths'. That is to say that there cannot be any *aggregate* of truths besides which there is no other truth, because the proposition that enunciates that this aggregate is made up of all truths presents already a truth that is not contained in this aggregate."[10]

We do not know, however, of any other passages in which these questions were taken up again.

As far as the second possible aim of a discussion of the antinomies problem in the Bolzanian system is concerned, besides the more or less interesting or amusing exercise of a reconstruction in it of the most familiar antinomies[11] for instance, the one of the ideas that do not refer to themselves – it does not seem completely devoid of interest to highlight a rather ticklish point concerning what seems one of the most suggestive and still somehow unexplored ideas of the Bolzanian heritage: the idea that the logical complexity of a proposition should be analysed in terms of attribution of semantic properties to the ideas and the propositions involved.

One of the most delicate problems connected with this idea seems to arise in the context of the interaction between variation and symbolization. In that idea, indeed, as we have seen, the symbolization (i.e. the treatment as objects of lectological things) plays an essential role, but its relations with the variation are never well clarified. A particularly critical point seems to be the following: given an idea *a*,

[9] *Phil. Tagebücher, 1827–44*; [BGA II B 18/2 p. 29] : "Zur Logik. |Ein merkwürdiges Beyspiel eines sich selbst widersprechenden Begriffes ist der des 'Alls der Dinge' oder der Summe aller Etwas; denn zu dieser Summe müßte auch diese Summe selbst, wenn sie etwas | wäre, als ein Theil gehören. Ein Gleiches gilt von dem Begriffe: das All der wirklichen Dinge | weil dieses All selbst etwas Wirkliches sein müßte.—Nicht also ist es mit dem Begriffe: Das All der Substanzen."

[10] "Genau so ist es auch mit dem Begriffe 'das All der Wahrheiten'. Es kann nämlich kein *Inbegriff* von Warheiten geben, außer dem keine andere Wahrheit da wäre, weil der Satz, der aussagt, daß dieser Inbegriff sich aus allen Wahrheiten zusammensetzt, stellt gleich eine Wahrheit dar, welche in diesem Inbegriff nicht enthalten is."

[11] See e.g. Morscher [1987] or Casari [1994].

a proposition p containing a, and another idea c, we may of course consider the proposition $p[a/c]$ (i.e. the proposition obtained by varying in p that a into c). Let us now consider the ideas of the two propositions at issue (i.e. $\lceil p \rceil$ and $\lceil p[a/c] \rceil$). By virtue of our (essential!) assumption $l_1 < \lceil l_2 \rceil \leftrightarrow l_1 \leq l_2$, also $\lceil p \rceil$ contains the a as a proper part and we may therefore consider the idea obtained from it by variation of a into c (i.e. the idea $\lceil p \rceil [a/c]$). The question is whether the $\lceil p[a/c] \rceil$ and the $\lceil p \rceil [a/c]$ are the same idea, or at least equivalent That they are the same idea, may be discussed. That, however, they are not even equivalent seems markedly conflictual, as we will immediately show, with the whole Bolzanian construction. It must, however, be said that we cannot see how such an equivalence could be proven using only the assumptions made up to now.

Let us illustrate the point by means of a quite simple example: the symbolization of the assertion that, *with respect to the idea a, from the proposition p the negation of the proposition q is derivable.* On the basis of the definitions, this may be thought to be obtainable through the following steps:

1. $\forall c(W(p[a/c]) \rightarrow \neg W(q[a/c]))$
2. $\forall c(W(p[a/c]) \rightarrow W(\boxminus(q[a/c]))$
3. $\forall c(W(p[a/c]) \rightarrow W(h(\lceil q[a/c] \rceil, \sim \overline{w}))$
4. $\forall c(W(p[a/c]) \rightarrow W(h(\lceil q \rceil [a/c], \sim \overline{w}))$
5. $\forall c(W(p[a/c]) \rightarrow W(h(\lceil q \rceil, \sim \overline{w})[a/c])$
6. $\forall c(W(p[a/c]) \rightarrow W((\boxminus q)[a/c])$
7. $p \underset{a}{\Rightarrow} \boxminus q$

Clearly the step from line 3 to line 4 requires just the equivalence between $\lceil q[a/c] \rceil$ and $\lceil q \rceil [a/c]$, whereas the step from line 4 to line 5 requires the equivalence between $r[b[a/c]]$and $(r[b])[a/c]$. The latter, which does not really involve the symbolic level, seems to be much less problematic; as to the former, it does not seem possible to avoid the postulation of the new general axiom

$$\lceil p \rceil [a/c] \equiv \lceil p[a/c] \rceil \qquad \text{AG59}$$

APPENDIX F
Real numbers

As remarked, the treatment of probability in our logical framework requires the availability in it of some systematization of the real numbers. To this end, we may proceed as follows.

We enrich the language with the two unary predicates $\mathcal{R}$ [*being a real number*] and $\mathcal{Z}$ [*being an integer*], the binary predicate $\leq$ [*being smaller or equal to*], the two individual constants 0 [*zero*] and 1 [*one*], the two unary functors – [*opposite of*] and $^{-1}$ [*inverse of*], and the two binary functors + [*sum of*] and $\cdot$ [*product of*], and we assume first of all the respective *closure axioms*

$\mathcal{R}(0)$	C1	$\mathcal{R}(1)$	C2
$\mathcal{R}(x) \rightarrow \mathcal{R}(-x)$	C3	$\mathcal{R}(x) \wedge x \neq 0 \rightarrow \mathcal{R}(x^{-1})$	C4
$\mathcal{R}(x) \wedge \mathcal{R}(y) \rightarrow \mathcal{R}(x + y)$	C5	$\mathcal{R}(x) \wedge \mathcal{R}(y) \rightarrow \mathcal{R}(x \cdot y)$	C6
$\mathcal{Z}(x) \rightarrow \mathcal{R}(x)$	C7		

We then agree on using, as special variables for $\mathcal{R}$, the Greek letters 'μ', 'ν', 'ρ',... and we accept the axioms ensuring that $\mathcal{R}$ is a *field* in the form

$\mu + (\nu + \rho) = (\mu + \nu) + \rho$	K1	$\mu \cdot (\nu \cdot \rho) = (\mu \cdot \nu) \cdot \rho$	K5
$\mu + \nu = \nu + \mu$	K2	$\mu \cdot \nu = \nu \cdot \mu$	K6
$\mu + 0 = \mu$	K3	$\mu \cdot 1 = \mu$	K7
$\mu + -\mu = 0$	K4	$\neg\mu = 0 \rightarrow \mu \cdot \mu^{-1} = 1$	K8
$\mu \cdot (\nu + \rho) = (\mu \cdot \nu) + (\mu \cdot \rho)$	K9	$\neg 0 = 1$	K10

Moreover, we accept that $\mathcal{R}$ in not only a field, but an *ordered* field (i.e. it has a linear ordering that is compatible with the sum and the product of positive factors); i.e.

$$\mu \leq \mu \quad \text{KO1} \qquad \mu \leq \nu \wedge \nu \leq \rho \rightarrow \mu \leq \rho \quad \text{KO2}$$

$$\mu \leq \nu \wedge \nu \leq \mu \rightarrow \mu = \nu \quad \text{KO3} \qquad \mu \leq \nu \vee \nu \leq \mu \quad \text{KO4}$$

$$\mu \leq \nu \rightarrow \mu + \rho \leq \nu + \rho \quad \text{KO5} \qquad 0 \leq \rho \rightarrow (\mu \leq \nu \rightarrow \mu \cdot \rho \leq \nu \cdot \rho) \quad \text{KO6}$$

But real numbers are not only an ordered field (such as are also the rational numbers), they also satisfy the *continuity principle of Dedekind*, asserting that *every non-empty set that has an upper bound also has a least upper bound.* We therefore accept also the schema

$$\exists \mu \alpha[\mu] \wedge \exists \nu \forall \mu(\alpha[\mu] \rightarrow \mu \leq \nu) \rightarrow \\ \exists \rho(\forall \mu(\alpha[\mu] \rightarrow \mu \leq \rho) \wedge \forall \nu(\forall \mu(\alpha[\mu] \rightarrow \mu \leq \nu) \rightarrow \rho \leq \nu)) \quad \text{KC}$$

Finally, the integers, which, as established above, are all real, constitute also a *discrete subordering* of the real numbers. Hence,

$$\mathcal{Z}(0) \quad \text{AR1} \qquad \mathcal{Z}(\mu) \rightarrow \mathcal{Z}(\mu + 1) \wedge \mathcal{Z}(\mu + -1) \quad \text{AR2}$$

$$\mathcal{Z}(\mu) \wedge \mathcal{Z}(\nu) \wedge \mu \leq \nu \wedge \nu \leq \mu + 1 \rightarrow \nu = \mu \vee \nu = \mu + 1 \quad \text{AR3}$$

REFERENCES

Works of Bolzano

1969ff. [BGA] *Bolzano, Bernard: Gesamtausgabe*, Herausgegeben von E. Winter, J. Berg, F. Kambartel, J. Louzil, E. Morscher & B. van Rootselaar, Stuttgart–Bad Cannstatt, Frommann.

1810 *Beyträge zu einer begründeteren Darstellung der Mathematik. Erste Lieferung*, Prag, Widtmann; repr. 1981, *Acta historiae rerum naturalium necnon technicarum* 12, pp. 83–250; English transl. 2004, Steve Russ, *The Mathematical Works of Bernard Bolzano*, Oxford, Oxford University Press, pp. 82–137.

1834 [RW] *Lehrbuch der Religionswissenschaft, ein Abdruck der Vorlesungen eines ehemaligen Religionslehers an einer katholiscen Universität, von einigen seiner Schüler gesammmelt und heraugegeben*, Sulbach.

1837 [WL] *Dr. Bolzanos Wissenschaftlehre. Versuch einer ausführlichen und größtentheils neuen Darstellung der Logik mir steter Rücksicht auf deren bisherige Bearbeiter. Heraugegeben von mehren seiner Freunde. Mit einer Vorrede des Dr. Ch. A. Heinroth*, Sulzbach; 2nd edn by A. Höfler & W. Schultz, 1928, Leipzig, Meiner.

2014 [TS] English transl. of WL by Rolf George & Paul Rusnock, *Theory of Science*, Vols. 1–4, Oxford, Oxford University Press.

1838 *Athanasia oder Gründe für die Unsterblichkeit der Seele*, 2nd edn, Sulzbach; repr. 1970, Frankfurt/M, Minerva.

1841 [WR] *Bolzano's Wissenschaftslehre und Religionswissenschaft in einer beurtheilenden Uebersicht. Eine Schrift für Alle, die dessen wichtigsten Ansichten kennen zu lernen wünschen*, Sulzbach.

1851 *Dr. Bernard Bolzano's Paradoxien des Unendlichen*, heraugegeben von Fr. Přihonsky, Leipzig, Reclam; repr. by A. Höfler, with Notes by H. Hahn, 1921, Leipzig, Reclam; repr. with Notes by B. van Rootselaar, 1975, Hamburg, Meiner.

1935 *Der Briefwechsel B. Bolzano's mit F. Exner*, heraugegeben von E.Winter, *Bernard Bolzano's Schriften*, Bd. 4, Prag, Kögl. Böhm. Gesell. d. Wissenschaften.

Works on Bolzano

BAR-HILLEL, Yehoshua,

1950 "Bolzano's Definition of Analytic Proposition", *Theoria*, 16, pp. 91–117.

1952 "Bolzano's Propositional Logic", *Archiv für Mathematische Logik und Grundlagenforschung*, 1, pp. 65–98.

BENOIST, Jocelyn,
2003 "Propriété et détermination: Sémantique et ontologie chez Bernard Bolzano", *Philosophiques* 30/1, pp. 137–48.
BENTHEM van, Johann,
1985 "The Variety of Consequence, According to Bolzano", *Studia Logica*, 44, No. 4, pp. 389–403.
2003 "Is There Still Logic in Bolzano's Key?", Morscher [2003], pp. 11–34.
BERG, Jan
1962 *Bolzano's Logic*, Stockholm, Almqvist & Wiksell.
1967 "Bolzano als Logiker", *Bernard Bolzano. Ein Denker und Erzieher im Östrerreichischen Vormärz*, Winter, E. (ed.), Wien, Sitzungsberichte der phil.-hist. Kl. der Österreichischen Akademie der Wissenshaften, 252/5, pp. 95–120.
1974 "Was ist ein Satz nach Bolzano" *Bolzano-Symposion. "Bolzano als Logiker"*, Wien, Sitzungsberichte der phil.-hist. Kl. der Österreichischen Akademie der Wissenshaften, 393/5, Bd. 19, pp. 21–30.
1977 "Bolzano's Contribution to Logic and Philosophy of Mathematics", Gandy, R. O. & Hyland, J. M. (eds.) *Logic Colloquium 76*, Amsterdam, North-Holland, pp.141–71.
1987 "Bolzano and Situation Semantics: Variations on a Theme of Variation", *Philosophia Naturalis*, 24, No. 4, pp. 373–7.
1987a "Is Russell's Antinomy Derivable in Bolzano's Logic?", *Philosophia Naturalis*, 24, No. 4, pp. 406–13.
1987b "Bolzano on Induction", *Philosophia Naturalis*, 24, No. 4, pp. 442–6.
1990 "Zur logischen und mathematischen Ontologie. Geneseologie und Resultatismus in der Analyse der Grundlagen der Bolzanoschen Zahlenlehre", Spalt D. (ed.), *Rechnen mit dem Unendlichen*, Basel, Birkhäuser, pp. 123–155.
1992 "The Connection between Bolzano's Logic of Variation and his Theory of Probability", *Bolzano's Wissenschaftslehre 1837–1987. An International Workshop*, Firenze, Olschki, pp. 107–120.
1992a *Ontology Without Ultrafilters and Possible Worlds*, St. Augustin, Academia Verlag.
BERG, Jan & MORSCHER, Edgar (eds.)
1992 *Bolzano-Forschung 1989–1991*, St. Augustin, Academia Verlag.
1999 *Bolzano-Forschung 1992–1998*, St. Augustin, Academia Verlag.
2002 *Bernard Bolzanos Bibliothek*, Teil I, St. Augustin, Academia Verlag.
2002a *Bernard Bolzanos Bibliothek*, Teil II, St. Augustin, Academia Verlag.
BERKA, Karel
1992 "Bolzano's Lehre vom natürlichen Schliessen", *Bolzano's Wissenschaftslehre 1837–1987. An International Workshop*, Firenze, Olschki, pp. 141–62.
BETTI, Arianna
2010 "Explanation in Metaphysics and Bolzano's Theory of Ground and Consequence", *Logique et Analyse*, 211, pp. 281–316.
BOUVERESSE, Jacques,
2000 "Sur les representations sans objet", *Les Etudes Philosophiques*, 4, pp. 519–34.
BUHL, Günter,
1961 *Ableitbarkeit und Abfolge in der Wissenschaftslehre Bolzanos*, Kantstudien, Ergänzungshefte No. 83.
CASARI, Ettore
1985 "L'universo logico bolzaniano", *Rivista di filosofia*, 76, pp. 339–66.

1986 "Bemerkungen über die Bolzanosche Wissenschaftslehre", *Logik und Grundlagenforschung. Festkolloquium zum 100. Geburtstag von H. Scholz*, Münster i. W., Aschendorff, pp. 53–66.

1989 "Remarks on Bolzano's Modalities", Corsi, G., Mangione C. & Mugnai M. (eds.), *Atti del convegno internazionale di storia della logica. Le teorie della modalità (1987)*, Bologna, CLUEB, pp. 319–22.

1989 "Una fonte dimenticata? La teoria bolzaniana del significato", *Rivista di Filosofia*, 80, pp. 319–49.

1992 "An Interpretation of some Ontological and Semantical Notions in Bolzano's Logic", *Bolzano's Wissenschaftslehre 1837–1987. An International Workshop*, Firenze, Olschki, pp. 55–105.

1997 "Sull'origine dell' «oggettivo» in Bolzano", Fabris, A., Fioravanti, G & Moriconi, E. (eds.), *Logica e teologia. Studi in onore di Vittorio Sainati*, Pisa, ETS, pp. 93–115.

2006 "Some Remarks on Bolzano's Notion of a Quality", Ballo, Edoardo & Franchella, M. (eds.), *Logic and Philosophy in Italy, Some Trends and Perspectives*, Monza, Polimetrica, pp. 185–201.

CELLUCCI, Carlo

1992 "Bolzano and Multiple-Conclusion Logic", *Bolzano's Wissenschaftslehre 1837–1987. An International Workshop*, Firenze, Olschki, pp. 179–189.

CENTRONE, Stefania

2010 "Functions in Frege, Bolzano and Husserl", *History and Philosophy of Logic*, 31, pp. 315–36.

2010a "Bolzano und Leibniz über Klarheit und Deutlichkeit", *Archiv für Geschichte der Philosophie* 92, pp. 256–89.

2011 "Begründungen bei Bolzano und Husserl", *Zeitschrift für philosophische Forschung*, 65, pp. 5–27.

2015 *Studien zu Bolzano*, St. Augustin, Academia Verlag.

CENTRONE, S. & KÜNNE, W.

2011 "Bolzanos Zeichentheorie. Eine Untersuchung zu §285 der Wissenschaftslehre", *Grazer philosophische Studien* 83, pp. 171–98.

CHRISTIAN, Kurt

1974 "Logische Wahrheit, logische Folge und Abfolge im Bolzanoschen Sinne", *Bolzano-Symposion. "Bolzano als Logiker"*, Wien, Sitzungsberichte der phil.-hist. Kl. der Ö.A.d.W., 393/5/12, Bd. 19, pp. 47–62.

1981 (ed.) *Bernard Bolzano. Leben und Wirkung*, Wien, Sitzungsberichte der phil.-hist. Kl. der Österreichische Akademie der Wissenschaften, 391/38.

COFFA, Alberto

1991 *The Semantic Tradition from Kant to Carnap*, Cambridge, Cambridge University Press.

DANEK, Jaromir

1975 *Les projets de Leibniz et de Bolzano: deux sources de la logique contemporaine*, Québec, Presses de l'Universite Laval.

DORN, Georg J. W.,

1987 "Zu Bolzanos Wahrscheinlichkeitslehre", *Bolzano-Studien. Philosophia Naturalis*, 24, No. 4, pp. 423–441.

FØLLESDAL, Dagfinn

2001 "Bolzano, Frege and Husserl on Reference and Object", Floyd, Juliet & Shieh, Sanford (eds.), *Future Pasts. The Analytic Tradition in Twentieth-Century Philosophy*, Oxford, Oxford University Press, pp. 67–80.

FRÉCHETTE, Guillaume

2010 *Gegenstandslose Vorstellungen—Bolzano und seine Kritiker*, St. Augustin, Academia Verlag.

GANTHALER, Heinrich & NEUMAIER, Otto (eds.)

1997 *Bolzano und die österreichische Geistesgeschichte*, St. Augustin, Academia Verlag.

GEORGE, Ralf

1983 "Bolzano's Consequence, Relevance, and Enthymemes", *Journal of Philosophical Logic*, 12, No. 3: 299–318.

1986 "Bolzano's Concept of Consequence", *Journal of Philosophy*, 83, No. 10, pp. 558–564.

1987 "Bolzano on Time", *Bolzano-Studien. Philosophia Naturalis*, 24, No. 4, pp. 452–468.

1991 "Concepts of Consequence", *Bolzano's Wissenschaftslehre* 1837–1987. *An International Workshop*, Firenze, Olschki, pp. 3–26.

2003 "Intuitions", *Philosophiques* 30, No. 1, pp. 19–46.

HAFNER, Johannes,

1999 "Bolzano's Criticism of Indirect Proofs", *Revue d'histoire des sciences*, 52, Nos. 3–4, pp. 385–98.

HAILPERIN, Theodore

1996 *Sentential Probability Logic: Origins, Development, Current Status, and Technical Applications*, Bethlehem, PA, Lehigh University Press.

HIECKE, Alexander & NEUMAIER, Otto (eds.)

2003 *Philosophie im Geiste Bolzanos, anläßlich des 222. Geburtstage von Bernard Bolzano. Edgar Morscher gewidmet*, St. Augustin, Academia Verlag.

KONZELMANN ZIV, Anita

2010 *Kräfte, Wahrscheinlichkeit und 'Zuversicht' —Bernard Bolzanos Erkenntnislehre*, St. Augustin, Academia Verlag.

KRAUSE, Andrej

2004 *Bolzanos Metaphysik*, Freiburg–München, K. Alber Verlag.

KRICKEL, Frank

1995 *Teil und Inbegriff*, St. Augustin, Academia Verlag.

KÜNNE, Wolfgang

1997 "Propositions in Bolzano and Frege", *Grazer Philosophische Studien*, 53, pp. 203–40.

1998 "Substanzen und Adhärenzen. Zur Ontologie in Bolzanos *Athanasia*", *Philosophiegeschichte und logische Analyse*, 1, pp. 233–50.

2003 "Bernard Bolzano's Wissenschaftslehre and Polish Analytical Philosophy between 1894 and 1935", Kijania-Placek, K. (ed.) *Philosophy and Logic. In Search of the Polish Tradition. Essays in Honour of Jan Wolenski on the Occasion of his 60th Birthday*, Dordrecht, Kluwer, pp. 179–92.

2006 "Analyticity and Logical Truth from Bolzano to Quine", *The Austrian Contribution to Analytical Philosophy*, Textor, M. (ed)., New York, Routledge, pp. 184–249.

2008 *Versuche über Bolzano/Essays on Bolzano*, St. Augustin, Academia Verlag.

2012 "Über Lügner, 'Lügner' und harmlosen Selbstbezug. Bolzano vs. Savonarola und die Geschichte einer Antinomie", Reboul, A. (ed.), *Philosophical Papers Dedicated to Kevin Mulligan*, Available online: http://www.philosophie.ch/kevin/festschrift/.

KÜNNE, W. and CENTRONE, S.

2011 "Bolzanos Zeichentheorie. Eine Untersuchung zu §285 der Wissenschaftslehre", *Grazer philosophische Studien* 83, pp. 171–98.

LAPOINTE, Sandra

2000 "Analyticité, universalité et quantification chez Bolzano", *Les Études Philosophiques*, No. 4, pp. 455–70.

LOUZIL, Jaromir

1974 "Bolzanos Begriff der Anschauung", *Bolzano-Symposion."Bolzano als Logiker"*, Wien, Sitzungsberichte der phil.-hist. Kl. der Österreichischen Akademie der Wissenshaften, 393/5, Bd. 19, pp. 63–76.

1975 (Ed.) *Kleine Wissenschaftslehre. Auszug au den ersten zwei Bänden der Wissenschaftslehre B. Bolzanos, verfasst von einem seiner Schüler*, Österreichische Akademie der Wissenschaften, Wien.

MANCOSU, Paolo

1999 "Bolzano and Cournot on Mathematical Explanation", *Revue d'histoire des sciences*, 52, Nos. 3–4, pp. 429–55.

MORSCHER, Edgar

1971a *Das logische An-sich bei Bernard Bolzano*, Salzburg, Pustet Verlag.

1974 "'Philosophische Logik' bei Bernard Bolzano", *Bolzano-Symposion "Bolzano als Logiker"*, Wien, Sitzungsberichte der phil.-hist. Kl. der Österreichischen Akademie der Wissenschaften, 393/5, Bd. 19, pp. 77–105; repr. Morscher [2007], pp. 169–98.

1974a "Ist Existenz ein Prädikat? Historische Bemerkungen zu einer philosophischen Frage", *Zeitschrift für philosophische Forschung*, 28, pp. 120–132; repr. Morscher [2007], pp. 135–48.

1981 "Bolzanos Wissenschaftslehre", Christian K. (ed.), *Bernard Bolzano. Leben und Wirkung*, Wien, Sitzungsberichte der phil.-hist. Kl. der Österreichischen Akademie der Wissenshaften, 391/38, pp. 99–126; repr. Morscher [2007], pp. 19–46.

1986 "Was Existence Ever a Predicate?", *Grazer Philosophische Studien* 25/26, pp. 269–84.

1987 "'Hintertürln' zur Paradoxien in Bolzanos Logik", *Philosophia Naturalis* 24, pp. 414–22; repr. Morscher [2007], pp. 159–68;

1987 "Bolzanos Syllogistik", *Bolzano-Studien. Philosophia Naturalis*, 24, No. 4, 447–51; repr. Morscher [2007], pp. 101–6.

1989 "Zu Bolzanos Lösung der Lügner-Paradoxie", Gombocz, W. L., Rutte H. & Sauer W. (eds.), *Traditionen und Perspectiven der analytischen Philosophie. Festschrift für Rudolf Haller*, Wien, Hölder-Pichler-Tempsky, pp. 89–96; repr. Morscher [2007], pp. 149–58.

1997 "Bolzanos Method of Variation: Three Puzzles", *Grazer Philosophische Studien*, 53, pp. 139–65.

1999 "Logische Allgemeingültigkeit", Morscher (ed.) [1999], pp. 179–206.

2003 "Sind alle wahren logischen Sätze logisch wahr?", Morscher (ed.) [2003], pp. 57–82; repr. Morscher [2007], pp. 75–100.

2003a "La définition bolzanienne de l'analyticité logique", *Philosophiques*, 30: 149–69.

2006 "The Great Divide within Austrian Philosophy: The Synthetic A Priori", Textor [2006], pp. 250–63.

2007 *Studien zur Logik Bernard Bolzanos*, St. Augustin, Academia Verlag.

2008 *Bernard Bolzano's Life and Work*, St. Augustin, Academia Verlag.

MORSCHER, Edgar (ed.)

1999 *Bernard Bolzanos geistiges Erbe für das 21. Jahrhundert*, St. Augustin, Academia Verlag.

2003 *Bernard Bolzanos Leistungen in Logik, Mathematik und Physik*, St. Augustin, Academia Verlag.

MUGNAI, Massimo

1992 "Leibniz and Bolzano on the 'Realm of truths'", *Bolzano's Wissenschaftslehre 1837–1987, An International Workshop*, Firenze, Olschki, pp. 207–20.

PALAGY, Melchior

1902 *Kant und Bolzano. Eine kritische Parallele*, Halle a. S., Niemayer; Italian transl. by Guidetti & Melandri, 1993, *Kant e Bolzano. Un confronto critico*, Ferrara, Spazio Libri.

PRETI, Giulio

1935 "I fondamenti della logica formale pura nella 'Wissenschaftslehre' di B. Bolzano e nelle 'Logische Unterschungen' di E. Husserl", *Sophia*; repr. *Saggi filosofici*, 1976, Firenze, La Nuova Italia, Vol. I, pp. 11–31.

PŘIHONSKY, Franz

1850 *Neuer Anti-Kant oder Prüfung der Kritik der reinen Vernunft nach den in Bolzano's Wissenschaftslehre niedergelegten Begriffen*, Bautzen, Weller.

PROUST, Joelle

1999 "Bolzano's Theory of Representation", *Revue d'histoire des sciences*, 52, Nos. 3–4, pp. 363–383.

ROOTSELAAR van, Bob

1992 "Axiomatics in Bolzanos Logico-Mathematical Research", *Bolzano's Wissenschaftslehre 1837–1987, An International Workshop*, Firenze, Olschki, pp. 221–230.

RUSNOCK, Paul

2003 "Qu'est-ce que la représentation? Bolzano et la philosophie autrichienne", *Philosophiques* 30, pp. 67–81.

2003 "La théorie des intuitions chez Bolzano", Fisette, D. & Lapointe, S. (eds.), *Aux origines de la phénoménologie: Husserl et le contexte des recherches logiques*, Québec, Presses de l'Université Laval, pp. 111–23.

RUSNOCK, Paul & GEORGE, Ralf

2004 "Bolzano as Logician", Gabbay, D. & Woods, J. (eds), *The Rise of Modern Logic: From Leibniz to Frege*, Amsterdam, North-Holland, pp. 177–205.

RUSS, Steve,

2004 *The Mathematical Works of Bernard Bolzano*, Oxford, Oxford University Press.

SCHENK, Günter

1981 "Über Bolzanos Theorie der Beziehungen zwischen 'Vorstellungen an sich' ", Schuffenhauer [1981], pp. 119–45.

SCHNIEDER, Benjamin

1964 *Substanz und Adhärenz, Bolzanos Ontologie des Wirklichen*, St. Augustin, Academia Verlag.

1986 "Bolzano sur la structure des propositions et le rôle sémantique des propriétés", *Philosophiques* 30, No. 1, pp. 83–103.

1992 "Particularised Attributes: An Austrian Tale", Textor (ed.) [2006], London, Routledge, pp. 130–58.

SCHOLZ, Heinrich

1995 "Die Wissenschaftslehre Bolzanos: Eine Jahrhundert-Betrachtung (1937)", *Abhandlungen der Friesschen Schule*, N.S., Bd. 6, pp. 399–472; repr. in *Mathesis Universalis*, 1961, Basel–Stuttgart, B. Schwabe, pp. 219–67.

SCHUFFENHAUER, Werner (ed.)

1981 *Bernard Bolzano 1781–1848. Studien und Quellen*, Berlin, Akademie-Verlag.

SEBESTIK, Jan

1964 "Bernard Bolzano et son mémoire sur le théorème fondamental de l'analyse", *Revue d'histoire des sciences*, 17, pp. 129–164.

1986 "La classe universelle et auto-appartenance chez Bernard Bolzano", *Eleuteria*, Athenes, pp. 72–84.

1992 *Logique et mathématique chez Bernard Bolzano*, Paris, Vrin.

1992a "The Construction of Bolzano's Logical System", *Bolzano's Wissenschaftslehre 1837–1987. An International Workshop*, Firenze, Olschki, pp. 163–77.

1995 "Twardowski entre Bolzano et Husserl: théorie de la representation", *Cahiers de la philosophie ancienne et du langage*, 1, pp. 61–85; German transl. *Grazer philosophische Studien*, 35, pp. 175ff.

1997 "Bolzano, Exner and the Origins of Analytical Philosophy", *Grazer philosophische Studien*, 53, pp. 33–59.

1999 "Forme, variation et déductibilité dans la logique de Bolzano", *Revue d'histoire des sciences*, 52, No. 3–4, pp. 479–506.

2003 "Husserl Reader of Bolzano", Fisette D. (ed.), *Husserl's Logical Investigations Reconsidered*, Dordrecht, Kluwer, pp.71–92.

SIEBEL, Mark

1996 *Der Begriff der Ableitbarkeit bei Bolzano*, St. Augustin, Academia Verlag.

2003 "La notion bolzanienne de déductibilité", *Philosophiques* 30, No. 1, pp. 171–89.

2004 "Bolzano Urteilslehre", *Archiv für Geschichte der Philosophie*, 86, pp. 56–87.

SIMONS, Peter

1987 "Bolzano, Tarski, and the Limits of Logic", *Bolzano-Studien. Philosophia Naturalis*, 24, No. 4, pp. 378–405.

1997 "Bolzano on Collections", *Grazer philosophische Studien*, 53, pp. 87–108.

"Austrian Philosophers on Truth", Textor [2006], pp. 159–83.

SPALT, Detlev (ed.)

1990 *Rechnen mit dem Unendlichen*, Basel, Birkhäuser.

STARCENKO, Anatoli A. & TJGNIBEDINA, Olga S.

1981 "Besonderheiten der wahrscheinlichkeitslogischen Forschungen B. Bolzanos", Schuffenhauer [1981], pp. 146–160.

STELZNER, Werner,

2002 "Compatibility and Relevance: Bolzano and Orlov", *Logic and Logical Philosophy*, 10, pp. 137–71.

SUNDHOLM, Göran

2000 "When and Why, Did Frege Read Bolzano?", *Logica Yearbook 1991*, Prague, Filosofia.

TATZEL, Arnim

2002 "Bolzano's Theory of Ground and Consequence", *Notre Dame Journal of Formal Logic*, 43, pp. 1–25; French transl. 2003, "La théorie bolzanienne du fondement et de la conséquence", *Philosophiques* 30, No. 1, pp. 191–217.

TEXTOR, Markus

1996 *Bolzanos Propositionalismus*, Berlin, W. de Gruyter.

TEXTOR, Markus (ed.)

2006 *The Austrian Contribution to Analytic Philosophy*, London, Routledge.

THOMPSON, Paul B.

1981 "Bolzano's deducibility and Tarski's logical consequence", *History and Philosophy of Logic*, 2, pp. 11–20.

TWARDOWSKI, Kasimierz

1894 *Zur Lehre von Inhalt und Gegenstand von Vorstellungen*, Wien; ed. by R. Haller, München 1982, Philosophia Verlag; English transl. and introd. by R. Grossmann, 1997, *On the content and object of presentations*, The Hague, Martinus Nijhoff.

WINTER, Eduard

1935 *Der Briefwechsel B. Bolzano's mit F. Exner (1833–1844), Spisy Bernarda Bolzana/Bernard Bolzano Schriften*, IV, Prague.

1944 *Der Bolzanoprozeß. Dokumente zur Geschichte der Prager Karlsuniversität im Vormärz*, Brünn–München–Wien, R. Roherer.

1956 *Der böhmischer Vormärz in Briefen B. Bolzanos an F. M. Prihonský (1824–1848),* Deutsche Akademie der Wissenschaften, Berlin.

1975 *Robert Zimmermanns philosophische Propädeutik und die Vorlagen aus der Wissenschaftslehre Bernard Bolzanos,* Wien, Österreichische Akademie der Wissenschaften.

1981 *Begriffe B. Bolzanos, gesammelt 1821 von Florian Werner, eingeleitet und vermehrt von Eduard Winter,* Schuffenhauer [1981], pp. 187–277.

WINTER, Eduard & WINTER, Maria

1970 *Der Bolzanokreis 1824–1833 in Briefen von Anna Hoffmann, Michael Josef Fesl, Franz Schneider und Franz Přihonsky,* Wien, Österreichische Akademie der Wissenschaften.

AUTHOR INDEX

A

Ackermann, Wilhelm 5

B

Bar-Hillel, Yehoshua 339
Benoist, Jocelyn 339
Benthem van, Johann 340
Berg, Jan 191, 201, 213, 235, 339, 340
Berka, Karel 340
Bernini, Sergio xi
Betti, Arianna xi, 340
Bouveresse, Jacques 340
Brentano, Franz 77
Bruni, Riccardo xi
Buhl, Günter 340
Burali-Forti, Cesare 125

C

Cantini, Andrea xi
Carnap, Rudolf 111
Casari, Ettore 340–1
Catalan, Eugène Charles 247
Cellucci, Carlo 341
Centrone, Stefania xi, 341, 342
Christian, Kurt 341
Cicero 303
Coffa, Alberto 341

D

Dalla Chiara, Maria Luisa xi
Danek, Jaromir 341
Dedekind, Richard 257
De Morgan, Augustus 118
Diller, Justus xi
Dorn, Georg J. W. 341

E

Exner, František 237, 243, 244, 339, 345

F

Frege, Gottlob 301, 341, 342
Fossati, Lorenzo xi
Føllesdal, Dagfinn 341
Fréchette, Guillaume 341

G

Ganthaler, Heinrich 341
Gentzen, Gerhard 109, 263
George, Rolf x, xi, 339, 342, 344
Geymonat, Ludovico v
Grayg, John 121

H

Hafner, Johannes 342
Hailperin, Theodore 342
Haller, Rudolf xi, 343
Hartley, David 296
Hasenjäger, Gisbert v
Hermes, Hans v
Hiecke, Alexander 342
Hilbert, David viii
Hume, David 296
Husserl, Edmund vii, viii, 22, 77, 238, 276, 305, 341, 344, 345

J

Jeffreys, Harold 111

K

Kambartel, Friedrich xi, 339
Kant, Immanuel 80, 321
Kaulbach, Friedrich xi
Keynes, John Maynard 111
Krickel, Frank xi
Konzelmann Ziv, Anita 342
Krause, Andrej 18, 342
Künne, Wolfgang xi, 333, 341, 342

L

Lapointe, Sandra 342, 344
Leibniz, Gottfried Wilhelm 3, 17
Louzil, Jaromir 339, 342

M

Mancosu, Paolo xi, 343
Maaß, Johann Gerhard 26
Minari, Pierluigi xi
Morscher, Edgar xi, 335, 339, 340, 342, 343
Mugnai, Massimo xi, 341, 343

N

Neumaier, Otto 341, 342

P

Palagy, Melchior 343
Paoli, Francesco xi
Pico, della Mirandola 245, 246
Preti, Giulio v, 344
Přihonsky, Franz 344, 346
Proust, Joelle 344

Q

Quine, Willlard van Orman 342

R

Rigamonti, Gianni xi
Rootselaar Van, Bob 344
Rusnock, Paul x, xi, 339, 344
Russ, Steve 339
Russell, Bertrand 325

S

Savonarola, Gerolamo 331, 332, 333, 342
Schenk, Günter 344
Schnieder, Benjamin 235, 239, 344
Scholz, Heinrich xi, 340, 344
Schuffenhauer, Werner 344, 345, 346
Sebestik, Jan xi, 333, 344
Segner, Johann Andreas 247
Siebel, Mark 282, 345
Simons, Peter 345
Spalt, Detlev 340, 345
Starcenko, Anatoli A. 345
Steinman, Laurence xi
Stelzner, Werner 345

Stumpf, Carl 77
Sundholm, Göran 345

T

Taber, Daniel xi
Tatzel, Arnim 252, 345
Textor, Markus 18, 342, 343, 344, 345
Thompson, Paul B. 345
Tjgnibedina, Olga S. 345
Twardowski, Kasimierz 22, 345

V

Vecerka, Karel 333

W

Winter, Eduard 339, 340, 345
Winter, Maria 346
Wolff, Christian 3

SUBJECT INDEX

A
Abstraction 42–3
Ackermann's Axiom 5
Additionability 93, 97, 103
Adherence 8, 43, 249, 267, 275–8, 280, 282
Aggregate 15, 18, 23, 35, 265, 315
  complete 321
  concrete 319, 329
  Existence of 325
  Idea of 322–3
  Quality of 17
  relatively to 317–9, 321–6
Alethic Modalities 263, 272
Antinomies 321, 331, 335
Approval 281
  Degree of 293
Asserting ix, 4, 75, 121–2, 127, 131–2, 135, 182, 218, 232, 235, 255, 267, 272–3, 278, 283, 287, 333, 336, 338
Attributing 55, 62, 64, 70, 75, 124, 127–8, 162, 236, 245, 247, 259, 336
Awareness 281

B
Baralip 202
Barbara 202
Barocco 202
Baroco 202, 206
Bearer x, 276, 283
Believing 294–5
Belonging ix, 1, 4, 6, 8–9, 11, 18, 38, 40–1, 46, 54, 56–8, 63–5, 71, 120, 124, 131, 135, 140, 143, 164, 166, 202, 233, 236–7, 239–49, 276, 291, 293, 304, 329, 334
Bocardo 205
Bolzanian definition 234
Bolzano's Conjecture 126–7, 132
Bolzano's Thesis 132, 136, 140, 143, 146, 152, 154, 160–2, 166, 169, 175
Bramantip 202
Burali-Forti's Paradox 125

C
Calentes 202
Camenes 202
Camestres 204
Causality x, 3, 267, 270–1
Cause 4, 267, 270–1, 282–4
  complete 271
Celarent 202
Cesare 203–4
Chain 36–7
Clearness 280–1
Compatibility 20, 24–6, 28, 30–2, 34–6, 40, 46, 48, 61, 82–97, 99–100, 104, 106, 108–11, 116–7, 140, 143–4, 149, 153, 158, 166–7, 170, 174, 195, 211, 214–6, 218
Complexity ix, 70, 263–5, 335
Component 320
Concatenation 25, 33, 99–100, 144, 167, 213
  mediate 35
  omnilateral 34, 213
Concretization 38–41, 43, 46, 48, 59, 143–4, 166–7, 177, 184–5, 228
Conjunction 131, 136–7, 213–4, 254, 257
  iterated 28
  material 161–2
  probability of 115–9
Connective 258
  formal 110, 149, 170, 217
  material 136
Consecutivity x, 3, 250, 252–6, 258–63, 266–7, 285, 287
Consequence 3, 250, 258–9, 267
  complete 251–4, 256
  partial 251–2, 257, 259, 272
Content 4, 21, 26, 74, 128, 275–6
Contingency 54, 272–4
Contradictoriness 26, 35, 61, 100–3, 106, 213, 223–4, 332, 334
Contrariety 26, 35, 61, 100–2, 105, 132, 140, 159
Copula 54, 62, 126, 301

D
Darapti 203–4
Darii 202, 205
Datisi 205
Denoting 235, 237, 296–8, 300, 303
Dependence 77, 115, 259, 260–1, 263–4
Derivability x, 88–90, 92, 95–6, 110–2, 144, 149, 152, 167, 173, 215, 218–9, 222, 261–3, 285, 287
  adequate 94
  exact 94–5, 174
  exactly measured 94
  unilateral 98
Dimaris 202
Dimatis 202
Disamis 205
Dissonance 24, 84
Disjunction ix, 69, 71, 114, 118, 141, 224
  formal 107
  material 68, 161–2
Disparateness 25–6, 33, 61, 153, 159
Distinctness 280–1

E
Effect 4, 267, 270–1, 281
  complete 271

Enunciating x, 70, 110, 121, 135, 193, 210–5, 222–3, 225, 231, 258, 269–70, 273, 278, 280, 285, 302–3, 335
 a Situation 239, 241–3
Epistemology i, 10
Epsilon Calculus viii, 5, 11, 37, 307
Equivalence 20, 25–6, 32, 33, 42, 50, 61, 75–6, 96–8, 100, 102–3, 113, 128, 144, 153, 158, 167, 174, 212, 221, 327
Error 279–80, 289–9, 292
Etiology 251
Existence 2, 4, 237, 267–273, 283, 289
Extension 4–5, 22, 24, 32–3, 53, 74–5, 192, 209, 213, 327, 334

F

Fact 239–42
Falsity ix, 3, 63–4, 67–71, 78, 80–1, 84, 86, 100, 102, 105–8, 115, 119, 128, 135–6, 141, 143–4, 149, 166–7, 170, 174, 185–6, 191, 203, 216–8, 222, 224–7, 231, 238, 243, 279–80, 289–91, 293–4, 332–3
 analytic 79–80
 synthetic 79–80
Felapton 203
Ferio 202, 205
Ferison 205
Festino 205
Formulating 277, 280–1, 286–91, 293–4
Fresiso 202
Fresison 205

H

Handeling 62
Having a Quality ix–x, 1–2, 6, 38–9, 41–2, 44, 53–7, 59, 63, 65–6, 80, 126–8, 140–2, 165, 175, 177, 180, 183, 192, 196, 234–5, 241, 264
Having as Content (Material) 22, 275–7, 296–7
Hermeneutics 302

I

Idea x, 2, 9, 10, 20, 29, 34–5, 38, 41, 52, 76, 124, 210–1
 abstract 39–40, 175
 collective 323
 composite 20–1, 189, 322
 concrete 39, 41
 distributive 323
 imaginary 22, 238
 Object 3, 5–6, 8, 12, 27, 30, 42–3, 52, 63, 135, 151, 154
 objectless 21–2, 74, 76, 191, 243, 335
 objectual 4, 21, 24, 26, 127, 129, 131, 240, 242–3
 of Propositions 133, 151, 162, 270
 Quality 3, 5–6, 8, 12, 23–4, 38, 42–3, 45–52, 59, 63, 128–9, 134–5, 236, 238
 simple 20–1, 26, 43, 45, 263, 268, 322
 singular 23, 43, 45, 193
 Sub-idea 263
 symbolic 128, 130
 universal 23
Impossibility 238, 272–4
Inclusion 20, 25–6, 32, 61, 75, 153, 157–8, 174, 211–2, 214, 315
Incompatibility 24, 26, 30–1, 34–6, 46, 61, 84–9, 91, 94–5, 97, 100–1, 106, 108, 112, 119, 143–4, 149, 153–4, 159, 166–7, 170, 179–80, 215, 221, 238
 r–Incompatibility 211
Inference x, 173–200, 202–206, 210–25, 229–31, 263, 266, 285–6, 290, 292
Integration 83, 85, 106–7, 109–10, 174
 absolute 106
 conditioned 106, 145, 167, 231
 formal 82, 107, 109, 143–4, 149, 166, 170, 227–9
 material 68, 71, 137, 225, 228
Intensity 281–2
Intention 297, 304
Interpretation 303
Intuition 264, 272, 280, 289, 326

J

Judgement vii, viii, x, 4, 10, 276–8, 280, 289–92, 294
 cognitive Properties of 280–1
 confused 281
 correct 280
 distinct 281
 Formation of 282–8
 indistinct 281
 logical Properties of 278–80
 perceptive 288–9
 physical Properties of 280–1

K

Keeping as true 281–2
Knowing 294–5
Knowledge vii, viii, x, 4, 43, 45, 273, 275, 277, 279, 282, 289, 294

L

Law of association of the representations 206
Lectology 2–4, 8–10, 12, 20, 38, 52, 63, 70, 127, 129, 134, 140, 151, 154, 268–9, 272, 274, 276, 336
Level
 elementary x, 4, 7, 9, 20
 higher 4, 7–10, 43, 233
 symbolic x, 4, 9, 21, 29, 70, 124
  extended 128
  restricted 128
Liveliness 282
Logical
 Objectivity vii–viii
 Subjectivity vii–viii

M

Meaning 297, 304–5
 immediately 299–302
 in a broad sense 297–300
 in a strict sense 297–300
 mediately 299–302
Multiplicity 328
 abstract 329
 concrete 329

N

Necessity 272–4
Negation 27, 52–6, 59, 67, 71, 73, 87, 90–2, 94–5, 98, 100, 102–4, 113–5, 131, 135, 185, 188, 192, 197–8, 206, 213–4, 217, 220–2, 224, 228, 273–4, 336

O

Object ix, x, 1–11, 15–9, 22–3, 28–9, 37–8, 43, 55, 62, 70, 124–5, 128, 133, 233, 235–7, 239–44, 249, 269, 273, 278, 297–300, 316–23, 325–7, 328, 333–5
  Musilian 1, 7
Ontology vii, 1–2, 6–8, 17
Ordered-tuple 15–6

P

Part ix, 2, 5, 8, 11, 13, 21, 43, 63, 74–5, 77, 128, 130, 134, 183, 272, 281, 316–23, 329, 333–4
  distant 319
  homogeneous 321
  inhomogeneous 321
  integrant 36, 322
  proper 2, 7, 10, 154, 212–3, 268, 336
  proximate 319
Perceiving 297
Possibility 272–4
Power 281, 293
Predicate 5, 8, 10–1, 16, 19, 21, 28, 29, 39, 44, 46, 50, 63, 67, 83, 124, 152, 251, 276, 278, 337
  Idea 52, 54, 62–4, 96, 99, 126, 177, 182, 184–5, 191, 218, 238, 243, 265, 288, 332
Premise 90–2, 94–5, 98–9, 103, 114, 116–9, 121, 173–4, 176, 182–4, 202–11, 214–5, 221–3, 231, 265–7, 282, 285, 290–2
  existential 188, 193–5, 197, 199
  inexistential 188, 190, 198–9, 201
  universal 177, 179, 185–6, 195, 198, 202
Probability 54, 111–20, 283, 286–7, 290–4
  alleged 291
  objective 291
  of Events 120–3
  real 291
  subjective 291
Producing 296–7, 300–1, 305
Proposition x, 3–6, 8, 10, 12, 29, 38
  affirmative 55
  analytic 79, 81
  Aristotelian universal with existential Import 66
  atomic 62–4
  complex 70
  conceptual 45
  in itself vii–x, 2–3
  intuitive 45
  perceptive or empirical 45
  synthetic 79
  traditional categorical 132

Q

Quality ix–x, 1, 3–9, 11, 15, 17–9, 40–1, 43, 56, 62–3, 70, 124–7, 129, 133, 140–1, 154, 163, 233–44, 249, 267, 293, 328–30
  empty or vacuous 40, 233–9
  Idea ix, 2–3, 8, 12, 20–1, 23–4, 38–9, 42–3, 45–52, 55–9, 63, 76, 128–9, 134–5, 141, 164, 175
Quantifier, restricted 11

R

Reason x, 3–5, 250–2, 256–7, 259–60, 262–7, 270, 283
  complete 251, 253–4
  partial 251, 272, 283
Real Numbers 337–8
Referring 9
Representation
  cognitive Properties of 280–1
  logical Properties of 279
  objective (or in itself) *see* Idea
  physical Properties of 280–1, 293
  subjective (or Presentation) viii, x, 4, 10, 22, 237, 275, 277–8, 280, 288, 296–305
Representing x, 22, 278, 280, 297

S

Semiotics 296, 302, 304
Sign 4, 296–305
Sense x, 127, 145, 297–8, 301, 303
Situation 10, 239–43, 278, 280, 302
Something 1, 37
Speaking 303–4
Stimulating 237, 297–302
Subject 52–4, 62–3, 77, 80, 96, 99, 126, 130, 173, 175, 177, 180, 182–5, 191, 218, 238, 241, 243, 265, 288, 302, 332
  thinking 291
Subordination 25–6, 32–3, 61, 96, 98–9, 101, 113, 117–8, 144, 153, 159, 167, 186, 212, 288
Substance 8, 43, 249, 267, 271, 335
Substitutivity 26, 90, 97, 100, 103, 105, 113, 212, 247, 283
Sum 23, 52, 316, 322–3, 334–5, 337
  of Quality Ideas 45–48, 50–1, 184, 191, 239
Superfluous 74, 90
Supplement 68

T

Totality 231, 212–3, 242, 282, 292, 333
  abstract 330
  concrete 329
Trope ix–x, 234, 244
Trust viii, 281, 293
  Degree of 281, 287, 290, 293–5
Truth viii–x, 3–4, 24, 45, 53, 63–5, 67–71, 78, 80, 85, 89, 97, 110, 127–8, 131,

Truth (*continued*) 133, 135, 137, 141–4, 147, 149, 160–2, 166–7, 170, 174, 177–8, 192, 218, 225, 233, 238–43, 250–8, 262–5, 272, 280, 283, 285, 291, 295, 331–2, 335
  analytic 79–81
  auxiliary 259–61
  in itself vii, viii, 3
  synthetic 79–80, 237

U

Unity 328
  abstract 328
  concrete 328

V

Validity 76, 78–80, 112, 289–90
  comparative 111
  degree of 78–9, 293
Variation ix, 2, 12–3, 20, 74–8, 80–1, 84, 88, 107, 221, 332, 335–6